# Collins

# Edexcel GCSE
# Maths
## 4th Edition

Foundation Student Book

Kevin Evans
Keith Gordon
Brain Speed
Michael Kent

# Contents

# How to use this book

Welcome to Collins *Edexcel GCSE Maths 4th Edition Foundation Student Book*. You will find a number of features in the book that will help you with your course of study.

## Chapter overview

See what maths you will be doing, what skills you will learn and how you can build on what you already know.

## About this chapter

Maths has numerous everyday uses. This section puts the chapter's mathematical skills and knowledge into context, historically and for the modern world.

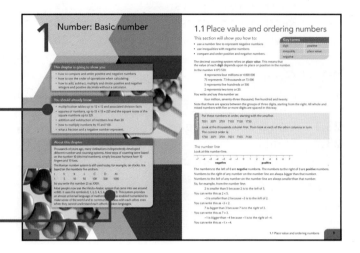

## This section will show you …

Detailed learning objectives show you the skills you will learn in that section.

## Key terms and glossary

Learn the important words you need to know. The explanations for the words in bold in the text can be found in the glossary at the back of the book.

## Examples

Understand the topic before you start the exercise by reading the examples in blue boxes. These take you through questions step by step.

## Exercises

Once you have worked through the examples you will be ready to tackle the exercises. There are plenty of questions, carefully designed to provide you with enough practice to become fluent.

## Hints and tips

These are provided where extra guidance can save you time or help you out.

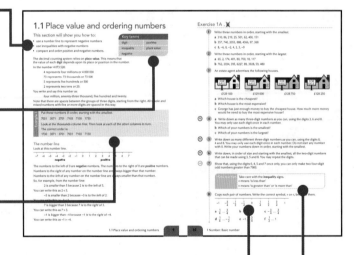

## Colour-coded questions

The questions in the exercises and the review questions are colour-coded, to show you how difficult they are. Most exercises start with more accessible questions and progress through intermediate to more challenging questions.

## Mathematical skills

As you progress you will be expected to absorb new ways of thinking and working mathematically. Some questions are designed to help you develop a *specific* skill. Look for the icons:

**(MR)** Mathematical reasoning – you need to apply your skills and draw conclusions from mathematical information.

**(CM)** Communicate mathematically – you need to show how you have arrived at your answer by using mathematical arguments.

**(PS)** Problem solving and making connections – you need to devise a strategy to answer the question, based on the information you are given.

**(EV)** Evaluate and interpret – your answer needs to show that you have considered the information you are given and commented upon it.

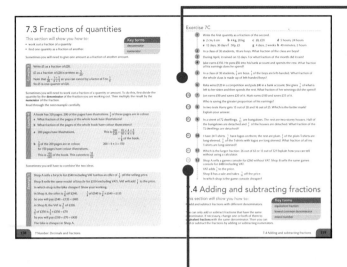

## Worked exemplars

Develop your mathematical skills with detailed commentaries walking you through how to approach a range of questions.

## Ready to progress?

Review what you have learnt from the chapter with this colour-coded summary to check you are on track throughout the course.

## Review questions

Practise what you have learnt in all of the previous chapters and put your mathematical skills to the test. Questions range from accessible through to more challenging.

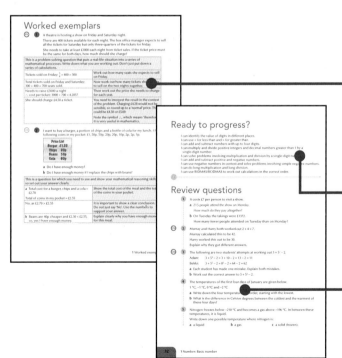

# 1 Number: Basic number

## This chapter is going to show you:

- how to compare and order positive and negative numbers
- how to use the order of operations when calculating
- how to add, subtract, multiply and divide positive and negative integers and positive decimals without a calculator.

## You should already know:

- multiplication tables up to $12 \times 12$ and associated division facts
- squares of numbers, up to $15 \times 15 = 225$ and the square roots of the square numbers up to 225
- addition and subtraction of numbers less than 20
- how to multiply numbers by 10 and 100
- what a fraction and a negative number represent.

## About this chapter

Thousands of years ago, many civilisations independently developed different number and counting systems. Most ways of counting were based on the number 10 (decimal numbers), simply because humans have 10 fingers and 10 toes.

The Roman number system is still used today, for example, on clocks. It is based on the numbers five and ten.

| I | V | X | L | C | D | M |
|---|---|---|---|---|---|---|
| 1 | 5 | 10 | 50 | 100 | 500 | 1000 |

So you write the number 23 as: XXIII.

Most people now use the Hindu–Arabic system that came into use around AD900. It uses the symbols 0, 1, 2, 3, 4, 5, 6, 7, 8 and 9. This system provides an almost universal language of mathematics. It has enabled humankind to make sense of the world and to communicate ideas with each other, even when they cannot understand each other's spoken languages.

# 1.1 Place value and ordering numbers

This section will show you how to:

- use a number line to represent negative numbers
- use inequalities with negative numbers
- compare and order positive and negative numbers.

The decimal counting system relies on **place value**. This means that the value of each **digit** depends upon its place or position in the number.

In the number 4 073 520:

> 4 represents four millions or 4 000 000
>
> 73 represents 73 thousands or 73 000
>
> 5 represents five hundreds or 500
>
> 2 represents two tens or 20.

You write and say this number as:

> four million, seventy-three thousand, five hundred and twenty.

Note that there are spaces between the groups of three digits, starting from the right. All whole and mixed numbers with five or more digits are spaced in this way.

---

**Example 1**

Put these numbers in order, starting with the *smallest*.

7031  3071  3701  7103  7130  1730

Look at the thousands column first. Then look at each of the other columns in turn.

The correct order is:

1730  3071  3701  7031  7103  7130

---

## The number line

Look at this number line.

The numbers to the left of 0 are **negative** numbers. The numbers to the right of 0 are **positive** numbers.

Numbers to the *right* of any number on the number line are always *bigger* than that number.

Numbers to the *left* of any number on the number line are always *smaller* than that number.

So, for example, from the number line:

> 2 is *smaller* than 5 because 2 is to the *left* of 5.

You can write this as 2 < 5.

> −3 is *smaller* than 2 because −3 is to the *left* of 2.

You can write this as −3 < 2.

> 7 is *bigger* than 3 because 7 is to the *right* of 3.

You can write this as 7 > 3.

> −1 is *bigger* than −4 because −1 is to the *right* of −4.

You can write this as −1 > −4.

# Exercise 1A

1. Write these numbers in order, starting with the *smallest*.

   a 310, 86, 219, 25, 501, 62, 400, 151

   b 357, 740, 2053, 888, 4366, 97, 368

   c 8, −6, 0, −2, 4, 2, 3, −9

2. Write these numbers in order, starting with the *largest*.

   a 65, 2, 174, 401, 80, 700, 18, 117

   b 762, 2034, 395, 6227, 89, 3928, 59, 480

3. An estate agent advertises the following houses.

| £129 100 | £129 000 | £128 750 | £128 250 |

   a Which house is the cheapest?

   b Which house is the most expensive?

   c George has just enough money to buy the cheapest house. How much more money would he need to buy the most expensive house?

**(PS)** 4. a Write down as many three-digit numbers as you can, using the digits 3, 6 and 8. You may only use each digit once in each number.

   b Which of your numbers is the smallest?

   c Which of your numbers is the largest?

**(PS)** 5. Write down as many different three-digit numbers as you can, using the digits 0, 4 and 8. You may only use each digit once in each number. Do not start any number with 0. Write your numbers down in order, starting with the smallest.

**(PS)** 6. Write down, in order of size and starting with the smallest, all the two-digit numbers that can be made using 3, 5 and 8. You may repeat the digits.

**(MR)** 7. Show that, using the digits 0, 4, 5 and 7 once only, you can only make two four-digit odd numbers greater than 7000.

> **Hints and tips** Take care with the **inequality** signs.
> < means 'is less than'
> > means 'is greater than' or 'is more than'

8. Copy each pair of numbers. Write the correct symbol, > or <, between them.

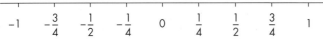

$$-1 \qquad -\frac{3}{4} \qquad -\frac{1}{2} \qquad -\frac{1}{4} \qquad 0 \qquad \frac{1}{4} \qquad \frac{1}{2} \qquad \frac{3}{4} \qquad 1$$

   a $\frac{1}{4} \ldots \frac{3}{4}$      b $-\frac{1}{2} \ldots 0$      c $-\frac{3}{4} \ldots \frac{3}{4}$

   d $\frac{1}{4} \ldots -\frac{1}{2}$      e $-1 \ldots \frac{3}{4}$      f $\frac{1}{2} \ldots 1$

**9** Nick has these number cards.

   **a** Copy this blank calculation. Write numbers into the boxes to give the largest possible total.

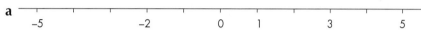

   **b** Copy this blank calculation. Write numbers into the boxes to give the smallest possible difference.

**10** Copy each pair of numbers. Write the correct symbol, > or <, between them.

   **a** −1 ... 3      **b** 3 ... 2      **c** −4 ... −1      **d** −5 ... −4

   **e** 1 ... −6      **f** −3 ... 0      **g** −2 ... −1      **h** 2 ... −3

   **i** 5 ... −6      **j** 3 ... 4      **k** −7 ... −5      **l** −2 ... −4

**11** Copy and complete these number lines by filling in all the missing numbers.

   **a**
   −5      −2      0   1      3      5

   **b**
   −8      −4      0   2      6

   **c**
   −9  −6      0   3   6      12

   **d**
   −100      −40      0   20      60

**12** Copy and complete the weather report below. Use these temperatures to fill the gaps.

   2 °C      −2 °C      −4 °C      6 °C

   The hottest place today is Edinburgh, with a temperature of ____, while in Eastbourne a ground frost has left the temperature just below zero, at ____. In Bristol it is even colder, at ____. Finally, in Tenby the temperature is just above freezing, at ____.

**13** Mark the numbers $-4\frac{1}{2}$, $+3\frac{3}{4}$ and $-\frac{1}{4}$ on a copy of this number line.

   −6  −5  −4  −3  −2  −1  0  1  2  3  4  5  6

**14** The sign ⩾ means 'greater than or equal to'.

   The sign ⩽ means 'less than or equal to'.

   $N$ is a whole number.

   Write down all the possible values that $N$ can have in each case.

   **a** $N \geqslant 7$ and $N < 12$      **b** $N > 5$ and $N \leqslant 10$      **c** $N \geqslant 2$ and $N \leqslant 8$

# 1.2 Order of operations and BIDMAS

## This section will show you how to:

- work out the answers to problems with more than one mathematical operation.

You should already know the BIDMAS (or BODMAS) rule. It tells you the order in which you *must* do the operations in calculations involving more than one operation, for example $9 \div 3 + 4 \times 2$.

| | | | | |
|---|---|---|---|---|
| **B** | Brackets | **B** | Brackets |
| **I** | Indices (Powers) | **O** | pOwers |
| **D** | Division | **D** | Division |
| **M** | Multiplication | **M** | Multiplication |
| **A** | Addition | **A** | Addition |
| **S** | Subtraction | **S** | Subtraction |

So, to work out $9 \div 3 + 4 \times 2$:

| | | | |
|---|---|---|---|
| first, divide | $9 \div 3 = 3$ | giving | $3 + 4 \times 2$ |
| then multiply | $4 \times 2 = 8$ | giving | $3 + 8$ |
| then add | $3 + 8 = 11$ | | |

**Example 2**

Work out $60 - 5 \times 3^2 + (4 \times 2)$.

| | | | |
|---|---|---|---|
| First, work out the brackets. | $(4 \times 2) = 8$ | gives | $60 - 5 \times 3^2 + 8$ |
| Next, work out the index (power). | $3^2 = 9$ | gives | $60 - 5 \times 9 + 8$ |
| Then multiply. | $5 \times 9 = 45$ | gives | $60 - 45 + 8$ |
| Now add. | $60 + 8 = 68$ | gives | $68 - 45$ |
| Finally, subtract. | $68 - 45 = 23$ | | |

So $60 - 5 \times 3^2 + (4 \times 2) = 23$

## Exercise 1B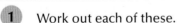

**1** Work out each of these.

    **a** $2 \times 3 + 5 =$     **b** $6 \div 3 + 4 =$     **c** $5 + 7 - 2 =$

    **d** $4 \times 6 \div 2 =$     **e** $2 \times 8 - 5 =$     **f** $3 \times 4 + 1 =$

    **g** $3 \times 4 - 1 =$     **h** $3 \times 4 \div 1 =$     **i** $12 \div 2 + 6 =$

    **j** $12 \div 6 + 2 =$     **k** $3 + 5 \times 2 =$     **l** $12 - 3 \times 3 =$

**2** Work out each of these.
Remember to work out the brackets first.

    **a** $2 \times (3 + 5) =$     **b** $6 \div (2 + 1) =$     **c** $(5 + 7) - 2 =$

    **d** $5 + (7 - 2) =$     **e** $3 \times (4 \div 2) =$     **f** $3 \times (4 + 2) =$

    **g** $2 \times (8 - 5) =$     **h** $3 \times (4 + 1) =$     **i** $3 \times (4 - 1) =$

    **j** $3 \times (4 \div 1) =$     **k** $12 \div (2 + 2) =$     **l** $(12 \div 2) + 2 =$

**3** Copy each of these, put a loop round the part that you do first and then work out the answer.

The first one has been done for you.

**a** $(3 \times 3) - 2 = 7$  **b** $3 + 2 \times 4 =$  **c** $9 \div 3 - 2 =$

**d** $9 - 4 \div 2 =$  **e** $5 \times 2 + 3 =$  **f** $5 + 2 \times 3 =$

**g** $10 \div 5 - 2 =$  **h** $10 - 4 \div 2 =$  **i** $4 \times 6 - 7 =$

**j** $7 + 4 \times 6 =$  **k** $6 \div 3 + 7 =$  **l** $7 + 6 \div 2 =$

**4** Work out each of these.

**a** $6 \times 6 + 2 =$  **b** $6 \times (6 + 2) =$  **c** $6 \div 6 + 2 =$

**d** $12 \div (4 + 2) =$  **e** $12 \div 4 + 2 =$  **f** $2 \times (3 + 4) =$

**g** $3 \times 5 + 5 =$  **h** $6 \times 2 + 7 =$  **i** $6 \times (2 + 7) =$

**j** $12 \div 3 + 3 =$  **k** $12 \div (3 + 3) =$  **l** $14 - 7 \times 1 =$

**m** $(14 - 7) \times 1 =$  **n** $12 - 6 \div 3 =$  **o** $(12 - 6) \div 3 =$

**5** Copy each number sentence and put in brackets, where necessary, to make the answer correct.

**a** $3 \times 4 + 1 = 15$  **b** $6 \div 2 + 1 = 4$  **c** $6 \div 2 + 1 = 2$

**d** $4 + 4 \div 4 = 5$  **e** $4 + 4 \div 4 = 2$  **f** $16 - 4 \div 3 = 4$

**g** $20 - 10 \div 2 = 15$  **h** $3 \times 5 + 5 = 30$  **i** $6 \times 4 + 2 = 36$

**j** $12 \div 3 + 3 = 7$  **k** $24 \div 8 - 2 = 1$  **l** $24 \div 8 - 2 = 4$

**6** Three dice are thrown.
They give scores of three, one and four.

A class uses the numbers to make up these questions.
Work out the answers.

**a** $3 + 4 + 1 =$  **b** $3 + 4 - 1 =$  **c** $4 \times 3 + 1 =$

**d** $(4 - 1) \times 3 =$  **e** $4 \times 3 \times 1 =$  **f** $(3 - 1) \times 4 =$

**g** $(4 + 1) \times 3 =$  **h** $4 \times (3 + 1) =$  **i** $1 \times (4 - 3) =$

 **7** Jack says that $5 + 6 \times 7$ is equal to 77.

Is he correct?

Explain your answer.

 **8** This is Micha's homework.

| | |
|---|---|
| *a* $2 + 3 \times 4 = 20$ | *b* $8 - 4 \div 4 = 7$ |
| *c* $6 + 3 \times 2 = 12$ | *d* $7 - 1 \times 5 = 30$ |
| *e* $2 \times 7 + 2 = 16$ | *f* $9 - 3 \times 3 = 18$ |

Copy the questions in which she has made mistakes and work out the correct answers.

**9** Add ÷, ×, + or − signs to make each number sentence correct.

   **a** $2 \ldots 3 \ldots 5 = 11$      **b** $2 \ldots 3 \ldots 5 = 16$      **c** $2 \ldots 3 \ldots 5 = 17$

   **d** $5 \ldots 3 \ldots 2 = 4$      **e** $5 \ldots 3 \ldots 2 = 13$      **f** $5 \ldots 3 \ldots 2 = 30$

(CM) **10** Which is smaller?     $4 + 5 \times 3$   or   $(4 + 5) \times 3$

    Show your working.

(PS) **11** Here is a list of numbers, some signs and one pair of brackets.

    2    5    6    18    −    ×    =    (  )

    Use *all* of them to make a correct calculation.

(PS) **12** Here is a list of numbers, some signs and one pair of brackets.

    3    4    5    8    −    ÷    =    (  )

    Use *all* of them to make a correct calculation.

(MR) **13** Jamil has a piece of pipe that is 10 m long.

    He wants to use his calculator to work out how much pipe will be left when he cuts off three pieces, each of length 1.5 m.

    Which calculations would give him the correct answer?

    $10 - 3 \times 1.5$     $10 - 1.5 + 1.5 + 1.5$     $10 - 1.5 - 1.5 - 1.5$

**14** Say whether each statement is true or false.

   **a** $6 \times 5 - 2 = 28$      **b** $8 \times (6 - 3) + 4 = 49$      **c** $36 - 12 \times 2 = 48$

   **d** $3 + 7 \times 3 = 24$      **e** $25 - (6 + 2) = 21$      **f** $29 - (6 - 2) = 25$

   **g** $60 \div (6 + 4) = 14$      **h** $6 - 2^2 = 16$      **i** $9 - 3^2 = 0$

   **j** $16 + 3 - 2 \times 5 = 9$

(PS) **15** You can use four 4s to make the numbers from 1 to 5.

   **a** Copy each calculation and insert a mathematical symbol +, −, × or ÷ into each box (☐) to make it true.

> **Hints and tips** The symbol + is used 7 times, the symbol − is used once, the symbol × is used twice and the symbol ÷ is used 5 times.

    **i** $(4 \square 4) \square (4 \square 4) = 1$      **ii** $(4 \square 4) \square (4 \square 4) = 2$

    **iii** $(4 \square 4 \square 4) \square 4 = 3$      **iv** $(4 \square 4) \square 4 \square 4 = 4$

    **v** $(4 \square 4 \square 4) \square 4 = 5$

   **b** Use four 4s and some of the symbols (, ), +, −, × and ÷ to make the numbers 5 to 9.

   **c** Use four 4s and some of the symbols (, ), +, −, ×, ÷ and √ to make the number 10.

> **Hints and tips** Remember that the square root of 4 ($\sqrt{4}$) = 2.

(MR) **16** Copy this diagram and write the digits 1 to 9 in the correct places so that the rows and columns give the totals shown. The numbers 8 and 5 have been put in to help you.

| 8 | ÷ |  | + |  | = | 8 |
|---|---|---|---|---|---|---|
| × |  | + |  | − |  |  |
|  | + |  | − |  | = | 10 |
| − |  | ÷ |  | × |  |  |
|  | + |  | − | 5 | = | 5 |
| = |  | = |  | = |  |  |
| 9 |  | 7 |  | 1 |  |  |

# 1.3 The four rules

This section will show you how to:

- use the four rules of arithmetic with integers and decimals.

## Addition with positive numbers

Remember these rules, when you are adding positive whole numbers or decimals.

- The answer will always be larger than the bigger number.
- If one or more of the numbers is a decimal, line up the decimal points.
- Fill any blanks with zeros.
- Always start by adding the numbers in the right-hand column and work from right to left.
- When the total for a column is more than nine, carry a digit into the next column on the left, as shown in Example 3.
- Always write down the carried digit, as a reminder to include it in the addition for the next column.

**Example 3**

Add these numbers.    **a**   $167 + 25 + 344$    **b**   $22.96 + 117.3$

**a**
```
    1 6 7
      2 5
  + 3 4 4
  -------
    5 3 6
    1 1
```

**b**
```
    2 2 . 9 6
  + 1 1 7 . 3 0
  -----------
    1 4 0 . 2 6
        1 1
```

## Subtraction with positive numbers

Remember these rules, when you are subtracting positive whole numbers or decimals.

- When you are subtracting from the bigger number, the answer will always be smaller than the bigger number.
- If one or more of the numbers is a decimal, line up the decimal points.
- Fill any blanks with zeros.
- Always start by subtracting in the right-hand column and work from right to left.
- When you have to take a bigger number from a smaller number in a column, you must 'borrow' from the column to the left and write a 1 before the smaller number, as shown in Example 4.

**Example 4**

Subtract these numbers.    **a**   $874 - 215$    **b**   $32.7 - 5.63$

**a**
```
    8 ⁶7̶ ¹4
  - 2 1 5
  -------
    6 5 9
```

**b**
```
    ²3̶ ¹2 . ⁶7 ¹0
  - 0 5 . 6 3
  -----------
    2 7 . 0 7
```

## Multiplication with positive numbers

Remember these rules, when you are multiplying a positive whole number or decimal greater than 1 by a positive single-digit whole number.

- Always write down the whole number or a decimal you are multiplying first.
- The answer will always be larger than the bigger number.
- There will be the same number of decimal places in the answer as there are in the question.
- When the total for a column is more than nine, carry a digit into the next column on the left, as shown in Example 5.
- Always write down the carried digit, as a reminder to include it in the working for the next column.

**Example 5**

**a** Multiply 231 by 4.

**b** Multiply 5.43 by 6.

In the first multiplication of part **a** ($3 \times 4 = 12$), you need to carry a digit into the next column on the left, as you do for addition.

**a**
$$\begin{array}{r} 2\,1\,3 \\ \times \quad 4 \\ \hline 8\,5\,2 \\ \phantom{8\,5}{}_{1} \end{array}$$

The same thing happens in part **b**.

**b**
$$\begin{array}{r} 5\,.\,4\,3 \\ \times \quad\quad 6 \\ \hline 3\,2\,.\,5\,8 \\ \phantom{3}{}_{2}\phantom{2.5}{}_{1} \end{array}$$

## Division with positive numbers

Remember these rules, when you are dividing a positive whole number or a decimal greater than 1 by a positive single-digit whole number.

- The answer will always be smaller than the bigger number.
- You must start the division at the left-hand side.

**Example 6**

**a** Divide 417 by 3.    **b** Divide 5.08 by 4.

**a** $417 \div 3$ is set out as:

$$\begin{array}{r} 1\,3\,9 \\ 3\,\overline{)\,4\,^{1}1\,^{2}7} \end{array}$$

- First, divide 3 into 4 to get 1 and remainder 1. Note where to put the 1 and the remainder 1.
- Then, divide 3 into 11 to get 3 and remainder 2. Note where to put the 3 and the remainder 2.
- Finally, divide 3 into 27 to get 9 with no remainder, giving the answer 139.

**b** $5.08 \div 4$ is set out as:

$$\begin{array}{r} 1\,.\,2\,7 \\ 4\,\overline{)\,5\,.\,^{1}0\,^{2}8} \end{array}$$

Follow the steps in part **a** to work through part **b**.

# Exercise 1C

**1**  Copy and complete each addition.

**a**      3 6 5
      + 3 4 8
     ―――――

**b**      9 5
      + 5 6
     ―――――

**c**      4 8 7 2
      + 1 5 0 9
     ―――――

**d**      3 1 7
         4 1 6
      + 2 3 5
     ―――――

**e**      2 8 7
      + 3 3 5
     ―――――

**2**  Copy and complete each addition.

**a** 128 + 518

**b** 563 + 85 + 178

**c** 3086 + 58 + 674

**d** 34.7 + 40.8

**e** 8.5 + 18.52 + 65.9

**f** 75.9 + 0.43 + 9

**3**  Copy and complete each subtraction.

**a**      6 3 7
      − 1 8 7
     ―――――

**b**      9 0 8
      − 3 4 5
     ―――――

**c**      9 5 4
      − 4 7 2
     ―――――

**d**      5 7 2
      − 1 5 8
     ―――――

**e**      7 3 2
      − 4 4 7
     ―――――

**f**      6 2 5 4
      − 3 3 6 2
     ―――――

**g**      8 0 4 3
      − 3 6 2 6
     ―――――

**h**      8 4 3 2
      − 4 6 6 5
     ―――――

**i**      8 0 3 4
      − 3 9 4 7
     ―――――

**j**      5 3 7 5
      − 3 5 4 7
     ―――――

**4**  Copy and complete each subtraction.

**a** 354 − 226

**b** 285 − 256

**c** 663 − 329

**d** 50.6 − 32.8

**e** 65.4 − 3.77

**f** 7.33 − 0.448

**5**  The distance from Cardiff to London is 152 miles.

The distance from Cardiff to Edinburgh is 406 miles.

**a** I travel from London to Cardiff and then from Cardiff to Edinburgh. How far have I travelled altogether?

**b** How much further is it to travel from Cardiff to Edinburgh than from Cardiff to London?

**(EV)** **6**  Jon is checking the addition of two numbers.

His answer is 843.

One of the numbers is 591.

What should the other number be?

**(EV)** **7**  Luisa is checking this subtraction.

614 − 258

How do you know that her answer of 444 is incorrect, without doing the whole calculation?

**8** A two-digit number is subtracted from a three-digit number.

The answer is 154.

Work out one pair of possible values for the numbers.

**9** Copy and complete each multiplication.

a    2 3
   ×   5
   ———

b    3 4
   ×   6
   ———

c    4 2
   ×   7
   ———

d    5 3
   ×   4
   ———

e    8 5
   ×   5
   ———

f    5 0
   ×   3
   ———

g    2 0 0
   ×     4
   ———

h    3 2 0
   ×     3
   ———

i    3 4 0
   ×     4
   ———

j    2 5 3
   ×     6
   ———

**10** Set each multiplication in columns and complete it.

a $42 \times 7$

b $74 \times 5$

c $48 \times 6$

d $20.8 \times 4$

e $3.09 \times 7$

f $6.3 \times 4$

g $5.48 \times 3$

h $64.3 \times 5$

i $8 \times 3.75$

**11** Copy and complete each division.

a $438 \div 2$

b $634 \div 2$

c $945 \div 3$

d $63.6 \div 6$

e $2.97 \div 3$

f $84.7 \div 7$

g $75.6 \div 3$

h $8.46 \div 6$

i $5.76 \div 4$

**12** Dean, Sean and Andy are doing a charity cycle ride from Chester to Southend. The distance from Chester to Southend is 235 miles.

a How many miles do all three travel in total?

b Dean is sponsored £6 per mile.

Sean is sponsored £5 per mile.

Andy is sponsored £4 per mile.

How much money do they raise altogether?

**13** The 235-mile charity cycle ride in question **12** must be completed in five days.

a Suppose the cyclists travelled the same distance on each of the five days. How many miles would they travel each day?

b In fact, the cyclists travel 60 miles on the first day and 50 miles on the second day.

The number of miles covered each day should be fewer than for the previous day.

Copy and complete this plan for the remaining days so that they finish the ride on time.

|       | Mileage |
|-------|---------|
| Day 3 |         |
| Day 4 |         |
| Day 5 |         |

**14** Answer these questions by completing a suitable multiplication.

  **a** How many days are there in 17 weeks?

  **b** How many hours are there in four days?

  **c** Eggs are packed in boxes of six. How many eggs altogether are there in 24 boxes?

  **d** Joe bought five boxes of matches. Each box contained 42 matches.
  How many matches did Joe buy altogether?

  **e** A box of toffees holds 35 sweets.
  What is the total number of sweets in six boxes?

**15** Answer these questions by completing a suitable division.

  **a** How many weeks are there in 91 days?

  **b** How long will it take me to save £111, if I save £3 a week?

  **c** A rope, 21.5 m long, is cut into five equal pieces.
  How long is each piece?

  **d** Granny has a bottle of 144 tablets.
  How many days will they last if she takes four each day?

  **e** I share a box of 360 sweets equally among eight children.
  How many sweets does each child receive?

(MR) **16** This is part of the 38 times table.

| ×1 | ×2 | ×3 | ×4 | ×5 |
|----|----|-----|-----|-----|
| 38 | 76 | 114 | 152 | 190 |

Show how you can use this table to work out these multiplications.

  **a** $9 \times 38$     **b** $52 \times 38$     **c** $105 \times 38$

## Arithmetic with negative numbers

The operations of addition and subtraction can be illustrated on a thermometer scale.

Adding a positive number moves the marker *up* the thermometer scale.

$-2 + 6 = 4$

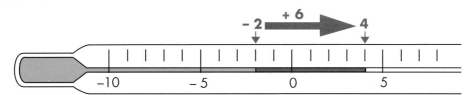

Subtracting a positive number moves the marker *down* the thermometer scale.

$3 - 5 = -2$

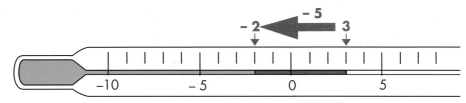

**Example 7**

The temperature at midnight was 2 °C but then it fell by five Celsius degrees. What was the new temperature?

Falling five Celsius degrees means the calculation is 2 – 5, which is equal to –3.
So, the new temperature is –3 °C.

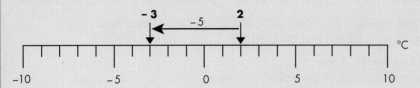

**Example 8**

The temperature is –4 °C. It then falls by five Celsius degrees. What is the new temperature?

The temperature falling five Celsius degrees means the calculation is –4 – 5, which is equal to –9.

So, the new temperature is –9 °C.

# Exercise 1D

1   Use a thermometer scale to help you find the answers to these calculations.

   **a** $-4 + 3 =$  **b** $-6 + 5 =$  **c** $-3 + 5 =$  **d** $-5 + 2 =$

   **e** $-1 - 3 =$  **f** $-2 - 4 =$  **g** $-5 - 1 =$  **h** $3 - 4 =$

   **i** $2 - 7 =$  **j** $1 - 5 =$  **k** $-3 + 7 =$  **l** $5 - 6 =$

2   Copy and complete these calculations *without* using a thermometer scale.

   **a** $5 - 9 =$  **b** $3 - 7 =$  **c** $-2 - 8 =$  **d** $-5 + 7 =$

   **e** $-10 - 22 =$  **f** $-13 - 17 =$  **g** $17 - 25 =$  **h** $-25 + 35 =$

   **i** $-23 - 13 =$  **j** $31 - 45 =$  **k** $-19 + 31 =$  **l** $199 - 300 =$

3   Work these out.

   **a** $8 + 3 - 5 =$   **b** $-2 + 3 - 6 =$   **c** $-1 + 3 + 4 =$

   **d** $-2 - 3 + 4 =$   **e** $-1 + 1 - 2 =$   **f** $-4 + 5 - 7 =$

   **g** $-3 - 3 - 3 =$   **h** $-3 + 4 - 6 =$   **i** $-102 + 45 - 23 =$

   **j** $8 - 10 - 5 =$   **k** $9 - 12 + 2 =$   **l** $99 - 100 - 46 =$

(MR)  4   Write down the calculations shown by these diagrams.

   **a**        **b**        **c**

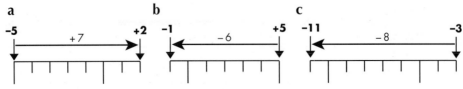

**5** At 5:00 am the temperature in London was −4 °C.

At 11:00 am the temperature was 3 °C.

**a** By how many Celsius degrees did the temperature rise?

**b** At 5:00 am, the temperature in Brighton was two Celsius degrees lower than in London.

What was the temperature is Brighton at 5:00 am?

 **6** Here are five numbers.

    4    7    8    2    5

**a** Use two of the numbers to make a calculation with an answer of −6.

**b** Use three of the numbers to make a calculation with an answer of −1.

**c** Use four of the numbers to make a calculation with an answer of −18.

**d** Use all five of the numbers to make a calculation with an answer of −12.

 **7** A submarine is 1600 feet below sea level.

A radar system can detect submarines down to 900 feet below sea level.

To avoid detection, the submarine captain keeps the submarine 200 feet below the level of detection.

How many feet can the submarine climb and stay safe from detection?

## Adding and subtracting negative numbers

To *subtract a negative number* treat the − − as a +

For example: 4 − (−2) = 4 + 2 = 6

To *add a negative number* treat the + − as a −

For example: 3 + (−5) = 3 − 5 = −2

You can use the **(−)** key when working with negative numbers on a calculator.

---

**Example 9**

Use your calculator to work out −3 + 7.

Press:

The answer should be 4.

---

**Example 10**

Use your calculator to work out −6 − (−2).

Press:

The answer should be −4.

---

## Exercise 1E

**1** Work out each of these. Then use a calculator to check your answers.

**a** 2 − (−4) =       **b** 4 − (−3) =       **c** 3 − (−5) =       **d** 5 − (−1) =

**e** 6 − (−2) =       **f** 8 − (−2) =       **g** −1 − (−3) =       **h** −4 − (−1) =

**i** 4 + (−2) =       **j** 2 + (−5) =       **k** 3 + (−2) =       **l** 1 + (−6) =

**m** 5 + (−2) =       **n** 4 + (−8) =       **o** −2 + (−1) =       **p** −6 + (−2) =

**2** Write down the answer to each calculation.
Then use a calculator to check your answers.

**a** $-2 + 9 =$   **b** $-6 + -2 =$   **c** $-1 + -4 =$   **d** $-8 + -3 =$

**e** $5 - -6 =$   **f** $3 - -3 =$   **g** $6 - -2 =$   **h** $3 - -5 =$

**i** $-5 - -3 =$   **j** $-2 - -1 =$   **k** $-4 - 5 =$   **l** $2 - 7 =$

**m** $-3 + 8 =$   **n** $-4 + -5 =$   **o** $1 - -7 =$   **p** $-5 - -5 =$

**3** The temperature at midnight was 4 °C.
Find the temperature after it *fell* by:

**a** 1 Celsius degree   **b** 4 Celsius degrees   **c** 7 Celsius degrees

**d** 9 Celsius degrees   **e** 15 Celsius degrees.

**4** What is the *difference* between the following temperatures?

**a** 4 °C and −6 °C   **b** −2 °C and −9 °C   **c** −3 °C and 6 °C

**5** Rewrite this list in order of size, starting with the smallest.

1     −5     3     −6     −9     8     −1     2

**6** What number do you *add to* 5 to get:

**a** 7   **b** 2   **c** 0

**d** −2   **e** −5   **f** −15?

**7** What number do you *subtract from* 4 to get:

**a** 2   **b** 0   **c** 5

**d** 9   **e** 15   **f** −4?

**8** What number do you *add to* −5 to get:

**a** 8   **b** −3   **c** 0

**d** −1   **e** 6   **f** −7?

**9** What number do you *subtract from* −3 to get:

**a** 7   **b** 2   **c** −1

**d** −7   **e** −10   **f** 1?

**10** Use a calculator to work these out.

**a** $-7 + -3 - -5 =$   **b** $6 + 7 - 7 =$   **c** $-3 + -4 - -7 =$

**d** $-4 + 5 - 7 =$   **e** $-4 + -6 - -8 =$   **f** $103 - -102 - -7 =$

**g** $-45 + -56 - -34 =$   **h** $-3 + 4 - -6 =$   **i** $102 + -45 - 32 =$

**11** The thermometer in a car is inaccurate by up to two Celsius degrees.

An ice alert warning comes when the thermometer measures the temperature as 3 °C or below.

Will the alert come on for all negative temperatures?

Give a reason for your answer.

**12** Give the outputs of each function machine.

a  $-4, -3, -2, -1, 0$ → [+ 3] → ?, ?, ?, ?, ?

b  $-4, -3, -2, -1, 0$ → [- 5] → ?, ?, ?, ?, ?

c  $-5, -4, -3, -2, -1, 0$ → [+ 3] → ?, ?, ?, ?, ?, ? → [- 2] → ?, ?, ?, ?,

d  $-5, -4, -3, -2, -1, 0$ → [- 7] → ?, ?, ?, ?, ?, ? → [- 2] → ?, ?, ?, ?,

e  $-3, -2, -1, 0, 1, 2, 3$ → [- 5] → ?, ?, ?, ?, ?, ? → [+ 3] → ??, ?, ?, ?,

f  $-3, -2, -1, 0, 1, 2, 3$ → [- 7] → ?, ?, ?, ?, ?, ? → [+ 9] → ??, ?, ?, ?,

**13** Write down the number missing from the box to make each number sentence true.

a  $2 + -6 = \square$

b  $4 + \square = 7$

c  $-4 + \square = 0$

d  $5 + \square = -1$

e  $3 + 4 = \square$

f  $\square - -5 = 7$

g  $2 + -2 = \square$

h  $\square - 2 = -2$

i  $-2 + -4 = \square$

j  $2 + 3 + \square = -2$

k  $-2 + -3 + -4 = \square$

l  $\square - 5 = -1$

m  $-3 - \square = 0$

n  $-6 + -3 = \square$

o  $\square - 3 - -2 = -1$

**(PS)** **14** You have these cards.

$-9$ $-8$ $-4$ $0$ $+1$ $+3$ $+5$

a  Which card should you choose to make the answer to this sum as large as possible?

$+6 + \square = \ldots$

What is the answer to the calculation?

b  Which card should you choose to make the answer to this sum as small as possible?

$+6 + \square = \ldots$

What is the answer to the calculation?

c  Which card should you choose to make the answer to this subtraction as large as possible?

$+6 - \square = \ldots$

What is the answer to the calculation?

d  Which card should you choose to make the answer to to this subtraction as small as possible?

$+6 - \square = \ldots$

What is the answer to the calculation?

**15** You have these cards.

$$-9 \quad -7 \quad -5 \quad -4 \quad 0 \quad +1 \quad +2 \quad +4 \quad +7$$

**a** Which cards should you choose to make the answer to this calculation as large as possible?

$+5 + \square - \square = \ldots$

What is the answer to the calculation?

**b** Which cards should you choose to make the answer to this calculation as small as possible?

$+5 + \square - \square = \ldots$

What is the answer to the calculation?

**c** Which cards should you choose to make the answer to this number sentence zero? Give all possible answers.

$\square + \square = 0$

 **16** Two single-digit numbers have a sum of 5.

One of the numbers is negative.

The other number is even.

What are the two numbers, when the even number is as large as possible?

 **17** Two numbers have a sum of –12 and a difference of 3.

Work out the numbers.

## Multiplying and dividing with negative numbers

These are the rules for multiplying and dividing with negative numbers.

- When the signs of the numbers are the *same*, the answer is *positive*.
- When the signs of the numbers are *different*, the answer is *negative*.

$$2 \times 4 = 8 \qquad 12 \div -3 = -4 \qquad -2 \times -3 = 6 \qquad -12 \div -3 = 4$$

A common error is to confuse, for example, $-3^2$ and $(-3)^2$.

$$-3^2 = -3 \times 3 = -9 \qquad \text{but} \qquad (-3)^2 = -3 \times -3 = +9$$

---

**Example 11**

Work out each of these.

**a** $(-2)^2$  **b** $(-2)^2 + (-6)^2$  **c** $(-6)^2 - (-2)^2$  **d** $(-2--6)^2$

**a** $(-2)^2 = -2 \times -2$

$= +4$

**b** $(-2)^2 + (-6)^2 = +4 + -6 \times -6$

$= 4 + 36$

$= 40$

**c** $(-6)^2 - (-2)^2 = 36 - 4$

$= 32$

**d** $(-2--6)^2 = (-2+6)^2$

$= (4)^2$

$= 16$

---

# Exercise 1F

**1** Write down the answers.

    **a** $-3 \times 5$      **b** $-2 \times 7$      **c** $-4 \times 6$      **d** $-2 \times -3$      **e** $-7 \times -2$

    **f** $-12 \div -6$      **g** $-16 \div 8$      **h** $24 \div -3$      **i** $16 \div -4$      **j** $-6 \div -2$

    **k** $4 \times -6$      **l** $5 \times -2$      **m** $6 \times -3$      **n** $-2 \times -8$      **o** $-9 \times -4$

    **p** $24 \div -6$      **q** $12 \div -1$      **r** $-36 \div 9$      **s** $-14 \div -2$      **t** $100 \div 4$

    **u** $-2 \times -9$      **v** $32 \div -4$      **w** $5 \times -9$      **x** $-21 \div -7$      **y** $-5 \times 8$

**2** Write down the answers.

    **a** $-3 + -6$      **b** $-2 \times -8$      **c** $2 + -5$      **d** $8 \times -4$      **e** $-36 \div -2$

    **f** $-3 \times -6$      **g** $-3 - -9$      **h** $48 \div -12$      **i** $-5 \times -4$      **j** $7 - -9$

    **k** $-40 \div -5$      **l** $-40 + -8$      **m** $4 - -9$      **n** $5 - 18$      **o** $72 \div -9$

    **p** $-7 - -7$      **q** $8 - -8$      **r** $6 \times -7$      **s** $-6 \div -1$      **t** $-5 \div -5$

    **u** $-9 - 5$      **v** $4 - -2$      **w** $4 \div -1$      **x** $-7 \div -1$      **y** $-4 \times 0$

**(MR) 3** What number do you multiply by $-3$ to get:

    **a** 6          **b** $-90$          **c** $-45$          **d** 81          **e** 21?

**(MR) 4** What number do you divide $-36$ by to get:

    **a** $-9$          **b** 4          **c** 12          **d** $-6$          **e** 9?

**5** Work out the value of each calculation.

    **a** $-6 + (4 - 7)$          **b** $-3 - (-9 - -3)$          **c** $8 + (2 - 9)$

> **Hints and tips** Work out the calculations in brackets first.

**6** Work out the value of each expression.

    **a** $4 \times (-8 \div -2)$          **b** $-8 - (3 \times -2)$          **c** $-1 \times (8 - -4)$

**7** What do you get if you divide $-48$ by:

    **a** $-2$          **b** $-8$          **c** 12          **d** 24?

**(PS) 8** Write down six different multiplications that give the answer $-12$.

**(PS) 9** Write down six different divisions that give the answer $-4$.

**10** Write down the answers.

    **a** $-3 \times -7$      **b** $3 + -7$      **c** $-4 \div -2$      **d** $-7 - 9$      **e** $-12 \div -6$

    **f** $-12 - -7$      **g** $5 \times -7$      **h** $-8 + -9$      **i** $-4 + -8$      **j** $-3 + 9$

    **k** $-5 \times -9$      **l** $-16 \div 8$      **m** $-8 - -8$      **n** $6 \div -6$      **o** $-4 + -3$

    **p** $-9 \times 4$      **q** $-36 \div -4$      **r** $-4 \times -8$      **s** $-1 - -1$      **t** $2 - 67$

**11**  **a** Work out $6 \times -2$.

    **b** The average temperature drops by two Celsius degrees every day for six days. How much has the temperature dropped altogether?

**(CM)**     **c** The temperature drops by six Celsius degrees for each of the next three days. Write down the calculation to work out the total change in temperature over these three days.

**(EV) 12** Put these calculations in order, from the one with the smallest answer to the one with the biggest answer.

$-5 \times 4$     $-20 \div 2$     $-16 \div -4$     $3 \times -6$

**13** Work these out.

**a** $(-4)^2$       **b** $(-4)^2 + (-5)^2$       **c** $(-5)^2 - (-2)^2$       **d** $(-3 + -5)^2$

## Long multiplication with integers

The three most common methods of long multiplication (without a calculator) are:

- the **grid method** (or box method), see Example 12
- the **column method** (or traditional method), see Example 13
- the **partition method**, see Example 14.

Whichever method you use, make sure your working is clear.

---

**Example 12**

Work out $243 \times 68$.

Use the grid method. Split the two numbers into hundreds, tens and units.

Write them in a grid. Multiply all the pairs of numbers.

| $\times$ | 200 | 40 | 3 |
|---|---|---|---|
| **60** | 12 000 | 2400 | 180 |
| **8** | 1600 | 320 | 24 |

Add the separate answers to find the total.

```
   1 2 0 0 0
     2 4 0 0
       1 8 0
     1 6 0 0
       3 2 0
 +      2 4
   1 6 5 2 4
       1 1
```

So $243 \times 68 = 16\,524$

---

Note the use of carried figures in the previous example, to help with the calculation.
Always write carried figures much smaller than the other numbers, so that you don't confuse them with the main calculation.

---

**Example 13**

Work out $357 \times 24$.

Use the column method. This is perhaps the method that is most commonly used.

```
      3 5 7     357 multiplied by 4.
  ×     2 4     357 multiplied by 20.
    1 4 2 8
      2 2
    7 1 4 0     Write down the 0 first, then multiply by 2.
      1 1
    8 5 6 8     The two results are added together.
```

So $357 \times 24 = 8568$

---

Example 14

Work out $358 \times 74$.

Use the partition method. Set out a grid.

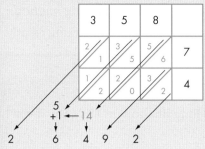

- Write the larger number along the top and the smaller number down the right-hand side.
- Multiply each possible pair in the grid, putting each answer into the two halves of the box, as shown.
- Starting on the right, add the numbers along each diagonal. If a total is larger than 9 (in this example there is a total of 14), split the number and put the 1 in the next column on the left, ready to be added in that diagonal.
- When you have completed the totalling, the number you are left with is the answer to the multiplication.

So $358 \times 74 = 26\,492$

# Exercise 1G

**1** Use your preferred written method to complete these multiplications.

| | | | |
|---|---|---|---|
| **a** $357 \times 34$ | **b** $724 \times 63$ | **c** $714 \times 42$ | **d** $898 \times 23$ |
| **e** $958 \times 54$ | **f** $676 \times 37$ | **g** $239 \times 81$ | **h** $437 \times 29$ |
| **i** $2504 \times 56$ | **j** $4037 \times 23$ | **k** $8009 \times 65$ | **l** $2070 \times 38$ |

**2** A supermarket had a delivery of 125 crates of soup. Each crate holds 48 cans of soup. The supermarket shelves have space for 2500 cans. How many cans will be left in the store room?

**(MR) 3** Greystones Primary School has 12 classes, each of which has 26 students. The school hall will hold 250 students. Can all the students fit in the hall for an assembly?

**4** Suhail lives 450 m from his school. He walks to school and back each day.

Suhail claims that in a school term of 64 days he walks more than the distance of a marathon, to and from school. A marathon is 42.1 km. Is he correct?

**(PS) 5** There are seven columns on one page of a newspaper. In each column there are 172 lines, and in each line there are 50 letters. The newspaper has the equivalent of 20 pages of print, excluding photographs and adverts. Are there more than a million letters in the paper?

**(PS) 6** A tank of water was emptied into casks. Each cask held 81 litres. 71 casks were filled and there were 68 litres left over. How much water was there in the tank to start with?

**(CM) 7** Joy is doing a sponsored walk. She managed to get 18 people to sponsor her, each for 35p per kilometre. She walked a total of 48 km. Did Joy reach her fundraising target of £400?

## Long division

There are several methods for **long division**. The most popular method is probably the *Italian method*, also called the *DMSB method* (divide, multiply, subtract and bring down) shown in Example 15. Example 15 shows a method of repeated subtraction, which is also called the *chunking method*.

You can use any method you are confident with as long as you always show your working clearly.

---

**Example 15**

Work out $840 \div 24$.

Jot down the appropriate times table before you start the long division. In this case, it will be the 24 times table.

| 1 | 2 | 3 | 4 | 5 | 6 | 7 | 8 | 9 |
|---|---|---|---|---|---|---|---|---|
| 24 | 48 | 72 | 96 | 120 | 144 | 168 | 192 | 216 |

```
        3 5
  24 ) 8 4 0
        7 2
      1 2 0
      1 2 0
            0
```

Start with: How many 24s in 8?

There are none, of course, so move on to 84.                    D

Look at the 24 times table to find the biggest number that is less than 84.

This is 72, which is $3 \times 24$.                              M

Take 72 from 84 and bring down the 0.                          S, B

Look again at the 24 times table to find that $5 \times 24 = 120$.   D, M

Because $120 - 120 = 0$, you have finished.                     S

So $840 \div 24 = 35$

---

If you are very confident, you can use the same method as for short division, without writing down all the numbers. It will look like this.

```
        3 5
  24 ) 8 4¹²0
```

Notice that you place the remainder from 84 in front of the 0 to make it 120.

---

**Example 16**

Work out $1655 \div 35$.

Jot down some of the multiples of 35 that may be useful.

$1 \times 35 = 35$   $2 \times 35 = 70$   $5 \times 35 = 175$   $10 \times 35 = 350$   $20 \times 35 = 700$

```
    1 6 5 5
  -   7 0 0      20 × 35    From 1655, subtract a large multiple of 35, such as 20 × 35 = 700.
      9 5 5
  -   7 0 0      20 × 35    From 955, subtract a large multiple of 35, such as 20 × 35 = 700.
      2 5 5
  -   1 7 5       5 × 35    From 255, subtract a multiple of 35, such as 5 × 35 = 175.
        8 0
  -     7 0       2 × 35    From 80, subtract a multiple of 35, such as 2 × 35 = 70.
        1 0      ‾‾‾‾
                    47      Once you have found the remainder of 10, you cannot subtract any
```
more multiples of 35. Add up the multiples to see how many times you have subtracted 35.

So, $1655 \div 35 = 47$ remainder 10

Sometimes, as here, you will not need a whole multiplication table. You could jot down only those parts of the table that you will need. This is very helpful if you are working *without* a calculator.

---

Example 17

Naseema is organising a coach trip for 640 people. Each coach will carry 46 people. How many coaches should she book?

You need to divide the number of people (640) by the number of people in a coach (46).

```
        1 3
46 ) 6 4 0
     4 6
     1 8 0
     1 3 8
       4 2
```

Start by dividing 64 by 46 … which gives 1 remainder 18.

Now divide 180 by 46 … which gives 3 remainder 42.

You have come to the end of the whole-number division to give 13 remainder 42.

This tells Naseema that she needs 14 coaches to take all 640 passengers as every passenger will need a seat.

There will be 46 – 42 = 4 spare seats.

## Exercise 1H

**1** Solve these by long division.

a $525 \div 21$   b $480 \div 32$   c $925 \div 25$   d $645 \div 15$

e $621 \div 23$   f $576 \div 12$   g $1643 \div 31$   h $728 \div 14$

i $832 \div 26$   j $2394 \div 42$   k $829 \div 22$   l $780 \div 31$

m $895 \div 26$   n $873 \div 16$   o $875 \div 24$   p $225 \div 13$

q $759 \div 33$   r $1478 \div 24$   s $756 \div 18$   t $1163 \div 43$

**2** 3600 football supporters want to go to an away game by coach. Each coach can hold 53 passengers.
How many coaches will they need altogether?

**3** Pencils costs 26p each. I have £10.
How many pencils can I buy?

**(PS) 4** Kirsty is collecting a set of 40 toy animals. Each animal costs 45p. Each month she spends up to £5 of her pocket money on these animals. How many months will it take her to buy the full set?

**(PS) 5** Amina is saving up to buy a ticket for a concert. The price of a ticket is £25. She is paid 75p per hour to look after her little sister.
How many hours must Amina spend looking after her sister to pay for the ticket?

**(MR) 6** *Skateboarder* magazine costs £2.20 at a newsagent's. A yearly subscription for the magazine costs £21.
How much cheaper is the magazine each month, if it is bought by subscription?

**7** Martyn has a paper round. He does two rounds each day from Monday to Friday, which take him 2 hours in total each day. He does one round on Saturday and Sunday, which takes him 90 minutes each day. He gets paid £3.25 per hour on Monday to Friday and £4.75 per hour on Saturday and Sunday.

a How much does he earn each week?

He saves £25 a week.

b How many weeks will it take him to save enough to buy a bicycle that costs £320?

**(PS) 8** How you can tell, without doing the long division, that the answer to 803 ÷ 22 will not be a whole number?

**9** You are given that $32 \times 44 = 1408$.

Write down the value of:

**a** $1408 \div 22$ **b** $704 \div 32$ **c** $2816 \div 128$.

## Long multiplication with decimals

As with any calculations with decimals, you must put each digit in its correct column and keep the decimal points in line.

**Example 18**

Work out $4.27 \times 34$.

```
      4 . 2 7
  ×       3 4
    1 7 . 0 8
          1 2
  1 2 8 . 1 0
        2
  1 4 5 . 1 8
        1
```

So, $4.27 \times 34 = 145.18$

## Exercise 1I

**1** Work out each of these.

**a** $3.72 \times 24$ **b** $5.63 \times 53$ **c** $1.27 \times 52$ **d** $4.54 \times 37$

**e** $67.2 \times 35$ **f** $12.4 \times 26$ **g** $62.1 \times 18$ **h** $81.3 \times 55$

**i** $5.67 \times 82$ **j** $0.73 \times 35$ **k** $23.8 \times 44$ **l** $99.5 \times 19$

> **Hints and tips** Remember to keep the decimal points in line in the working.

**2** Find the total cost of each purchase.

**a** 18 ties at £12.45 each

**b** 25 shirts at £8.95 each

**c** 13 pairs of tights at £2.30 a pair

> **Hints and tips** When the answer is an amount of money in pounds, you must write it with two decimal places, for example, you should write £224.1 as £224.10.

**3** Theo says that he can make multiplications easier by doubling one number and halving the other number. He says he still gets the same answer.

For example, to work out $8.4 \times 12$ he calculates $16.8 \times 6$.

Use his method to work out $2.5 \times 14$.

**4** A party of 24 scouts and their leader visited a zoo. Each scout's ticket cost £2.15 and the leader's ticket cost £2.60.

What was the total cost to get into the zoo?

**5** **a** Merry the market gardener bought 35 trays of seedlings.

Each tray cost her £3.45.

What was the total cost of the seedlings?

**b** There were 20 seedlings in each tray. Merry sold them at 85p each.

How much profit did she make per tray?

# Worked exemplars

  **1**   A theatre is hosting a show on Friday and Saturday night.

There are 400 tickets available for each night. The box office manager expects to sell all the tickets for Saturday but only three-quarters of the tickets for Friday.

She needs to take at least £3000 each night from ticket sales. If the ticket price must be the same for both days, how much should she charge?

| This is a problem-solving question that puts a real-life situation into a series of mathematical processes. Write down what you are working out. Don't just put down a series of calculations. | |
|---|---|
| Tickets sold on Friday: $\frac{3}{4} \times 400 = 300$ | Work out how many seats she expects to sell on Friday. |
| Total tickets sold on Friday and Saturday: 300 + 400 = 700 seats sold. | Now work out how many tickets she expects to sell on the two nights together. |
| Needs to raise £3000 a night ∴ cost per ticket: 3000 ÷ 700 = 4.2857 | Then work out the price she needs to charge for each seat. |
| She should charge £4.50 a ticket. | You need to interpret the result in the context of the problem. Charging £4.28 would not be sensible, so round up to a 'normal' price. This could be £4.50 or £5.00<br><br>Note the symbol ∴, which means 'therefore'. It is very useful in mathematics. |

  **2**   I want to buy a burger, a portion of chips and a bottle of cola for my lunch. I have the following coins in my pocket: £1, 50p, 50p, 20p, 20p, 10p, 2p, 2p, 1p.

**Price List**
Burger £1.20
Chips   90p
Beans   50p
Cola    60p

**a** Do I have enough money?

**b** Do I have enough money if I replace the chips with beans?

| This is a question for which you need to use and show your mathematical reasoning skills, so set out your answer clearly. | |
|---|---|
| **a** Total cost for a burger, chips and a cola = £2.70<br><br>Total of coins in my pocket = £2.55 | Show the total cost of the meal and the total of the coins in your pocket. |
| No, as £2.70 > £2.55 | It is important to show a clear conclusion. Do not just say 'No'. Use the numbers to support your answer. |
| **b** Beans are 40p cheaper and £2.30 < £2.55, so, yes I have enough money. | Explain clearly why you have enough money for this meal. |

# Ready to progress?

I can identify the value of digits in different places.
I can use < for less than and > for greater than.
I can add and subtract numbers with up to four digits.
I can multiply and divide positive integers and decimal numbers greater than 1 by a single-digit number.
I can solve problems involving multiplication and division by a single-digit number.
I can add and subtract positive and negative numbers.
I can use negative numbers in context and solve problems involving simple negative numbers.
I can do long multiplication and long division.
I can use BIDMAS/BODMAS to work out calculations in the correct order.

# Review questions

**1** It costs £7 per person to visit a show.

   **a** 215 people attend the show on Monday.

     How much do they pay altogether?

   **b** On Tuesday the takings were £1372.

     How many fewer people attended on Tuesday than on Monday?

**CM** **2** Murray and Harry both worked out $2 + 4 \times 7$.

   Murray calculated this to be 42.

   Harry worked this out to be 30.

   Explain why they got different answers.

**CM** **3** The following are two students' attempts at working out $3 + 5^2 - 2$.

   Adam:     $3 + 5^2 - 2 = 3 + 10 - 2 = 13 - 2 = 11$

   Bekki:     $3 + 5^2 - 2 = 8^2 - 2 = 64 - 2 = 62$

   **a** Each student has made one mistake. Explain both mistakes.

   **b** Work out the correct answer to $3 + 5^2 - 2$.

**4** The temperatures of the first four days of January are given below.

   1 °C, –1 °C, 0 °C and –2 °C

   **a** Write down the four temperatures, in order, starting with the lowest.

   **b** What is the difference in Celsius degrees between the coldest and the warmest of these four days?

**5** Nitrogen freezes below –210 °C and becomes a gas above –196 °C. In between these temperatures, it is liquid.

   Write down one possible temperature where nitrogen is:

   **a** a liquid           **b** a gas           **c** a solid (frozen).

**(PS) 6** A coach costs £345 to hire.

There are 55 seats on the coach.

Work out a sensible seat price assuming at least 45 seats are sold.

**(PS) 7** Mark went into these three shops and bought one item from each shop.

In total he spent £43.97.

| Music store | | Clothes store | | Book store | |
|---|---|---|---|---|---|
| CDs: | £5.98 | Shirt: | £12.50 | Magazine: | £2.25 |
| DVDs: | £7.99 | Jeans | £32.00 | Pen: | £3.98 |

What did he buy?

**(MR) 8** Cans of soup are sold in packs of five for £3.25 and packs of eight for £5. In which pack size is the cost of one can cheaper?

**(PS) 9** **a** Mike took his wife and four children to a theme park. The tickets were £13.25 for each adult and £5.85 for each child. What was the total cost of the tickets for Mike and his family?

**b** After lunch, the children ask for ice creams but their mother only has £5.

A large ice cream costs £1.60 and a small ice cream costs £1.20.

Which size should she buy? Give a reason for your answer.

**(PS) 10** Mary wanted to lay a path straight down the garden, which is 10 m long. She bought nine paving stones, each 1.35 m long. Has Mary bought too many paving stones? Show all your working.

**(CM) 11** Some multiplications can be made into easier multiplications by doubling one number and halving the other number. You can do this several times.

For example, $2.5 \times 6.4 = 5 \times 3.2 = 10 \times 1.6 = 16$.

Use this method to work out $1.25 \times 16.8$.

**12** A party of 25 students and their three teachers went to the theatre. The cost of a student ticket was £4.65 and the cost of a teacher ticket was £5.60. What was the total cost of the tickets?

**(PS) 13** **a** A shopkeeper bought 45 trays of eggs. Each tray cost £3.15. What was the total cost of the trays of eggs?

**b** There were 24 eggs on each tray which are sold at £1.45 for 6. How much profit did the shopkeeper make if he sold all the eggs?

**(PS) 14** Here are four number cards, showing the number 2745.

$$\boxed{2}\ \boxed{7}\ \boxed{4}\ \boxed{5}$$

Using all four cards once only, write down:

**a** the largest possible number

**b** the smallest possible number

**c** the missing numbers from this problem.

$$\boxed{\phantom{0}}\,\boxed{7} \times 2 = \boxed{\phantom{0}}\,\boxed{\phantom{0}}$$

# 2 Geometry and measures: Measures and scale drawings

## This chapter is going to show you:

- how to convert from one metric unit to another
- how to convert from one imperial unit to another
- how to convert from imperial units to metric units
- how to read and draw scale drawings
- how to read map scales
- how to draw nets of 3D shapes and identify 3D shapes from their nets
- how to draw plans and elevations.

## You should already know:

- the basic units used for measuring length, mass and capacity
- how to multiply or divide numbers by 10, 100 or 1000
- the names of common 3D shapes
- how to measure lines accurately.

## About this chapter

Simon Stevin was a Dutch mathematician, born in Bruges (now in Belgium) in 1548. In 1585 he published a small pamphlet in Dutch, in which he suggested using a decimal system for all measurement. He stated that the introduction of decimal coins, measures and weights everywhere was only a question of time.

In the UK we continued to use imperial units, such as feet, inches, ounces and pounds, for many years. Your grandparents will probably still remember them. In the later part of the 20th century, the British Government started trying to make sure that, in Britain, we use metric units as much as possible. This is now common practice, apart from using the mile.

The metric system is now used in many parts of the world, including the UK. The USA still uses imperial measurements. This chapter shows you the relationships between metric and imperial measurements, including how to use them on scale diagrams.

Sometimes we need to use scale drawings to represent large distances, such as on maps and in satellite navigation systems. It is also useful to represent a 3D shape using plans and elevations. This chapter introduces both these ideas.

# 2.1 Systems of measurement

## This section will show you how to:

- convert from one metric unit to another
- convert from one imperial unit to another.

Two systems of measurement are used in Britain now: the **imperial** system and the **metric** system.

The imperial system has a lot of awkward conversions, such as 12 inches = 1 foot. The metric system is based on powers of 10 (10, 100, 1000 and so on) so calculations are more straightforward.

It will be many years before all the units of the imperial system disappear, so it helps to know units in both systems.

### Key terms

| | |
|---|---|
| centilitre (cl) | ounce (oz) |
| foot (ft) | pound (lb) |
| gallon (gal) | stone (st) |
| imperial | ton (T) |
| inch (in) | tonne (t) |
| metric | yard (yd) |

| System | Unit | How to estimate it |
|---|---|---|
| | **Length** | |
| Metric system | 1 centimetre | The distance across a fingernail |
| | 1 metre | A long stride for an average person |
| | 1 kilometre | The distance you can walk in about twelve minutes |
| Imperial system | 1 inch | The length of the top joint of an adult's thumb |
| | 1 foot | The length of an A4 sheet of paper |
| | 1 yard | The distance from your nose to your fingertips when you stretch out your arm |
| | **Mass** | |
| Metric system | 1 gram | A 1p coin has a mass of about 4 grams |
| | 1 kilogram | A bag of sugar |
| | 1 tonne | A saloon car |
| Imperial system | 1 pound | A jar full of jam |
| | 1 stone | A large cat |
| | 1 ton | A saloon car |
| | **Volume/Capacity** | |
| Metric system | 1 millilitre | A full teaspoon is about 5 millilitres |
| | 1 centilitre | The amount of ink in three biros |
| | 1 litre | A full carton of orange juice |
| Imperial system | 1 pint | A full glass bottle of milk |
| | 1 gallon | A half-full bucket of water |

## Metric units

You should already know the relationships between these metric units.

| Length | Mass |
|---|---|
| 10 millimetres = 1 centimetre | 1000 grams = 1 kilogram |
| 1000 millimetres = 100 centimetres<br>= 1 metre | 1000 kilograms = 1 **tonne** |
| 1000 metres = 1 kilometre | |
| **Capacity / Volume** | |
| 10 millilitres = 1 **centilitre** | 1000 litres = 1 metre³ |
| 1000 millilitres = 100 centilitres<br>= 1 litre | 1 millilitre = 1 centimetre³ |

Note the relationship between the units of capacity and volume.

$$1 \text{ litre} = 1000 \text{ cm}^3 \qquad \text{which means} \qquad 1 \text{ ml} = 1 \text{ cm}^3$$

Since the metric system is based on powers of 10, you can multiply or divide by powers of 10 to change between units for the same type of measure.

**Note:**

* To change *large* units to *smaller* units, always *multiply*.

* To change *small* units to *larger* units, always *divide*.

---

**Example 1**

Change:

**a** 1.2 m to centimetres

**b** 0.62 cm to millimetres

**c** 3 m to millimetres

**d** 75 cl to millilitres.

**a** $1.2 \times 100 = 120$ cm

**b** $0.62 \times 10 = 6.2$ mm

**c** $3 \times 1000 = 3000$ mm

**d** $75 \times 10 = 750$ ml

---

**Example 2**

Change:

**a** 732 cm to metres

**b** 410 mm to centimetres

**c** 840 mm to metres

**d** 450 cl to litres.

**a** $732 \div 100 = 7.32$ m

**b** $410 \div 10 = 41$ cm

**c** $840 \div 1000 = 0.84$ m

**d** $450 \div 100 = 4.5$ litres

---

## Exercise 2A

**1** Estimate the approximate metric length, mass or capacity of each of the following.

**a** The length and mass of this book

**b** The length of your school hall

**c** The capacity of a glass milk bottle

**d** The length, width and mass of a brick

**e** The diameter and mass of a 10p coin

**f** The distance from your school to Manchester

**g** The mass of a cat

**h** The amount of water in one raindrop

**i** The dimensions of the room you are in

**j** Your own height and mass

**2** Bob is going to put up some bunting from the top of each lamp post in his street. He has three ladders: a 2 metre, a 3.5 metre and a 5 metre ladder.

Bob is slightly below average height for an adult male. He estimates that the lamp posts are about three times his height.

Which of the ladders should he use? Give a reason for your choice.

**3** The distance from Bournemouth to Basingstoke is shown on a website as 88 kilometres.

Why is this unit used instead of metres or centimetres?

**4** Copy and complete these statements.

a  3.4 m = … mm      b  13.5 cm = … mm      c  0.67 m = … cm

d  7.03 km = … m      e  2.4 l = … ml        f  5.9 l = … cl

g  8.4 cl = … ml      h  5.2 m$^3$ = … l      i  3.75 t = … kg

j  0.94 cm$^3$ = … l   k  15.2 kg = … g       l  0.19 cm$^3$ = … ml

**5** Copy and complete these statements.

a  125 cm = … m       b  82 mm = … cm        c  550 mm = … m

d  2100 m = … km      e  4200 g = … kg       f  5750 kg = … t

g  2580 ml = … l      h  340 cl = … l        i  600 kg = … t

j  630 ml = … cl      k  35 ml = … cm$^3$     l  1035 l = … m$^3$

**6** Sarif wanted to buy two pieces of 1.5 cm by 2 cm wood, each 2 m long. In his local store the types of wood were described as:

2000 mm × 15 mm × 20 mm

200 mm × 15 mm × 20 mm

200 mm × 150 mm × 2000 mm

1500 mm × 2000 mm × 20 000 mm

Should he choose any of these? If so, which one?

**7** 1 litre is equivalent to 1000 millilitres (ml) or 1000 cubic centimetres (cm$^3$).

Given that 1 centimetre is equal to 10 millimetres, write down the number of cubic millimetres in 1 litre.

**8** a  How many millimetres are there in 1 kilometre?

b  How many square millimetres are there in a square kilometre?

| Hints and tips | The answer is not 1 000 000. |

## Imperial units

Examples of everyday imperial measures include miles for distances by road, pints for milk, gallons for petrol and pounds for the mass of babies (in conversation), feet and inches for people's heights and ounces for the mass of food ingredients in a food recipe.

To answer the questions below you will need these imperial conversions.

| Length | 12 **inches** | = 1 **foot** |
|---|---|---|
| | 3 feet | = 1 **yard** |
| | 1760 yards | = 1 mile |
| Mass | 16 **ounces** | = 1 **pound** |
| | 14 pounds | = 1 **stone** |
| | 2240 pounds | = 1 **ton** |
| Capacity | 8 pints | = 1 **gallon** |

Remember

- To change *large* units to *smaller* units, always *multiply*.
- To change *small* units to *larger* units, always *divide*.

**Example 3**

Change:

**a** 4 feet to inches    **b** 5 gallons to pints

**c** 36 feet to yards    **d** 48 ounces to pounds.

**a** $4 \times 12 = 48$ inches    **b** $5 \times 8 = 40$ pints

**c** $36 \div 3 = 12$ yards    **d** $48 \div 16 = 3$ pounds

## Exercise 2B

**1** Copy and complete these statements.

**a** 2 feet = … inches   **b** 4 yards = … feet   **c** 2 miles = … yards

**d** 5 pounds = … ounces  **e** 4 stone = … pounds  **f** 3 tons = … pounds

**g** 5 gallons = … pints   **h** 4 feet = … inches   **i** 1 yard = … inches

**j** 10 yards = … feet    **k** 4 pounds = … ounces  **l** 60 inches = … feet

**m** 5 stone = … pounds   **n** 36 feet = … yards   **o** 1 stone = … ounces

**2** Copy and complete these statements.

**a** 8800 yards = … miles   **b** 15 gallons = … pints   **c** 1 mile = … feet

**d** 96 inches = … feet    **e** 98 pounds = … stones  **f** 56 pints = … gallons

**g** 32 ounces = … pounds  **h** 15 feet = … yards   **i** 11 200 pounds = … tons

**j** 1 mile = … inches    **k** 128 ounces = … pounds **l** 72 pints = … gallons

**m** 140 pounds = … stones  **n** 15 840 feet = … miles  **o** 1 ton = … ounces

 **3** Andrew's teacher set up an old fashioned shop, selling in pounds and ounces. He was asked to buy a two-pound bag of sugar from the shop. The only bags of sugar were:

8-ounce bags, 16-ounce bags, 32-ounce bags and 40-ounce bags

Which bag should Andrew buy?

 **4** How many square inches are there in a square mile?

**5** **a** Copy and complete the table by writing a *sensible* imperial unit for each measurement.

| The length of the Amazon river | 4000 |
|---|---|
| The height of a mature apple tree | 25 |
| The mass of a hen's egg | 2 |
| The amount of diesel in the tank of a bus | 120 |

**b** Change 9 feet to inches.

**c** Change 24 pints to gallons.

# 2.2 Conversion factors

## This section will show you how to:

- use approximate conversion factors to change between imperial units and metric units.

Here are some conversions between imperial units and metric units.

Here are some conversion factors between imperial units and metric units. The symbol ≈ means 'is approximately equal to'.

**Length**
- 1 inch ≈ 2.5 centimetres
- 1 foot ≈ 30 centimetres
- 1 mile ≈ 1.6 kilometres
- 5 miles ≈ 8 kilometres

**Mass**
- 1 pound ≈ 450 grams
- 2.2 pounds ≈ 1 kilogram

**Capacity**
- 1 pint ≈ 570 millilitres
- 1 gallon ≈ 4.5 litres
- 1.75 or $1\frac{3}{4}$ pints ≈ 1 litre

---

**Example 4**

Use the conversion factors above to work out the following.

**a** Change 5 gallons into litres.

**b** Change 45 miles into kilometres.

**c** Change 5 pounds into kilograms.

**a** $5 \times 4.5 \approx 22.5$ litres

**b** $45 \times 1.6$ kilometres ≈ 72 kilometres

**c** $5 \div 2.2 \approx 2.3$ kilograms (1 dp)

**Note:** You should round an answer when it has several decimal places, as it is only an approximation.

**1** Copy and complete these conversions using the conversion factors above.

**a** 8 inches = … cm    **b** 6 kg = … pounds    **c** 30 miles = … km

**d** 15 gallons = … litres    **e** 5 pints = … ml    **f** 45 litres = … gallons

**g** 30 cm = … inches    **h** 80 km = … miles    **i** 11 pounds = … kg

**j** 1710 ml = … pints    **k** 100 miles = … km    **l** 56 kg = … pounds

**m** 40 gallons = … litres    **n** 200 pounds = … kg    **o** 1 km = … yards

**p** 1 foot = … cm    **q** 1 stone = … kg    **r** 1 yard = … cm

 **2** Which is heavier, a tonne or a ton? Show your working clearly.

(CM) **3** Which is longer, a metre or a yard? Show your working clearly.

**4** The mass of 1 cm³ of water is about 1 gram.

**a** What is the mass of 1 litre of water:

**i** in grams    **ii** in kilograms?

**b** What is the approximate mass of 1 gallon of water:

**i** in grams    **ii** in kilograms?

 (PS) **5** In France I saw a road sign that read: 'Paris 216 km'. I was travelling on a road with a speed limit of 80 km/h.

**a** Approximately how many miles was I from Paris?

**b** What was the approximate speed limit in miles per hour?

**c** How long would it take me to get to Paris if I travelled at the maximum speed all the way? Give your answer in hours and minutes.

(PS) **6** While on a cycling holiday in France, Tom had to cover 200 km in one day. He knew that, at home, his average speed was 30 mph.

How long would he expect the journey to take, with no stops?

(PS) **7** A cowboy's 'ten-gallon' hat could actually hold only 1 gallon of water.

How many cubic inches could a 'ten-gallon' hat hold?

# 2.3 Scale drawings

This section will show you how to:

- read and draw scale drawings
- use a scale drawing to make estimates.

| Key terms | |
|---|---|
| estimate | scale drawing |
| ratio | scale factor |

### Scale drawings

A **scale drawing** is an accurate representation of a real object. It is usually smaller than the original object but can be larger, for example, drawings of miniature electronic circuits and very small watch movements.

In a scale drawing:

- all the measurements must be in proportion to the corresponding measurements on the original object
- all the angles must be equal to the corresponding angles on the original object.

To work out the measurements for a scale drawing, multiply all the actual measurements by a common **scale factor**, usually referred to as a scale. (You will learn more about scale factors in a later chapter.)

Scales are often given as **ratios**, for example, 1 cm : 1 m. When the units in a ratio are the *same*, they are normally not given. For example, a scale of 1 cm : 1000 cm is written as 1 : 1000.

**Note:** When making a scale drawing, take care to express *all* measurements in the *same* unit.

---

**Example 5**

The diagram shows the front of a kennel.

It is drawn to a scale of 1 : 30. Work out:

**a** the actual width of the front

**b** the actual height of the doorway.

1 : 30

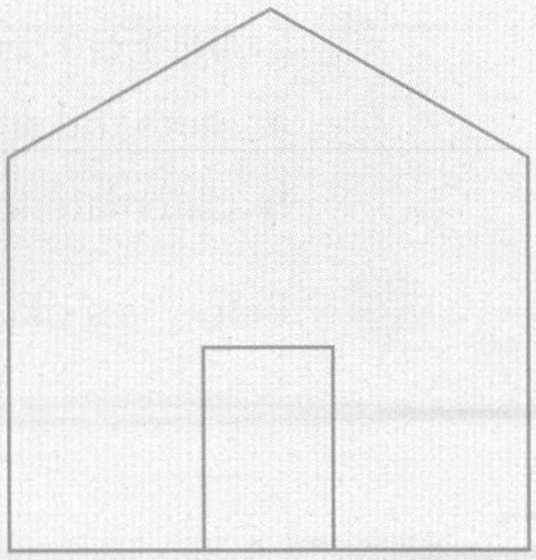

The scale of 1 : 30 means that a measurement of 1 cm on the diagram represents a measurement of 30 cm on the actual kennel.

**a** So the actual width of the front is:      4 cm × 30 = 120 cm

**b** The actual height of the doorway is:      1.5 cm × 30 = 45 cm

---

Map scales are often expressed as ratios, such as 1 : 50 000 or 1 : 200 000.

The first ratio means that 1 cm on the map represents 50 000 cm or 500 m in the real situation. The second ratio means that 1 cm represents 200 000 cm or 2 km.

---

**Example 6**

Work out the actual distances between the following towns represented on this map.

**a** Bowden and Goldagel          **b** Goldagel and St Jidd          **c** Trentham and St Jidd

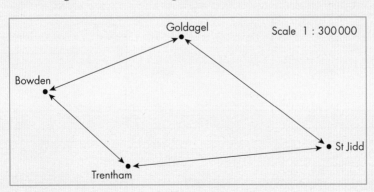

The map is drawn to a scale of 1 : 300 000.

300 000 cm = 3000 m = 3 km so the scale means that a distance of 1 cm on the map represents a distance of 3 km on the land.

So, the actual distances are:

**a** Bowden to Goldagel: 4 × 3 km = 12 km

**b** Goldagel to St Jidd: 5 × 3 km = 15 km

**c** Trentham to St Jidd: 5.5 × 3 km = 16.5 km

**PS** **1** Look at this plan of a garden.

| | | |
|---|---|---|
| | Onions | Soft fruits |
| Apple trees | Lawn | Potatoes |

Scale: 1 cm represents 10 m

**a** State the actual dimensions of each plot of the garden.

**b** Calculate the actual area of each plot.

**2** Below is a plan for a mouse mat.

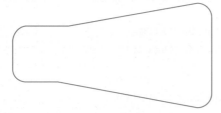

Scale: 1 cm represents 6 cm

**a** How long is the actual mouse mat?

**b** How wide is the narrowest part of the mouse mat?

**3** This plan of a kitchen garden is drawn to a scale of 1 : 200.

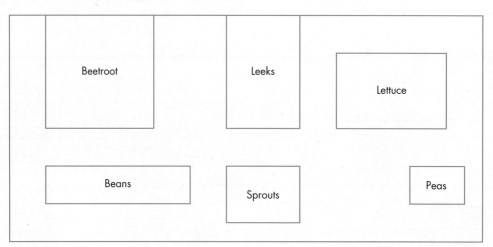

What are the actual dimensions of each of these plots?

**a** Beetroot        **b** Beans        **c** Sprouts

**d** Leeks        **e** Lettuce        **f** Peas

**4** The diagram shows a sketch of a garden.

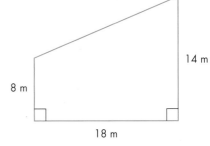

**a** Make an accurate scale drawing of the garden.

Use a scale of 1 cm to represent 2 m.

**b** Amelia wants to plant flowers on the sloping edge shown on the diagram. Use your scale drawing to work out how many she can plant if they are 0.5 m apart.

 **5** This map is drawn to a scale of 1 : 300 000.

Work out the actual distance between each pair of places to the nearest tenth of a kilometre.

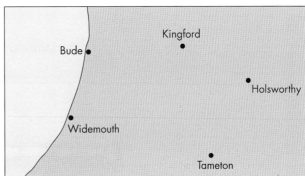

**a** Widemouth to Holsworthy

**b** Kingford to Tameton

**c** Widemouth to Tamenton

**d** Bude to Holsworthy

**e** Bude to Kingford

**f** Bude to Widemouth

 **6** This map is drawn to scale.

The approximate direct distance from Carlisle to Scarborough is 232 km.

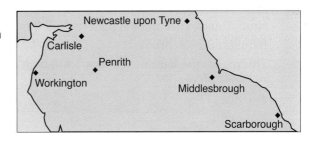

**a** Show that the scale of the map is 1 : 4 000 000.

**b** What is the direct distance from:

 **i** Workington to Newcastle

 **ii** Newcastle to Middlesbrough?

 **7** This map is drawn to a scale of 1 : 2 000 000.

Work out the direct distance, to the nearest 5 kilometres, from:

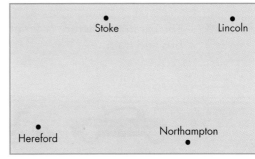

**a** Stoke to Northampton

**b** Stoke to Lincoln

**c** Lincoln to Hereford.

 **8** The Hiran Minar in India, shown in this scale drawing, is 30 m high. Which of the following is the correct scale for the drawing?

**a** 1 : 5000

**b** 1 : 6000

**c** 1 : 500

**d** 1 : 600

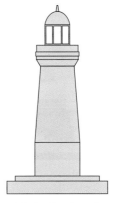

**9** Find a paper map of Britain.

    **a** Write down the scale on the map . Then use the scale to work out the direct distance in miles between:

       **i** Manchester and London         **ii** Edinburgh and Bristol.

## Sensible estimates for scales

The average height of a man is 1.78 m. For a sensible **estimate**, we would usually say that the height of a man is about 1.8 m. You can use this information to estimate the lengths or heights of other objects.

**Example 7**

Look at the picture.

Estimate, in metres, the height of the lamppost and the length of the bus.

Assume the man is about 1.8 m tall. The lamppost is about three times as high as he is.

**Note:** You can use tracing paper to mark off the height of the man and then measure the other lengths against this.

This makes the lamppost about 5.4 m high, or close to 5 m high. As it is an estimate, there is no need for an exact value.

The bus is about four times as long as the man is tall so the bus is about 7.2 m long, or close to 7 m long.

## Exercise 2E

**1** The car in the picture is 4 metres long. Use this to estimate the length of the bicycle, bus and train.

**2** Estimate the greatest height and length of the whale.

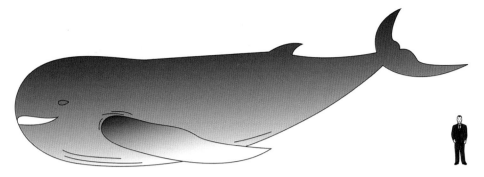

**3** Estimate the following.

**a** the height of the traffic lights

**b** the width of the road

**c** the height of the flagpole

(MR) **4** This illustration shows Joel standing next to a statue.

Joel's height is 1.5 m. How could he use this information to estimate the height of the statue?

(EV) **5** Ben looked at the diagram and said: 'I think the Tyrannosaurus Rex is about 7 metres tall. If it lay down, it would be about 10 metres long.'

Is Ben correct? How do you know?

# 2.4 Nets

This section will show you how to:

- draw nets of some 3D shapes
- identify a 3D shape from its net.

A **net** is a flat shape that you can fold into a **3D shape**.

**Example 8**

Sketch a net for each of these shapes.

**a** cube

**b** square-based pyramid

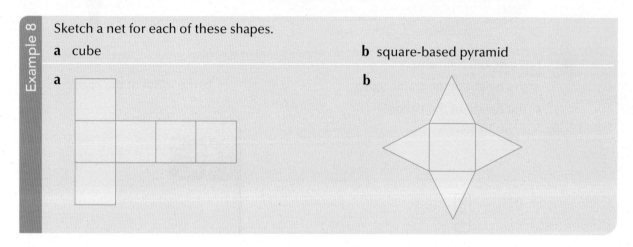

**a**

**b**

## Exercise 2F

**1** Six nets are shown below.

Which of these nets will **not** make a cube?

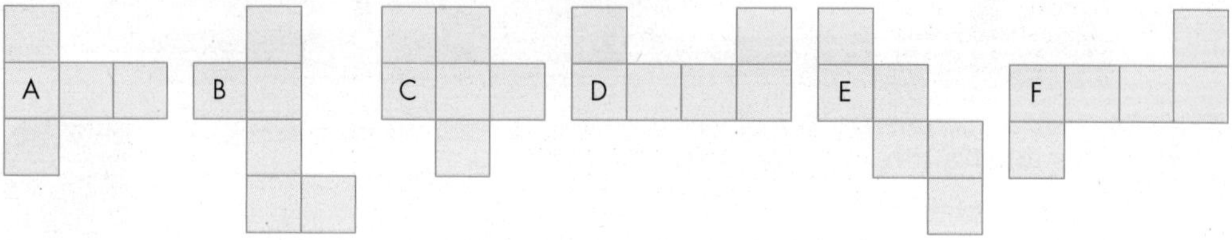

A     B     C     D     E     F

**2** Draw, on squared paper, an accurate net for each of these cuboids.

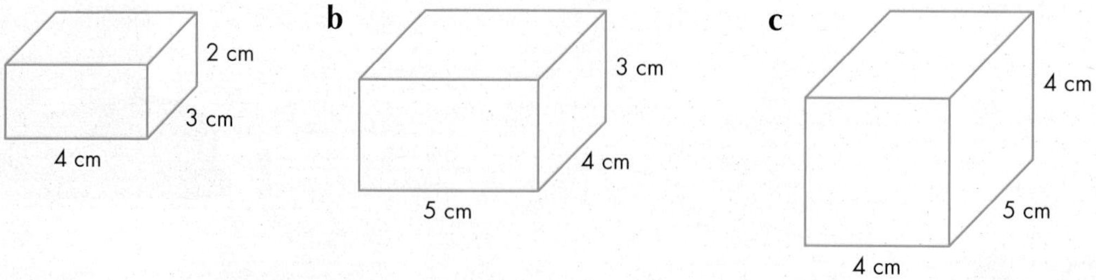

**a** 2 cm, 3 cm, 4 cm

**b** 3 cm, 4 cm, 5 cm

**c** 4 cm, 5 cm, 4 cm

**EV** **3** Jenny is making an open box from card.
This is a sketch of the box.

Jenny has a piece of card that measures 15 cm
by 21 cm. Can she make the box using only
this card?

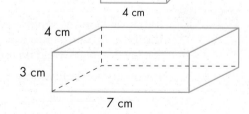

4 cm

3 cm

7 cm

**4** The shape on the right is a triangular prism. Its ends are isosceles triangles and its other faces are rectangles. Draw an accurate net for this prism on squared paper.

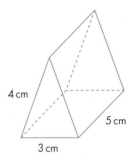

4 cm

5 cm

3 cm

**5** Sketch the nets of these shapes.

**a**

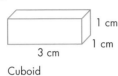

1 cm

1 cm

3 cm

Cuboid

**b**

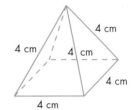

4 cm

4 cm

4 cm

4 cm

4 cm

Square-based pyramid

**c**

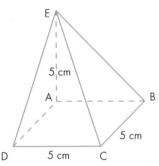

E

5 cm

A

B

D

5 cm

C

5 cm

Square-based pyramid, with point E directly above point A

**d**

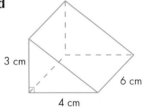

3 cm

6 cm

4 cm

Right-angled triangular prism

**PS** **6** Here is one possible net for a cube.

Draw as many different nets for a cube as you can.

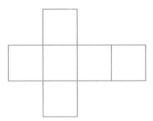

| Hints and tips | There are 11 altogether. How many can you find? |

**MR** **7** Which of these are nets for a square-based pyramid?

**a**

**b**

**c**

**CM** **8** Emma was asked to make a net for a square-based pyramid with base length 5 cm and height 6 cm. She was given a piece of card measuring 15 cm by 15 cm. Show that this card is not big enough to make the net.

# 2.5 Using an isometric grid

## This section will show you how to:

- read from and draw on isometric grids
- interpret diagrams to draw plans and elevations.

| Key terms | |
| --- | --- |
| elevation | isometric grid |
| plan | |

### Isometric grids

It can be difficult to draw a 3D shape on a flat (2D) surface so that it looks like the original 3D shape. One method is to use an **isometric grid** (a grid of equilateral triangles).

Below are two drawings of the same cuboid, one on squared paper, the other on isometric paper.

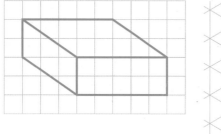
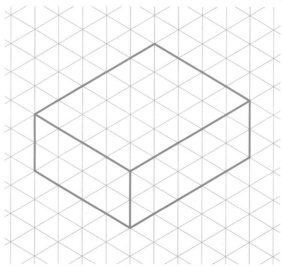

The cuboid measures 5 × 4 × 2 units. You can read these dimensions from the isometric drawing but not from the drawing on squared paper.

You can use a triangular dot grid instead of an isometric grid but you *must* make sure that it is the **correct way round** – as shown here.

### Plans and elevations

A **plan** is the view of a 3D shape when you see it from above.

An **elevation** is the view of a 3D shape when you see it from the front or from another side.

Example 9

The 3D shape is drawn on a triangular dot grid. Draw its plan, front elevation and side elevation on squared paper.

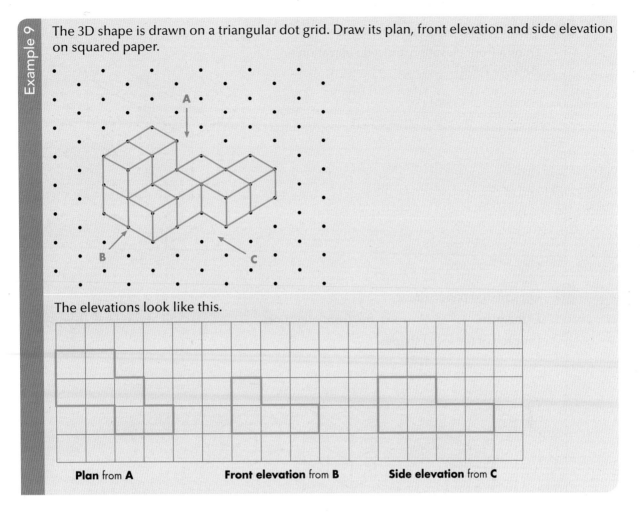

The elevations look like this.

**Plan** from **A**          **Front elevation** from **B**          **Side elevation** from **C**

Note that when drawing plans and elevations, you show the edges of shapes that can be seen.

## Exercise 2G

**1** Draw each of these cuboids on an isometric grid.

**a**  2 cm  3 cm  4 cm

**b**  5 cm  4 cm  2 cm

**c**  3 cm  4 cm  5 cm

**2** The diagram shows an L-shaped prism.

Draw a front elevation and side elevation on squared paper.

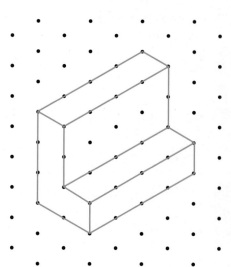

**3** On squared paper, draw:

**i** the plan          **ii** the front elevation          **iii** the side elevation

for each of these 3D shapes.

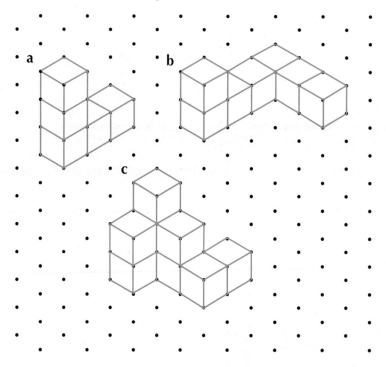

**4** This drawing shows the plan view of a solid made from five cubes.

Draw the solid on an isometric grid.

**(PS)** **5** Here are the plan, front elevation and side elevation of a 3D shape.

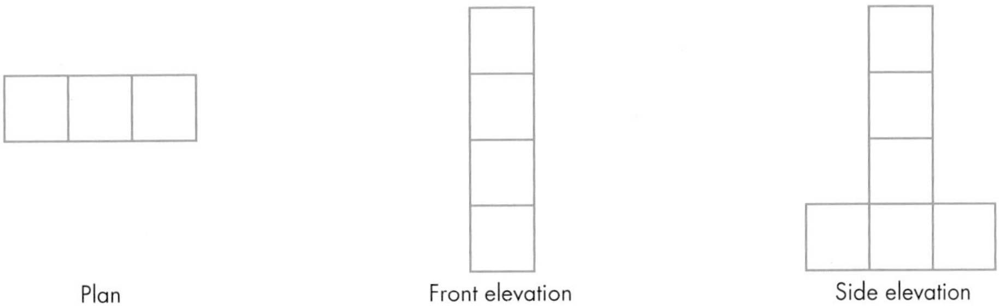

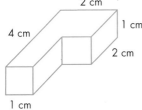

Plan             Front elevation            Side elevation

Draw the 3D shape on an isometric dotted grid.

**(EV)** **6** The diagram shows a toy brick.

2 cm
1 cm
4 cm
2 cm
1 cm

**a** Draw an accurate diagram of the brick on an isometric grid.

**b** Leila said: 'I bought 88 bricks in a box measuring 12 cm by 9 cm by 5 cm.'

Could she be correct? Show working to support your answer.

**(MR)** **7** The diagram shows the plan view, front and side elevations of a solid shape. Draw the shape on an isometric grid.

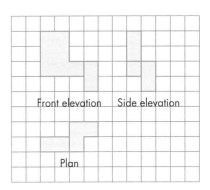

Front elevation     Side elevation

Plan

**8** Copy and complete the drawing of a cuboid measuring 4 cm by 2 cm by 3 cm.

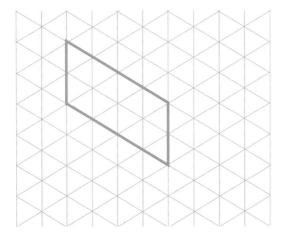

# Worked exemplars

  **1** Two towns are 60 km apart.

Pete knows that 8 kilometres are about 5 miles. He said: 'If I drive at 30 mph, it will take me about one and a quarter hours to drive from one town to the other.'

Comment on Pete's statement.

| | |
|---|---|
| This is an evaluation question where you are asked to check and comment on a statement to see whether or not it is correct. | |
| 30 mph = 30 ÷ 5 × 8 kph<br><br>     = 48 kph<br><br>Time to drive 60 km: 60 ÷ 48 = 1.25 hours | You need to work out the time it will take Pete to drive 60 km at 30 mph. The method shown converts the imperial speed to metric speed first. Note that 0.25 in the answer 1.25 means one quarter of an hour.<br><br>An alternative method is to convert the distance travelled to an imperial measure and then calculate the speed. Both methods give the same answer of 1.25 hours. |
| So Pete is correct, the time taken is one and a quarter hours. | You need to say that your answer means Pete's statement is correct. |

 **2** A café ordered 94 pints of milk. They receive 49 litres.
1 gallon ≈ 4.5 litres ≈ 8 pints

Does the cafe have more or less than they ordered?

Show your working.

| | |
|---|---|
| This is a problem-solving question so you need to identify a strategy to solve the problem. You will usually need to make connections between different parts of mathematics, in this case equivalent imperial units and imperial to metric conversions. | |
| So 1 litre ≈ 8 ÷ 4.5<br><br>     ≈ 1.78 pints<br><br>49 litres ≈ 1.78 × 49<br><br>     ≈ 87.22 pints | Use the given relationship between imperial units and metric units to work out how many pints there are in 1 litre.<br><br>Then calculate the equivalent number of pints to 49 litres. |
| They received less than the 94 pints ordered. | Finish with a clear statement showing that less milk has been delivered than ordered. |

3 Alec is planning a holiday in Madrid. He looks at the map of Spain shown below.

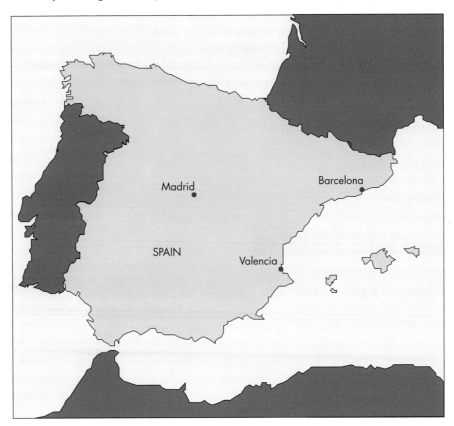

Alec knows the direct distance from Madrid to Barcelona is 460 km.

Show that the direct distance from Valencia to Barcelona is 305 km.

| This is a mathematical reasoning question. So you need to apply your knowledge and demonstrate chains of reasoning to achieve a given result. | |
|---|---|
| On the map, the direct distance between Madrid and Barcelona is 4.6 cm.<br><br>If 4.6 cm represents 460 km, 1 cm represents 100 km. | First work out the scale of the map using the information given and the measurement from Madrid to Barcelona on the map. |
| On the map, the direct distance from Valencia to Barcelona is 3.05 cm, so the direct distance from Valencia to Barcelona is 3.05 × 100 km = 305 km. | Then apply the scale factor to the map measurement from Valencia to Barcelona. |

# Ready to progress?

I can convert from one metric unit to another.
I can convert from one imperial unit to another.
I can use conversion factors to change between imperial units and metric units.
I can solve problems using metric units and imperial units.
I can use ratios when drawing scale drawings.
I can draw nets for 3D shapes.
I can draw cuboids on isometric paper.

I can draw plans and elevations of 3D shapes.

# Review questions

1. Copy and complete this table. Write a sensible metric unit for each measurement.

| | |
|---|---|
| The length of a garden | ............... |
| The mass of a puppy | ............... |
| The length of a brick | ............... |

2. Two villages are 40 km apart.

   a Change 40 km into metres.

   b Given that 8 km is about the same as 5 miles, how many miles is 40 km?

3. a Write down a sensible metric unit for measuring:

   i the distance from Newcastle to Bristol      ii the amount of water in a reservoir.

   b Change:

   i 9 centimetres to millimetres      ii 3000 grams to kilograms.

4. (EV) Brian is driving through Belgium. He sees the speed limit on the motorway is 130 kilometres per hour.

   Brian, remembering that 8 km is about the same as 5 miles, says: 'That's more than 80 miles per hour.'

   Comment on Brian's statement.

5. (PS) The diagram shows the dimensions of a bookcase. The thickness of all the wood used is 30 mm.

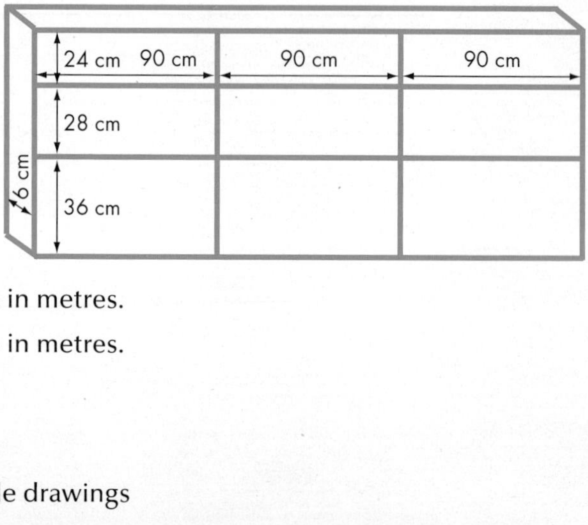

   24 cm   90 cm     90 cm     90 cm

   28 cm

   6 cm

   36 cm

   a Calculate the height of the bookcase in metres.

   b Calculate the length of the bookcase in metres.

**6** A model of the Funkturm in Berlin is made to a scale of 4 millimetres to 1 metre.

The actual width of the base of the Funkturm is 40 metres.

**a** Work out the width of the base of the model.

The height of the model is 60 centimetres.

**b** Work out the actual height of the Funkturm.

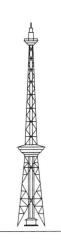

**7** The diagram shows a pyramid with a square base.

The length of each side of the base is 2 cm.

The length of each sloping edge is 2 cm.

Draw an accurate net of the pyramid on a centimetre-square grid.

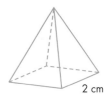

2 cm

**8** The distance from Sheffield to Birmingham is shown on a website as 140 kilometres. Why are kilometres used instead of centimetres?

**9** Oliver and Evie go on a self-catering holiday in Portugal with their gran.

They travel 12 km to the nearest supermarket for supplies.

Gran asks them to get 5 lbs of potatoes, 8 oz of butter and 4 pts of milk.

1 lb = 16 ounces ≈ 450 grams     and   1 litre ≈ 1.75 pints

**a** Gran asks Evie how far it is to the supermarket, in miles. What answer should Evie give her?

**b** The supermarket only sells goods in metric units. Convert Gran's shopping list into metric units.

**10** The diagram shows a solid shape made from seven one-centimetre cubes.

On squared paper, draw the plan, the front elevation and the side elevation.

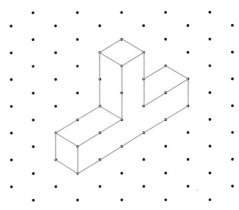

**11** Here are the front elevation, side elevation and plan of a 3D shape.

Draw a sketch of the shape.

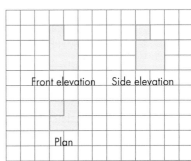

Front elevation     Side elevation

Plan

# 3 Statistics: Charts, tables and averages

## This chapter is going to show you:

- how to use tally charts and frequency tables to collect and organise data
- how to represent data on various types of diagram
- how to draw a line graph to show trends in data
- how to work out the mode, median, mean and range of small sets of data
- how to decide which is the best average to use
- how to use an average and the range to compare sets of data
- how to draw and interpret stem-and-leaf diagrams.

## You should already know:

- how to use a tally for recording data
- how to read information from charts and tables.

## About this chapter

William Playfair, a Scottish engineer, was probably the first person to work out how to represent statistics in diagrams. He invented the line graph and bar chart in 1786 and the pie chart in 1801.

Florence Nightingale was also a pioneer in presenting information visually. She was born in 1820 and was very good at mathematics from an early age. In 1859, she was elected the first female member of the Royal Statistical Society.

She developed a form of the pie chart now known as the 'polar area diagram' or the 'Nightingale rose diagram'. It illustrated the numbers of patient deaths, per month, in military field hospitals. She called these diagrams 'coxcombs' and used them in her reports on the conditions of medical care in the Crimean War to parliament and to civil servants, who may not have understood traditional statistical reports.

Since the latter part of the twentieth century, statistical graphs have become an important way of analysing information. Computer-generated graphs are shown nearly every day on TV, in newspapers and in magazines. In this chapter, you will learn how to draw and interpret a range of these graphs.

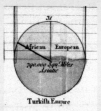

William Playfair pioneered charts and graphs such as this pie chart.

Florence Nightingale was a nurse and hospital reformer. She used charts and graphs in her work.

# 3.1 Frequency tables

This section will show you how to:

- use tally charts and frequency tables to collect and represent data
- use grouped frequency tables to collect and represent data.

**Key terms**

class interval

data collection sheet

experiment

frequency

frequency table

grouped frequency table

observation

sample

tally chart

When you study statistics you need to collect and organise data, represent it on diagrams and then interpret it.

To collect data for a simple survey, you can use a **tally chart**, also known as a **data collection sheet**. You can use tally charts to gather information about, for example:

- how people travel to work
- how students spend their free time
- the amount of time people spend watching TV.

Example 1 shows you how to use tally marks to collect data. Notice that you draw every fifth mark through the previous four. This means you can use the five times table to count the tally marks in each row of the chart. This gives the **frequency** of each data value. You can use these frequencies to produce a **frequency table**, which is useful when you are making statistical calculations.

**Example 1**

Sophie carried out a survey to find out how students travelled to school. She asked a **sample** of students and recorded her results in a tally chart like this one.

| Method of travel | Tally | Frequency |
|---|---|---|
| Walk | ٴ ٴ ٴ ٴ ٴ ||| | 28 |
| Car | ٴ ٴ || | 12 |
| Bus | ٴ ٴ ٴ ٴ ||| | 23 |
| Bicycle | ٴ | 5 |
| Taxi | || | 2 |

Draw a frequency table for Sophie's data and interpret her results.

The frequency table looks like this.

Adding together all the frequencies, you can see that 70 students took part in the survey. Comparing the frequencies, you can see that more students travelled to school on foot than by any other method of transport.

| Method of travel | Frequency |
|---|---|
| Walk | 28 |
| Car | 12 |
| Bus | 23 |
| Bicycle | 5 |
| Taxi | 2 |

There are three methods of collecting data.

- **Taking a sample:** For example, to find out which television programmes students watch, take a sample from the whole school population with equal numbers of boys and girls chosen at random from each year group. A good sample size for this would be 50.
- **Observation:** For example, to find how many vehicles use a certain road each day, count and record the numbers of vehicles passing a point at different times of the day.
- **Experiment:** For example, to find out how often a six occurs when you throw a dice, throw the dice 50 times or more and record each score.

Example 2

Andrew wanted to find out the most likely number of heads you obtain when you throw two coins.

He carried out an **experiment** by throwing two coins 50 times.

His tally chart looked like this.

| Number of heads | Tally | Frequency |
|---|---|---|
| 0 | ЖЖ ЖЖ II | 12 |
| 1 | ЖЖ ЖЖ ЖЖ ЖЖ ЖЖ II | 27 |
| 2 | ЖЖ ЖЖ I | 11 |

Draw a frequency table for Andrew's data and interpret his results.

From the frequency table, you can see that a single head appeared the highest number of times.

| Number of heads | Frequency |
|---|---|
| 0 | 12 |
| 1 | 27 |
| 2 | 11 |

## Grouped data

Many surveys produce a lot of data with a wide range of values. It is sensible to put such data into groups before trying to draw up a frequency table. These groups of data are called classes or **class intervals**.

Once you have grouped your data, you can draw up and complete a **grouped frequency table**.

Example 3

These are the marks for 36 students in a Year 10 mathematics examination.

| 31 | 49 | 52 | 79 | 40 | 29 | 66 | 71 | 73 | 19 | 51 | 47 |
|---|---|---|---|---|---|---|---|---|---|---|---|
| 81 | 67 | 40 | 52 | 20 | 84 | 65 | 73 | 60 | 54 | 60 | 59 |
| 25 | 89 | 21 | 91 | 84 | 77 | 18 | 37 | 55 | 41 | 72 | 38 |

a Draw a tally chart to show the data. Use class intervals of 1–20, 21–40 and so on.

b Draw a grouped frequency table to show the data.

c Which class interval has the greatest frequency?

a Draw the tally chart, as shown, and write in the headings.

Use a tally mark to indicate each student's score, writing it in the class to which it belongs. For example, 81, 84, 84, 89 and 91 belong to the class 81–100, giving five tally marks, as shown in the table.

Finally, count the tally marks for each class and write the result in the column headed 'Frequency'. The table is now complete.

| Marks | Tally | Frequency |
|---|---|---|
| 1–20 | III | 3 |
| 21–40 | ЖЖ III | 8 |
| 41–60 | ЖЖ ЖЖ I | 11 |
| 61–80 | ЖЖ IIII | 9 |
| 81–100 | ЖЖ | 5 |

*(continued)*

**b** Now put the data into a grouped frequency table.

**c** From the grouped frequency table, you can see that the group of marks scored by the highest number of students obtained is 41–60. This is the class with the greatest frequency.

| Marks | Frequency |
|---|---|
| 1–20 | 3 |
| 21–40 | 8 |
| 41–60 | 11 |
| 61–80 | 9 |
| 81–100 | 5 |

## Exercise 3A

**1** Ethan listed the numbers of goals scored by Burnley Rangers in their last 20 matches. These are his results.

0 1 1 0 2      0 1 3 2 1

0 1 0 3 2      1 0 2 1 1

**a** Draw a frequency table for his data.

**b** What was the most frequent score?

**c** How many goals did the team score in total in the 20 matches?

**2** Amelia was doing a geography project about the weather. She recorded the daily midday temperatures in June, to the nearest degree.

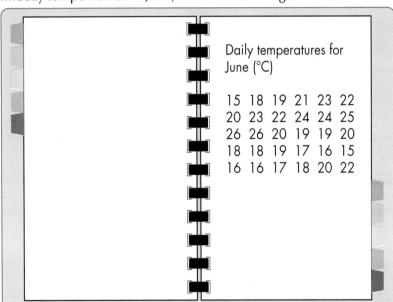

Daily temperatures for June (°C)

15 18 19 21 23 22
20 23 22 24 24 25
26 26 20 19 19 20
18 18 19 17 16 15
16 16 17 18 20 22

**a** Copy and complete the tally chart for her data.

**b** Draw a grouped frequency table to show her data.

**c** Which class interval has the most temperatures recorded in it?

**d** Describe a possible weather pattern for the month.

| Temperature (°C) | Tally | Frequency |
|---|---|---|
| 14–16 | | |
| 17–19 | | |
| 20–22 | | |
| 23–25 | | |
| 26–28 | | |

(MR) **3** In a game, Mitesh uses a six-sided dice.
He decides to keep a record of his scores to find out whether the dice is fair.
These are his scores.

| 2 | 4 | 2 | 6 | 1 | | 5 | 4 | 3 | 3 | 2 |
| 3 | 6 | 2 | 1 | 3 | | 5 | 4 | 3 | 4 | 2 |
| 1 | 6 | 5 | 1 | 6 | | 4 | 1 | 2 | 3 | 4 |

a Draw a frequency table for his data.

b How many times does Mitesh throw the dice during the game?

c Do you think the dice is fair? Explain why.

**4** These are the heights, to the nearest centimetre, of a random sample of 32 Year 10 students.

| 172 | 158 | 160 | 175 | | 180 | 167 | 159 | 180 |
| 167 | 166 | 178 | 184 | | 179 | 156 | 165 | 166 |
| 184 | 175 | 170 | 165 | | 164 | 172 | 154 | 186 |
| 167 | 172 | 170 | 181 | | 157 | 165 | 152 | 164 |

a Draw a tally chart for the data, using class intervals 151–155, 156–160, …

b Draw a grouped frequency table for the data.

c Which class interval has the highest frequency?

(MR) **5** For each of these surveys, decide whether the data should be collected by:

i sampling      ii observation      iii experiment.

a The number of people using a new superstore

b How people will vote in a forthcoming election

c The number of times a person scores a bullseye in a game of darts

d Where people go for their summer holidays

e The number of times a double six is obtained when two dice are thrown

f The frequency of a bus service on a particular route

g The number of times a drawing pin lands point up when it is dropped

(PS) **6** Over a seven-week period, Olivia used a stopwatch to time the interval between putting fresh food into her hamster's hutch and the hamster finding it.

The times she recorded, to the nearest second, are given below.

| 7 | 30 | 14 | 27 | 8 | 31 | 8 | | 28 | 10 | 41 | 51 | 37 | 15 | 21 |
| 37 | 16 | 38 | 23 | 20 | 9 | 11 | | 55 | 9 | 33 | 8 | 35 | 45 | 35 |
| 25 | 25 | 49 | 23 | 43 | 55 | 45 | | 8 | 13 | 9 | 39 | 12 | 57 | 16 |
| 37 | 26 | 32 | 19 | 48 | 29 | 37 | | | | | | | | |

How would you put this data into a grouped frequency table?

(MR) **7** Max was doing a survey to find the ages of people attending a school football match.

He said that he would draw a grouped frequency table for the ages, with class intervals of 10–20, 20–30, 30–40, 40–50 and 50–60.

Why may these class intervals not be suitable?

# 3.2 Statistical diagrams

This section will show you how to:

- draw pictograms to represent statistical data
- draw bar charts and vertical line charts to represent statistical data.

**Key terms**

bar chart

composite bar chart

dual bar chart

key

pictogram

vertical line chart

When you collect data in a survey, you can present it in pictorial or diagrammatic form. This helps people to understand it. There are plenty of examples of this in newspapers, in magazines and on TV, where a large range of visual aids are used to show statistical information.

## Pictograms

A **pictogram** is a frequency table in which the frequency of each type of data is shown by a repeated symbol. The symbol itself may represent a single item or a number of items. The **key** tells you how many items the symbol represents.

---

**Example 4**

The pictogram shows the number of phone calls made by Georgia from her mobile phone during a week.

How many calls did Georgia make in the week?

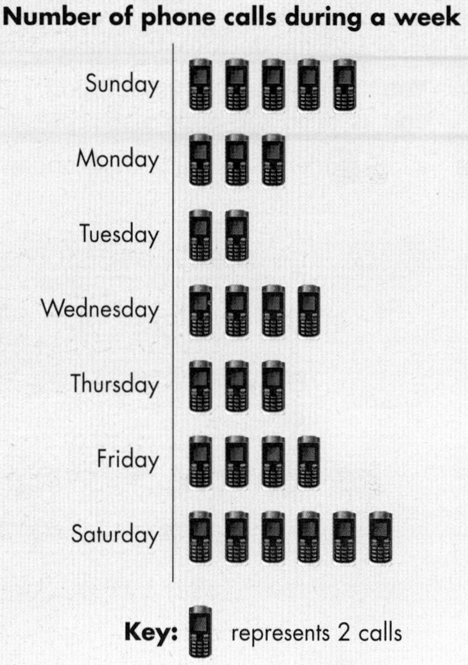

**Number of phone calls during a week**

**Key:** 📱 represents 2 calls

From the pictogram, adding up the calls each day gives 10 + 6 + 4 + 8 + 6 + 8 + 12 = 54.

So Georgia made a total of 54 calls.

---

Pictograms are straightforward to draw and read when they contain only whole symbols. However, because it can be difficult to draw fractions of a symbol, the frequencies may only be approximations, as shown in Examples **5** and **6**.

Example 5

The pictogram shows the numbers of Year 10 students who were late for school during a week.

**The number of Year 10 students
who were late for school**

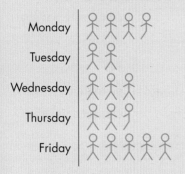

Monday

Tuesday

Wednesday

Thursday

Friday

**Key:** represents 5 students

How many students were late on:

**a** Monday          **b** Thursday?

How can you decide what the fractions of the symbols represent?

You have to make an assumption, such as each 'limb' of the symbol represents one student and its 'head and body' also represents one student.

**a** 19 students were late on Monday.      **b** 13 students were late on Thursday.

Example 6

This pictogram shows the numbers of delayed trains over one weekend.

**The number of delayed trains over a weekend**

Saturday

Sunday

**Key:** represents 10 trains

Give one reason why the pictogram is difficult to interpret.

The last symbol for the number of delayed trains on Sunday is a fraction of 10. However, it is not easy to work out what this fraction represents.

## Bar charts and vertical line charts

A **bar chart** is a series of bars or blocks of the *same* width, drawn either vertically or horizontally from an axis.

The heights or lengths of the bars always represent frequencies.

The bars are separated by narrow gaps of equal width, which helps you to read the chart.

A **vertical line chart** is similar to a bar chart, but has vertical lines instead of bars.

Example 7

The frequency table shows the average monthly rainfall over a six-month period.

| Month | January | February | March | April | May | June |
|---|---|---|---|---|---|---|
| Rainfall (mm) | 10 | 26 | 18 | 36 | 14 | 4 |

a Draw a bar chart to show the data.

b Draw a vertical line chart to show the data.

c How much rain fell in January?

d What is the difference between the amounts of rain that fell in the driest and wettest months?

a The bar chart looks like this.

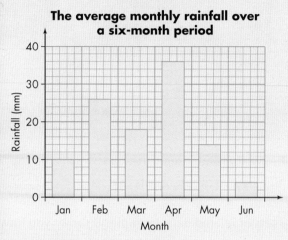

b The vertical line chart looks like this.

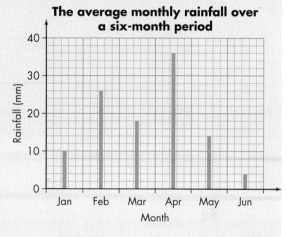

c 10 mm of rain fell in January.

d The driest month was June (4 mm) and wettest month was April (36 mm).

So the difference is 32 mm.

Example 8

The grouped frequency table shows the marks of 24 students in a test. Draw a bar chart to represent the data.

| Marks | 1–10 | 11–20 | 21–30 | 31–40 | 41–50 |
|---|---|---|---|---|---|
| Frequency | 2 | 3 | 5 | 8 | 6 |

The bar chart looks like this.

Note that:

• both axes are labelled

• the class intervals are written under the middle of each bar

• the bars are separated by equal spaces.

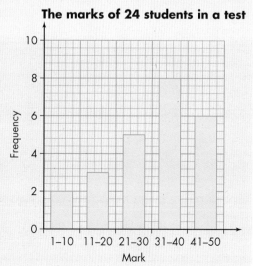

3.2 Statistical diagrams        63

You can use a multiple or **dual bar chart** to compare two or more sets of related data. Example 9 shows you how to do this.

Example 9

This dual bar chart shows the average daily maximum temperatures for England and Turkey over a five-month period.

In which month was the difference between the average temperatures in England and Turkey the greatest?

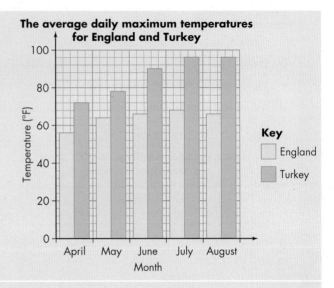

The largest difference is where the tops of the two bars are the furthest apart. This is in August.

**Note:** You must always include a key to identify the two different sets of data.

You can also use a **composite bar chart** to compare sets of related data. Example 10 shows you how to do this.

Example 10

This composite bar chart shows the numbers of visitors to a museum over a three-month period.

**a** How many visitors, in total, went to the museum over the three months?

**b** How many of the visitors over the three months were children?

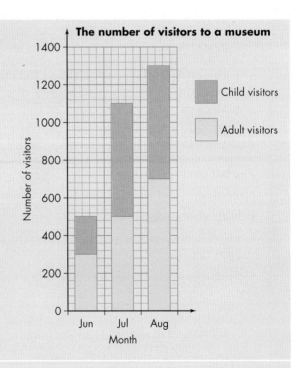

**a** 500 + 1100 + 1300 = 2900 visitors

**b** 200 + 600 + 600 = 1400 children

## Exercise 3B

**1** Mr Weeks, a milkman, recorded the numbers of pints of milk he delivered to 10 flats on a Friday morning. Draw a pictogram for his data.

Use a key of 1 symbol = 1 pint.

| Flat 1 | Flat 2 | Flat 3 | Flat 4 | Flat 5 | Flat 6 | Flat 7 | Flat 8 | Flat 9 | Flat 10 |
|--------|--------|--------|--------|--------|--------|--------|--------|--------|---------|
| 2 | 3 | 1 | 2 | 4 | 3 | 2 | 1 | 5 | 1 |

**EV**

**2** This pictogram is from a Suntours brochure. It shows the average daily hours of sunshine for five months in Tenerife.

**a** Write down the average daily hours of sunshine for each month.

**b** Which month had the most sunshine?

**c** Give a reason why pictograms are useful in holiday brochures.

**The average daily hours of sunshine in Tenerife**

**Key** ☀ represents 2 hours

**MR**

**3** The pictogram shows the amounts of money collected by six students doing a sponsored walk for charity.

**a** Who raised the most money?

**b** How much money did the six students raise altogether?

**c** Robert also took part in the walk and raised £33. Explain why it would be difficult to include him on the pictogram.

**The amount of money collected by students in a sponsored walk**

| Anthony | £ £ £ £ £ |
| Ben | £ £ £ £ £ £ |
| Emma | £ £ £ £ £ |
| Leanne | £ £ £ £ |
| Reena | £ £ £ £ £ £ |
| Simon | £ £ £ £ £ £ £ |

**Key** £ represents £10

**MR**

**4** Millie carried out a fitness survey. She waited outside a fitness centre and asked a sample of people, as they came out, which activity they had taken part in. She then drew this vertical line chart to show her data.

**a** Which was the most popular activity?

**b** How many people took part in Millie's survey?

**c** Give a reason why her data would be better drawn as a bar chart.

**d** Is a fitness centre a good place to carry out a survey on fitness? Explain your answer.

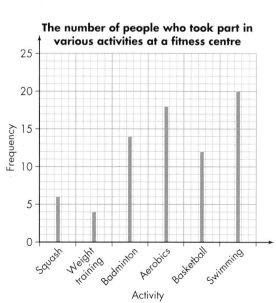

**The number of people who took part in various activities at a fitness centre**

**5** The table shows the number of accidents at a dangerous crossroads over a six-year period.

| Year | 2010 | 2011 | 2012 | 2013 | 2014 | 2015 |
|---|---|---|---|---|---|---|
| Number of accidents | 6 | 8 | 7 | 9 | 6 | 4 |

**a** Draw a pictogram to represent the data.

**b** Draw a bar chart to represent the data.

**c** Draw a vertical line chart to represent the data.

**d** You want to write to your council to suggest traffic lights are put in at the crossroads. Which diagram would you use? Explain why.

 **6** Emily did a survey on the time it took for students in her class to get to school on a particular morning. She wrote down their times, to the nearest minute.

| 15 | 23 | 36 | 45 | 8 | 20 | 34 | 15 | 27 | 49 |
|---|---|---|---|---|---|---|---|---|---|
| 10 | 60 | 5 | 48 | 30 | 18 | 21 | 2 | 12 | 56 |
| 49 | 33 | 17 | 44 | 50 | 35 | 46 | 24 | 11 | 34 |

**a** Draw a grouped frequency table for Emily's data. Use class intervals 1–10, 11–20, …

**b** Draw a bar chart to illustrate her data.

**c** What conclusions could Emily draw from the bar chart?

**7** This diagram shows the minimum and maximum temperatures for one day in August in five cities, to the nearest Celsius degree.

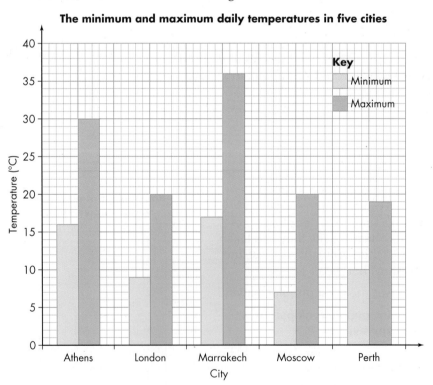

**The minimum and maximum daily temperatures in five cities**

Carlos says that, in most cities, the minimum temperature is about half the maximum temperature.

Is Carlos correct? Give reasons to explain your answer.

**8** The dual bar chart shows the average annual temperatures, in degrees Celsius, for England and Scotland.

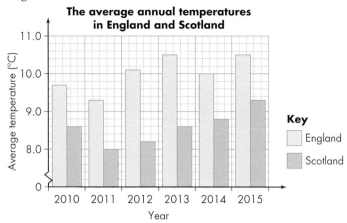

Leo says that the chart shows that the temperature in England is always more than double the temperature in Scotland.

Explain why Leo is wrong.

**9** The table shows the numbers of points Richard and George were each awarded in eight rounds of a general knowledge quiz.

| Round | 1 | 2 | 3 | 4 | 5 | 6 | 7 | 8 |
|---|---|---|---|---|---|---|---|---|
| Richard | 7 | 8 | 7 | 6 | 8 | 6 | 9 | 4 |
| George | 6 | 7 | 6 | 9 | 6 | 8 | 5 | 6 |

Draw a dual bar chart to illustrate the data.

**10** On a sports afternoon, students chose to play basketball, badminton or volleyball.

This composite bar chart represents the data.

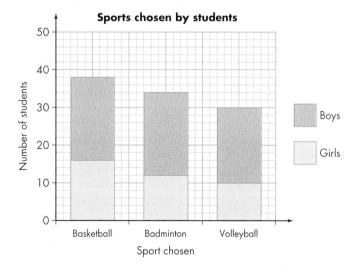

**a** How many students took part altogether?

**b** How many girls chose basketball?

**c** How many boys chose either badminton or volleyball?

**11** These bar charts have been drawn without scales.

The first bar chart shows the numbers of bags of crisps in a café at the start of the day.

The second bar chart shows how many of each flavour were left at the end of the day.

**The number of bags of crisps in a café at the start and at the end of a day**

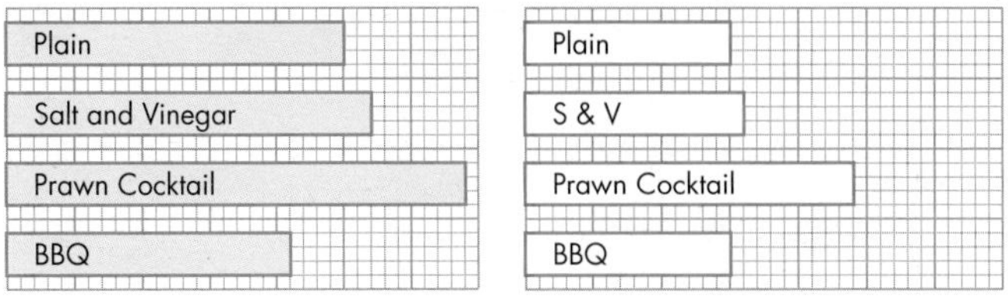

| Plain |
| Salt and Vinegar |
| Prawn Cocktail |
| BBQ |

| Plain |
| S & V |
| Prawn Cocktail |
| BBQ |

40 bags of prawn cocktail crisps were sold.

Each bag of crisps costs 60p.

How much money did the café take from selling crisps?

# 3.3 Line graphs

This section will show you how to:

* draw a line graph to show trends in data.

**Key terms**

line graph

time series graph

trend

In statistics, you can use **line graphs** to show how data changes over a period of time. Line graphs in which the horizontal axis represents time are often called **time series graphs**.

Line graphs can indicate **trends**. For example, they can show whether Earth's temperature is increasing as the concentration of carbon dioxide builds up in the atmosphere or whether a firm's profit margin is falling year on year.

As for any graph, you should always draw line graphs on graph paper.

 The line graph shows the outside temperature at a weather station, taken at hourly intervals. Estimate the temperature at 3:30 pm.

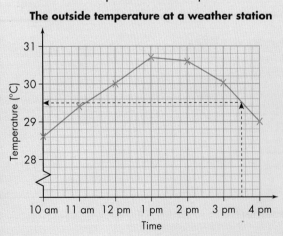

The outside temperature at a weather station

At 3:30 p.m. the temperature is approximately 29.5 °C.

Note that the temperature axis starts at 28 °C rather than 0 °C. This allows you to use a scale that makes it easy to plot the points and then to read the graph. The points are joined with straight lines so you can estimate the temperatures between the readings.

The line graph shows a company's annual profit over a five-year period. Between which two years was the increase in profits greatest?

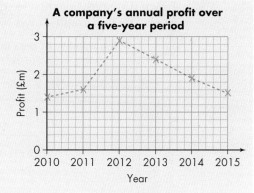

**A company's annual profit over a five-year period**

The greatest increase in profits is where the line is steepest. This is between 2011 and 2012.

On this graph, the values between the plotted points have no meaning because the company's profit was calculated at the end of every year. In cases like this, the lines are often dashed. Although the trend appears to be that profits fell after 2012, it would not be sensible to predict what would happen after 2015.

## Exercise 3C

**1** The line graph shows the value of a bank's shares on seven consecutive trading days.

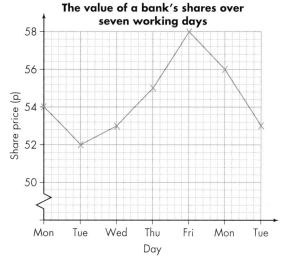

**The value of a bank's shares over seven working days**

a On which day did the share price have its lowest value? What was that value?

b By how much did the share price rise from Wednesday to Thursday?

c On which day was the greatest rise in the share price from the previous day?

d Mr Hardy sold 500 shares on Friday. He originally bought the shares at 40p each. How much profit did he make?

**2** The table shows the population of a town, rounded to the nearest thousand, at each census.

| Year | 1951 | 1961 | 1971 | 1981 | 1991 | 2001 | 2011 |
|------|------|------|------|------|------|------|------|
| Population (1000s) | 12 | 14 | 15 | 18 | 21 | 25 | 23 |

a Draw a line graph for the data.

b From your graph, estimate the population in 1976.

c Between which two consecutive censuses did the largest population increase occur?

d Can you predict the population for 2021? Give a reason for your answer.

**3** When plotting a graph to show the summer midday temperatures in Spain, Abbass decided to start his graph at the temperature 20 °C.

Explain why he may have made this decision.

**4** Trevor was studying ants. He counted the ants in an ants' nest at the end of each week. The line graph shows his results.

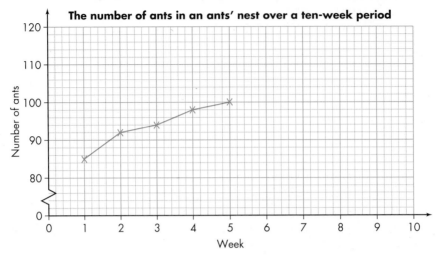

**The number of ants in an ants' nest over a ten-week period**

**a** At the end of week 6, he counted 104 ants. At the end of week 10, he counted 120 ants.

 **i** Copy the graph and plot the points for week 6 and week 10.

 **ii** Complete the graph with straight lines.

**b** Use your graph to estimate the number of ants at the end of week 8.

**5** The table shows some estimated numbers of tourists worldwide.

| Year | 1980 | 1985 | 1990 | 1995 | 2000 | 2005 | 2010 |
|---|---|---|---|---|---|---|---|
| Number of tourists (millions) | 100 | 150 | 220 | 280 | 290 | 320 | 340 |

**a** Draw a line graph for the data.

**b** Use your graph to estimate the number of tourists in 1975.

**c** In which five-year period did tourism increase the most?

**d** Comment on the trend in the data. Give a possible reason for it.

**6** The table shows the maximum and minimum daily temperatures, to the nearest Celsius degree, for London, over a week.

| Day | Sunday | Monday | Tuesday | Wednesday | Thursday | Friday | Saturday |
|---|---|---|---|---|---|---|---|
| Maximum (°C) | 12 | 14 | 16 | 15 | 16 | 14 | 10 |
| Minimum (°C) | 4 | 5 | 7 | 8 | 7 | 4 | 3 |

**a** Draw two line graphs, on the same axes, to show the maximum and minimum temperatures.

**b** Find the smallest and greatest differences between the maximum and minimum temperatures.

**7** Jack records the mass of a puppy at the end of each week, for five weeks after it is born.

| Week | 1 | 2 | 3 | 4 | 5 |
|---|---|---|---|---|---|
| Mass (g) | 850 | 920 | 940 | 980 | 1000 |

**a** Draw a line graph for the data.

**b** Jack uses the data to estimate the mass of the puppy after 8 weeks.

 Give a reason why his answer may not be accurate.

# 3.4 Statistical averages

This section will show you how to:

- work out the mode, median, mean and range of small sets of data
- decide which is the best average to use to represent a data set.

You often see the word **average** used to describe or compare sets of data, for example:

- the average rainfall in Britain
- the average score of a batsman
- the average weekly wage
- the average mark in an examination.

In each case, a single **representative** or typical value (the average) is used to stand for the whole set of values. An average is extremely useful because it enables you to compare one set of data with another set by comparing just two values – their averages.

There are several ways of expressing an average, but the most commonly used averages are the mode, the median and the mean.

| Key terms |
| --- |
| average |
| categorical |
| consistency |
| modal |
| outlier |
| range |
| representative |
| spread |
| stem-and-leaf diagram |

## The mode

The mode is the value that occurs most often in a set of data. It is the value with the highest frequency.

The mode is very easy to find and it can be applied to non-numerical (**categorical**) data. For example, you could find the **modal** style of shirts sold in a particular month.

---

**Example 13**

These are the numbers of goals that Suhail scored in 12 school football matches.

   1  2  1  0  1  0  0  1  2  1  0  2

What is the mode of his scores?

The number that occurs most often in this list is 1. So, the mode is 1.

You can also say that the modal score or modal value is 1.

---

**Example 14**

Alicia asked her friends how many books they had each taken out of the school library during the previous month. These are their responses.

   2  1  3  4  6  4  1  3  0  2  6  0

What is the mode?

This data has no mode, because no number occurs more than any of the others.

---

# Exercise 3D

**1** Write down the mode for each set of data.

Hints and tips It helps to put the data in order, or group all the same values together.

**a** 3, 4, 7, 3, 2, 4, 5, 3, 4, 6, 8, 4, 2, 7

**b** 47, 49, 45, 50, 47, 48, 51, 48, 51, 48, 52, 48

**c** −1, 1, 0, −1, 2, −2, −2, −1, 0, 1, −1, 1, 0, −1, 2, −1, 2

**d** $\frac{1}{2}, \frac{1}{4}, 1, \frac{1}{2}, \frac{3}{4}, \frac{1}{4}, 0, 1, \frac{3}{4}, \frac{1}{4}, 1, \frac{1}{4}, \frac{3}{4}, \frac{1}{4}, \frac{1}{2}$

**e** 100, 10, 1000, 10, 100, 1000, 10, 1000, 100, 1000, 100, 10

**f** 1.23, 3.21, 2.31, 3.21, 1.23, 3.12, 2.31, 1.32, 3.21, 2.31, 3.21

**2** Write down the modal category for each set of data.

**a** red, green, red, amber, green, red, amber, green, red, amber

**b** rain, sun, cloud, sun, rain, fog, snow, rain, fog, sun, snow, sun

**c** α, γ, α, β, γ, α, α, γ, β, α, β, γ, β, β, α, β, γ, β

**d** E, I, M, M, I, E, M, P, M, P, O, E, J, P, M, O, P

**3** Evie did a survey to find the shoe sizes of students in her class. The bar chart illustrates her data.

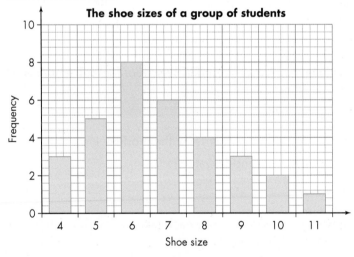

**a** How many students are there in Evie's class?

**b** What is the modal shoe size?

**c** Evie decided to draw a bar chart to show the shoe sizes of the boys and the girls separately. Do you think that the mode for the boys and the mode for the girls will be the same as the mode for the whole class? Explain your answer.

**4** The frequency table shows the marks that a class obtained in a spelling test.

| Mark | 3 | 4 | 5 | 6 | 7 | 8 | 9 | 10 |
|---|---|---|---|---|---|---|---|---|
| Frequency | 1 | 2 | 6 | 5 | 5 | 4 | 3 | 4 |

**a** Write down the modal mark.

**b** Do you think the modal mark is representative of this data? Explain your answer.

3 Statistics: Charts, tables and averages

**5** The grouped frequency table shows the number of emails each household in Orchard Street received during one day.

| Number of emails | 0–4 | 5–9 | 10–14 | 15–19 | 20–24 | 25–29 | 30–34 | 35–39 |
|---|---|---|---|---|---|---|---|---|
| Frequency | 9 | 12 | 14 | 11 | 10 | 8 | 4 | 2 |

**a** Draw a bar chart to illustrate the data.

**b** How many households are there in Orchard Street?

**c** How many households received 20 or more emails?

**d** How many households did not receive any emails during the week? Explain your answer.

**e** Write down the modal group for this data.

> **Hints and tips** You cannot find the mode of the data in a grouped frequency table. Instead, you need to find the modal group, which is the group with the highest frequency.

**6** The table shows the eye colours of the students in a class.

| | Blue | Brown | Green |
|---|---|---|---|
| Boys | 4 | 8 | 1 |
| Girls | 8 | 5 | 2 |

**a** How many students are there in the class?

**b** What is the modal eye colour for:

   **i** the boys         **ii** the girls         **iii** the whole class?

**c** After two students join the class, the modal eye colour for the whole class is blue. Which of these statements is true?

   Both students had green eyes.

   Both students had brown eyes.

   Both students had blue eyes.

   You cannot tell what their eye colours were.

**7** Here is a large set of raw data.

| | | | | | | | | |
|---|---|---|---|---|---|---|---|---|
| 5 | 6 | 8 | 2 | 4 | 8 | 9 | 8 | 1 |
| 3 | 4 | 2 | 7 | 2 | 4 | 6 | 7 | 5 |
| 3 | 8 | 9 | 1 | 3 | 1 | 5 | 6 | 2 |
| 5 | 7 | 9 | 4 | 1 | 4 | 3 | 3 | 5 |
| 6 | 8 | 6 | 9 | 8 | 4 | 8 | 9 | 3 |
| 4 | 6 | 7 | 7 | 4 | 5 | 4 | 2 | 3 |
| 4 | 6 | 7 | 6 | 5 | 5 | | | |

**a** What problems may you have if you try to find the mode by counting individual numbers?

**b** Explain a method that would make finding the mode more efficient and accurate.

**c** Use your method to find the mode of the data.

**8** Explain why the mode is often referred to as the 'shopkeeper's average'.

## The median

The median is the middle value of a list of values when they are put in order of size.

The advantage of using the median as an average is that half the data-values are below the median value and half are above it. Therefore, the average is affected only slightly by the presence of any **outliers**. Outliers are particularly high or low values that are not typical of the data as a whole.

**Example 15**

Work out the median for this list of numbers.

  2, 3, 5, 6, 1, 2, 3, 4, 5, 4, 6

Put the list in numerical order.  1, 2, 2, 3, 3, **4**, 4, 5, 5, 6, 6

There are 11 numbers in the list, so the middle of the list is the 6th number.

Therefore, the median is 4.

**Example 16**

Work out the median for this list of numbers.

  7, 3, 5, 1, 6, 7, 2, 3, 4, 5, 4, 6

Put the list in numerical order.  1, 2, 3, 3, 4, **4, 5**, 5, 6, 6, 7, 7

There are 12 numbers in the list, so the middle value lies between the middle pair.

These are the 6th and 7th values, which are 4 and 5.

The median is the number that is *halfway between* 4 and 5. So the median is $4\frac{1}{2}$.

## Exercise 3E

 **1** Find the median for each set of data.

> Hints and tips   Remember to put the data in order before finding the median.
> If there is an even number of data-values, the median is halfway
> between the middle pair.

**a** 7, 6, 2, 3, 1, 9, 5, 4, 8            **b** 26, 34, 45, 28, 27, 38, 40, 24, 27, 33, 32, 41, 38

**c** 4, 12, 7, 6, 10, 5, 11, 8, 14, 3, 2, 9      **d** 12, 16, 12, 32, 28, 24, 20, 28, 24, 32, 36, 16

**e** 10, 6, 0, 5, 7, 13, 11, 14, 6, 13, 15, 1, 4, 15    **f**  −1, −8, 5, −3, 0, 1, −2, 4, 0, 2, −4, −3, 2

**g** 5.5, 5.05, 5.15, 5.2, 5.3, 5.35, 5.08, 5.9, 5.25

 **2** The list shows the amounts of money spent by a group of 15 sixth-formers on lunch in the school's cafeteria.

£2.30, £2.20, £2, £2.50, £2.20, £3.50, £2.20, £2.25, £2.20, £2.30, £2.40, £2.20, £2.30, £2, £2.35

**a** Write down the mode for the data.

**b** Work out the median for the data.

**c** Which is the better average to use? Explain your answer.

 **3** **a** Work out the median of 7, 4, 3, 8, 2, 6, 5, 2, 9, 8, 3.

**b** Use your answer to part **a** to find the median for each of these sets.

> Hints and tips   Look for a connection between the original data and the new data.
> For example, in **i**, the numbers are each 10 more than those in **a**.

**i**  17, 14, 13, 18, 12, 16, 15, 12, 19, 18, 13      **ii**  217, 214, 213, 218, 212, 216, 215, 212, 219, 218, 213

**iii** 12, 9, 8, 13, 7, 11, 10, 7, 14, 13, 8          **iv**  14, 8, 6, 16, 4, 12, 10, 4, 18, 16, 6

**4** The chart shows the age, height and mass of each of the seven players in a netball team.

| | Amina | Bella | Dolly | Elisa | Grace | Lily | Martha |
|---|---|---|---|---|---|---|---|
| Age (years) | 13 | 15 | 12 | 11 | 11 | 15 | 14 |
| Height (cm) | 161 | 165 | 162 | 158 | 154 | 168 | 169 |
| Mass (kg) | 45 | 49 | 43 | 40 | 46 | 55 | 52 |

**a** Work out the median age of the team. Which player has the median age?

**b** Work out the median height of the team. Which player has the median height?

**c** Work out the median mass of the team. Which player has the median mass?

**d** Who would you choose as the average player in the team? Give a reason for your answer.

**5** These are seven cheque payments.

£10, £10, £12, £15, £80, £100, £200

Explain why the median is not a good average to use for this set of payments.

**6** Here is list of numbers.

4, 4, 5, 8, 10, 11, 12, 15, 15, 16, 20

**a** Write down four more numbers to make the median 12.

**b** Write down six more numbers to make the median 12.

**c** What are the fewest number of values you can add to the list to make the median 4?

**7** A list contains seven even numbers. The largest number is 24. The smallest number is half the largest. The mode is 14 and the median is 16. Two of the numbers add up to 42. What are the seven numbers?

**8** The bar chart shows the marks that Mrs Woodhead gave her students for a mental arithmetic test.

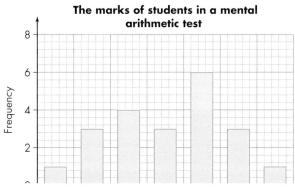

**The marks of students in a mental arithmetic test**

**a** How many students are there in Mrs Woodhead's class?

**b** What is the modal mark?

**c** Copy and complete this frequency table.

| Mark | 12 | 13 | 14 | 15 | 16 | 17 | 18 |
|---|---|---|---|---|---|---|---|
| Frequency | 1 | 3 | | | | | |

**d** How could you work out the median mark from the frequency table?

## The mean

The mean of a set of data is the sum of all the values in the set *divided by* the total number of values in the set. That is:

$$\text{mean} = \frac{\text{sum of all values}}{\text{total number of values}}$$

The advantage of using the mean as an average is that it takes into account all the values in the set of data. The disadvantage is that it may be affected by outliers. It is the most commonly used average.

**Example 17**

Find the mean of 4, 8, 7, 5, 9, 4, 8, 3.

The sum of all the values is $4 + 8 + 7 + 5 + 9 + 4 + 8 + 3 = 48$.

The total number of values is 8.

Therefore, mean = $\frac{48}{8}$

$= 6$

**Example 18**

These are the ages of 11 players in a football team.

21, 23, 20, 27, 25, 24, 25, 30, 21, 22, 28

What is the mean age of the team?

The sum of all the ages is 266.

The total number in the team is 11.

Therefore, mean age = $\frac{266}{11}$

$= 24.1818\ldots$

$= 24.2$ (1 dp)

When the answer is not exact, it is usual to round the mean to 1 decimal place.

**Example 19**

The mean mass of eight members of a rowing crew is 89 kg.

When the cox is included, the mean mass is 85 kg.

What is the mass of the cox?

The eight crew members have a total mass of $8 \times 89 = 712$ kg.

With the cox, the total mass is $9 \times 85 = 765$ kg.

So the cox has a mass of $765 - 712 = 53$ kg.

## Exercise 3F

1. Work out the mean for each set of data.

   **a** 7, 8, 3, 6, 7, 3, 8, 5, 4, 9

   **b** 47, 3, 23, 19, 30, 22

   **c** 42, 53, 47, 41, 37, 55, 40, 39, 44, 52

   **d** 1.53, 1.51, 1.64, 1.55, 1.48, 1.62, 1.58, 1.65

   **e** 1, 2, 0, 2, 5, 3, 1, 0, 1, 2, 3, 4

**2** Calculate the mean for each set of data. Give your answers correct to 1 decimal place.

   **a** 34, 56, 89, 34, 37, 56, 72, 60, 35, 66, 67

   **b** 235, 256, 345, 267, 398, 456, 376, 307, 282

   **c** 50, 70, 60, 50, 40, 80, 70, 60, 80, 40, 50, 40, 70

   **d** 43.2, 56.5, 40.5, 37.9, 44.8, 49.7, 38.1, 41.6, 51.4

   **e** 2, 3, 1, 0, 2, 5, 4, 3, 2, 0, 1, 3, 4, 5, 0, 3, 1, 2

**3** The table shows the marks that 10 students obtained in their Year 10 examinations in mathematics, English and science.

| Student | Anna | Blake | Chloe | Dexter | Evan | Freya | George | Hassan | Imogen | Jasmine |
|---|---|---|---|---|---|---|---|---|---|---|
| Maths | 45 | 56 | 47 | 77 | 82 | 39 | 78 | 32 | 92 | 62 |
| English | 54 | 55 | 59 | 69 | 66 | 49 | 60 | 56 | 88 | 44 |
| Science | 62 | 58 | 48 | 41 | 80 | 56 | 72 | 40 | 81 | 52 |

   **a** Work out the mean mark for mathematics.

   **b** Work out the mean mark for English.

   **c** Work out the mean mark for science.

   **d** Which student obtained marks closest to the mean in all three subjects?

   **e** How many students were above the mean mark in all three subjects?

**(PS)** **4** Heather kept a record of the amount of time she spent on her homework over 10 days.

$\frac{1}{2}$ hour, 20 minutes, 35 minutes, $\frac{1}{4}$ hour, 1 hour, $\frac{1}{2}$ hour, $1\frac{1}{2}$ hours, 40 minutes, $\frac{3}{4}$ hour, 55 minutes

Calculate the mean time, in minutes, that Heather spent on her homework over this period.

**(EV)** **5** These are the weekly wages of 10 people working in an office.

£350   £200   £180   £200   £350   £200   £240   £480   £300   £280

   **a** Write down the modal wage.

   **b** Work out the median wage.

   **c** Calculate the mean wage.

   **d** Which of the three averages best represents the office staff's wages? Give reasons for your answer.

**6** The ages of five people in a group of walkers are 38, 28, 30, 42 and 37.

   **a** Calculate the mean age of the group.

   **b** Steve, who is 41, joins the group. Calculate the new mean age of the group.

**(MR)** **7** **a** Calculate the mean of 3, 7, 5, 8, 4, 6, 7, 8, 9 and 3.

   **b** Calculate the mean of 13, 17, 15, 18, 14, 16, 17, 18, 19 and 13. What do you notice?

   **c** Write down, without calculating, the mean for each of these sets of data.

     **i** 53, 57, 55, 58, 54, 56, 57, 58, 59, 53

     **ii** 103, 107, 105, 108, 104, 106, 107, 108, 109, 103

     **iii** 4, 8, 6, 9, 5, 7, 8, 9, 10, 4

**8** Two families enter a quiz.

| Speed family | Age | Roberts family | Age |
|---|---|---|---|
| Brian | 59 | Frank | 64 |
| Kath | 54 | Mary | 62 |
| James | 34 | David | 34 |
| June | 34 | Jim | 32 |
| John | 30 | Tom | 30 |
| Joseph | 24 | Helen | 30 |
| Joy | 19 | Evie | 16 |

To enter the quiz, each family had to choose four members with a mean age from 35 to 36. Choose two teams, one from each family, with this mean age.

 **9** Asif had scored 315 runs in nine games of cricket.

What is the least number of runs he needs to score in the next match to increase his average score?

 **10** The mean age of a group of eight walkers is 42. Joanne joins the group and the mean age changes to 40. How old is Joanne?

## Which average to use?

An average must be truly representative of a set of data. So, when you have to find an average, it is important to choose the appropriate type of average for the particular set of data. If you use the wrong average, your results could be distorted and give misleading information.

This table shows the advantages and disadvantages of each type of average. You can use it to help you decide on an appropriate average for a data set.

| | Mode | Median | Mean |
|---|---|---|---|
| Advantages | Very easy to find<br>Not affected by outliers<br>Can be used for non-numerical data | Easy to find for ungrouped data<br>Not affected by outliers | Easy to find<br>Uses all the values<br>The total for a given number of values can be calculated from it |
| Disadvantages | Does not use all the values<br>May not exist | Does not use all the values<br>Often not understood | Outliers can distort it<br>Has to be calculated |
| Used for | Non-numerical data<br>Finding the most likely value | Data with outliers | Data with values that are spread in a balanced way |

# Exercise 3G 🖩

**1**  **a** For each set of data, work out the mode, the median and the mean.

    **i** 6, 10, 3, 4, 3, 6, 2, 9, 3, 4    **ii** 6, 8, 6, 10, 6, 9, 6, 10, 6, 8    **iii** 7, 4, 5, 3, 28, 8, 2, 4, 10, 9

    **b** Decide on the best average to represent each set of data. Give a reason for your answers.

**(MR)** **2** These are the ages of the members of a hockey team.

    29  26  21  24  26  28  35  23  29  28  29

    **a** Work out:

      **i** the modal age      **ii** the median age      **iii** the mean age.

    **b** Which is the best average to use? Give a reason for your answer.

**(MR)** **3** A newsagent sold the following numbers of copies of *The Evening Star* on 12 consecutive evenings during a promotion exercise.

    65  73  75  86  90  112  92  87  77  73  68  62

    **a** Work out the mode, the median and the mean for the sales.

    **b** The publisher asked the newsagent to report the average number of sales during the promotion. Which of the three averages would you advise the newsagent to use? Explain why.

**(MR)** **4** Decide which average you would use for each of these. Give reasons for your answers.

    **a** The average mark in an examination

    **b** The average pocket money for a group of 16-year-old students

    **c** The average shoe size for all the girls in Year 10

    **d** The average height for all the artistes on tour with a circus

    **e** The average hair colour for students in your school

    **f** The average mass of all newborn babies in a hospital's maternity ward.

**(CM)** **5** A pack of matches consisted of 12 boxes. These are the numbers of matches in the individual boxes.

    34  31  29  35  33  30  31  28  29  35  32  31

    On the box it stated: 'Average contents 32 matches'. Is this correct? Explain your answer.

**(EV)** **6** The table shows the annual salaries for a firm's employees.

    **a** What is:

| | |
|---|---|
| **Chairperson** | £83 000 |
| **Managing director** | £65 000 |
| **Floor manager** | £34 000 |
| **Skilled worker 1** | £28 000 |
| **Skilled worker 2** | £28 000 |
| **Machinist** | £20 000 |
| **Computer engineer** | £20 000 |
| **Secretary** | £20 000 |
| **Office junior** | £8 000 |

      **i** the modal salary

      **ii** the median salary

      **iii** the mean salary?

    **b** The management has suggested a pay rise of 6% for everyone. The shop floor workers want a pay rise of £1500 for everyone.

      **i** Which one of the suggestions is more likely to cause problems for the firm? Why?

      **ii** What difference would each suggestion make to the modal, median and mean salaries?

**7** Mr Brennan, a mathematics teacher, told each student their individual test mark. He told the whole class the modal mark, the median mark and the mean mark.

    **a** Which average would tell a student whether they were in the top half or the bottom half of the class?

    **b** Which average really tells the students nothing?

    **c** Which average allows a student to gauge how well they have done, compared with everyone else?

**8** Three players were hoping to be chosen for the basketball team. The table shows their scores for the last few games they played. The teacher said they would be selected by their best average score. By which average would each boy choose to be selected? Give reasons for your answers.

| Tom | 16, 10, 12, 10, 13, 8, 10 |
| David | 16, 8, 15, 25, 8 |
| Mohammed | 15, 2, 15, 3, 5 |

**9** Here are two responses to the question: *What is the average pay at a factory with 100 employees?*

    Boss: 'The average is £43 295.'

    A worker: 'The average is £18 210.'

    Explain how they can both be correct.

## The range

The **range** for a set of data is the difference between the highest and lowest values.

    range = highest value – lowest value

The range is *not* an average. It shows the **spread** of the data. You can use it when you compare two or more sets of similar data, for example, to comment on their **consistency**.

---

**Example 20**

Rachel's marks in 10 mental arithmetic tests were 4, 4, 7, 6, 6, 5, 7, 6, 9 and 6.

Adil's marks in the same tests were 6, 7, 6, 8, 5, 6, 5, 6, 5 and 6.

Compare their results.

Rachel's mean mark is $60 \div 10 = 6$ and her range is $9 - 4 = 5$.

Adil's mean mark is $60 \div 10 = 6$ and his range is $8 - 5 = 3$.

Although their means are the same, Adil has a smaller range. This shows that Adil's results are more consistent.

When calculating the range, you must check for outliers at the start or end of the data, as they will give an unrepresentative value. This is a value that is very different from all the others.

---

## Exercise 3H

**1** Work out the range for each set of data.

    **a** 3, 8, 7, 4, 5, 9, 10, 6, 7, 4

    **b** 62, 59, 81, 56, 70, 66, 82, 78, 62, 75

    **c** 1, 0, 4, 5, 3, 2, 5, 4, 2, 1, 0, 1, 4, 4

    **d** 3.5, 4.2, 5.5, 3.7, 3.2, 4.8, 5.6, 3.9, 5.5, 3.8

    **e** 2, –1, 0, 3, –1, –2, 1, –4, 2, 3, 0, 2, –2, 0, –3

**2** The table shows the maximum and minimum temperatures at midday for five cities in England during a week in August.

| | Bristol | Leeds | London | Oxford | Truro |
|---|---|---|---|---|---|
| Maximum temperature (°C) | 28 | 25 | 26 | 27 | 24 |
| Minimum temperature (°C) | 23 | 22 | 24 | 20 | 21 |

**a** Write down the range of the temperatures for each city.

**b** What do the ranges tell you about the weather for England during the week?

**3** The table shows the amounts taken by the school equipment shop over a three-week period.

| | Monday | Tuesday | Wednesday | Thursday | Friday |
|---|---|---|---|---|---|
| Week 1 (£) | 32 | 29 | 36 | 30 | 28 |
| Week 2 (£) | 34 | 33 | 25 | 28 | 20 |
| Week 3 (£) | 35 | 34 | 31 | 33 | 32 |

**a** Calculate the mean amount taken each week.

**b** Find the range for each week.

**c** What can you say about the total amounts taken for each of the three weeks?

**4** In a golf tournament, the club captain had to choose either Leila or Shona to play in the first round. In the previous eight rounds, their scores were as follows.

Leila's scores: 75, 92, 80, 73, 72, 88, 86, 90

Shona's scores: 80, 87, 85, 76, 85, 79, 84, 88

**a** Calculate the mean score for each golfer.

**b** Work out the range for each golfer.

**c** Which golfer would you choose to play in the tournament? Explain why.

**5** Daniel has a choice of two buses to catch to school: the 50 or the 50A. Over a month, he kept a record of the number of minutes each bus was late on ten different occasions, when it set off from the bus stop near his home.

Number 50:  4, 2, 0, 6, 4, 8, 8, 6, 3, 9    Number 50A:  3, 4, 0, 10, 3, 5, 13, 1, 0, 1

**a** For each bus:

**i** calculate the mean number of minutes it was late

**ii** work out the range of the minutes it was late.

**b** Which bus would you advise Daniel to catch? Give a reason for your answer.

**6** The table gives the ages and heights of 10 children in Emma's family.

**a** Emma wants to take some of the children in her family to a theme park for her birthday. She wants to invite as many of them as possible but does not want the range of ages to be more than 5. Who should she invite?

| Name | Age (years) | Height (cm) |
|---|---|---|
| Jake | 9 | 121 |
| Isaac | 4 | 73 |
| Zoe | 8 | 93 |
| Lewis | 10 | 118 |
| Eliza | 3 | 66 |
| Andrew | 6 | 82 |
| Oliver | 4 | 78 |
| Beatrice | 2 | 69 |
| Zac | 9 | 87 |
| Niamh | 7 | 82 |

**b** This is a sign at the theme park.

Isaac is the shortest person who can go on the ride and Zac is the tallest.

What are the smallest and largest possible missing values on the sign?

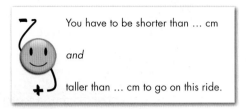

You have to be shorter than … cm

*and*

taller than … cm to go on this ride.

**7** A school quiz has four teams: Year 10, Year 11, the sixth form and teachers.

    **a** The age range of one team is 20 and the mean age is 34.

       Who would you expect to be in this team? Explain your answer.

    **b** Another team has an age range of 1 and a mean age of $15\frac{1}{2}$.

       Who would you expect to be in this team? Give a reason to explain your answer.

**8**  **a** Write down a set of five numbers that has a range of 5 *and* a mean of 5.

    **b** Write down a set of five numbers that has a range of 5, a mode of 5, a median of 5 *and* a mean of 5.

## Stem-and-leaf diagrams

Here is a list of the ages of 20 people.

    23, 13, 34, 44, 26, 12, 41, 31, 20, 18, 19, 31, 48, 32, 45, 14, 12, 27, 31, 19

Another way of working out the median and range for discrete data is to use a **stem-and-leaf diagram**.

First put the ages in order.

    12, 12, 13, 14, 18, 19, 19, 20, 23, 26, 27, 31, 31, 31, 32, 34, 41, 44, 45, 48

The tens digits will be the 'stem' and the units digits will be the 'leaves'.

Here is the completed stem-and-leaf diagram.

A stem-and-leaf diagram gives a better idea of how the data is distributed.

Remember to include a key.

| 1 | 2 2 3 4 8 9 9 |
| 2 | 0 3 6 7 |
| 3 | 1 1 1 2 4 |
| 4 | 1 4 5 8 |

**Key**   1 | 2   represents an age of 12

---

**Example 21**

Here are the exam marks for 15 students.

    45, 62, 58, 58, 61, 49, 61, 47, 52, 58, 48, 56, 65, 46, 54

**a** Draw a stem-and-leaf diagram for the data.

**b** Work out the range of the marks.

**c** Work out the median mark.

**a** The marks in order are:

    45, 46, 47, 48, 49, 52, 54, 56, 58, 58, 58, 61, 61, 62, 65

The stem-and-leaf diagram looks like this.

| 4 | 5 6 7 8 9 |
| 5 | 2 4 6 8 8 8 |
| 6 | 1 1 2 5 |

**Key**   4 | 5   represents 45 marks

**b** The range = 65 – 45 = 20 marks.

**c** The median is the eighth value in the list, which is 56 marks.

# Worked exemplars

  **1**  The dual bar chart shows the number of people who attended an evening class lesson.

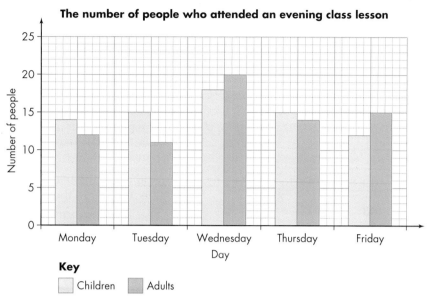

**a** How many adults attended the lesson on Tuesday?

**b** On which day did most children attend the lesson?

**c** How many people attended the lesson on Friday?

**d** Louise said: 'More adults than children attended the five lessons.'

Is she correct? Give a reason for your answer, clearly showing your working.

| This question requires you to interpret and draw conclusions from mathematical information. | |
|---|---|
| **a** 11 | Make sure you read the correct value carefully. |
| **b** Wednesday | This is the day with the highest bar for the children. Make sure you give the day. The answer is not 20. |
| **c** 27 | Add together the heights of the bars for Friday. |
| **d** (see table below) | Lay out your working as shown in the table. |

**d**

|   | Mon | Tue | Wed | Thu | Fri | Total |
|---|---|---|---|---|---|---|
| **C** | 14 | 15 | 18 | 15 | 12 | 74 |
| **A** | 12 | 11 | 20 | 14 | 15 | 72 |

No, because 74 children attended the five lessons but only 72 adults attended.

**2** These are the masses, in kilograms, of a rowing boat crew.

91, 81, 89, 91, 79, 85, 87, 45

**a** Write down the modal mass.  **b** Work out the median mass.

**c** Calculate the mean mass.  **d** Which average best describes the data? Explain your answer.

| In part **d** you need to evaluate your answers to parts **a** to **c** in relation to the set of data. First, work out the answers for parts **a** to **c**. | |
|---|---|
| **a** 91 kg | The mode is the most common mass. |
| **b** Masses in order:<br><br>45, 79, 81, 85, 87, 89, 91, 91<br><br>Median = 86 kg | Start by putting the masses in order.<br><br>The middle pair is 85 and 87, so the median is the mass that is halfway between them. |
| **c** Mean = $\dfrac{91 + 81 + 89 + 91 + 79 + 85 + 87 + 45}{8}$<br><br>$= \dfrac{648}{8}$<br><br>$= 81$ kg | Add up all the masses and divide by the number in the crew.<br><br>This gives $648 \div 8$. |
| **d** The median as it avoids using the outlier mass of 45 kg. | This is an evaluation question so you need to interpret the three averages from parts **a** to **c** in the context of the given problem. |

 **3** Roger and Brian keep a record of their scores in eight games of darts.

| Game number | 1 | 2 | 3 | 4 | 5 | 6 | 7 | 8 |
|---|---|---|---|---|---|---|---|---|
| Roger | 45 | 60 | 142 | 74 | 48 | 54 | 89 | 64 |
| Brian | 37 | 180 | 120 | 46 | 72 | 80 | 97 | 48 |

**a** What is the range of scores for each player?

**b** What is the mean score for each player?

**c** Which player is the more consistent and why?

**d** Who would you say is the better player and why?

| This question requires you to show your skills in mathematical reasoning, which means that you should show how you reach your answer. | |
|---|---|
| **a** Roger's range is 142 – 45 = 97<br><br>Brian's range is 180 – 37 = 143 | The range is the difference between the highest value and the smallest value. |
| **b** Roger's mean is 72 and Brian's mean is 85. | Add up all the scores for each player and divide by the number of games.<br><br>Roger's mean = 576 ÷ 8 = 72<br><br>Brian's mean = 680 ÷ 8 = 85 |
| **c** Roger, because his range is smaller. | You are drawing a conclusion from the information you have worked out. |
| **d** Brian, because he has a higher mean score. | You are drawing a conclusion from the information you have worked out. |

# Ready to progress?

I can draw frequency tables for sets of data.
I can draw and interpret pictograms, bar charts and vertical line charts.
I can draw line graphs to predict trends.
I can work out the mode, median and mean for a set of data.
I can work out the range for a set of data.

I know how to choose the best average to represent a set of data.
I can use the range and an average to compare sets of data.

# Review questions

1   Lucy threw a dice 24 times.

These are her scores.

3   5   3   4   1   2   4   5
6   2   3   4   3   1   4   3
2   3   5   5   3   4   2   1

**a** Copy and complete the frequency table.

| Score | Tally | Frequency |
|-------|-------|-----------|
| 1 | | |
| 2 | | |
| 3 | | |
| 4 | | |
| 5 | | |
| 6 | | |

**b** Draw a vertical line chart to show the data.

2   Percy wants to plant a conifer hedge. He buys 10 plants from a local garden centre.

These are the heights, in centimetres, of the 10 plants.

61, 84, 63, 58, 81, 43, 78, 80, 68, 84

**a** Write down the modal height.

**b** Work out the median height.

**c** Calculate the mean height.

**d** Which average best represents the height of the plants? Give a reason for your answer.

3   These are the ages of the members of a school hockey team.

15   15   17   16   16   14   16   16   16   15   16

Which average best represents the age of the team? Give a reason for your answer.

(EV) **4** The dual bar chart shows information about the amounts of time, in minutes, that Hannah and Josh spent on their computers on four days.

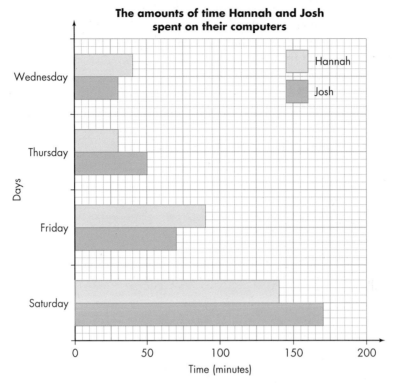

**The amounts of time Hannah and Josh spent on their computers**

**a** Work out the mean time for:

  **i** Hannah   **ii** Josh.

**b** Write down the range of the times for:

  **i** Hannah   **ii** Josh.

**c** Compare the amount of time that Hannah and Josh spent on their computers.

**5** The table shows the percentage sales for three products sold over a four-week period in a park kiosk.

|  | Ice cream | Chocolate bars | Crisps |
|---|---|---|---|
| Week 1 | 36% | 36% | 28% |
| Week 2 | 32% | 28% | 40% |
| Week 3 | 42% | 26% | 32% |
| Week 4 | 34% | 32% | 34% |

Draw a composite bar chart to illustrate the data.

(MR) **6** This back-to-back stem-and-leaf diagram shows the marks for a group of boys and girls in a history test.

```
        Boys                      Girls
   9  6  6  5   │ 3 │ 0  5  7  9
9  6  6  2  0   │ 4 │ 2  2  3  8  8  8
      5  4  4  3 │ 5 │ 1  1  5
```

**Key** Boys: 2 │ 4 represents 42 marks
Girls: 3 │ 5 represents 35 marks

Hints and tips  Read the boys' marks from right to left.

a Write down the range of marks for the boys.

b Write down the range of marks for the girls.

c Work out the median mark for the boys.

d Work out the median mark for the girls.

e What overall conclusion can you draw from this data?

(MR) **7** The line graph shows a company's sales figures.

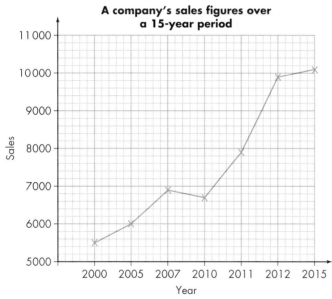

**A company's sales figures over a 15-year period**

Give three reasons why this graph is misleading.

(PS) **8** Finley took eight mental arithmetic tests.

These are his marks.

10   7   15   15   7   6   14   14

a Work out the median.

b Work out the mean.

Finley took another test.

The mean for all nine tests is the same as the mean for the first eight tests.

c Work out his mark in the ninth test.

(PS) **9** A list of nine numbers has a mean of 7.6. What number must be added to the list to give a new mean of 8?

# 4 Geometry and measures: Angles

## This chapter is going to show you:

- how to calculate angles on a line and around a point
- how to calculate angles in a triangle and in any polygon
- how to calculate angles in parallel lines
- how to calculate interior and exterior angles in polygons
- how to use bearings.

## You should already know:

- how to use a protractor to measure an angle
- the meaning of the terms 'acute', 'obtuse', 'reflex', 'right' and how to use these terms to describe angles
- the names and angle properties of quadrilaterals
- how to use three-letter notation to describe any angle
- what a polygon is and the names of polygons with up to ten sides
- that a diagonal is a line joining two non-adjacent vertices of a polygon
- the meaning of the terms 'parallel' and 'perpendicular' in relation to lines.

## About this chapter

Ancient civilisations used right angles to survey and construct buildings, but not everything can be measured in right angles. They needed a smaller, more useful unit. The ancient Babylonians chose a unit angle that led to the development of the degree, which we still use today.

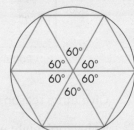

Historians think the ancient Babylonians believed the 'circle' of the year consisted of 360 days. The ancient Babylonians also knew that the side of a regular hexagon drawn in a circle is equal to the radius of the circle. This may have led to the division of the full circle (360 'days') into six equal parts, each part consisting of 60 'days', and so giving a full circle 360 units. They divided one angle of an equilateral triangle into 60 equal parts, which we now call degrees.

This chapter will show you the connections between various shapes and their angles. Angles help us construct so many things, from tables to skyscrapers. It is essential that you understand them: they literally shape our world.

# 4.1 Angle facts

This section will show you how to:

- calculate angles on a straight line
- calculate angles around a point
- use vertically opposite angles.

**Key terms**

angles around a point

angles on a straight line

vertically opposite angles

## Angles on a line

The **angles on a straight line** add up to 180°.

Draw an example for yourself (and measure $a$ and $b$) to show that the statement is true.

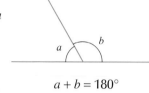

$$a + b = 180°$$

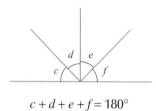

$$c + d + e + f = 180°$$

## Angles around a point

The sum of the **angles around a point** is 360°.

Again, check this for yourself by drawing an example and measuring the angles.

Sometimes you will need to use equations to solve angle problems, as shown in the next examples.

**Note:** Unless you are told otherwise, diagrams in exercises are not drawn accurately.

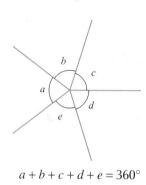

$$a + b + c + d + e = 360°$$

**Example 1**

Work out the value of $x$ in the diagram.

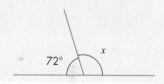

Angles on a straight line add up to 180°.

Therefore,  $x + 72° = 180°$

$x = 180° - 72°$

So $x = 108°$.

**Example 2**

Work out the value of $x$ in the diagram.

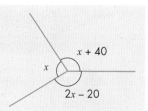

These angles are around a point, so they must add up to 360°.

Therefore,  $x + x + 40° + 2x - 20° = 360°$

$4x + 20° = 360°$

$4x = 340°$

So $x = 85°$.

## Vertically opposite angles

**Vertically opposite angles** are equal.

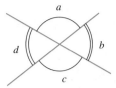

$a = c$ and $b = d$.

**Example 3**

Work out the size of angle $x$ in the diagram.

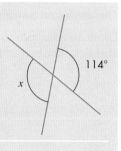

The two angles are vertically opposite, so $x = 114°$.

## Exercise 4A

**1** Work out the size of the angle marked $x$ in each of these examples.

**a**

132°
$x$

**b**

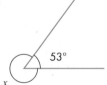

53°
$x$

**c**

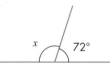

$x$
72°

**d**

$x$
38°

**e**

78°
$x$ 43°

**f**

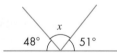

$x$
48° 51°

> **Hints and tips** Never measure angles in questions like these as diagrams are not always drawn accurately. Always calculate angles unless you are told to measure them.

**2** Write down the value of $x$ in each of these diagrams.

**a**

$x$
82°

**b**

105°
$x$

**c**

$x$
75°

(CM) **3** In the diagram, angle ABD is 45° and angle CBD is 125°.

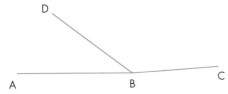

Decide if ABC is a straight line. How did you decide?

(PS) **4** Calculate the value of $x$ in each of these examples.

**a** 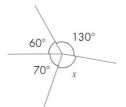  **b** 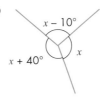  **c**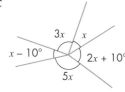

(PS) **5** Calculate the value of $x$ in each of these examples.

**a** 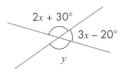  **b** 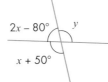  **c**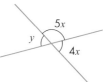

(PS) **6** Calculate the value of $x$ and $y$ in each of these examples. Calculate $x$ first each time.

**a**   **b**   **c**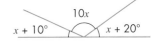

(CM) **7** Ella has a collection of tiles. They are all equilateral triangles and are all the same size.

She says that six of the tiles will fit together and leave no gaps.

Show that Ella is correct.

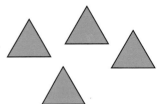

Hints and tips   All the angles in an equilateral triangle are 60°.

(PS) **8** Work out the value of $y$ in the diagram.

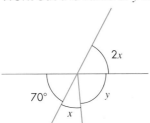

Hints and tips   Remember that the diagrams are not drawn accurately. You must calculate the angles.

# 4.2 Triangles

## This section will show you how to:

- recognise and calculate the angles in different sorts of triangle.

**Example 4**

Calculate the size of angle $a$ in the triangle opposite.

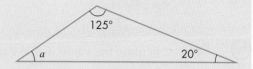

Angles in a triangle add up to $180°$.

Therefore, $a + 20° + 125° = 180°$

$$a + 145° = 180°$$

So $a = 35°$.

## Special triangles

### Scalene triangle

A **scalene triangle** is a triangle in which each side is a different length.

### Equilateral triangle

An **equilateral triangle** is a triangle with all its sides equal. Therefore, all three angles are $60°$.

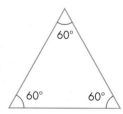

### Isosceles triangle

An **isosceles triangle** is a triangle with two equal sides and, therefore, with two equal angles (at the foot of the equal sides).

Notice how to mark the equal sides and equal angles.

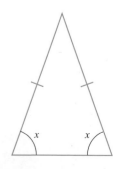

### Right-angled triangle

A **right-angled triangle** has an angle of 90°.

$a + b = 90°$

A right-angled triangle may also be scalene or isosceles.

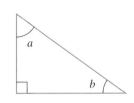

### Obtuse-angled triangle

An **obtuse-angled triangle** is a triangle with an obtuse angle (more than 90°).

An obtuse-angled triangle may also be scalene or isosceles.

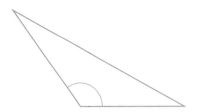

### Acute-angled triangle

An **acute-angled triangle** is a triangle with *all* its angles less than 90°.

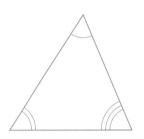

## Exercise 4B 🖩

**1** Work out the size of the angle marked with a letter in each of these triangles.

**a**
60°
50°
$a$

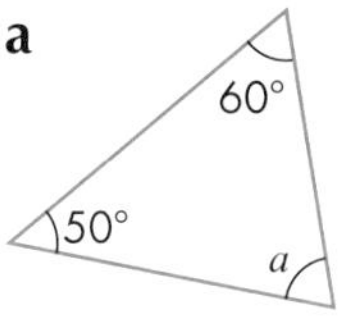

**b**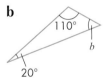
110°
$b$
20°

**c**
70°
$c$
30°

**d**
69°
51°
$d$

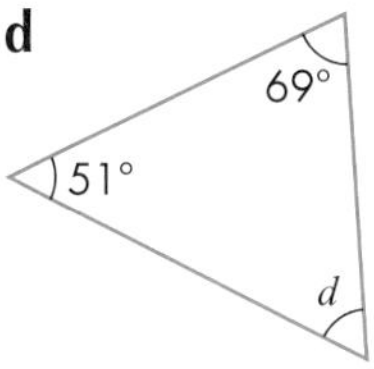

**e**
67°
$e$
38°

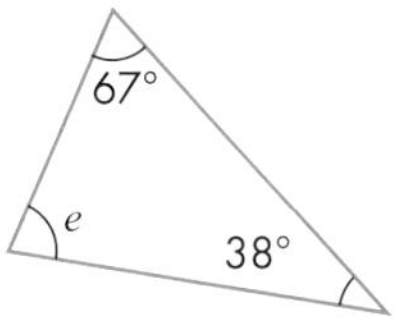

**f**
39°
$f$
32°

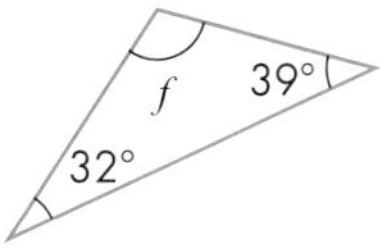

**g**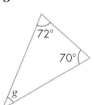
72°
70°
$g$

**h**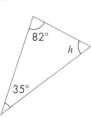
82°
$h$
35°

> **Hints and tips** Remember that the diagrams are not drawn accurately. You must calculate the angles.

**2** Which of these sets of angles form the three angles of a triangle? How do you know?

**a** 35°, 75°, 80°  **b** 50°, 60°, 70°  **c** 55°, 55°, 60°

**d** 60°, 60°, 60°  **e** 35°, 35°, 110°  **f** 102°, 38°, 30°

**3** Each set of angles form the three interior angles of a triangle. Work out the value of the angle given by a letter in each case.

**a** 20°, 80°, $a$  **b** 52°, 61°, $b$  **c** 80°, 80°, $c$

**d** 25°, 112°, $d$  **e** 120°, 50°, $e$  **f** 122°, 57°, $f$

**4**
  **a** Sketch a scalene obtuse-angled triangle.

  **b** Sketch a scalene acute-angled triangle.

  **c** Sketch a scalene right-angled triangle.

  **d** Write down the angle sum of any triangle.

**5** In the triangle on the right, all the angles are the same.

  **a** What is the size of each angle?

  **b** What is the name of a special triangle like this?

  **c** What is special about the sides of this triangle?

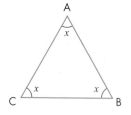

**6** In the triangle on the right, two of the angles are the same.

  **a** Work out the size of the lettered angles.

  **b** What is the name of a special triangle like this?

  **c** What is special about the sides AC and AB of this triangle?

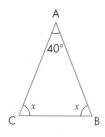

**7** Work out the size of the angle marked with a letter in each of these diagrams.

  **a**

  **b**
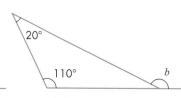

  **c**
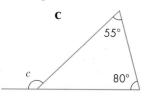

**8** A town planner has drawn this diagram to show three paths in a park but has missed out the angle marked $x$.

Work out the value of $x$.

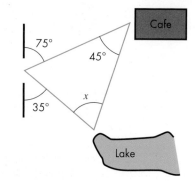

**(EV)** **9** Joe and Hannah looked at triangle DEF shown opposite.

Joe said: 'It's a right-angled triangle.'

Hannah said: 'It's an isosceles triangle.'

Comment on each of the statements.

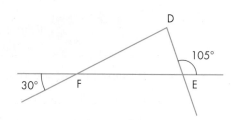

CM **10** The diagram shows three intersecting straight lines.
Show that the angle labelled $a$ is an acute angle.
Give reasons for each stage of your working.

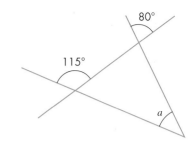

CM **11** Show that $x = a + b$.

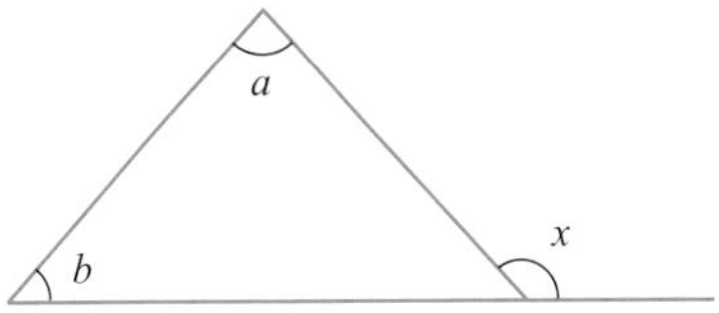

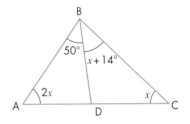

PS **12** ABC is a triangle.
Point D is on AC.
Work out the size of angle ABD.

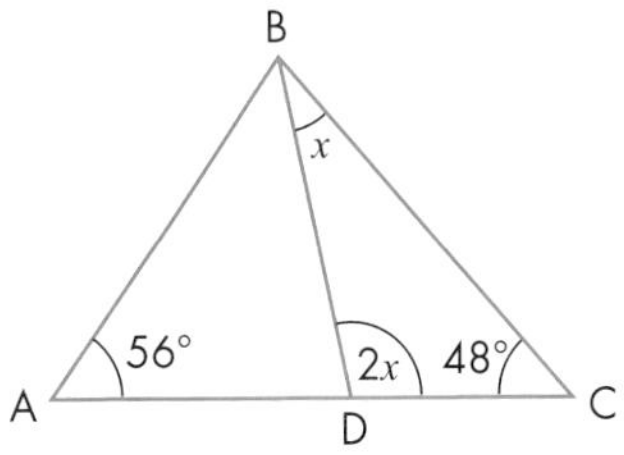

MR **13** ABC is a triangle.
Point D is on AC.
Work out the size of angle BDA.

# 4.3 Angles in a polygon

## This section will show you how to:

- calculate the sum of the interior angles in a polygon.

**Key terms**

interior angle

polygon

### Angle sums from triangles

Working through Exercise 4C will show you how you can use triangles to help work out the angle sum of **polygons**.

## Exercise 4C

**1** **a** Draw a quadrilateral (a four-sided shape).

**b** Draw in a diagonal to make it into two triangles.

**c** Copy and complete this statement.

The sum of the angles in a quadrilateral is equal to the sum of the angles in … triangles, which is … $\times$ 180° = …°.

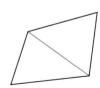

**2**   **a** Draw a pentagon (a five-sided shape).

   **b** Draw in two diagonals to make it into three triangles.

   **c** Copy and complete this statement.

   The sum of the angles in a pentagon is equal to the sum of
   the angles in … triangles, which is … × 180° = …°.

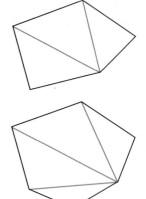

**3**   **a** Draw a hexagon (a six-sided shape).

   **b** Draw in three diagonals to make it into four triangles.

   **c** Copy and complete this statement.

   The sum of the angles in a hexagon is equal to the sum of the
   angles in … triangles, which is … × 180° = …°.

**4**   Complete the table below. Use the number pattern to carry on the angle sum up to a
   decagon (ten-sided shape).

| Shape | Number of sides | Triangles | Angle sum |
|---|---|---|---|
| triangle | 3 | 1 | 180° |
| quadrilateral | 4 | 2 | |
| pentagon | 5 | 3 | |
| hexagon | 6 | 4 | |
| heptagon | 7 | | |
| octagon | 8 | | |
| nonagon | 9 | | |
| decagon | 10 | | |

**5**   Using the number pattern, copy and complete this statement.

   The number of triangles in a 20-sided shape is …, so the sum of the angles in a
   20-sided shape is … × 180° = …°.

## $n$-sided polygon

For an $n$-sided polygon, the sum of the **interior angles** is $180(n - 2)°$.

The interior angles are the angles inside the shape.

**Example 5**

Calculate the size of angle $a$ in the quadrilateral opposite.

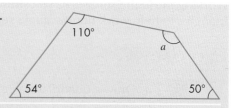

Angles in a quadrilateral add up to 360°.

Therefore, $a + 50° + 54° + 110° = 360°$

$$a + 214° = 360°$$

$$\text{So } a = 146°.$$

## Exercise 4D 🖩

**1** Work out the size of the angle marked with a letter in each of these quadrilaterals.

**a**

**b**

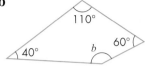

**c**

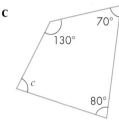

**d**

**e**

**f**

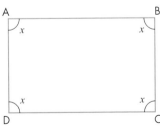

> **Hints and tips**  Remember, the sum of the interior angles of a quadrilateral is 360°.

**CM** **2** Which of these sets of angles form the four interior angles of a quadrilateral? How do you know this is true?

**a** 135°, 75°, 60°, 80°  **b** 150°, 60°, 80°, 70°  **c** 85°, 85°, 120°, 60°

**d** 80°, 90°, 90°, 110°  **e** 95°, 95°, 60°, 110°  **f** 102°, 138°, 90°, 30°

**3** Each set of angles form the four interior angles of a quadrilateral. Calculate the value of the lettered angle in each case.

**a** 120°, 80°, 60°, $a$  **b** 102°, 101°, 90°, $b$  **c** 80°, 80°, 80°, $c$

**d** 125°, 112°, 83°, $d$  **e** 120°, 150°, 50°, $e$  **f** 122°, 157°, 80°, $f$

**4** In this quadrilateral, all the angles are the same.

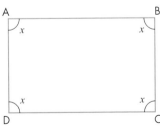

**a** What size is each angle?

**b** What is the name of a special quadrilateral like this?

**c** Is there another quadrilateral with four equal angles? What is it called?

**5** Work out the size of the angle marked with a letter in each of the polygons below.
You may find the table you completed in Exercise 4C question **4** useful.

**a**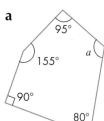

**b**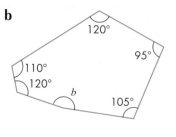

**c**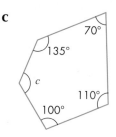

**d**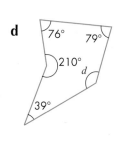

**e**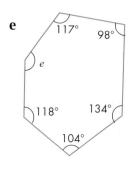

**f**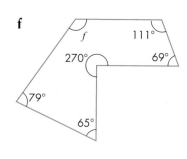

> **Hints and tips** Remember, the sum of the interior angles of an $n$-sided polygon is $180(n-2)°$.

**(PS)** **6** Anna is drawing this logo for a school magazine.

It is made up of four equilateral triangles that are all the same size.

She needs to know the sizes of the six angles so that she can draw it accurately.

What are the sizes of the six angles?

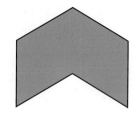

> **Hints and tips** First mark the four equilateral triangles on a copy of the diagram.

**(MR)** **7** This quadrilateral is made from two isosceles triangles.
They are both the same size.

Work out the value of $y$ in terms of $x$.

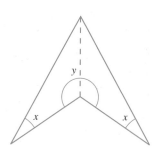

**(PS)** **8** The four angles in a quadrilateral are $2x$, $x + 20$, $2x - 10$ and $3x + 30$.

What is the size of the smallest angle in the quadrilateral?

# 4.4 Regular polygons

This section will show you how to:

- calculate the exterior angles and the interior angles of a regular polygon.

A polygon is regular if all its interior angles are equal and all its sides have the same length.

Here are three regular polygons.

Square          Pentagon          Hexagon
4 sides          5 sides            6 sides

A square is a regular four-sided shape that has an angle sum of $360°$, so each angle is $360° \div 4 = 90°$.

A regular pentagon has an angle sum of $540°$, so each angle is $540° \div 5 = 108°$.

| Shape | Number of sides | Angle sum | Each angle |
|---|---|---|---|
| square | 4 | $360°$ | $90°$ |
| pentagon | 5 | $540°$ | $108°$ |
| hexagon | 6 | $720°$ | $720 \div 6 = 120°$ |

## Interior and exterior angles of regular shapes

Look again at these three regular polygons.

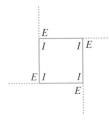 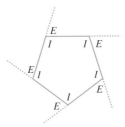 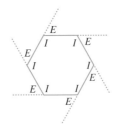

At each vertex, there is an interior angle, $I$, and an **exterior angle**, $E$. Notice that: $I + E = 180°$.

Clearly, the exterior angles of a square are each $90°$. So, the sum of the exterior angles of a square is $4 \times 90° = 360°$.

You can calculate the exterior angle of a regular pentagon as follows. You know from the previous table that the interior angle of a regular pentagon is $108°$.

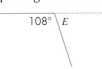

So the exterior angle is $180° - 108° = 72°$.

Therefore, the sum of the exterior angles is $5 \times 72° = 360°$.

| Regular polygon | Number of sides | Interior angle | Exterior angle | Sum of exterior angles |
|---|---|---|---|---|
| square | 4 | 90° | 90° | $4 \times 90° = 360°$ |
| pentagon | 5 | 108° | 72° | $5 \times 72° = 360°$ |
| hexagon | 6 | 120° | 60° | $6 \times 60° = 360°$ |

From this table, you can see that the sum of the exterior angles is always 360°.

You can use this information to calculate the exterior angle and the interior angle for any regular polygon.

For an $n$-sided regular polygon, the exterior angle is given by $E = \frac{360°}{n}$ and the interior angle is given by $I = 180° - E$.

Calculate the size of the exterior and interior angle of a regular 12-sided polygon (a regular dodecagon).

$E = \frac{360°}{12}$

$\quad = 30°$

$I = 180° - 30°$

$\quad = 150°$

## Exercise 4E

**1** Copy and complete the table below.

| Regular polygon | Number of sides | Interior angle sum | Each interior angle |
|---|---|---|---|
| octagon | 8 | | |
| nonagon | 9 | | |
| decagon | 10 | | |

**2** Copy and complete the table below for regular polygons.

| Regular polygon | Number of sides | Interior angle | Exterior angle |
|---|---|---|---|
| square | 4 | 90° | 90° |
| pentagon | 5 | 108° | 72° |
| hexagon | 6 | 120° | |
| octagon | 8 | | |
| nonagon | 9 | | |
| decagon | 10 | | |

**3** Each diagram shows an interior angle of a regular polygon. For each one, work out:

　**i** the exterior angle 　　**ii** the number of sides

　**iii** the sum of the interior angles of the polygon.

**a** 135°　　**b** 160°　　**c** 165°　　**d** 144°

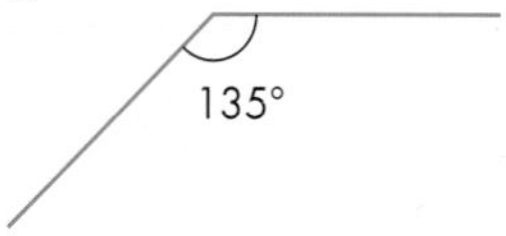

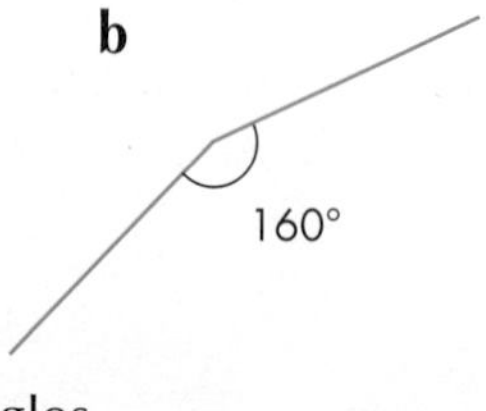

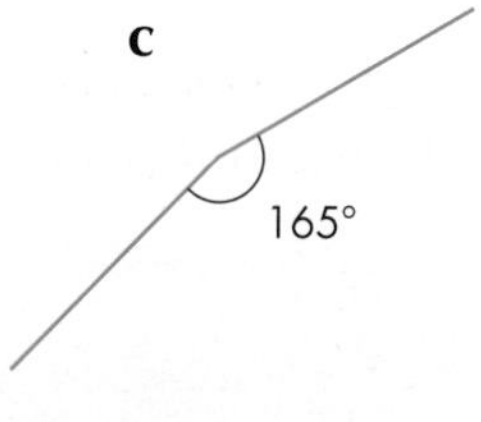

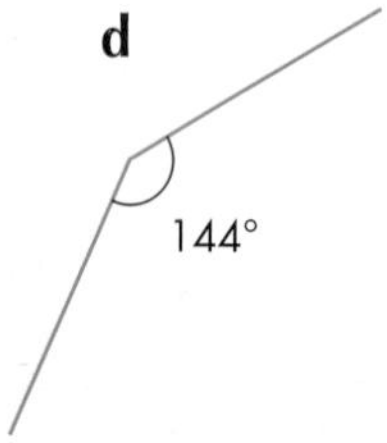

**4** Each diagram is an exterior angle of a regular polygon. For each one, work out:

    **i** the interior angle    **ii** the number of sides

    **iii** the sum of its interior angles.

**a**            **b** 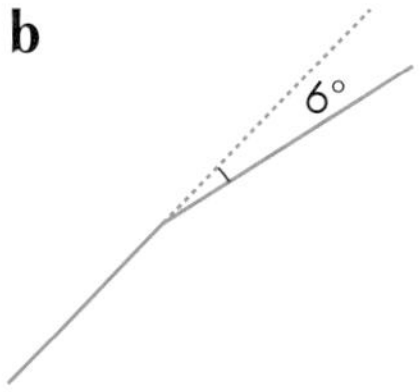            **c** 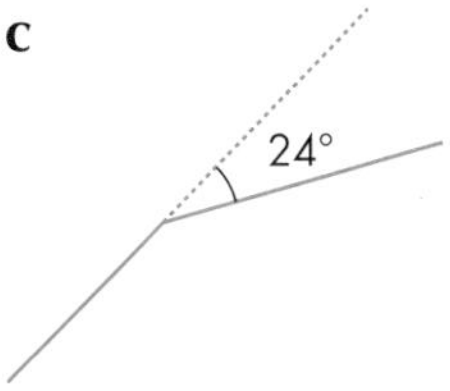            **d**

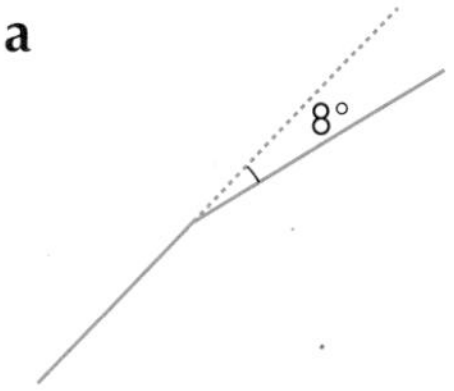

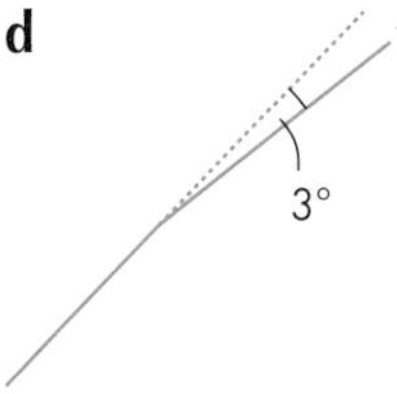

    8°            6°           24°            3°

> **Hints and tips** Remember that the angle sum is (number of sides − 2) × 180°.

**5** Show why each of these cannot be the interior angle of a regular polygon.

**a**            **b** 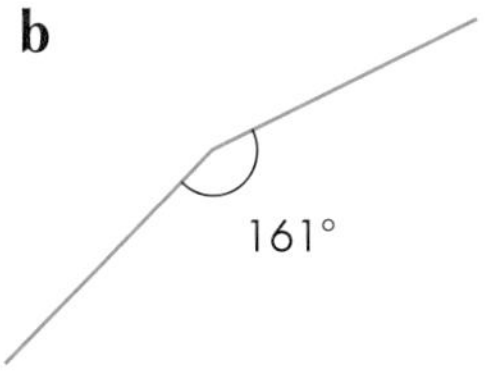            **c** 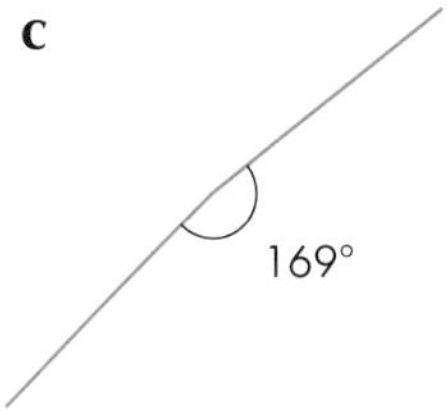            **d**

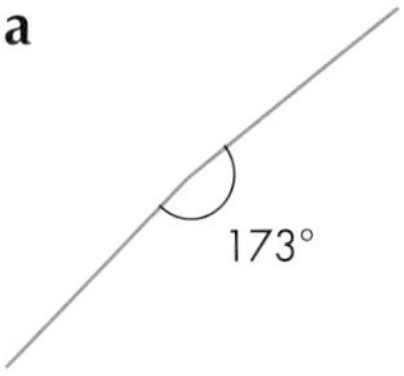

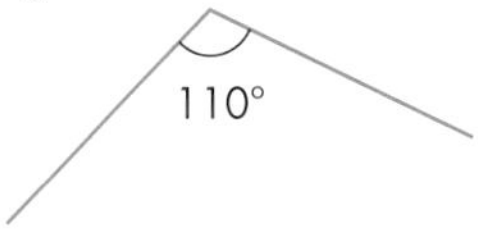

   173°       161°       169°      110°

**6** Show why each of these cannot be the exterior angle of a regular polygon.

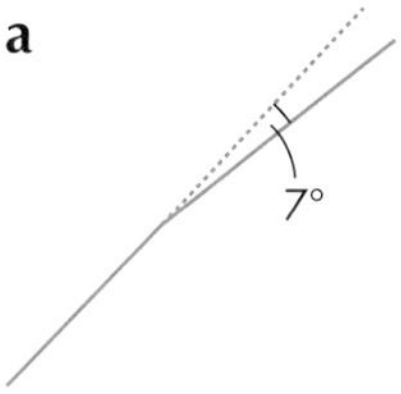

**a**            **b** 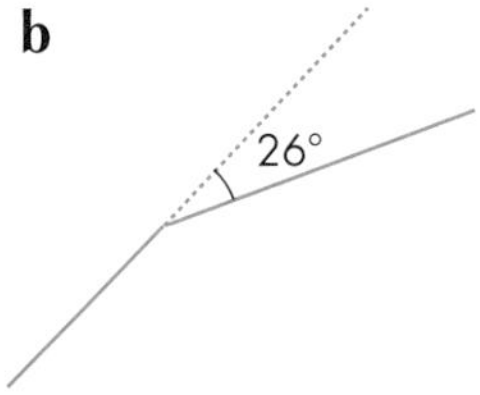            **c** 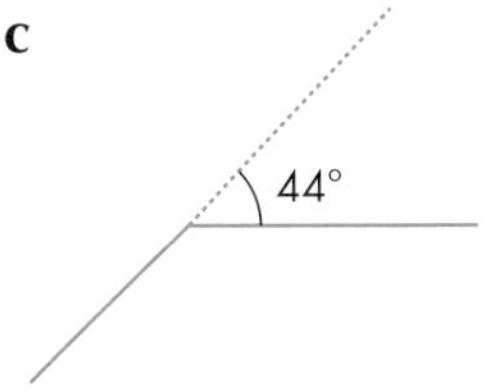            **d**

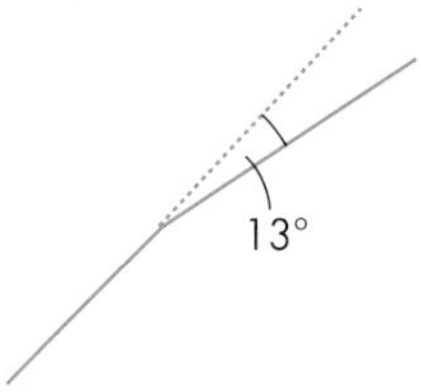

   7°        26°       44°         13°

**7** Draw a sketch of a regular octagon and join each vertex to the centre.

Calculate the value of the angle at the centre (marked $x$).

What connection does this have with the exterior angle?

Is this true for all regular polygons?

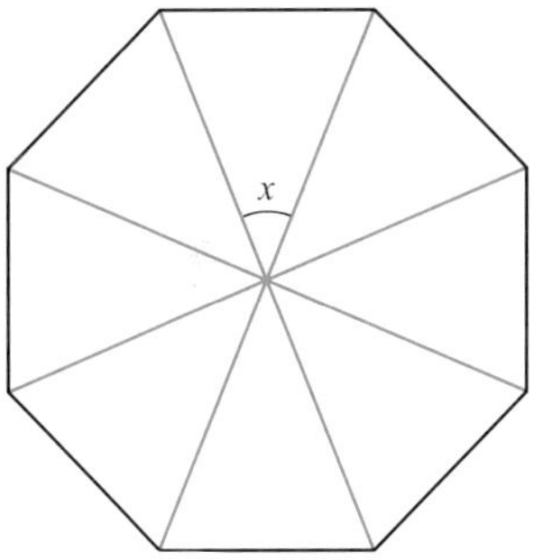

**8** A joiner is making tables so that the shape of each one is half a regular octagon, as shown in the diagram.

What are the sizes of each angle on the table top?

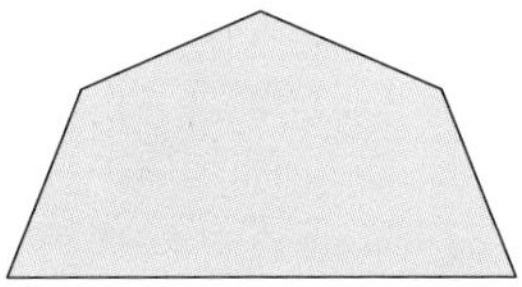

**9** This star shape has ten sides that are equal in length.

Each reflex interior angle is 240°.

Work out the size of each acute interior angle.

> **Hints and tips** Work out the sum of the interior angles of a decagon first.

**10** The diagram shows part of a regular polygon.

Each interior angle is 144°. How many sides does the polygon have?

144°

 **11** Calculate the angles of a pentagon whose interior angles are in the ratio 2 : 2 : 3 : 4 : 4.

 **12** Joe measured all the angles in a polygon and got 987°, but he forgot to measure one angle. What was the size of the missing angle?

# 4.5 Angles in parallel lines

This section will show you how to:

- calculate angles in parallel lines.

By drawing a pair of parallel lines with a line through them, you can check the following results for yourself.

| Key terms |
| --- |
| allied angles |
| alternate angles |
| corresponding angles |

| | | |
| --- | --- | --- |
| Angles like these 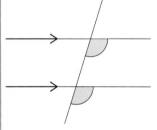 | Angles like these 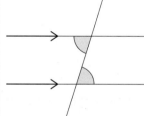 | Angles like these 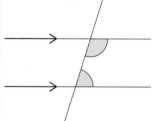 |
| are called **corresponding angles**. Corresponding angles are equal. | are called **alternate angles**. Alternate angles are equal. | are called **allied angles** or co-interior angles. Allied angles add up to 180°. |

**Example 7**

State the size of each lettered angle in the diagram and give a reason.

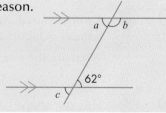

62°

$a = 62°$ (alternate angle to 62°)

$b = 118°$ (allied angle to 62° or angles on a line with $a$)

$c = 62°$ (vertically opposite angle to 62° or corresponding angle to $a$)

# Exercise 4F

**1** Copy and complete these statements to make them true.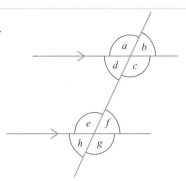

   **a** Angles $h$ and ... are corresponding angles.

   **b** Angles $d$ and ... are alternate angles.

   **c** Angles $e$ and ... are allied angles.

   **d** Angles $b$ and ... are corresponding angles.

   **e** Angles $c$ and ... are allied angles.

   **f** Angles $c$ and ... are alternate angles.

**(CM)**

**2** State the sizes of the lettered angles in each diagram and give a reason.

**a** 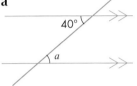     **b** 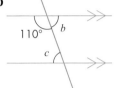     **c**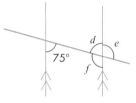

**d** 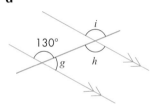     **e** 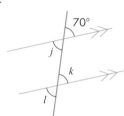     **f**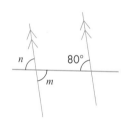

**g** 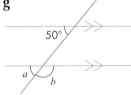     **h** 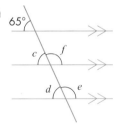     **i**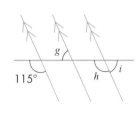

**j** 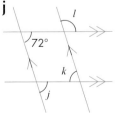     **k** 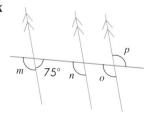     **l**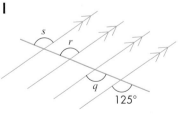

**3** State the sizes of the lettered angles in these diagrams.

**a** 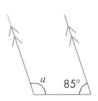     **b**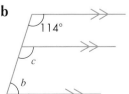

(PS) **4** Calculate the values of *x* and *y* in each diagram.

**a**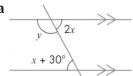

**b**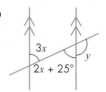

**c**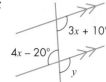

**d**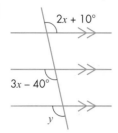

**e**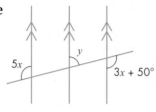

**f**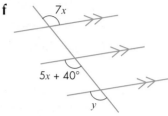

(PS) **5** A company makes signs in the shape of a chevron, like the one shown.

This sign has one line of symmetry.

What is the size of angle *x*?

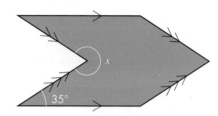

(CM) **6** In the diagram, AE is parallel to BD.

Work out the size of angle *a*.

Describe clearly how you calculated your answer.

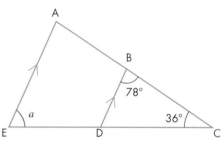

(EV) **7** Lizzie is writing out a solution to this question.

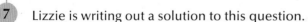
The line *XY* crosses the parallel lines *AB* and *CD* at *P* and *Q*.
Work out the size of angle *DQY*.
Give reasons for your answer.

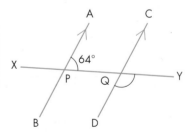

This is her solution.

   Angle PQD = 64° (corresponding angles)

   So angle DQY = 124° (angles on a line = 190°)

Lizzie has made a number of errors in her solution. Find her errors and write out a correct solution for the question.

(CM) **8** Use this diagram to prove that the three angles in a triangle add up to 180°.

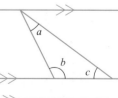

(CM) **9** Prove that *p* + *q* + *r* = 180°.

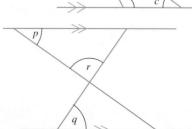

# 4.6 Special quadrilaterals

This section will show you how to:

- use angle properties in quadrilaterals.

## Parallelogram

- A parallelogram has opposite sides parallel.
- Its opposite sides are equal.
- Its diagonals **bisect** each other (i.e. cut each other in half).
- Its opposite angles are equal. That is:

    angle BAD = angle BCD

    angle ABC = angle ADC

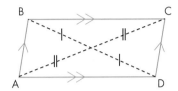

## Rectangle

- A rectangle is a parallelogram with all its angles equal.
- All four angles are right angles.
- Its opposite sides are equal in length.

## Rhombus

- A rhombus is a parallelogram with all its sides equal.
- Its diagonals bisect each other at right angles.
- Its diagonals also bisect the angles.

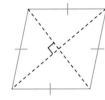

## Square

- A square is a rhombus with all its angles equal (90°).

## Kite

- A kite is a quadrilateral with two pairs of equal adjacent sides.
- Its longer diagonal bisects its shorter diagonal at right angles.
- The opposite angles between the sides of different lengths are equal.

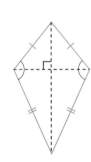

## Trapezium

- A trapezium has two parallel sides.
- The sum of the interior angles at the ends of each non-parallel side is 180°. That is:

    angle BAD + angle ADC = 180°

    angle ABC + angle BCD = 180°

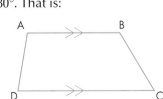

# Exercise 4G 🖩

1    Calculate the size of the lettered angles in each trapezium.

**a**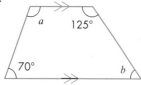

**b**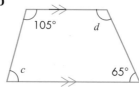

**c**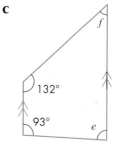

| Hints and tips | Remember that the diagrams are not drawn accurately. You must calculate the angles. |
|---|---|

2    Calculate the size of the lettered angles in each parallelogram.

**a**

**b**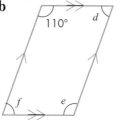

**c**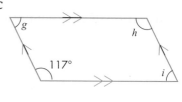

3    Calculate the size of the lettered angles in each rhombus.

**a**

**b**

**c**

4    Calculate the size of the lettered angles in each kite.

**a**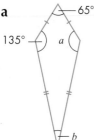

**b**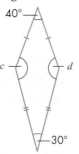

**c**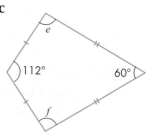

5    Calculate the size of the lettered angles in each of these shapes.

**a**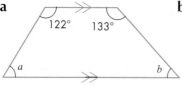

**b**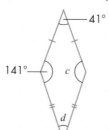

**c**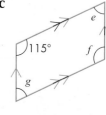

**6** Calculate the value of $x$ in each of these shapes.

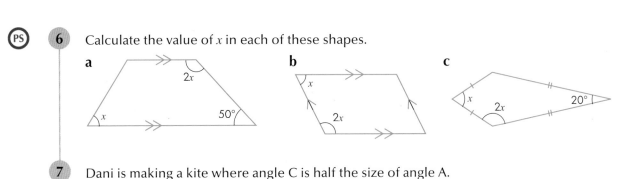

a

b

c

**7** Dani is making a kite where angle C is half the size of angle A.

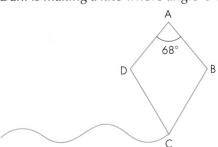

Work out the size of angles B and D.

**8** David says that a parallelogram is a special type of rectangle.

Marie says that he is wrong and that a rectangle is a special type of parallelogram.

Who is correct? Give a reason for your answer.

**9** The diagram shows a quadrilateral ABCD.

**a** Calculate the size of angle B.

**b** What special name is given to the quadrilateral ABCD? Give a reason for your answer.

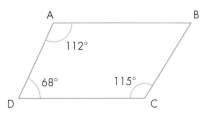

**10** A parallelogram has a pair of allied angles where one is three times as large as the other. What is the size of that largest angle?

**11** A kite has every angle either 30° larger or 30° smaller than another one in the kite. Sketch two different kites possible from this data and state the size of the smallest angle in the kite.

**12** A quadrilateral ABCD has interior angles of size $x$, $2x$, $3x$ and $4x$ respectively at A, B, C and D. What type of quadrilateral is this?

Give a reason for your answer.

> **Hints and tips** Remember that the sum of the angles in a quadrilateral is 360°.

# 4.7 Bearings

This section will show you how to:

- use a bearing to specify a direction.

The **bearing** of a point B from a point A is the angle through which you turn *clockwise* as you change direction from due north to the direction of B.

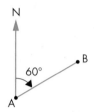

For example, in the diagram the bearing of B from A is 060°.

**Key terms**

bearing

three-figure bearing

As a bearing can have any value from 0° to 360°, you give all bearings in three figures. This is known as a **three-figure bearing**.

Here are three more examples of bearings.

D is on a bearing of 048° from C

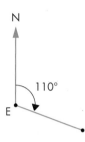

F is on a bearing of 110° from E

H is on a bearing of 330° from G

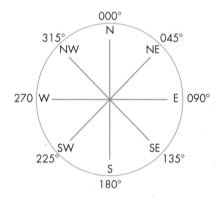

These are eight bearings you should be familiar with. Clockwise from N, they are north, northeast, east, southeast, south, southwest, west and northwest.

---

**Example 8**

A, B and C are three towns.

**a** Write down the bearing of B from A and the bearing of C from A.

**b** Use the scale to work out the actual distances between:

   **i** A and B      **ii** A and C.

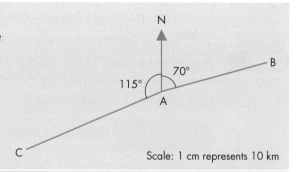

Scale: 1 cm represents 10 km

**a** The bearing of B from A is 070°.

   The bearing of C from A is 360° − 115° = 245°.

**Remember:** A bearing is always measured clockwise from the north line.

**b i** On the diagram AB is 3 cm, so the actual distance between A and B is 30 km.

  **ii** On the diagram AC is 4 cm, so the actual distance between A and C is 40 km.

---

1. This map is drawn to a scale of 1 cm to 2 km.

   By measuring distances and angles, work out the bearings and distances of each of the following.

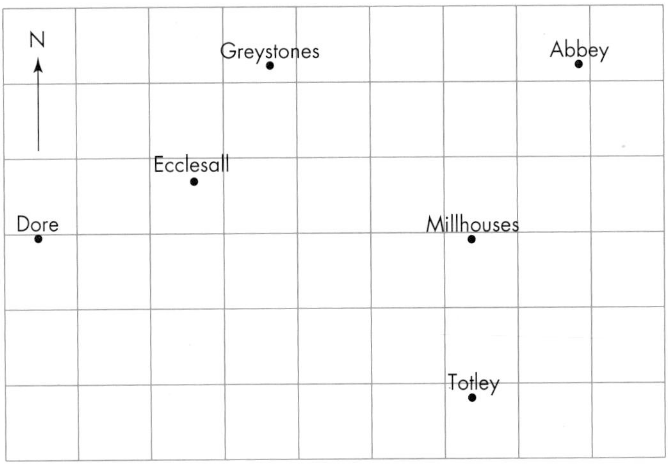

   **a** Totley from Dore
   **b** Dore from Ecclesall
   **c** Millhouses from Dore
   **d** Greystones from Abbey
   **e** Millhouses from Greystones
   **f** Totley from Millhouses

2. Draw sketches to show these situations.

   **a** Castleton is on a bearing of 170° from Hope.

   **b** Bude is on a bearing of 310° from Wadebridge.

3. A is due north of C. B is due east of A. B is on a bearing of 045° from C.

   **a** Sketch the layout of the three points, A, B and C.

   **b** D is due south of B.

   Al said that the bearing of A from D is 030°.

   How do you know that Al must be wrong?

4. Captain Bird decided to sail his ship around the four sides of a square kilometre.

   **a** Assuming he started sailing due north, write down the three bearings he should follow in order to complete the square in a clockwise direction.

   **b** Assuming he started sailing on a bearing of 090°, write down the three bearings he should follow in order to complete the square in an anticlockwise direction.

5. Draw diagrams to solve the following problems.

   **a** The three-figure bearing of A from B is 070°. Work out the three-figure bearing of B from A.

   **b** The three-figure bearing of P from Q is 145°. Work out the three-figure bearing of Q from P.

   **c** The three-figure bearing of X from Y is 324°. Work out the three-figure bearing of Y from X.

**6** The diagram shows a port P and two harbours X and Y on the coast.

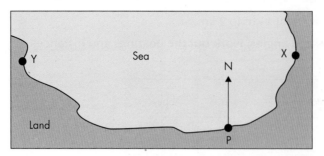

a A fishing boat sails to X from P.

What is the three-figure bearing of X from P?

b A yacht sails to Y from P.

What is the three-figure bearing of Y from P?

c Point X is 15 km away from point Y.

Show that it is 11.8 km from point Y to point P.

**7** The diagram shows the position of Kim's house (H) and the college (C).

Scale: 1 cm reprsents 200 m

a Use the diagram to work out the actual distance from Kim's house to the college.

b Measure and write down the three-figure bearing of the college from Kim's house.

c The supermarket (S) is 600 m from Kim's house on a bearing of 150°.

Mark S on a copy of the diagram.

**8** Trevor is flying a plane on a bearing of 072°.

He is instructed by a control tower to turn and fly due south towards an airport.

Through what angle does he need to turn?

**9** Apple Bay (A), Broadside (B) and Caverly (C) are three villages in a bay.

The villages lie on the vertices of a square.

The bearing of B from A is 030°.

Work out the bearing of Apple Bay from Caverly.

**10** Bryony set sail from Port Terry on a bearing of 036°. After sailing 5 km, she changed course on a bearing due east. After sailing a further 5 km, she changed course to sail due south. After sailing a further 10 km, Bryony sailed straight back to Port Terry on a bearing of 300°. Work out the length of the final part of Bryony's journey.

# Worked exemplar

 **1** ABC is a triangle. D is a point on AB such that BC = BD.

**a** Work out the value of $x$.

**b** Work out the value of $y$.

**c** Is it true that AD = DC? Give a reason for your answer.

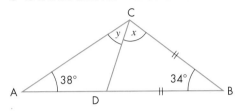

> In this question you are required to communicate mathematically. You need to show clearly how you have found the missing angles and explain your final response to part **c**.

| | |
|---|---|
| **a** Triangle BCD is isosceles, so angle BDC is equal to $x$.<br><br>Angles in a triangle = 180°<br><br>Therefore, $x + x + 34° = 180°$<br>$2x = 146°$<br>$x = 73°$ | First, make an equation in $x$ from your knowledge that angles in a triangle add up to 180°. Then solve the equation. |
| **b** **Method 1**<br>Angle ADC = $180° - 73°$<br>$\quad = 107°$ (*angles on a line*)<br>$y + 38° + 107° = 180°$ (*angles in a triangle*)<br>$\quad y + 145° = 180°$<br>$\quad\quad y = 35°$<br><br>**Method 2**<br>Angle ACB = $180° - (38° + 34°)$<br>$\quad = 108°$ (*angles in a triangle*)<br>$y + x$ = Angle ACB<br>$\quad = 108°$<br>$\quad y = 108° - 73°$<br>$\quad y = 35°$ | To work out angle $y$, you need to show how you are using the given angles and the found angle $x$. You should show the mathematical reasoning used at each stage. There are two ways of working out $y$ here. Both are acceptable. |
| **c** No, $y$ is not 38° so triangle ACD is not an isosceles triangle. No two sides of the triangle are equal. | Clearly state your explanation about ACD not being isosceles. The answer 'No' alone, is not enough. |

# Ready to progress?

I can calculate angles on a line or at a point.
I can calculate angles in triangles, quadrilaterals and polygons.

I can calculate interior and exterior angles in polygons.
I can use bearings.

# Review questions

**(PS)** **1** The diagram shows three angles on a straight line.

What is the value of $x$?

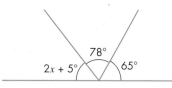

**2** Look at the diagram.

**a i** Write down the value of $x$.

**(CM)**     **ii** Give a reason for your answer.

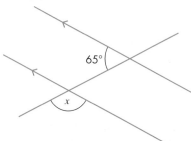

**(EV)** **b** What is wrong with this diagram?

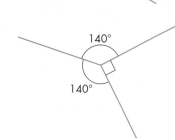

**(PS)** **3** Look at the triangle PQR.

**a** What is the size of the angle at P?

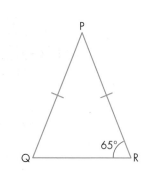

**b** The diagram has been extended to point T as shown.

What is the size of the angle at T?

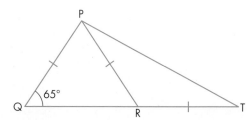

**CM** **4** Why do the interior angles of a pentagon add up to 540°?

**MR** **5** A quadrilateral has three angles of $x$, $3x$, $5x$ and a right angle.

What is the size of the largest angle in the quadrilateral?

**CM** **6** Look at the diagram. Why is angle QTS 66°?

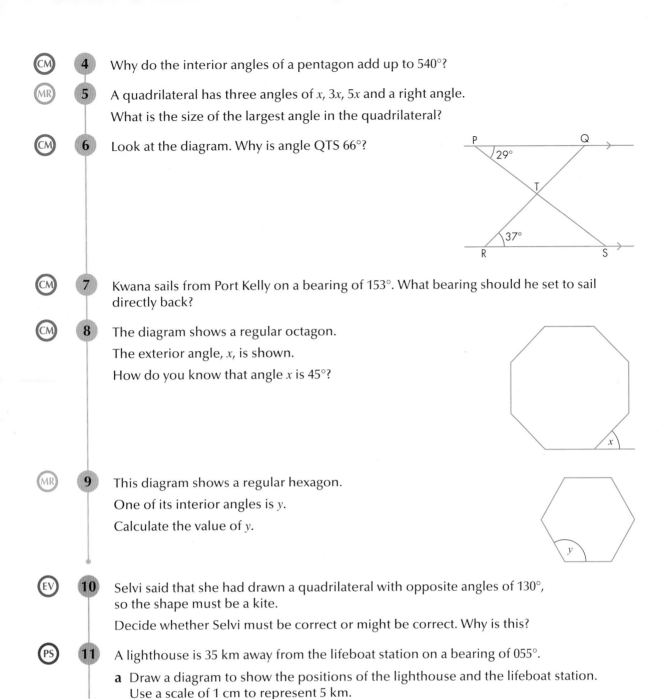

**CM** **7** Kwana sails from Port Kelly on a bearing of 153°. What bearing should he set to sail directly back?

**CM** **8** The diagram shows a regular octagon.

The exterior angle, $x$, is shown.

How do you know that angle $x$ is 45°?

**MR** **9** This diagram shows a regular hexagon.

One of its interior angles is $y$.

Calculate the value of $y$.

**EV** **10** Selvi said that she had drawn a quadrilateral with opposite angles of 130°, so the shape must be a kite.

Decide whether Selvi must be correct or might be correct. Why is this?

**PS** **11** A lighthouse is 35 km away from the lifeboat station on a bearing of 055°.

**a** Draw a diagram to show the positions of the lighthouse and the lifeboat station. Use a scale of 1 cm to represent 5 km.

**b** A lifeboat sails from the station on a bearing of 075° until it is due south of the lighthouse. How far away from the lighthouse is the lifeboat at this point?

# 5 Number: Number properties

- how to find multiples and factors
- what prime numbers are
- how to break a number down into its prime factors
- how to work out the lowest common multiple of two numbers
- how to work out the highest common factor of two numbers
- how to work out squares and square roots
- how to use a calculator for basic calculations.

**You should already know:**

- the multiplication tables up to 12 × 12.

**About this chapter**

In everyday life you meet numbers in many different situations, such as wages, shopping, taking out loans and budgeting for the future. It is important that you have a good grasp of numbers, and their properties, so that you can make sense of the everyday mathematics that you will use without even realising.

When people started counting, they used their fingers. The numbers 1, 2, 3, 4, … are therefore called the natural numbers or counting numbers. As time went on, the number system became more complex and expanded to include zero, negative numbers, decimals and fractions. This chapter concentrates on the counting numbers and their properties.

# 5.1 Multiples of whole numbers

This section will show you how to:

- find multiples of whole numbers
- recognise multiples of numbers.

**Key terms**

multiple

multiplication table

When you multiply any two whole numbers together, the answer is a **multiple** of both of those numbers. For example, when you multiply 5 and 7:

$5 \times 7 = 35$

35 is a multiple of 5 and a multiple of 7. Here are some other multiples of 5 and 7.

| Multiples of 5 | 5 | 10 | 15 | 20 | 25 | 30 | 35 | ... | |
| Multiples of 7 | 7 | 14 | 21 | 28 | 35 | 42 | ... | ... | |

Multiples are the answers that appear in **multiplication tables**.

| × | 2 | 3 | 4 | 5 | 6 |
|---|---|---|---|---|---|
| 2 | 4 | 6 | 8 | 10 | (12) |
| 3 | 6 | 9 | (12) | 15 | 18 |
| 4 | 8 | (12) | 16 | 20 | 24 |
| 5 | 10 | 15 | 20 | 25 | 30 |
| 6 | (12) | 18 | 24 | 30 | 36 |

Notice that 12 is a multiple of 2, 3, 4 and 6 as well as being a multiple of 1 and 12.

## Recognising multiples

These 'tricks' will help you to recognise the multiples of 2, 3, 4, 5, 6 and 9.

- Multiples of 2 are always even numbers, so end in 0, 2, 4, 6 or 8. For example:

    12   34   96   1938   370

- Multiples of 3 are always made up of digits that add up to a multiple of 3. For example, you can recognise that:

| 15 is a multiple of 3 | because | $1 + 5 = 6$ | which is $2 \times \mathbf{3}$ |
| 72 is a multiple of 3 | because | $7 + 2 = 9$ | which is $3 \times \mathbf{3}$ |
| 201 is a multiple of 3 | because | $2 + 0 + 1 = 3$ | which is $1 \times \mathbf{3}$. |

- Multiples of 4 always give an even number when divided by 2. For example, you can recognise that:

| 64 is a multiple of 4 | because | $64 \div 2 = 32$ | which is even |
| 212 is a multiple of 4 | because | $212 \div 2 = 106$ | which is even |
| 500 is a multiple of 4 | because | $500 \div 2 = 250$ | which is even. |

> **Hints and tips**  Another quick check for multiples of 4 is to look at the last two digits of the number. If they give a multiple of 4, the whole number is a multiple of 4.

- Multiples of 5 always end in 5 or 0. For example:

    35   60   155   300

- Multiples of 6 are multiples of 2 and multiples of 3. For example, you can recognise that:

| 42 is a multiple of 6 | because it is even | and | $4 + 2 = 6$ | which is $2 \times \mathbf{3}$ |
| 54 is a multiple of 6 | because it is even | and | $5 + 4 = 9$ | which is $3 \times \mathbf{3}$. |

- Multiples of 9 are always made up of digits that add up to a multiple of 9. For example, you can recognise that:

| 63 is a multiple of 9 | because | $6 + 3 = 9$ | which is $1 \times$ **9** |
| 738 is a multiple of 9 | because | $7 + 3 + 8 = 18$ | which is $2 \times$ **9**. |

You can use your calculator to find out whether numbers are multiples of 7 or 8. For example, to find out whether 341 is a multiple of 7, check whether 341 gives a whole-number answer when it is divided by 7. Key in:

**3  4  1  ÷  7  =**

The answer is 48.714 286. This is a decimal number, not a whole number, so 341 is *not* a multiple of 7.

## Exercise 5A

1. Write out the first five multiples of each number.
   - **a** 3
   - **b** 7
   - **c** 9
   - **d** 11
   - **e** 16

   > **Hints and tips** Remember the first multiple is the number itself.

2. Write down the numbers from the box below that are multiples of:
   - **a** 2
   - **b** 3
   - **c** 5
   - **d** 9.

   | 111 | 254 | 255 | 108 | 73 |
   | 68 | 162 | 711 | 615 | 98 |
   | 37 | 812 | 102 | 75 | 270 |

   > **Hints and tips** Use the rules about recognising multiples.

3. Write down the numbers from the box below that are multiples of:
   - **a** 4
   - **b** 7
   - **c** 6.

   Use your calculator to help you.

   | 72 | 135 | 102 | 161 197 |
   | 132 | 78 | 91 | 216 514 |
   | 312 | 168 | 75 | 144 294 |

   > **Hints and tips** Don't waste time by testing whether odd numbers are multiples of even numbers.

4. Find the biggest number less than 100 that is a multiple of:
   - **a** 2
   - **b** 3
   - **c** 4
   - **d** 5
   - **e** 7
   - **f** 6.

5. Find the smallest number greater than 1000 that is a multiple of:
   - **a** 6
   - **b** 8
   - **c** 9.

6. Vishal is packing eggs into boxes of six. He has 50 eggs. Will all the boxes be full?
   Give a reason for your answer.

**MR** **7** A hotel orders taxis for a party of 20 people. Each taxi holds the same number of passengers. If all the taxis are full, how many people could there be in each taxi? Give two possible answers.

**8** Write down a number from the box below that is a multiple of:

    **a** 9         **b** 7         **c** 3 and 5.

| 6 | 8 | 12 | 15 | 18 | 28 |
|---|---|----|----|----|----|

**PS** **9** Find the smallest even number that is a multiple of 11 and a multiple of 3.

**PS** **10** How many numbers, between 1 and 100, are multiples of both 6 and 9? List them.

**MR** **11** The number 24 appears in eight multiplication tables: 1, 2, 3, 4, 6, 8, 12 and 24!

    **a** In which multiplication tables is the number 10?

    **b** In which multiplication tables is the number 30?

    **c** In which multiplication tables is the number 100?

    **d** Which number less than 100 is in the most multiplication tables?

# 5.2 Factors of whole numbers

## This section will show you how to:

- identify the factors of a number.

<table>
<tr><td>Key terms</td></tr>
<tr><td>factor</td></tr>
<tr><td>factor pair</td></tr>
</table>

A **factor** is any whole number that divides into another whole number exactly.

The factors of 20 are:     1   2   4   5   10   20.

The factors of 12 are:     1   2   3   4   6   12.

To recognise factors, it is important to know your multiplication tables.

Remember these factor facts.

- 1 is always a factor of any number and so is the number itself.

- When you find one factor, there is always another factor that goes with it, unless the factor is multiplied by itself to give the number. For example, look at the number 20.

    $1 \times 20 = 20$     so 1 and 20 are both factors of 20

    $2 \times 10 = 20$     so 2 and 10 are both factors of 20

    $4 \times 5 = 20$     so 4 and 5 are both factors of 20

    These are called **factor pairs**.

You may need to use your calculator to find the factors of large numbers.

**Example 1**

Find the factors of 32.

Look for the factor pairs of 32.

$1 \times 32 = 32$     $2 \times 16 = 32$     $4 \times 8 = 32$

So, the factors of 32 are {1, 2, 4, 8, 16, 32}.

**Hints and tips** The curly brackets, { }, are used to show a set of numbers.

Example 2

Find the factors of 36.

Look for the factor pairs of 36.

$1 \times 36 = 36$     $2 \times 18 = 36$     $3 \times 12 = 36$     $4 \times 9 = 36$     $6 \times 6 = 36$

Notice that 6 is a repeated factor so you only count it once.

So, the factors of 36 are {1, 2, 3, 4, 6, 9, 12, 18, 36}.

## Exercise 5B

**1**   Find the factors of each number.

   **a**  10        **b**  28        **c**  18        **d**  17        **e**  25
   **f**  40        **g**  30        **h**  45        **i**  24        **j**  16

(MR)  **2**   How many different ways can 24 chocolate bars be packed into boxes so that there are exactly the same number of bars in each box?

**3**   Use your calculator to find the factors of each of these numbers.

   **a**  120       **b**  150       **c**  144       **d**  180       **e**  169
   **f**  108       **g**  196       **h**  153       **i**  198       **j**  199

   | Hints and tips | Remember that when you find one factor it will give you another, unless it is a repeated factor such as $5 \times 5$. |
   |---|---|

**4**   For each number, work out the largest factor less than 100.

   **a**  110       **b**  201       **c**  145       **d**  117       **e**  130
   **f**  240       **g**  160       **h**  210       **i**  162       **j**  250

**5**   Work out the largest factor that each pair of numbers has in common.

   **a**  2 and 4        **b**  6 and 10        **c**  9 and 12        **d**  15 and 25
   **e**  9 and 15       **f**  12 and 21       **g**  14 and 21       **h**  25 and 30
   **i**  30 and 50      **j**  55 and 77

   | Hints and tips | Look for the largest number that has both numbers in its multiplication table. |
   |---|---|

(MR)  **6**   A designer is making a box to hold 12 Christmas decorations. The decorations are to be packed in layers.

   The designer wants the box to have a square base. Explain how the 12 decorations could be packed to make a square.

   | Hints and tips | How many decorations must be put in each layer for the box to have a square base? |
   |---|---|

(CM)  **7**   Look at these five numbers.

   | 18 | 21 | 27 | 32 | 36 |
   |---|---|---|---|---|

   Use factors to explain why 32 could be the odd one out.

(PS)  **8**   What is the largest odd number that is a factor of 40 and a factor of 60?

# 5.3 Prime numbers

This section will show you how to:

- identify prime numbers.

Look at the numbers 2, 3, 5, 7, 11 and 13.

What are their factors?

Each of these numbers has only two factors: itself and 1. They are all examples of **prime numbers**.

A prime number is a whole number that has only two factors: itself and 1.

**Note:** 1 is *not* a prime number, since it has only one factor – itself.

There is no easy rule that helps you to work out the prime numbers. Therefore, it is useful to know the first few prime numbers. Here are the prime numbers up to 50.

2, 3, 5, 7, 11, 13, 17, 19, 23, 29, 31, 37, 41, 43, 47

## Exercise 5C

**1** Write down all the prime numbers between 20 and 30.

**2** There is only one only prime number between 90 and 100. Write it down.

(MR) **3** Use the rules for recognising multiples to decide which of these numbers are not prime numbers.

| 462 | 108 | 848 | 365 | 711 |
|-----|-----|-----|-----|-----|

(PS) **4** When three different prime numbers are multiplied together, the answer is 105.

What are the three prime numbers?

(CM) **5** A shopkeeper has 31 identical bars of soap.

He is trying to arrange the bars on a shelf in rows, each with the same number of bars.

Is this possible?

Give a reason for your answer.

(MR) **6** $p$ is a prime number, $q$ is an even number and $r$ is an odd number.

Look at each expression and choose whether it will give an answer that is always even (E), always odd (O) or could be either (C).

**a** $pqr$            **b** $pr$            **c** $p + q + r$

(PS) **7** Three different prime numbers add up to 29.

**a** Explain why 2 cannot be one of the three prime numbers.

**b** Work out one possible set of three prime numbers that add up to 29.

(CM) **8** Explain why the difference of any two prime numbers greater than 2 will always be even.

(PS) **9** The difference between two two-digit prime numbers is a two-digit square number. There are 16 possible answers. Work out at least five of them.

# 5.4 Prime factors, LCM and HCF

This section will show you how to:

- identify prime factors
- identify the lowest common multiple (LCM) of two numbers
- identify the highest common factor (HCF) of two numbers.

The **unique factorisation theorem** states that every integer greater than 1 is either a prime number or can be written as the product of prime numbers. This gives the **prime factors** of the number.

For example, consider 110. Find two numbers that, when multiplied together, give that number, for example, $2 \times 55$. Are they both prime? No, 55 isn't. So take 55 and repeat the operation, to get $5 \times 11$. Are these both prime? Yes. So:

$110 = 2 \times 5 \times 11$

The prime factors of 110 are 2, 5 and 11.

This method is not very logical and you need to know your multiplication tables well to use it. There are, however, two methods that you can use to make sure you do not miss any of the prime factors. The next two examples show you how to use the first of these methods.

---

**Example 3**

Find the prime factors of 24.

Divide 24 by any prime number that goes into it. (2 is an obvious choice.)

Divide the answer (12) by a prime number. As 12 is even, 2 is again an obvious choice.

Repeat this process until you have a prime number as the answer.

$24 = 2 \times 2 \times 2 \times 3$

So the prime factors of 24 are 2 and 3.

| 2 | 24 |
|---|----|
| 2 | 12 |
| 2 | 6 |
|   | 3 |

---

**Example 4**

Find the prime factors of 96.

$96 = 2 \times 2 \times 2 \times 2 \times 2 \times 3$

So the prime factors of 96 are 2 and 3.

| 2 | 96 |
|---|----|
| 2 | 48 |
| 2 | 24 |
| 2 | 12 |
| 2 | 6 |
|   | 3 |

---

When 24 is expressed as $2 \times 2 \times 2 \times 3$ it is written as a **product of prime factors**. Another name for this is the **prime factorisation** of 24.

A quicker and neater way to write this answer is to use **index notation**, expressing the answer using powers.

In index notation, the prime factorisation of $24 = 2^3 \times 3$.

**Remember:** The small number 3 (the power) tells you how many times the factor 2 occurs in the prime factorisation of 24. The name for 2 to the power of 3 is 'two cubed'.

The prime factorisation of 96, or 96 written as the product of its prime factors, is $2 \times 2 \times 2 \times 2 \times 2 \times 3$ or $2^5 \times 3$.

This time, the power tells you how many times the factor 2 occurs in the prime factorisation of 96.

You can use powers for any prime factors.

## Factor trees

A **factor tree** is another way to find the prime factors of a number.

Start by splitting the number into a product of two factors. Then split these factors until you reach prime numbers at the end of each branch.

---

**Example 5**

Find the prime factors of 76.

Start with 76 and work down, splitting each factor until you reach prime numbers.

2, 2 and 19 are all prime numbers, so the prime factors of 76 are 2 and 19.

As a product of prime factors, $76 = 2 \times 2 \times 19$

$= 2^2 \times 19$

---

You can use this method for large numbers, which will have lots of branches, as shown in the next example.

---

**Example 6**

Find the prime factors of 420.

Start with 420 and work down, splitting each factor into pairs until you reach prime numbers.

The prime factors of 420 are 2, 3, 5 and 7.

As a product of prime factors, $420 = 2 \times 5 \times 2 \times 3 \times 7$

$= 2^2 \times 3 \times 5 \times 7$

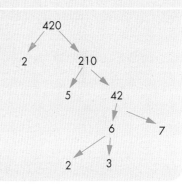

---

**Example 7**

Written as the product of its prime factors, 50 is $2 \times 5^2$.

Use this information to write each of these numbers as a product of prime factors in index form.

**a** 150          **b** 500          **c** 200

**a** 150 is $3 \times 50$     so     $150 = 2 \times 3 \times 5^2$.

**b** 500 is $10 \times 50$    so     $500 = 2^2 \times 5^3$.

**c** 200 is $4 \times 50$     so     $200 = 2^3 \times 5^2$.

---

# Exercise 5D

1 Copy and complete these factor trees and prime factorisations.

a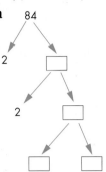

$84 = 2 \times 2 \,...\, ...$

b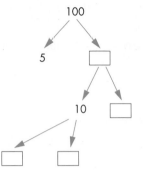

$100 = 5 \,...\, ...\, ...$

c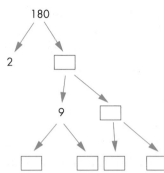

$180 = 2 \,...\, ...\, ...\, ...$

d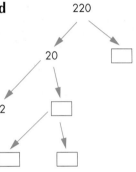

$220 = 2 \,...\, ...\, ...$

e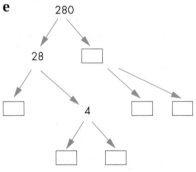

$280 = ...\, ...\, ...\, ...\, ...$

f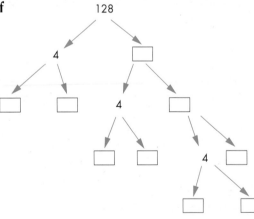

$128 = ...\, ...\, ...\, ...\, ...\, ...\, ...$

g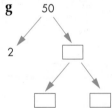

$50 = ...\, ...\, ...$

h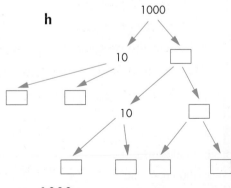

$1000 = ...\, ...\, ...\, ...\, ...\, ...$

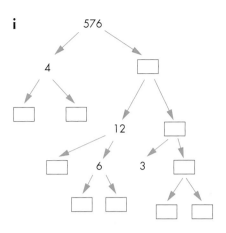

**i**

576 = ... ... ... ... ... ... ... ...

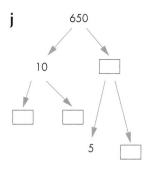

**j**

650 = ... ... ... ...

**2** In index notation, 100 is written as $2^2 \times 5^2$ and 540 is written as $2^2 \times 3^3 \times 5$. Rewrite your prime factorisations from question **1** using index notation.

**3** Write the numbers from 1 to 50 as a product of their prime factors. Use index notation. The first six have been done for you.

$1 = 1$  $2 = 2$  $3 = 3$  $4 = 2^2$  $5 = 5$  $6 = 2 \times 3$  ...

> **Hints and tips** Use your previous answers to help you. For example, $9 = 3^2$, so as $18 = 2 \times 9$, $18 = 2 \times 3^2$.

**4** **a** What is special about the numbers 2, 4, 8, 16, 32, ...?

**b** What are the next two terms in this series?

**c** What are the next three terms in the series 3, 9, 27, ...?

**d** Continue the series 4, 16, 64, ..., for three more terms.

**e** The series in parts **a** and **b** can be written in index notation as $2^1, 2^2, 2^3, 2^4, 2^5, 2^6, ...$

Rewrite the series in parts **c** and **d** in index notation.

**(MR)** **5** **a** Express 60 as a product of its prime factors.

**b** Write your answer to part **a** in index form.

**c** Use your answer to part **b** to write 120, 240 and 480 as products of prime factors in index form.

**(PS)** **6** $1001 = 7 \times 11 \times 13$

$1001^2 = 1\,002\,001$

$1001^3 = 1\,003\,003\,001$

**a** Write 1 002 001 as a product of prime factors, in index form.

**b** Write 1 003 003 001 as a product of prime factors, in index form.

**c** Write $1001^{10}$ as a product of prime factors, in index form.

**(CM)** **7** Harriet wants to share £40 equally among three of her grandchildren. Why is it not possible?

**(PS)** **8** **a** In prime factor form $385 = a \times b \times c$. Work out values for $a$, $b$ and $c$.

**b** In prime factor form $147 = x \times y^2$. Work out values for $x$ and $y$.

## Lowest common multiple

Numbers that appear in the multiplication tables of two (or more) numbers are common multiples of those numbers.

The **lowest common multiple** (LCM) of two numbers is the smallest number that appears in the multiplication tables of both numbers.

For example, the LCM of 3 and 5 is 15, the LCM of 2 and 7 is 14 and the LCM of 6 and 9 is 18.

Examples 8 and 9 show two ways of working out the LCM.

**Example 8**

Find the LCM of 18 and 24.

Write out the multiples of 18.   {18, 36, 54, ⑦2, 90, 108, …}

Write out the multiples of 24.   {24, 48, ⑦2, 96, 120, …}

You can see that 72 is the smallest (lowest) number that appears in both lists, so it is the lowest common multiple.

**Example 9**

Find the LCM of 42 and 63.

Write 42 in prime factor form.   $42 = 2 \times 3 \times 7$

Write 63 in prime factor form.   $63 = 3^2 \times 7$

Write down, in prime factor form, the smallest number that includes *all* the prime factors of 42 *and* 63.

$2 \times 3^2 \times 7$          (This includes $2 \times 3 \times 7$ and $3^2 \times 7$.)

Then work it out.

$$2 \times 3^2 \times 7 = 2 \times 9 \times 7$$
$$= 18 \times 7$$
$$= 126$$

The LCM of 42 and 63 is 126.

## Highest common factor

A number that divides exactly into two different numbers is a common factor of those two numbers.

The biggest number that divides exactly into both of them is their **highest common factor** (HCF).

For example, the HCF of 24 and 18 is 6, the HCF of 45 and 36 is 9 and the HCF of 15 and 22 is 1.

**Example 10**

Find the HCF of 28 and 16.

Write out the factors of 28.     {1, 2, ④, 7, 14, 28}

Write out the factors of 16.     {1, 2, ④, 8, 16}

You can see that 4 is the biggest (highest) number that appears in both lists, so it is the highest common factor.

Example 11

**a** Find the HCF of 48 and 120.

**b** Write 48 and 120 as products of prime factors.

**c** What is the connection between the answers to **a** and **b**?

**a** List the factors of 48.      {1, 2, 3, 4, 6, 8, 12, 16, ⃝24, 48}

List the factors of 120.      {1, 2, 3, 4, 5, 6, 8, 10, 12, 15, 20, ⃝24, 30, 40, 60, 120}

The HCF of 48 and 120 is 24.

**b**   $48 = 2 \times 2 \times 2 \times 2 \times 3$

$= 2^4 \times 3$

$120 = 2 \times 2 \times 2 \times 3 \times 5$

$= 2^3 \times 3 \times 5$

**c** As the product of prime factors, $24 = 2 \times 2 \times 2 \times 3$ or $2^3 \times 3$. These are the common factors of the prime factorisations of 48 and 120.

## Exercise 5E

**1** Find the LCM of each pair of numbers.

**a** 4 and 5      **b** 7 and 8      **c** 2 and 3      **d** 4 and 7

**e** 2 and 5      **f** 3 and 5      **g** 3 and 8      **h** 5 and 6

**2** What connection is there between each pair of numbers and its LCM in question **1**?

**3** Find the LCM of each pair of numbers.

**a** 4 and 8      **b** 6 and 9      **c** 4 and 6      **d** 10 and 15

 **4** Does the connection you found in question **2** apply to the numbers in question **3**? If not, why not?

**5** Find the LCM of each pair of numbers.

**a** 24 and 56      **b** 21 and 35      **c** 12 and 28      **d** 28 and 42

**e** 12 and 32      **f** 18 and 27      **g** 15 and 25      **h** 16 and 36

 **6** Cheese slices are in packs of eight.

Bread rolls are in packs of six.

What is the smallest number of each pack that can be bought to have the same number of cheese slices and bread rolls?

**7** Find the HCF of each pair of numbers.

**a** 24 and 56      **b** 21 and 35      **c** 12 and 28      **d** 28 and 42

**e** 12 and 32      **f** 18 and 27      **g** 15 and 25      **h** 16 and 36

**i** 42 and 27      **j** 48 and 64      **k** 25 and 35      **l** 36 and 54

**PS** **8** In prime factor form, $1250 = 2 \times 5^4$ and $525 = 3 \times 5^2 \times 7$.

    **a** Which of these are common multiples of 1250 and 525?

       **i** $2 \times 3 \times 5^3 \times 7$    **ii** $2^3 \times 3 \times 5^4 \times 7^2$    **iii** $2 \times 3 \times 5^4 \times 7$    **iv** $2 \times 3 \times 5 \times 7$

    **b** Which of these are common factors of 1250 and 525?

       **i** $2 \times 3$    **ii** $2 \times 5$    **iii** $5^2$    **iv** $2 \times 3 \times 5 \times 7$

**PS** **9** Two numbers have a HCF of 6 and a LCM of 72.

    What are the two numbers?

# 5.5 Square numbers

## This section will show you how to:

- identify square numbers
- use a calculator to find the square of a number.

**Key term**

square number

What is the next number in this sequence?

1, 4, 9, 16, 25, …

Write each number as:

$1 \times 1$, $2 \times 2$, $3 \times 3$, $4 \times 4$, $5 \times 5$, …

These factors can be represented by square patterns of dots.

From these patterns, you can see that the next pair of factors must be $6 \times 6 = 36$ and therefore 36 is the next number in the sequence.

$1 \times 1$    $2 \times 2$    $3 \times 3$    $4 \times 4$    $5 \times 5$

Because they form square patterns, the numbers 1, 4, 9, 16, 25, 36, … are called **square numbers**.

When you multiply any number by itself, the result is called the *square of the number* or the *number squared*. This is because the answer is a square number. For example:

    the square of 5 (or 5 squared) is $5 \times 5 = 25$

    the square of 6 (or 6 squared) is $6 \times 6 = 36$.

There is a short way to write the square of any number. For example:

    you can write 5 squared ($5 \times 5$) as $5^2$

    you can write 13 squared ($13 \times 13$) as $13^2$.

So you can write the sequence of square numbers, 1, 4, 9, 16, 25, 36, …, as:

$1^2$, $2^2$, $3^2$, $4^2$, $5^2$, $6^2$, …

You should learn the square numbers up to $15 \times 15$ ($= 225$).

## Exercise 5F

  **1** The square number pattern starts:

    1    4    9    16    25    …

    Copy and continue this pattern until you have written down the first 20 square numbers.

**(PS)** **2** Work out the answer to each of these number sentences.

$1 + 3 \quad =$

$1 + 3 + 5 \quad =$

$1 + 3 + 5 + 7 =$

Look carefully at the pattern of the three number sentences.

Write down the next three number sentences in the pattern.

**(MR)** **3** Draw one counter.

Draw more counters to your picture to make the next square number.

**a** How many extra counters did you add?

Draw more counters to your picture to make the next square number.

**b** How many extra counters did you add?

**c** Without drawing, how many more counters will you need to make the next square number?

**d** Describe the pattern of counters you are adding.

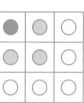

**4** Use your calculator to work out these numbers. Look for the $x^2$ key.

**a** $23^2$      **b** $57^2$      **c** $77^2$      **d** $123^2$      **e** $152^2$

**f** $3.2^2$      **g** $9.5^2$      **h** $23.8^2$      **i** $(-4)^2$      **j** $(-12)^2$

**5** **a** Write down the value of $13^2$.      **b** Write down the value of $14^2$.

**c** Estimate the value of $13.2^2$.

**6** Find the next three numbers in each of these number patterns. You may use your calculator.

| | 1 | 4 | 9 | 16 | 25 | 36 | 49 | 64 | 81 |
|---|---|---|---|---|---|---|---|---|---|
| **a** | 2 | 5 | 10 | 17 | 26 | 37 | ... | ... | ... |
| **b** | 2 | 8 | 18 | 32 | 50 | 72 | ... | ... | ... |
| **c** | 3 | 6 | 11 | 18 | 27 | 38 | ... | ... | ... |
| **d** | 0 | 3 | 8 | 15 | 24 | 35 | ... | ... | ... |
| **e** | 101 | 104 | 109 | 116 | 125 | 136 | ... | ... | ... |

**Hints and tips** They are all based on square numbers, so look for the connection with the square numbers on the top line.

**7** **a** Work out the value of each expression and the square number that appears after it. You may use your calculator.

$3^2 + 4^2$ and $5^2$

$5^2 + 12^2$ and $13^2$

$7^2 + 24^2$ and $25^2$

$9^2 + 40^2$ and $41^2$

$11^2 + 60^2$ and $61^2$

**(EV)** **b** Describe what you notice about your answers to part **a**.

**8** Jasper's bill for his internet is £12 each month for 12 months.

How much does he pay for the whole year?

**(PS)** **9** A builder is using flagstones to lay a patio. He buys enough flagstones for 15 rows, each with 15 flagstones in them. When he starts laying them, he realises he forgot to allow for the gaps between the flagstones so only lays 14 rows, each with 14 flagstones in them.

How many flagstones does he have left?

**(MR)** **10** 4 and 81 are square numbers and their sum is 85.

Find two different square numbers that have a sum of 85.

The following exercise will give you some practice with multiples, factors, square numbers and prime numbers.

## Exercise 5G

**1** Write out the first five multiples of each number.

   **a** 6       **b** 13       **c** 8       **d** 20       **e** 18

> Hints and tips  Remember, the first multiple is the number itself.

**2** Write down the square numbers up to 100.

**3** Write down the factors of each of these numbers.

   **a** 12       **b** 20       **c** 9       **d** 32       **e** 24

   **f** 38       **g** 13       **h** 42       **i** 45       **j** 36

**4** Write out the first three numbers that are multiples of both numbers in each pair.

   **a** 3 and 4    **b** 4 and 5    **c** 3 and 5    **d** 6 and 9    **e** 5 and 7

**(MR)** **5** In question **3**, every number had an even number of factors, except parts **c** and **j**. What sort of numbers are 9 and 36?

**(MR)** **6** The number in question **3**, part **g**, had only two factors. Why?

**7** Write down the prime numbers up to 20.

**(MR)** **8** Copy these number sentences and write out the *next four* sentences in the pattern.

$$1 = 1$$
$$1 + 3 = 4$$
$$1 + 3 + 5 = 9$$
$$1 + 3 + 5 + 7 = 16$$

**(PS)** **9** Here are four numbers.

   10     16     35     49

Copy and complete the table by putting each of the numbers in the correct box.

| | Square number | Factor of 70 |
|---|---|---|
| **Even number** | | |
| **Multiple of 7** | | |

**CM** **10** Arrange these four number cards to make a square number.

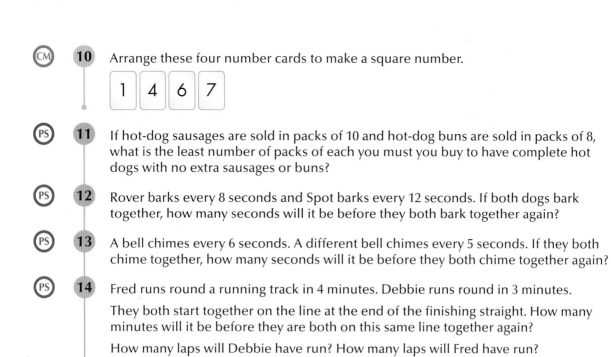

1  4  6  7

**PS** **11** If hot-dog sausages are sold in packs of 10 and hot-dog buns are sold in packs of 8, what is the least number of packs of each you must you buy to have complete hot dogs with no extra sausages or buns?

**PS** **12** Rover barks every 8 seconds and Spot barks every 12 seconds. If both dogs bark together, how many seconds will it be before they both bark together again?

**PS** **13** A bell chimes every 6 seconds. A different bell chimes every 5 seconds. If they both chime together, how many seconds will it be before they both chime together again?

**PS** **14** Fred runs round a running track in 4 minutes. Debbie runs round in 3 minutes.

They both start together on the line at the end of the finishing straight. How many minutes will it be before they are both on this same line together again?

How many laps will Debbie have run? How many laps will Fred have run?

**15** From this box, choose one number that fits each of these descriptions.

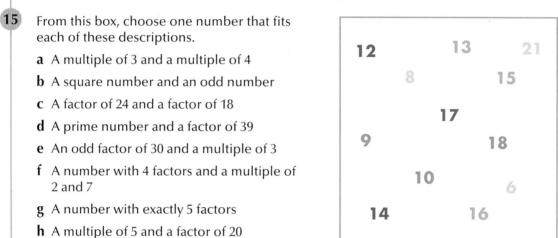

a A multiple of 3 and a multiple of 4

b A square number and an odd number

c A factor of 24 and a factor of 18

d A prime number and a factor of 39

e An odd factor of 30 and a multiple of 3

f A number with 4 factors and a multiple of 2 and 7

g A number with exactly 5 factors

h A multiple of 5 and a factor of 20

i An even number that is a factor of 36 and a multiple of 9

j A prime number that is one more than a square number

k A number with factors that, when written out in order, make a number pattern in which each number is twice the one before

l An odd number that is a multiple of 7

**MR** **16** These numbers are described as triangular numbers.

1, 3, 6, 10, 15

a Investigate why they are called triangular numbers.

b Write down the next five triangular numbers.

# 5.6 Square roots

This section will show you how to:

- recognise the square roots of square numbers up to 225
- use a calculator to find the square roots of any number.

The **square root** of a given number is a number that, when multiplied by itself, produces the given number.

For example, the square root of 9 is 3, since $3 \times 3 = 9$.

Numbers also have a negative square root, since $-3 \times -3$ also equals 9.

A square root is represented by the symbol $\sqrt{\phantom{x}}$. For example, $\sqrt{16} = 4$.

**Remember:** A square root is treated like a power or index, according to BIDMAS.

## Exercise 5H

**1** Write down the positive square root of each of these numbers.

| | | | | |
|---|---|---|---|---|
| **a** 4 | **b** 25 | **c** 49 | **d** 1 | **e** 81 |
| **f** 100 | **g** 64 | **h** 9 | **i** 36 | **j** 16 |
| **k** 121 | **l** 144 | **m** 400 | **n** 900 | **o** 169 |

**2** Write down both possible values of each square root.

| | | | | |
|---|---|---|---|---|
| **a** $\sqrt{25}$ | **b** $\sqrt{36}$ | **c** $\sqrt{100}$ | **d** $\sqrt{49}$ | **e** $\sqrt{64}$ |
| **f** $\sqrt{16}$ | **g** $\sqrt{9}$ | **h** $\sqrt{81}$ | **i** $\sqrt{1}$ | **j** $\sqrt{144}$ |

**3** Write down the value of each of these. You need only give positive square roots. You will need to use your calculator for some of them. Look for the $\sqrt{\square}$ key.

| | | | | |
|---|---|---|---|---|
| **a** $9^2$ | **b** $\sqrt{1600}$ | **c** $10^2$ | **d** $\sqrt{196}$ | **e** $6^2$ |
| **f** $\sqrt{225}$ | **g** $7^2$ | **h** $\sqrt{144}$ | **i** $5^2$ | **j** $\sqrt{441}$ |
| **k** $11^2$ | **l** $\sqrt{256}$ | **m** $8^2$ | **n** $\sqrt{289}$ | **o** $21^2$ |

**4** Write down the positive value of each of these.

| | | | | |
|---|---|---|---|---|
| **a** $\sqrt{576}$ | **b** $\sqrt{961}$ | **c** $\sqrt{2025}$ | **d** $\sqrt{1600}$ | **e** $\sqrt{4489}$ |
| **f** $\sqrt{10\,201}$ | **g** $\sqrt{12.96}$ | **h** $\sqrt{42.25}$ | **i** $\sqrt{193.21}$ | **j** $\sqrt{492.84}$ |

**5** Put these in order, from smallest value to largest value.

$3^2$ $\qquad$ $\sqrt{90}$ $\qquad$ $\sqrt{50}$ $\qquad$ $4^2$

**6** Between which two consecutive whole numbers does the square root of 20 lie?

**7** Use these number cards to make this calculation correct.

$$\boxed{1} \quad \boxed{2} \quad \boxed{3} \quad \boxed{4} \quad \boxed{8}$$

$$\sqrt{\boxed{\phantom{0}}\boxed{\phantom{0}}\boxed{\phantom{0}}} = \boxed{\phantom{0}}\boxed{\phantom{0}}$$

**8** Tebor is tiling a square wall in a kitchen.

Altogether he needs 225 square tiles.

How many tiles are there in each row?

 **9** **a** Work out $\frac{9}{25}$ as a decimal.

**b** Use a calculator to work out the square root of the answer to **a**.

**c** Convert the answer to part **b** to a fraction.

**d** Work out these square roots. Give your answers as fractions.

    **i** $\sqrt{\frac{4}{9}}$           **ii** $\sqrt{\frac{25}{49}}$           **iii** $\sqrt{\frac{64}{81}}$

 **10** How do you know that $8.7^2$ is between 64 and 81?

 **11** By choosing values for $a$ and $b$, decide whether each of these statements is true or false.

**a** $\sqrt{a + b} = \sqrt{a} + \sqrt{b}$ when $a > 0$ and $b > 0$

**b** $\sqrt{a \times b} = \sqrt{a} \times \sqrt{b}$

**c** $2\sqrt{a} = \sqrt{2a}$ when $a > 0$

**d** $3\sqrt{b} = \sqrt{9b}$

 **12** **a** Write down the prime factors of 324.

**b** Use your answer to part **a** to write down the square root of 324.

**c** Without using a calculator, work out the square root of 484.

 **13** **a** The square root of a number $N$, written as a product of its prime factors, is:

$\sqrt{N} = 2 \times 5 \times 13$

Work out the value of $N$.

**b** The square root of a number $M$, written as a product of its prime factors, is:

$\sqrt{M} = 2^2 \times 3 \times 5$

Work out the value of $M$.

# 5.7 Basic calculations on a calculator

## This section will show you how to:

- use some of the important keys when working on a calculator.

In this section, you will use a calculator to find answers to algebraic or geometric problems. The examples will show you how to use some of the **function keys** on the calculator. Remember that some functions will need you to press the **shift key** **shift** to make them work. When you have keyed in the calculation, press **=** to give the answer.

| Key terms |
| --- |
| function key |
| shift key |

Some calculators display answers to calculations as fractions. There is always a key to convert this to a decimal. It is usually acceptable to give an answer as a fraction or a decimal unless you are asked to round to a given accuracy.

Most scientific calculators can be set up to display the answers in the format you want.

These three angles are on a straight line. Find the value of $a$.

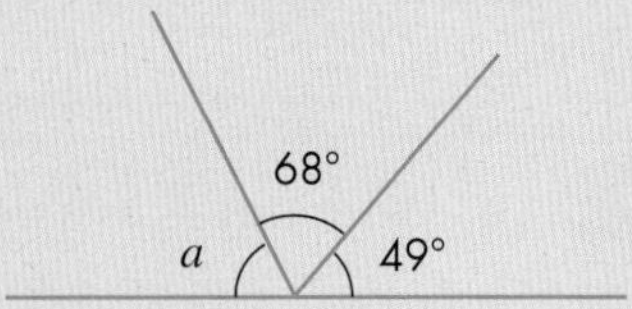

To find the size of the angle labelled $a$, subtract 68° and 49° from 180°.

You can do the calculation in two ways.

$$180 - 68 - 49 \quad \text{or} \quad 180 - (68 + 49)$$

Try keying each calculation into your calculator.

$$180 - 68 - 49$$

The display will show 63.

$$180 - (68 + 49)$$

Again, the display should show 63.

It is important that you can do this both ways.

A common error is to work out $180 - 68 + 49$, which will give the wrong answer. You must use the correct calculation or use brackets to combine parts of the calculation.

Work out the area of this trapezium, where $a = 12.3$, $b = 16.8$ and $h = 2.4$.

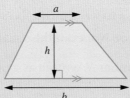

To work out the area of the trapezium, use the formula: $A = \frac{1}{2}(a + b)h$.

Remember, you should always substitute into a formula before working it out.

$$A = \frac{1}{2}(12.3 + 16.8)2.4$$

Between the brackets and the numbers at either end of the brackets there is an assumed multiplication sign, so the calculation is:

$$\frac{1}{2} \times (12.3 + 16.8) \times 2.4$$

$\frac{1}{2}$ can be keyed in lots of different ways

• as a division ![1 ÷ 2]

• as a decimal ![0 . 5]

• as a fraction using the fraction key   (and arrows ![arrows])

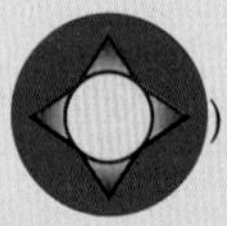

Key in the full calculation, using the fraction key:

The display should show 34.92 or $\frac{873}{25}$.

    5 Number: Number properties

You will learn more about areas of shapes in a later chapter.

Your calculator has a power key $x^2$ and a cube key $x^3$.

**Example 21**

Find the value of $4.5^3 - 2 \times 4.5$.

Try keying in: **4** **.** **5** **x³** **−** **2** **×** **4** **.** **5** **=**

The display should show 82.125 or $\frac{657}{8}$.

Most calculations involving circles will involve the number $\pi$ (said as pi), which has its own calculator button $\pi$ .

The decimal value of $\pi$ goes on for ever. Its approximate value is 3.14 but the value in a calculator is far more accurate and may be displayed as 3.1415926535 or $\pi$.

**Example 22**

Work these out.    **a** $\pi \times 3.2^2$    **b** $2 \times \pi \times 4.9$

Give your answers correct to 1 decimal place.

**a**  Try keying in: **π** **×** **3** **.** **2** **x²** **=**

The display should show 32.16990877 or $\frac{256}{25}\pi$. (Convert this to a decimal.)

This is 32.2 to 1 decimal place.

**b**  Try keying in: **2** **×** **π** **×** **4** **.** **9** **=**

The display should show 30.78760801 or $\frac{49}{5}\pi$. (Convert this to a decimal.)

This is 30.8 to 1 decimal place.

# Exercise 5I

For each question, try to key in the whole calculation in one step.

**1** Subtract these sets of numbers from 180.

    **a** 54 then 81             **b** 21 then 39 then 68      **c** 51 then 34 then 29

**2** Subtract these sets of numbers from 360.

    **a** 68, 92                 **b** 90, 121, 34           **c** 32, 46, 46

**3** **a** Subtract 68 from 180 and divide the answer by 2.

    **b** Subtract 46 from 360 and divide the answer by 2.

    **c** Subtract 52 from 180 twice.

    **d** Subtract 39 and 2 lots of 64 from 360.

**4** Work these out.

    **a** $(10 - 2) \times 180 \div 10$       **b** $180 - (360 \div 5)$

**5** Work these out.

**a** $\frac{1}{2} \times (4.6 + 6.8) \times 2.2$  **b** $\frac{1}{2} \times (2.3 + 9.9) \times 4.5$

**6** Work these out. Give your answers correct to 1 decimal place.

**a** $\pi \times 8.5$  **b** $2 \times \pi \times 3.9$  **c** $\pi \times 6.8^2$  **d** $\pi \times 0.7^2$

(PS) **7** At Sovereign garage, Jon bought 21 litres of petrol for £21.52.

At the Bridge garage he paid £15.41 for 15 litres.

At which garage is petrol cheaper?

(MR) **8** A teacher asked her class to work out this problem.

$\frac{2.3 + 8.9}{3.8 - 1.7}$

Abby keyed in:

( 2 • 3 + 8 • 9 ) ÷
3 • 8 − 1 • 7 =

Bobby keyed in:

2 • 3 + 8 • 9 ÷ 3 • 8 −
1 • 7 =

Col keyed in:

( 2 • 3 + 8 • 9 ) ÷
( 3 • 8 − 1 • 7 ) =

Donna keyed in:

2 • 3 + 8 • 9 ÷
( 3 • 8 − 1 • 7 ) =

They each rounded their answers to 3 decimal places.

Work out the answer that each of them found.

Who had the correct answer?

(MR) **9** Show that a speed of 31 metres per second is approximately 70 miles per hour.

You will need to know that 1 mile ≈ 1610 metres.

**10** Work these out.

**a** $3.4 \times 5.6 \times 8.8$  **b** $2 \times (3.4 \times 5.6 + 3.4 \times 8.8 + 5.6 \times 8.8)$

**11** Work these out, giving your answers correct to 2 decimal places.

**a** $\sqrt{(3.2^2 - 1.6^2)}$  **b** $\sqrt{(4.8^2 + 3.6^2)}$

**12** Work these out.

**a** $7.8^3 + 3 \times 7.8$  **b** $5.45^3 - 2 \times 5.45 - 40$

(MR) **13** Choose values for $a$ and $b$ and use your calculator to test whether or not these statements are true.

**a** $(a - b)(a + b) = a^2 - b^2$  **b** $(a + b)^2 = a^2 - 2ab + b^2$

# Worked exemplars

  A PE teacher is organising 20 students into equal-sized teams for a competition. How many can she have in each team?

| This is a mathematical reasoning question, so you need to construct a chain of reasoning to reach the required result. | |
| --- | --- |
| Factors of 20 are {1, 2, 4, 5, 10, 20}.<br>1 student alone would not make a team.<br>20 students in a team could not play in a competition.<br>So the teacher can have teams of 2, 4, 5 or 10. | The teams will need to be of equal size and there will have to be more than one student in a team so they can play each other. So, although 1 and 20 are factors, they cannot form part of the final answer as this does not address the team requirement. |

  Give a reason why each of these numbers could be the odd one out.

123    144    169

| This is a 'communicating mathematics' question, so make sure your answers are clear and that you use correct mathematical terms. | |
| --- | --- |
| 123 because it is not a square number.<br>144 because it is not odd.<br>169 because it is not a multiple of 3. | Answer the whole question by giving a reason for each number being the odd one out and not just identifying one of the numbers as the odd one out.<br>Each answer is for any valid reason, so there could be other solutions. |

  Jack is investigating writing year numbers as products of prime factors. He finds that:

$2015 = 5 \times 13 \times 31$

$2016 = 2^5 \times 3^2 \times 7$

2017 is a prime number

$2018 = 2 \times 1009$

**a** How can you tell from the prime factorisations that 2015 and 2016 do not have any common factors other than 1?

**b** Write down the factors of 2017?

**c** Is 42 a factor of 2016? Justify your answer.

| This question requires you to evaluate the results shown, so you need to show an understanding of what prime numbers, prime factors and factors are in each part respectively. | |
| --- | --- |
| **a** As there are no numbers common to the prime factorisations of 2015 and 2016, they cannot have any common factors. | This part requires that you show an understanding of what prime factors are and that two numbers cannot have any factors in common, if they have no common prime factors. |
| **b** 1 and 2017 | This question assesses that you know what a prime number represents. As 2017 is a prime number, it only has 2 factors. |
| **c** Write 42 as a product of prime factors. $42 = 2 \times 3 \times 7$<br>$2016 = 2 \times 3 \times 7 \times 2^4 \times 3$, so 42 is a factor of 2016. | As the prime factors of 42 appear in the prime factorisation of 2016, 42 must be a factor of 2016. |

# Ready to progress?

I can recognise multiples of the first ten whole numbers.
I can find factors of numbers less than 100.
I can write down any square number up to $15 \times 15 = 225$.
I can find the square root of any number, using a calculator.
I can recognise two-digit prime numbers.

I can work out the prime factors of numbers.
I can work out the LCM and HCF of two numbers.
I can use systematic counting strategies to identify arrangements.

# Review questions

**1**   **a** Write down the largest factor of 360 that is less than 100.

   **b** Write down the smallest factor of 315 that is greater than 100.

**2**   **a** Copy and complete the missing numbers in the pattern below.

|  |  |  | Last digit |
|---|---|---|---|
| $4^1$ | $= 4$ | $= 4$ | 4 |
| $4^2$ | $= 4 \times 4$ | $= 16$ | 6 |
| $4^3$ | $= 4 \times 4 \times 4$ | $= 64$ | |
| $4^4$ | $= 4 \times 4 \times 4 \times 4$ | $= 256$ | |
| $4^5$ | $=$ | $=$ | |

   **b** What will the last digit of $4^{17}$ be?

**3**   **a** Write down the first five multiples of 6.

   **b** Write down the factors of 12.

   **c** Write down a square number between 20 and 30.

   **d** Write down two prime numbers between 20 and 30.

**4**   Find the value of $3.7^2$.

**5**   Which two numbers in the box are prime numbers?

| 51 | 52 | 53 | 54 | 55 | 56 | 57 | 58 | 59 |
|---|---|---|---|---|---|---|---|---|

**6**   Write down the value of each of these.

   **a** $2^3$          **b** $\sqrt{64}$          **c** the cube of 5

**7**   John set up two computer virus checkers on his computer on 1 January.

   Checker A would scan every 8 days.

   Checker B would scan every 10 days.

   On what date will both checkers next be scanning on the same day?

**(PS) 8** Small pies are sold in packs of 4.

Bread sticks are sold in packs of 10.

What is the least number of each pack that Mary needs to buy, to have the same number of pies and bread sticks?

**9** **a** Express 90 as a product of its prime factors.

**b** Find the highest common factor (HCF) of 90 and 35.

**10** Find the lowest common multiple (LCM) of 45 and 70.

**(PS) 11** $A$ and $B$ are numbers written as the products of their prime factors.

$$A = 3 \times 5 \times 7 \qquad B = 2 \times 3^2 \times 5^2$$

**a** Find the highest common factor (HCF) of $A$ and $B$.

**b** Find the lowest common multiple (LCM) of $A$ and $B$.

**12** **a** Write down the value of $14^2$.

**(CM)** **b** How do you know that $35^2$ is not equal to 1220?

**(MR) 13** Two clocks, A and B, are both set at 12:00 midnight and started. Each clock chimes at every quarter hour.

Clock A keeps perfect time; clock B gains 5 minutes every hour.

**a** What time will it be on each clock when they next chime together?

**b** What time will it be on each clock when they next chime together, with both clocks showing a whole number of hours?

**(MR) 14** A blue light flashes every 10 seconds. A yellow light flashes every 15 seconds. A red light flashes every 25 seconds.

At the start, they all flash together. How many seconds will it take before:

**a** blue and yellow   **b** blue and red    **c** red and yellow   **d** all three flash together again?

**(CM) 15** Britain's most famous mathematician, Isaac Newton, lived between 1643 and 1726. He did not have a calculator so he needed a quick way of working out square roots.

This was his method to work out the square root of 20.

Guess a value for the root, $R$, say 4.5.

Work out $20 + R^2$: $\qquad\qquad\qquad\qquad$ $(20 + 4.5^2) = 40.25$

Divide $(20 + R^2)$ by $2 \times R$: $\qquad$ $(40.25) \div (2 \times 4.5) = 4.47222...$

On a calculator, the value of $\sqrt{20}$ is $= 4.4721...$ so Newton's method works almost straight away, correct to 3 decimal places.

If you want a more accurate answer, repeat the method with R = 4.47222...

**a** Use Newton's method to work out $\sqrt{60}$, taking $R$ as 7.7 and replacing 20 in the formula above with 60.

Compare your answer with the value of $\sqrt{60}$ on your calculator. For how many decimal places is the answer the same?

**b** Use Newton's method to work out $\sqrt{39}$, taking $R$ as 6.2 and replacing 20 in the formula above with 39.

Compare your answer with the value of $\sqrt{39}$ on your calculator. For how many decimal places is the answer the same?

**(MR) 16** The LCM and the HCF of two numbers, $x$ and $y$, are equal. What must be true about $x$ and $y$?

# 6 Number: Approximations

## This chapter is going to show you:

- how to round a number to a given accuracy
- how to estimate the answer to calculations by rounding
- how to find the limits of numbers rounded to a given accuracy
- how to work out the error interval due to rounding or truncation.

## You should already know:

- how to multiply and divide whole numbers.

## About this chapter

You have probably heard people ask for a 'ball park' figure. It comes from newspapers in the US reporting on crowds at baseball games. They generally give the numbers to the nearest thousand, for example: '40 000 fans watch the Red Sox beat the Yankees.' The number 40 000 is an approximate value, but close enough to give a good idea of the number of people in the 'ball park'.

If you think about it, you will probably realise that you talk in approximate numbers every day. For example, you may say that it takes you about 20 minutes to get to school or that your mobile phone costs about £30 a month, or a car journey will take about two and a half hours.

It is important that you can round numbers accurately. You also need to have an idea of what it actually means when numbers are given approximately. For example, when a hotel claims to be 'approximately 5 minutes' walk from the beach!' it is probably going to take you a bit longer than that!

# 6.1 Rounding whole numbers

This section will show you how to:

- round a whole number.

On a packet of mints you may see the contents given as, for example: 'Average contents: 30 mints'. If you know that this number is **rounded** to the nearest 10, then:

- the smallest number that is rounded up to 30 is 25
- the largest number that is rounded down to 30 is 34 (because 35 would be rounded up to 40).

So, there could actually be from 25 to 34 mints in the packet.

What about the number of runners in a marathon? If you know that the number 23 000 is rounded to the nearest 1000:

- the smallest number that is rounded up to 23 000 is 22 500
- the largest number that is rounded down to 23 000 is 23 499 (because 23 500 would be rounded up to 24 000).

So, there could actually be from 22 500 to 23 499 people in the marathon.

Do you think the cooking time on a pie is a rounded value? Does a cooking time of 30 minutes mean that you could cook it for any time between 25 and 35 minutes? This is different because it is generally unsafe to undercook food. Most cooks would bake the pie for at least 30 minutes. If it is cooked for too long it will burn, so 35 minutes is probably the maximum time.

Do you think the time it takes the teacher to drive to school is rounded? Does the journey always take between 35 and 45 minutes? Traffic conditions vary a lot, so 40 minutes is probably an average.

Most numbers used on a daily basis are estimates, sensible values or averages.

| | a | Round these numbers to the nearest 10. | | i | 53 | | ii | 67 | | iii | 125 |
| --- | --- | --- | --- | --- | --- | --- | --- | --- | --- | --- |
| | b | Round these numbers to the nearest 100. | | i | 489 | | ii | 821 | | iii | 1350 |
| | a | i | 50 | ii | 70 | iii | 130 | | | |
| | b | i | 500 | ii | 800 | iii | 1400 | | | |

Note that the rule for a 'halfway' value is to round up.

## Exercise 6A

**1**    Round each number to the nearest 10.

| a | 24 | b | 57 | c | 78 | d | 54 | e | 96 |
|---|----|---|----|---|----|---|----|---|----|
| f | 21 | g | 88 | h | 66 | i | 14 | j | 26 |
| k | 29 | l | 51 | m | 77 | n | 49 | o | 94 |

**2**    Round each number to the nearest 100.

| a | 240 | b | 570 | c | 780 | d | 504 | e | 967 |
|---|-----|---|-----|---|-----|---|-----|---|------|
| f | 112 | g | 645 | h | 358 | i | 998 | j | 1050 |
| k | 350 | l | 650 | m | 750 | n | 1020 | o | 1070 |

**3**    These three jars are on the shelf of a sweet shop.

Jar 1       Jar 2       Jar 3

Look at each number below and write down which jar it could be describing. (For example, could there be 76 sweets in jar 1?)

| a | 78 sweets | b | 119 sweets | c | 84 sweets | d | 75 sweets |
|---|-----------|---|------------|---|-----------|---|-----------|
| e | 186 sweets | f | 122 sweets | g | 194 sweets | h | 115 sweets |
| i | 81 sweets | j | 79 sweets | k | 192 sweets | l | 124 sweets |

m   Which of these numbers of sweets could not be in jar 1?    74   84   81   76

n   Which of these numbers of sweets could not be in jar 2?    124 126 120 115

o   Which of these numbers of sweets could not be in jar 3?    194 184 191 189

**4**    Round each number to the nearest 1000.

| a | 2400 | b | 5700 | c | 7806 | d | 5040 | e | 9670 |
|---|------|---|------|---|------|---|------|---|------|
| f | 1120 | g | 6450 | h | 3499 | i | 9098 | j | 1500 |
| k | 2990 | l | 5110 | m | 7777 | n | 5020 | o | 9400 |
| p | 3500 | q | 6500 | r | 7500 | s | 1020 | t | 1770 |

**5** Round each number to the nearest 10.

a 234    b 567    c 718    d 524    e 906
f 231    g 878    h 626    i 114    j 296
k 375    l 625    m 345    n 1012   o 1074

**6** Look at these signs.

**Welcome to Elsecar**
Population 800
(to the nearest 100)

**Welcome to Hoyland**
Population 1200
(to the nearest 100)

**Welcome to Jump**
Population 600
(to the nearest 100)

Which of these sentences could be true? Which must be false?

a There are 789 people living in Elsecar.

b There are 1278 people living in Hoyland.

c There are 550 people living in Jump.

d There are 843 people living in Elsecar.

e There are 1205 people living in Hoyland.

f There are 650 people living in Jump.

**7** The sign-maker who made the signs in question **6** is making a similar sign for Drayton. Drayton has a population of 1385.

Draw a diagram to show what she should write on the sign.

**8** The table shows the numbers of spectators in the crowds at ten Premier Division games on a weekend in October 2014.

| Match | Number of spectators |
| --- | --- |
| Burnley v Everton | 19 927 |
| Liverpool v Hull City | 44 591 |
| Man Utd v Chelsea | 75 327 |
| QPR v Aston Villa | 18 022 |
| Southampton v Stoke | 30 017 |
| Sunderland v Arsenal | 44 449 |
| Swansea v Leicester | 20 259 |
| Tottenham v Newcastle | 35 650 |
| West Brom v Crystal Palace | 24 738 |
| West Ham v Man City | 34 977 |

a Which match had the largest crowd?

b Which had the smallest crowd?

c Round all the numbers to the nearest 1000.

d Round all the numbers to the nearest 100.

**9** Give these times to the nearest 5 minutes.

    **a** 34 minutes    **b** 57 minutes    **c** 14 minutes    **d** 51 minutes

    **e** 8 minutes    **f** 13 minutes    **g** 44 minutes    **h** 32.5 minutes

    **i** 3 minutes    **j** 50 seconds

**(PS) 10** Matthew and Vikki are playing a game with whole numbers.

    **a** Matthew is thinking of a number. Rounded to the nearest 10, it is 380.

       What is the smallest number Matthew could be thinking of?

    **b** Vikki is thinking of a different number. Rounded to the nearest 100, it is 400. If Vikki's number is definitely smaller than Matthew's, what are the smallest and largest possible numbers that Vikki is thinking of?

**(PS) 11** The number of adults attending a comedy show is 80, to the nearest 10.

    The number of children attending is 50, to the nearest 10.

    Katie says that 130 adults and children attended the comedy show.

    Give an example to show that she may *not* be correct.

# 6.2 Rounding decimals

## This section will show you how to:

- round decimal numbers to a given accuracy.

The decimal number system carries on after the **decimal point** to include **decimal fractions**, with place values for tenths, hundredths, thousandths and so on.

| Key terms |
|---|
| decimal fraction |
| decimal place |
| decimal point |
| error interval |

The decimal point separates the decimal fraction from the whole-number part of the number.

For example, the number 25.374 is made up of:

| Tens | Units | | tenths | hundredths | thousandths |
|---|---|---|---|---|---|
| 10 | 1 | | $\frac{1}{10}$ | $\frac{1}{100}$ | $\frac{1}{1000}$ |
| 2 | 5 | . | 3 | 7 | 4 |

You use decimal notation to express amounts of money. For example:

    £32.67     means   3 × £10

                       2 × £1

                       6 × £0.10       (10 pence)

                       7 × £0.01       (1 penny)

When you write a number in decimal form, the positions of the digits to the right of the decimal point are called **decimal places** (dp). For example:

- 79.4 is written 'with one decimal place'

- 6.83 is written 'with two decimal places'

- 0.526 is written 'with three decimal places'.

When a rounded number is written with a zero, for example, 3.50 (2 dp), it means that the number, before rounding, was possibly 3.495 of 4.99. The zero shows that the rounded number is accurate to two decimal places.

These are the steps to round a decimal number to a given number of decimal places.

- Count along the decimal places from the decimal point and look at the digit after the one to be rounded.
- If the value of this digit is less than 5, just remove this and the digits after it.
- If the value of this digit is 5 or more, add 1 to round up the digit in the previous decimal place and then remove any digits after the one you have rounded.

When a number has been rounded, you need to know what the original number may have been. For example, when a number is given as '4.5 to one decimal place', then it could originally have been somewhere between 4.45 and 4.55. It could not have been exactly 4.55 as this would be round up to 4.6.

You can use inequalities to show this. You write:

4.45 ⩽ actual value < 4.55

This is the **error interval** due to rounding. It means that the original number could be from 4.45 to 4.55 but not 4.55.

> **Hints and tips**   The sign ⩽ means that the value 4.45 is included; the sign < means that the value 4.55 is not included.

## Exercise 6B

**1** Round each number to one decimal place.

| | | | |
|---|---|---|---|
| **a** 4.83 | **b** 3.79 | **c** 2.16 | **d** 8.25 |
| **e** 3.673 | **f** 46.935 | **g** 23.883 | **h** 9.549 |
| **i** 0.109 | **j** 0.599 | **k** 64.99 | **l** 50.999 |

> **Hints and tips**   Just look at the value of the digit in the second decimal place.

**2** Round each number to two decimal places.

| | | | |
|---|---|---|---|
| **a** 5.783 | **b** 2.358 | **c** 0.977 | **d** 33.085 |
| **e** 23.5652 | **f** 91.7895 | **g** 7.995 | **h** 2.3076 |
| **i** 5.9999 | **j** 3.5137 | **k** 96.508 | **l** 0.009 |

**3** Round each number to the number of decimal places (dp) indicated.

   **a** 4.568 (1 dp)         **b** 0.0832 (2 dp)       **c** 45.71593 (3 dp)

   **d** 94.8531 (2 dp)      **e** 602.099 (1 dp)     **f** 671.7629 (2 dp)

   **g** 7.1124 (1 dp)         **h** 6.90354 (3 dp)    **i** 13.7809 (2 dp)

   **j** 0.07511 (1 dp)        **k** 4.00184 (3 dp)    **l** 59.983 (1 dp)

**4** Round each number to the nearest whole number.

   **a** 8.7          **b** 9.2          **c** 6.5         **d** 3.28

   **e** 7.82        **f** 7.55       **g** 6.172     **h** 3.961

   **i** 1.514       **j** 46.78     **k** 153.9     **l** 342.5

(MR) **5** Belinda puts these items in her shopping basket:

        bread (£1.09), meat (£6.99), cheese (£3.91) and butter (£1.13).

   By rounding each price to the nearest pound (£), work out an estimate for the total cost of the items.

(MR) **6** Which of these numbers are correctly rounded values of 3.456?

       3      3.0      3.4      3.40      3.45      3.46      3.47      3.5      3.50

(PS) **7** When an answer is rounded to three decimal places, it is 4.728.

   Which of these could be the original answer?

       4.71        4.7275       4.7282       4.73

(PS) **8** A number has three decimal places. When it is rounded to two decimal places it is 6.45. When it is rounded to one decimal place, it is 6.4. Work out a possible value of the number.

**9** A number is given as 8.8 correct to one decimal place. Write down the error interval due to rounding.

(PS) **10** Jake had £8 in change, to the nearest pound. To the nearest 10p he had £8.50 in change. How much change could he have had?

(PS) **11** A rectangle is 6.5 cm long and 3.3 cm wide. Both of these values are given correct to one decimal place.

   **a** Write down the error interval due to rounding of the length.

   **b** Use error intervals of the length and width to show that the perimeter has an error interval of $19.4 \leqslant$ perimeter $< 19.8$.

   **c** Use error intervals for the length and width to write down the smallest possible value for the area.

(MR) **12** $\pi$ is a never-ending decimal. A commonly used under-estimate is 3.14. A commonly used over-estimate is 3.142. The formula for the circumference, $C$, of a circle with diameter, $d$, is $C = \pi d$. The diameter of a circle is measured as 10 cm, to the nearest centimetre. Work out:

   **a** the smallest possible under-estimate of the circumference

   **b** the greatest possible over-estimate of the circumference.

(PS) **13** A number is given correct to 3 decimal places. It is rounded to 2 decimal places. This answer is then rounded to 1 decimal place, to give the value 3.7.

   Work out the smallest possible value that the original number could have had.

# 6.3 Approximating calculations

This section will show you how to:

- identify significant figures
- round numbers to a given number of significant figures
- use approximation to estimate answers and check calculations
- round a calculation at the end of a problem, to give what is considered to be a sensible answer.

| Key terms |
|-----------|
| approximate |
| significant figure |

## Rounding to significant figures

You often use **significant figures** (sf) when you want to **approximate** a number that has lots of digits in it.

This table shows some numbers written to one, two and three significant figures.

| One sf | 8 | 50 | 200 | 90 000 | 0.000 07 | 0.003 | 0.4 |
|--------|---|-----|-----|--------|----------|--------|------|
| Two sf | 67 | 4.8 | 0.76 | 45 000 | 730 | 0.0067 | 0.40 |
| Three sf | 312 | 65.9 | 40.3 | 0.0761 | 7.05 | 0.003 01 | 0.400 |

These are the steps for rounding a number to one significant figure. They are very similar to those used for rounding to one decimal place.

- From the left, find the first non-zero digit. The next digit is the second digit.
- When the value of the second digit is less than 5, leave the first (non-zero) digit as it is.
- When the value of the second digit is equal to or greater than 5, add 1 to the first (non-zero) digit.
- If the original number is greater than 1, replace all other digits after the rounded digit, as far as the decimal point, with zeros. If there were some decimal values do not include them.
- If the original number is less than 1 (it starts 0), write down all the zeros before the rounded digit and delete everything after the rounded digit.

To round to two significant figures, you use the same method but use the third (non-zero) digit from the left.

This table shows some numbers rounded to one or two significant figures.

| Number | Rounded to | | Number | Rounded to | |
|--------|------|------|--------|------|------|
| | 1 sf | 2 sf | | 1 sf | 2 sf |
| 78 | 80 | 78 | 45 281 | 50 000 | 45 000 |
| 32 | 30 | 32 | 568 | 600 | 570 |
| 0.692 | 0.7 | 0.69 | 8054 | 8000 | 8100 |
| 1.894 | 2 | 1.9 | 7.867 | 8 | 7.9 |
| 998 | 1000 | 1000 | 99.8 | 100 | 100 |
| 0.436 | 0.4 | 0.44 | 0.0785 | 0.08 | 0.079 |

## Exercise 6C

**1** Round each number to one significant figure.

| | | | | |
|---|---|---|---|---|
| **a** 46 313 | **b** 57 123 | **c** 30 569 | **d** 94 558 | **e** 85 299 |
| **f** 54.26 | **g** 85.18 | **h** 27.09 | **i** 96.432 | **j** 167.77 |
| **k** 0.5388 | **l** 0.2823 | **m** 0.00584 | **n** 0.04785 | **o** 0.000876 |
| **p** 9.9 | **q** 89.5 | **r** 90.78 | **s** 199 | **t** 999.99 |

**2** Round each number to two significant figures.

| | | | | |
|---|---|---|---|---|
| **a** 56 147 | **b** 26 813 | **c** 79 611 | **d** 30 578 | **e** 14 009 |
| **f** 5876 | **g** 1065 | **h** 847 | **i** 109 | **j** 638.7 |
| **k** 1.689 | **l** 4.0854 | **m** 2.658 | **n** 8.0089 | **o** 41.564 |
| **p** 0.8006 | **q** 0.458 | **r** 0.0658 | **s** 0.9996 | **t** 0.00982 |

**3** Round each number to three significant figures.

| | | | | |
|---|---|---|---|---|
| **a** 64 523 | **b** 19 316 | **c** 4.5489 | **d** 4.0756 | **e** 14.396 |
| **f** 6.689 | **g** 1.072 | **h** 9428 | **i** 10.37 | **j** 9.6821 |

**4** Write down the smallest and the greatest numbers of sweets that can be found in each of these jars, when full.

**5** Write down the smallest and the greatest numbers of people that live in these towns.

Ayton          population 800 (to 1 sf)

Beeville          population 1000 (to 1 sf)

Charlestown          population 200 000 (to 1 sf)

**6** When a number with one decimal place is rounded to 2 sf it is 64 and rounded to 1 sf it is 60.

**a** What is the largest possible value of the number?

**b** What is the smallest possible value of the number?

**7** A joiner estimates that he has 20 pieces of skirting board in stock. This is correct to one significant figure. He uses three pieces and now has 10 left, still correct to one significant figure.

How many pieces could he have had to start with? Work out all possible answers.

**8** There are 500 fish in a pond, correct to one significant figure.

What is the smallest possible number of fish that could be taken from the pond so that there are 400 fish in the pond, correct to one significant figure?

**9** The organisers of a conference expect 2000 people (to 2 sf) to attend. The venue can seat 1850 (to 3 sf). In the event, 100 people had to stand. How many people attended the conference? Write down a range of values.

## Approximating calculations

How do you approximate the value of a calculation?

What would you actually do when you try to approximate an answer to a problem?

For example, what is the approximate answer to 35.1 × 6.58?

To find the approximate answer, round each number to one significant figure, then complete the calculation. So, in this case, the approximation is:

$35.1 \times 6.58 \approx 40 \times 7 = 280$

> **Hints and tips** The symbol ≈ means 'approximately equal to'.

The approximation for the division 89.1 ÷ 2.98, is 90 ÷ 3 = 30.

**Example 4**

Find an approximate answer to 24.3 ÷ 3.87.

Round each number in the calculation to 1 sf.   24.3 ÷ 3.87

≈ 20 ÷ 4 = 5

So 24.3 ÷ 3.87 ≈ 5.

A quick approximation will help prevent you from giving an answer that is much too big or too small.

## Exercise 6D

**1** Work out approximate answers to these calculations.

**a** 5435 × 7.31      **b** 5280 × 3.211      **c** 63.24 × 3.514 × 4.2

**d** 3508 × 2.79      **e** 72.1 × 3.225 × 5.23      **f** 470 × 7.85 × 0.99

**g** 354 ÷ 79.8      **h** 36.8 ÷ 1.876      **i** 5974 ÷ 5.29

Use a calculator to see how close your approximations are to the actual answers.

**2** Work out the approximate monthly pay for each annual salary given.

**a** £35 200      **b** £25 600

**c** £18 125      **d** £8420

**3** Work out each person's approximate annual pay.

**a** Kevin who earns £270 a week

**b** Malcolm who earns £1528 a month

**c** David who earns £347 a week

**(MR)** **4** A farmer bought 2713 kg of seed at a cost of £7.34 per kilogram. Estimate the total cost of this seed.

**5** By rounding, work out an approximate answer to each calculation.

**a** $\frac{573 + 783}{107}$    **b** $\frac{783 - 572}{24}$    **c** $\frac{352 + 657}{999}$    **d** $\frac{1123 - 689}{354}$

**e** $\frac{589 + 773}{658 - 351}$    **f** $\frac{793 - 569}{998 - 667}$    **g** $\frac{354 + 656}{997 - 656}$    **h** $\frac{1124 - 661}{355 + 570}$

**i** $\frac{28.3 \times 19.5}{97.4}$    **j** $\frac{78.3 \times 22.6}{3.69}$    **k** $\frac{3.52 \times 7.95}{15.9}$    **l** $\frac{11.78 \times 77.8}{39.4}$

**6** Work out an approximate answer for each calculation.

**a** 208 ÷ 0.378      **b** 96 ÷ 0.48      **c** 53.9 ÷ 0.58

**d** 14.74 ÷ 0.285      **e** 28.7 ÷ 0.621      **f** 406.9 ÷ 0.783

Use a calculator to see how close your approximations are to the actual answers.

**7** A litre of paint will cover an area of about 8.7 square metres (m²). Approximately how many 1-litre cans will I need to paint a room with a total surface area of 73 m²?

**8** By rounding, work out an approximate answer to each calculation.

a $\dfrac{84.7 + 12.6}{0.483}$  b $\dfrac{32.8 \times 71.4}{0.812}$  c $\dfrac{34.9 - 27.9}{0.691}$  d $\dfrac{12.7 \times 38.9}{0.42}$

(CM) **9** It took me 6 hours and 40 minutes to drive from Sheffield to Bude, a distance of 295 miles. My car uses petrol at the rate of about 32 miles per gallon. The petrol cost £3.51 per gallon.

a Approximately how many miles did I travel each hour?

b Approximately how many gallons of petrol did I use in driving from Sheffield to Bude?

c What was the approximate cost of all the petrol for my journey from Sheffield to Bude and back again?

(PS) **10** Kirsty puts magazines inside envelopes and sticks an address label on each one. She puts 178 magazines in envelopes and addresses them between 10:00 am and 1:00 pm. Approximately how many magazines will she be able to deal with in a week in which she works for 17 hours?

(EV) **11** An athlete runs 3.75 km every day. A marathon is 42.1 kilometres. The athlete claims the distance he ran was the equivalent of over three marathons a month. Is this claim true?

(MR) **12** 1 kg = 1000 g

A box full of magazines weighs 8 kg. One magazine weighs about 15 g. Approximately how many magazines are there in the box?

**13** An apple weighs about 280 g.

a What is the approximate mass of a bag containing a dozen apples?

b Approximately how many apples will there be in a sack weighing 50 kg?

**14** At the 2015 American Football final between the New England Patriots and the Seattle Seahawks, in Arizona, the average price of a ticket was $2670. The attendance at the game was 70 288.

A Estimate how much money was made from ticket sales.

B Each team gets 17.5% of the money made from ticket sales. Estimate how much each team made.

(MR) **15** The error interval of the answer, $A$, to the calculation $N \times 15$ is:

$112.5 \leqslant A < 127.5$

Work out the error interval of $N$. Give your answer in the form $X \leqslant N < Y$, where $X$ and $Y$ are decimal numbers.

**16** The lengths in this rectangle are measured to the nearest centimetre.

a Work out the error interval of the length.

b Work out the error interval of the width.

c Work out the error interval of the perimeter

d Work out the error interval of the area.

10 cm

5 cm

# Worked exemplars

 **1** The perimeter of a square is 36 cm.

This length is accurate to the nearest centimetre.

Work out the error interval due to rounding of the length of a side of the square.

| This is a problem-solving question, so you need to make connections between different topics in mathematics (in this case properties of squares and rounding) and to show your strategy clearly. | |
|---|---|
| Perimeter = 4 × length of one side | Write down the correct formula for the perimeter of a square. |
| Error interval of perimeter is: <br> 35.5 ⩽ perimeter < 36.5 | Write down the error interval of the perimeter. |
| Error interval of one side is $\frac{1}{4}$ of the error interval of the perimeter. | Show that the error interval of a side is a quarter of the error interval of the perimeter. |
| Error interval of one side is: <br> 8.875 ⩽ length of side < 9.125 | Write down the error interval of the side. |

  **2** Use estimation to put the expressions below in order of size, starting with the smallest.

**A** $\frac{9.7 \times 10.3}{7.2 - 2.1}$  **B** $88.8 \div 5.8$  **C** $\left(\frac{3.7 + 6.2}{1.9}\right)^2$

| This is a mathematical reasoning question. You need to show your working and your conclusion. | |
|---|---|
| A:  $\frac{10 \times 10}{7 - 2} = \frac{100}{5} = 20$ | Round all the numbers in each expression to 1 sf and work out the approximate value. |
| B:  $90 \div 6 = 15$ | |
| C:  $\left(\frac{4 + 6}{2}\right)^2 = \left(\frac{10}{2}\right)^2 = 5^2 = 25$ | |
| The order from smallest is: <br> B, A, C | Write down the order, starting with the smallest. |

  **3** The length of a piece of wood is measured as 2.3 metres to the nearest 10 cm. Which of these is the error interval for the length?

**A**  2.2 m ⩽ length < 2.3 m   **B**  1.3 m ⩽ length < 3.3 m
**C**  2.25 m ⩽ length < 2.35 m   **D**  2.29 m ⩽ length < 2.31 m

| This is a mathematical reasoning question as there is a mixture of units. The length is given in metres but the accuracy is measured in centimetres. | |
|---|---|
| 2.3 metres = 230 centimetres | Change the length to centimetres. |
| Error interval is 225 ⩽ length < 235 | An accuracy of 10 cm means that the length is within ± 5 cm. |
| 2.25 m ⩽ length < 2.35 m | Convert the error interval back to metres. |

# Ready to progress?

I can identify the number of decimal places.

I can round numbers to a given number of decimal places.
I can round decimals and whole numbers to a given number of significant figures.
I can use approximations to estimate the answer to calculations.

I can use inequality notation to specify the error interval due to rounding.

# Review questions

1.  **a** Write the number 5639 correct to the nearest 1000.

    **b** Write the number seven hundred and eighty-six correct to the nearest 10.

2.  **a** Write the number 3185 correct to the nearest hundred.

    **b** Write the number 5472 correct to the nearest thousand.

3.  **a** Write these numbers in figures.

    **i** thirty-six million      **ii** three thousand, six hundred

    **b** What number should go in each box to make these calculations correct?

    **i** $36 \times \square = 36\,000$      **ii** $36\,000\,000 \div \square = 360\,000$

4.  Ben Nevis is 1344 metres high.

    Mount Snowdon is 1085 metres high.

    **a** How much higher is Ben Nevis than Mount Snowdon?

    Give your answer to the nearest 10 metres.

    **b** Round the height of each mountain to the nearest 10 metres and then work out the difference between them.

    **c** Are your results for parts **a** and **b** the same? Explain why.

5.  The population of Plaistow in 2014 is given as 7700 correct to the nearest hundred.

    **a** What is the smallest possible population of Plaistow?

    **b** What is the largest possible population of Plaistow?

6.  A seal colony is estimated to have 2500 seals, correct to the nearest 100.

    **a** What is the smallest possible number of seals in the colony?

    **b** What is the largest possible number of seals in the colony?

    **c** Better counting methods now show the colony to be 2500 correct to the nearest 50. What is

    **i** the smallest      **ii** the largest number of seals now?

**7** How many decimal places are there in each number?

    **a** 6.82         **b** 9.705         **c** 0.009         **d** 10.0708

**8** How many significant figures are there in each number?

    **a** 58         **b** 6000         **c** 0.0092         **d** 204

    **e** 18.5         **f** 10.52         **g** 0.0106         **h** 300 000

**9** Round these numbers to the accuracy shown.

    **a** 67.493 (1 dp)     **b** 30.6 (2 sf)     **c** 2999 (2 sf)     **d** 3.99 (1 dp)

    **e** 23 (1 sf)         **f** 12.35 (1 dp)    **g** 13.567 (2 dp)   **h** 29.3764 (3 sf)

**(PS) 10** A bus has 52 seats. 10 people are allowed to stand. The bus has 40 passengers, counted to the nearest 10. At a bus stop there are 20 people measured to the nearest 10. No one gets off the bus. Can you be sure that everyone at the bus stop will get on the bus? Show how you decide.

**(PS) 11** This is a postcard. The lengths are accurate to the nearest half-centimetre.

This is an envelope. The lengths are accurate to the nearest centimetre.

Will every postcard fit in every envelope? Show how you decide.

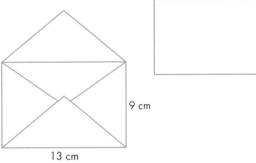

12.5 cm

8 cm

9 cm

13 cm

**(PS) 12** A number has three decimal places. When it is rounded to two decimal places it is 1.85. When it is rounded to one decimal place it is 1.9. Work out the range of values for the number.

**(PS) 13** A number is 200 when rounded to the nearest hundred and 240 when rounded to the nearest ten. What are the greatest and least possible values of the number?

**(MR) 14** Machine A makes square holes. Each hole has a side of 50 mm, measured to the nearest millimetre.

Machine B makes round pegs. Each peg has a diameter of 49 mm, measured to the nearest 2 millimetres.

    **a** Use inequalities to write down the error interval for the side of the square.

    **b** Will every round peg fit in every square hole? Show your working.

**15** The lengths in this right-angled triangle are measured to the nearest centimetre.

    **a** Work out the error interval of the perimeter.

    **b** Work out the error interval of the area.

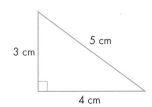

5 cm

3 cm

4 cm

# 7 Number: Decimals and fractions

## This chapter is going to show you:

- how to multiply and divide decimals
- how to convert between rational numbers, fractions and decimals
- how to find reciprocals of numbers and fractions
- how to work out one quantity as a fraction of another
- how to add, subtract, multiply and divide fractions, with and without a calculator.

## You should already know:

- how to cancel fractions to their simplest form
- how to convert a mixed number to an improper fraction and vice versa
- the multiplication tables up to $12 \times 12$
- how to multiply and divide whole numbers.

## About this chapter

Have you ever wondered why some numbers are written in different ways, either as fractions or as decimals?

The answer is that sometimes it is easier to use decimals and sometimes it is easier to use fractions. Fractions and decimals are just different ways of writing the same thing. For example, you usually use decimals when you write an amount of money but fractions when you talk about parts of an hour.

You need to have a basic understanding of decimals and fractions to make sense of the numbers you meet in your everyday life. For example:

- petrol prices up by 2.2 pence a litre
- $\frac{4}{5}$ of parents are happy with the local school.

# 7.1 Calculating with decimals

## This section will show you how to:

- multiply and divide with decimals.

## Multiplying two decimal numbers

Follow these steps to multiply two decimal numbers together.

**Step 1** Rewrite and complete the calculation without the decimal points.

**Step 2** Next, you need to find the position of the decimal point. The number of decimal places in the answer is the same as the total number of decimal places in the numbers you are multiplying.

Alternatively, you can round both of the decimal numbers to the nearest whole number and estimate the answer. This will give you the approximate size of the answer so you will know where to put the decimal point.

**Step 3** Put the decimal point back in.

> **Example 1**
>
> Evaluate $3.42 \times 2.7$.
>
> Rewrite and complete the calculation without decimal points.
>
> ```
>     3 4 2
>  ×    2 7
>   2 3 9 4
>      2 1
>   6 8 4 0
>   9 2 3 4
>     1 1
> ```
>
> Find the position of the decimal point.
>
> 3.42 has two decimal places (.42) and 2.7 has one decimal place (.7).
>
> So, there are three decimal places in the answer.
>
> Or, $3.42 \times 2.7 \approx 3 \times 3 = 9$.
>
> Therefore, $3.42 \times 2.7 = 9.234$

## Dividing one decimal number by another

Follow these steps to divide one decimal number by another decimal number.

- First, look at the number you are dividing by. Multiply it by a power of 10 (10, 100, 1000…) so that it becomes an integer (whole number).

- Next, multiply the number that you are dividing into by the same power of 10.

- Finally, work out this equivalent division.

> **Example 2**
>
> Complete these divisions.
>
> **a**  $0.24 \div 0.3$          **b**  $5.88 \div 0.8$
>
> **a**  Multiply both numbers by 10.          $0.24 \div 0.3 = 2.4 \div 3$
>
> ```
>        0 . 8
>   3 ) 2 . ²4
> ```
>
> **b**  Multiply both numbers by 10.          $5.88 \div 0.8 = 58.8 \div 8$
>
> ```
>          7 . 3 5
>   8 ) 5 ⁵8 . ²8 ⁴0
> ```
>
> Note that an extra zero has been put at the end of 58.8 to complete the calculation.

# Exercise 7A ✖

**1** Work these out.

   **a** $2.4 \times 0.2$     **b** $7.3 \times 0.4$     **c** $5.6 \times 0.2$     **d** $0.3 \times 0.4$

   **e** $0.14 \times 0.2$    **f** $0.3 \times 0.3$     **g** $0.24 \times 0.8$    **h** $5.82 \times 0.5$

**2** Evaluate each of these.

   **a** $2.4 \div 0.3$     **b** $6.3 \div 0.7$     **c** $3.6 \div 0.9$     **d** $3.48 \div 0.4$

   **e** $6.24 \div 0.8$    **f** $4.85 \div 0.5$    **g** $8.47 \div 0.7$    **h** $5.82 \div 0.6$

**(PS) 3** Jerome is making a jacket. He has £30 to spend and needs 3.4 m of cloth. The cloth costs £8.75 per metre.

Does Jerome have enough money for the cloth?

**(MR) 4** In cricket, a player's strike rate is the number of runs scored divided by the number of balls faced.

Copy and complete this table.

| Player | Runs scored | Balls faced | Strike rate |
|---|---|---|---|
| Ravi | 60 | 80 | 0.75 |
| Alistair | 45 | 50 | |
| Shane | 42 | | 0.6 |
| Charlotte | | 120 | 0.55 |

**(PS) 5** A carpenter needs some wood cut into 0.7-metre lengths.

How many 0.7-metre lengths can she cut from nine 2.4-metre lengths?

**6** For each calculation:

   **i** estimate the answer by first rounding each number to the nearest whole number

   **ii** calculate the exact answer and the difference between this and your estimate.

   **a** $4.8 \times 0.7$    **b** $2.4 \times 0.6$     **c** $15.3 \times 0.9$    **d** $20.1 \times 0.8$

**(MR) 7** **a** Use any method to work out $26 \times 22$.

   **b** Use your answer to part **a** to work out:

     **i** $2.6 \times 2.2$         **ii** $1.3 \times 1.1$         **iii** $2.6 \times 8.8$.

**(MR) 8** You are given that $6.96 \div 0.24 = 29$.

Use this information to work out the answers to these divisions.

   **a** $69.6 \div 2.4$    **b** $6.96 \div 0.29$    **c** $6.96 \div 58$    **d** $2.32 \div 29$

**9** The error interval of $a$ is $9.5 \leqslant a < 10.5$.

The error interval of $b$ is $3.5 \leqslant b < 4.5$.

   **a** Work out the error interval of $ab$.

**(CM)**    **b** Give a reason why the highest value of the error interval of $\frac{a}{b}$ is 3.

**(MR)**    **c** Work out the lowest value of the error interval of $\frac{a}{b}$.

# 7.2 Fractions and reciprocals

## This section will show you how to:

- recognise different types of fraction, reciprocal, terminating decimal and recurring decimal
- convert terminating decimals to fractions
- convert fractions to decimals
- find reciprocals of numbers or fractions.

### Fractions

When a fraction is converted to a decimal it will be either a **terminating decimal** or a **recurring decimal**.

A terminating decimal has a finite number of digits. For example, $\frac{1}{4} = 0.25$, $\frac{1}{8} = 0.125$.

In a recurring decimal, there is a digit, or block of digits, that repeat. For example, $\frac{1}{3} = 0.3333...$,

$\frac{2}{11} = 0.181\,818...$

> **Hints and tips**    A number that can be written as a fraction, for example, $\frac{1}{4}$ or $\frac{10}{3}$ is called a **rational number**.

You show recurring digits by putting a dot over the first and last digits of the group that repeats.

0.3333... becomes $0.\dot{3}$         0.181 818... becomes $0.\dot{1}\dot{8}$

0.123 123 123... becomes $0.\dot{1}2\dot{3}$      0.583 33 becomes $0.58\dot{3}$

0.618 181 8... becomes $0.6\dot{1}\dot{8}$       0.412 312 312 3... becomes $0.4\dot{1}2\dot{3}$

### Converting fractions into decimals

A fraction will convert to either a terminating decimal or a recurring decimal, depending on the denominator.

You may already know that $\frac{1}{3} = 0.333... = 0.\dot{3}$. This means that the 3s go on forever and the decimal never ends.

You can use a calculator to convert a fraction to a decimal, by dividing the numerator by the denominator. Be careful, because some calculators round the last digit so the number in the display may not appear as a true recurring decimal.

Use a calculator to check these recurring decimals.

$\frac{2}{11} = 0.181818... = 0.\dot{1}\dot{8}$     $\frac{4}{15} = 0.2666... = 0.2\dot{6}$     $\frac{8}{13} = 0.615\,384\,615\,384\,... = 0.\dot{6}15\,38\dot{4}$

## Converting terminating decimals into fractions

To convert a terminating decimal to a fraction, write it as a fraction with the digits after the decimal point as the numerator and a power of 10 as the denominator. The denominator is 10, 100, 1000, … depending on the number of decimal places in the number. Because a terminating decimal has a specific number of decimal places, you can use place value to work out exactly how many zeros there will be in the denominator. For example:

$$0.7 = \frac{7}{10} \qquad 0.23 = \frac{23}{100} \qquad 0.045 = \frac{45}{1000}$$

$$= \frac{9}{200}$$

$$2.34 = \frac{234}{100} \qquad\qquad\qquad 0.625 = \frac{625}{1000}$$

$$= \frac{117}{50} \qquad\qquad\qquad\qquad = \frac{5}{8}$$

$$= 2\frac{17}{50}$$

## Finding reciprocals of numbers or fractions

The **reciprocal** of any number is 1 divided by the number.

For example:

- the reciprocal of 2 is $1 \div 2$, which is $\frac{1}{2} = 0.5$
- the reciprocal of 0.25 is $1 \div 0.25 = 4$.

You can find the reciprocal of a fraction by **inverting** it. To do this, you swap the numerator and the denominator.

For example:

- the reciprocal of $\frac{2}{3}$ is $\frac{3}{2}$
- the reciprocal of $\frac{7}{4}$ is $\frac{4}{7}$.

## Exercise 7B

**1** Write each of these fractions as a terminating decimal or recurring decimal, as appropriate.

   **a** $\frac{1}{2}$        **b** $\frac{1}{3}$        **c** $\frac{1}{4}$        **d** $\frac{1}{5}$        **e** $\frac{1}{6}$

   **f** $\frac{1}{7}$        **g** $\frac{1}{8}$        **h** $\frac{1}{9}$        **i** $\frac{1}{10}$        **j** $\frac{1}{13}$

**2** There are several patterns to be found in recurring decimals. For example:

$$\frac{1}{7} = 0.142857142857142857142857\ldots$$

$$\frac{2}{7} = 0.285714285714285714285714\ldots$$

$$\frac{3}{7} = 0.428571428571428571428571\ldots$$

and so on.

   **a** Write down the decimals for these fractions, to 24 decimal places.

      **i** $\frac{4}{7}$        **ii** $\frac{5}{7}$        **iii** $\frac{6}{7}$

   **b** What do you notice?

**3** **a** Work these out.

  **i** $1 \div 2.5$    **ii** $1 \div 5$    **iii** $1 \div 10$

  **b** In the sequence 2.5, 5, 10, ... you double each term to find the next term.

  Use your answers to part **a** to explain how you can find the answers to $1 \div 20$ and $1 \div 40$ without using a calculator.

**4** Work out the ninths, $\frac{1}{9}, \frac{2}{9}, \frac{3}{9}, \dots \frac{8}{9}$, as recurring decimals.

Describe any patterns that you notice.

**5** Work out the elevenths, $\frac{1}{11}, \frac{2}{11}, \frac{3}{11}, \dots \frac{10}{11}$, as recurring decimals.

Describe any patterns that you notice.

**6** Write each fraction as a decimal. Use your results to write the list in order of size, smallest first.

  $\frac{4}{9}$    $\frac{5}{11}$    $\frac{3}{7}$    $\frac{9}{22}$    $\frac{16}{37}$    $\frac{6}{13}$

**7** Write this list of fractions in order of size, smallest first.

  $\frac{19}{60}$    $\frac{7}{24}$    $\frac{3}{10}$    $\frac{2}{5}$    $\frac{5}{12}$

**8** Convert these terminating decimals to fractions.

  **a** 0.125    **b** 0.34    **c** 0.725    **d** 0.3125

  **e** 0.89    **f** 0.05    **g** 2.35    **h** 0.21875

**9** Use a calculator to work out the reciprocal of each number.

  **a** 12    **b** 16    **c** 20    **d** 25    **e** 50

**10** Write down the reciprocal of each fraction.

  **a** $\frac{3}{4}$    **b** $\frac{5}{6}$    **c** $\frac{2}{5}$

  **d** $\frac{7}{10}$    **e** $\frac{11}{20}$    **f** $\frac{4}{15}$

**11** **a** Write the fractions and their reciprocals from question **10** as terminating decimals or recurring decimals, as appropriate.

  **b** Is it always true that the reciprocal of a terminating decimal is a recurring decimal?

**12** Multiply each fraction in question **10** by its reciprocal.

Comment on your results.

**13** Explain why zero has no reciprocal.

**14** **a** Work out the reciprocal of the reciprocal of 10.

  **b** Work out the reciprocal of the reciprocal of 2.

  **c** What do you notice?

**15** $x$ and $y$ are two positive numbers and $x$ is less than $y$.

Which of these statements is true?

  **a** The reciprocal of $x$ is less than the reciprocal of $y$.

  **b** The reciprocal of $x$ is greater than the reciprocal of $y$.

  **c** It is impossible to tell.

Give an example to support your answer.

# 7.3 Fractions of quantities

This section will show you how to:

- work out a fraction of a quantity
- find one quantity as a fraction of another.

Sometimes you will need to give one amount as a fraction of another amount.

**Example 3**

Write £5 as a fraction of £20.

£5 as a fraction of £20 is written as $\frac{5}{20}$.

Note that $\frac{5}{20} = \frac{1 \times 5}{4 \times 5}$ so you can cancel by a factor of 5 to $\frac{1}{4}$.

So £5 is one-quarter of £20.

Sometimes you will need to work out a fraction of a quantity or amount. To do this, first divide the quantity by the **denominator** of the fraction you are working out. Then multiply the result by the **numerator** of the fraction.

Read through the next example carefully.

**Example 4**

A book has 320 pages. 200 of the pages have illustrations. $\frac{3}{4}$ of these pages are in colour.

a What fraction of the pages of the whole book have illustrations?

b What fraction of the pages of the whole book have colour illustrations?

a 200 pages have illustrations.

This is $\frac{200}{320} = \frac{10 \times 4 \times 5}{10 \times 4 \times 8}$

$= \frac{5}{8}$ of the book.

b $\frac{3}{4}$ of the 200 pages are in colour.

$200 \div 4 \times 3 = 150$

So 150 pages have colour illustrations.

This is $\frac{150}{320}$ of the book. This cancels to $\frac{15}{32}$.

Sometimes you will have to combine the two ideas.

**Example 5**

Shop A sells a bicycle for £540 including VAT but has an offer of $\frac{1}{4}$ off the selling price.

Shop B sells the same model of bicycle for £350 (excluding VAT). VAT will add $\frac{1}{5}$ to the price.

In which shop is the bike cheaper? Show your working.

In Shop A, the offer is $\frac{1}{4}$ off £540.     $\frac{1}{4}$ of £540 is $\frac{1}{4} \times £540 = £135$

So you will pay £540 − £135 = £405

In Shop B, the VAT is $\frac{1}{5}$ of £350.

$\frac{1}{5}$ of £350 is $\frac{1}{5} \times £350 = £70$

So you will pay £350 + £70 = £420

The bike is cheaper in Shop A.

**1**   Write the first quantity as a fraction of the second.

     **a**   2 cm, 6 cm      **b**   4 kg, 20 kg      **c**   £8, £20          **d**   5 hours, 24 hours

     **e**   12 days, 30 days   **f**   50p, £3          **g**   4 days, 2 weeks   **h**   40 minutes, 2 hours

**2**   In a class of 30 students, 18 are boys. What fraction of the class are boys?

**3**   During April, it rained on 12 days. For what fraction of the month did it rain?

**4**   Jake earns £150. He puts £50 into his bank account and spends the rest. What fraction of his earnings does he spend?

**5**   In a class of 30 students, $\frac{3}{5}$ are boys. $\frac{1}{3}$ of the boys are left-handed. What fraction of the whole class is made up of left-handed boys?

**6**   Reka wins £120 in a competition and puts £40 in a bank account. She gives $\frac{1}{4}$ of what is left to her sister and then spends the rest. What fraction of her winnings did she spend?

**CM**   **7**   Jon earns £90 and saves £30 of it. Matt earns £100 and saves £35 of it.

     Who is saving the greater proportion of his earnings?

**CM**   **8**   In two tests Harry gets 13 out of 20 and 16 out of 25. Which is the better mark?

     Explain your answer.

**PS**   **9**   In a street of 72 dwellings, $\frac{5}{12}$ are bungalows. The rest are two-storey houses. Half of the bungalows are detached and $\frac{2}{7}$ of the houses are detached. What fraction of the 72 dwellings are detached?

**PS**   **10**   I have 24 T-shirts. $\frac{1}{6}$ have logos on them; the rest are plain. $\frac{2}{5}$ of the plain T-shirts are long-sleeved. $\frac{3}{4}$ of the T-shirts with logos are long sleeved. What fraction of all my T-shirts are long-sleeved?

**EV**   **11**   Which is the larger fraction: 26 out of 63 or 13 out of 32? Explain how you can tell without using a calculator.

**MR**   **12**   Shop A sells a games console for £360 without VAT. Shop B sells the same games console for £480 including VAT.

     VAT adds $\frac{1}{5}$ to the price.

     Shop B has a sale and takes $\frac{1}{10}$ off the price.

     In which shop is the game console cheaper?

# 7.4 Adding and subtracting fractions

## This section will show you how to:

- add and subtract fractions with different denominators.

| Key terms |
| --- |
| equivalent fraction |
| lowest common denominator |
| mixed number |

You can only add or subtract fractions that have the same denominator. If necessary, change one or both of them to **equivalent fractions** with the same denominator. Then you can add or subtract the fractions by adding or subtracting numerators.

Always look for the **lowest common denominator** of the fractions you are changing. This is the lowest common multiple (LCM) of both denominators.

**Example 6**

Work out $\frac{5}{6} - \frac{3}{4}$.

Find the lowest common denominator (LCM) of 4 and 6. This is 12.

Convert to equivalent fractions.
$$\frac{5}{6} - \frac{3}{4} = \frac{5}{6} \times \frac{2}{2} - \frac{3}{4} \times \frac{3}{3}$$
$$= \frac{10}{12} - \frac{9}{12}$$

Subtract 9 from 10.
$$= \frac{1}{12}$$

When you add and subtract **mixed numbers**, you can deal with the whole numbers and the fractions separately.

**Example 7**

Work these out.

**a** $2\frac{1}{3} + 3\frac{5}{7}$　　　　**b** $3\frac{1}{4} - 1\frac{3}{5}$

**a** $2\frac{1}{3} + 3\frac{5}{7} = 2 + 3 + \frac{1}{3} + \frac{5}{7}$
$$= 5 + \frac{7}{21} + \frac{15}{21}$$
$$= 5 + \frac{22}{21}$$
$$= 5 + 1\frac{1}{21}$$
$$= 6\frac{1}{21}$$

**b** $3\frac{1}{4} - 1\frac{3}{5} = 3 - 1 + \frac{1}{4} - \frac{3}{5}$
$$= 2 + \frac{5}{20} - \frac{12}{20}$$
$$= 2 - \frac{7}{20}$$
$$= 1\frac{13}{20}$$

## Exercise 7D

**1** Work these out.

**a** $\frac{1}{3} + \frac{1}{5}$　　　　**b** $\frac{1}{3} + \frac{1}{4}$　　　　**c** $\frac{2}{3} + \frac{1}{4}$

**d** $\frac{1}{5} - \frac{1}{10}$　　　　**e** $\frac{7}{8} - \frac{3}{4}$　　　　**f** $\frac{5}{6} - \frac{3}{4}$

 **2** Which is biggest: half of 96, one-third of 141, two-fifths of 120 or three-quarters of 68?

**3** Work these out.

**a** $3\frac{1}{3} + 1\frac{9}{20}$　　**b** $1\frac{1}{8} - \frac{5}{9}$　　**c** $\frac{7}{10} + \frac{3}{8} + \frac{5}{6}$　　**d** $1\frac{1}{3} + \frac{7}{10} - \frac{4}{15}$

**4** **a** In a class election, half of the students voted for Aminah, one-third voted for Jenet and the rest voted for Pieter. What fraction of the class voted for Pieter?

 **b** One of the numbers in the box is the number of students in the class in part **a**.

| 25 | 28 | 30 | 32 |

How many students are there in the class?

 **5** A one-litre bottle of milk is used to fill four glasses. Three of the glasses have a capacity of one-eighth of a litre. The fourth glass has a capacity of half a litre.

Priya likes milky coffee so she always has at least 10 cl of milk in her cup. Is there enough milk left in the bottle for Priya to have two cups of coffee?

**CM** **6** Mick has worked out this addition.

$1\frac{1}{3} + 2\frac{1}{4} = 3\frac{2}{7}$

His answer is incorrect. What mistake has he has made? Work out the correct answer.

**CM** **7** Write down what you would say to someone, in a telephone conversation, to explain how to find the answer to this calculation.

$\frac{1}{4} + \frac{2}{5}$

**PS** **8** There are 900 students in a school. $\frac{11}{20}$ of the students are boys. Of the boys, $\frac{2}{11}$ are left-handed. Of the girls, $\frac{2}{9}$ are left-handed. What fraction of all the students are left-handed? Show your working.

**9** There are 600 counters in a bag. They are red, blue or yellow. $\frac{3}{8}$ of the counters are red. $\frac{1}{5}$ of the counters are blue.

   **a** What fraction of the counters are yellow?

   **b** How many yellow counters are there in the bag?

**10** A small gym has 200 members. $\frac{27}{40}$ of the members are at least 40 years of age. $\frac{2}{5}$ of the members are female.

**CM**    **a** Use calculations to show that some of the female members are at least 40 years of age.

**MR**    **b** $\frac{5}{8}$ of the female members are at least 40 years old. How many of the men are aged less than 40?

# 7.5 Multiplying and dividing fractions

This section will show you how to:

- multiply proper fractions
- multiply mixed numbers
- divide by fractions.

| Key terms |
| --- |
| improper fraction |
| proper fraction |

## Multiplying fractions

Before you can multiply fractions, you must rewrite any mixed numbers as fractions with just a numerator and a denominator. In a **proper fraction**, the numerator is less than the denominator. In an **improper fraction**, the numerator is bigger than the denominator.

**Note:** The correct name for a fraction in which the numerator is bigger than the denominator is 'improper fraction'. These are sometimes called top-heavy fractions.

Follow these steps to multiply fractions.

**Step 1**   Convert any mixed numbers into improper fractions.

**Step 2**   Write down the multiplication and simplify the fractions by cancelling by any common factors in the numerators and the denominators, if possible.

**Step 3**   Multiply the numerators to obtain the numerator of the answer and multiply the denominators to obtain the denominator of the answer.

**Step 4**   If the answer is an improper fraction, convert it into a mixed number.

Example 8

Work these out.

**a** $\frac{4}{9} \times \frac{3}{10}$      **b** $2\frac{2}{5} \times 1\frac{7}{8}$

**a**   Simplify the fractions.

     2 is a factor of 4 and 10. 3 is a factor of 3 and 9.

     So cancel by 2 and 3.
$$\frac{{}^{2}\cancel{4}}{{}_{3}\cancel{9}} \times \frac{\cancel{3}^{1}}{\cancel{10}_{5}}$$

     Then multiply. $\qquad\qquad\qquad\qquad\qquad = \frac{2}{15}$

**b**   Convert the mixed numbers into improper fractions.
$$2\frac{2}{5} \times 1\frac{7}{8} = \frac{12}{5} \times \frac{15}{8}$$

     Simplify the fractions by cancelling by 4 and 5.
$$\frac{{}^{3}\cancel{12}}{{}_{1}\cancel{5}} \times \frac{\cancel{15}^{3}}{\cancel{8}_{2}}$$

     Then multiply. $\qquad\qquad\qquad\qquad\qquad = \frac{9}{2}$

$$= 4\frac{1}{2}$$

## Dividing fractions

Dividing by a fraction is equivalent to multiplying by the reciprocal of the fraction.

For example:    dividing by $\frac{2}{3}$ is the same as multiplying by $\frac{3}{2}$

              dividing by $\frac{1}{4}$ is the same as multiplying by 4.

Follow these steps to divide fractions.

**Step 1**   Convert any mixed numbers into improper fractions.

**Step 2**   Convert the division calculation into a multiplication calculation.

**Step 3**   Carry out the multiplication.

Example 9

Work these out.

**a** $\frac{5}{6} \div \frac{3}{4}$      **b** $2\frac{1}{2} \div 3\frac{1}{3}$

**a**   Convert the division calculation into a multiplication calculation. $\frac{5}{6} \div \frac{3}{4} = \frac{5}{6} \times \frac{4}{3}$

     Cancel any common factors. $\qquad\qquad\qquad\qquad\qquad = \frac{5}{{}_{3}\cancel{6}} \times \frac{\cancel{4}^{2}}{3}$

     Then multiply. $\qquad\qquad\qquad\qquad\qquad\qquad\qquad = \frac{10}{9}$

     Convert to a mixed number. $\qquad\qquad\qquad\qquad\quad = 1\frac{1}{9}$

**b**   Convert the mixed numbers into improper fractions. $2\frac{1}{2} \div 3\frac{1}{3} = \frac{5}{2} \div \frac{10}{3}$

     Convert the division calculation into a multiplication calculation.

$$= \frac{5}{2} \times \frac{3}{10}$$

     Cancel any common factors. $\qquad\qquad\qquad\qquad\qquad = \frac{{}^{1}\cancel{5}}{2} \times \frac{3}{\cancel{10}_{2}}$

     Then multiply. $\qquad\qquad\qquad\qquad\qquad\qquad\qquad = \frac{3}{4}$

# Exercise 7E

**1** Work these out. Give each answer as a mixed number or fraction in its simplest form.

**a** $\frac{1}{2} \times \frac{1}{3}$    **b** $\frac{3}{4} \times \frac{1}{2}$    **c** $\frac{14}{15} \times \frac{3}{8}$    **d** $\frac{6}{7} \times \frac{21}{30}$

**e** $1\frac{1}{4} \times \frac{1}{3}$    **f** $1\frac{3}{4} \times 1\frac{2}{3}$    **g** $3\frac{1}{4} \times 1\frac{1}{5}$    **h** $1\frac{1}{4} \times 2\frac{2}{3}$

**2** Work these out. Give each answer as a mixed number or fraction in its simplest form.

**a** $\frac{1}{4} \div \frac{1}{3}$    **b** $\frac{4}{5} \div \frac{3}{4}$    **c** $7\frac{1}{2} \div 1\frac{1}{2}$

**d** $1\frac{5}{12} \div 3\frac{3}{16}$    **e** $3\frac{3}{5} \div 2\frac{1}{4}$

 **3** Bilal eats one-quarter of a cake, and then half of what is left. How much cake is left uneaten?

 **4** You are given that 1 tonne = 1000 kilograms.

A dustbin lorry carries 12 tonnes of rubbish. Three-quarters of this is recycled.

Half of the remainder is sent for landfill and the rest is sent to an incinerator.

What fraction of the rubbish goes to landfill?

**5** Zahar made $12\frac{1}{2}$ litres of lemonade for a party. His glasses could each hold $\frac{5}{16}$ of a litre. How many of the glasses could he fill from the $12\frac{1}{2}$ litres of lemonade?

 **6** Which is larger: $\frac{3}{4}$ of $2\frac{1}{2}$ or $\frac{2}{5}$ of $6\frac{1}{2}$?

**7** If £5.20 is two-thirds of three-quarters of a sum of money, what is the total amount of money?

# 7.6 Fractions on a calculator

## This section will show you how to:

- use a calculator to add and subtract fractions
- use a calculator to multiply and divide fractions.

> **Key term**
>
> shift key

In this section, you will look at questions that require calculation with fractions within the context of another topic, such as algebra or geometry.

Remember that:

- a fraction with the numerator bigger than the denominator is an improper fraction
- a mixed number is made up of a whole number and a proper fraction.

For example:

- $\frac{14}{5} = 2\frac{4}{5}$

- $3\frac{2}{7} = \frac{23}{7}$.

When you use a calculator for work on fractions, the method of working is different from when you work 'on paper'. For example, you will not need to change the denominators to add or subtract, and you may not need to change mixed numbers to improper fractions to multiply or divide. Your calculator should do this for you.

> **Hints and tips** Not all models of calculator work the same way, so make sure you know how yours works. The keystrokes in this section are based on a standard calculator.

## Using a calculator to convert improper fractions to mixed numbers

Find the fraction key on your calculator. Remember, for some functions, you may need to use the **shift key**.

To key in a fraction, press [▢/▢].

Input the fraction so that it looks like this: $\frac{9}{5}$ or [ 9⌐5 ]

Now press the equals key [=] so that the fraction displays in the answer part of the screen.

Pressing shift and the key [S⇔D] will convert the fraction to a mixed number:

[ 1⌐4⌐5 ] [S⇔D]

This is the mixed number $1\frac{4}{5}$.

Pressing the equals key [=] again will convert the mixed number back to an improper fraction.

- Try to think of a way of converting an improper fraction to a mixed number without using a calculator.
- Test your idea, and then use your calculator to check it.

## Using a calculator to convert mixed numbers to improper fractions

To input a mixed number, press the shift key **shift** first and then the fraction key [▢/▢].

Pressing the equals key [=] will convert the mixed number to an improper fraction.

- Key in at least 10 improper fractions and convert them to mixed numbers.
- Remember to press the equals key [=] to change the mixed numbers back to improper fractions.
- Input at least 10 mixed numbers and convert them to improper fractions.
- Look at your results. Try to think of a way of converting a mixed number to an improper fraction without using a calculator.
- Test your idea, and then use your calculator to check it.

## Using a calculator to add and subtract fractions

> **Example 10**
>
> A water tank is half-full. One-third of the full capacity of the tank is poured out.
>
> What fraction of the tank is now full of water?
>
> The calculation is $\frac{1}{2} - \frac{1}{3}$
>
> Keying in the calculation gives:
>
> [▢/▢] [1] [▼] [2] [▶] [−] [▢/▢] [1] [▼] [3] [▶] [=]
>
> The display should show $\frac{1}{6}$.
>
> The tank is now one-sixth full of water.

**Example 11**

Work out the perimeter of a rectangle $1\frac{1}{2}$ cm long and $3\frac{2}{3}$ cm wide.

The formula for the perimeter of a rectangle is:

$$P = 2l + 2w$$

where $l = 1\frac{1}{2}$ cm and $w = 3\frac{2}{3}$ cm.

$$P = 2 \times 1\frac{1}{2} + 2 \times 3\frac{2}{3}$$

Keying in the calculation gives:

**2** **×** **shift** **▭** **1** **▶** **1** **▼** **2** **▶**
**+** **2** **×** **shift** **▭** **3** **▶** **2** **▼** **3** **▶** **=**

The display should show $10\frac{1}{3}$.

So the perimeter is $10\frac{1}{3}$ cm.

## Using a calculator to multiply and divide fractions

**Example 12**

Work out the area of a rectangle of length $3\frac{1}{2}$ m and width $2\frac{2}{3}$ m.

The formula for the area of a rectangle is:

area = length × width

Keying in the calculation, where length = $3\frac{1}{2}$ and width = $2\frac{2}{3}$ gives:

**shift** **▭** **3** **▶** **1** **▼** **2** **▶** **×**
**shift** **▭** **2** **▶** **2** **▼** **3** **▶** **=**

The display should show $9\frac{1}{3}$.

The area is $9\frac{1}{3}$ cm².

**Example 13**

Work out the average speed of a bus that travels $20\frac{1}{4}$ miles in $\frac{3}{4}$ hour.

The formula for the average speed is average speed = $\dfrac{\text{total distance travelled}}{\text{total time taken}}$

The total distance is $20\frac{1}{4}$ miles and the total time is $\frac{3}{4}$ hour.

Keying in the calculation gives:

**shift** **▭** **2** **0** **▶** **1** **▼** **4** **▶**
**÷** **▭** **3** **▼** **4** **▶** **=**

The display should show 27.

The average speed is 27 mph.

You will learn more about distance, speed and time in a later chapter.

# Exercise 7F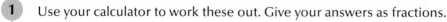

In this exercise, try to key in each calculation as one continuous set of operations, without writing down any intermediate values.

**1** Use your calculator to work these out. Give your answers as fractions.

   **a** $\frac{3}{4} + \frac{4}{5}$              **b** $\frac{4}{5} + \frac{9}{20}$             **c** $\frac{5}{8} + \frac{9}{16} + \frac{3}{5}$

   **d** $\frac{9}{20} - \frac{1}{12}$          **e** $\frac{7}{16} + \frac{3}{8} - \frac{1}{20}$       **f** $\frac{4}{5} + \frac{9}{16} - \frac{2}{3}$

**2** **a** What is the distance between Wickersley and Redbrook, using these roads?

   **b** How much further is it to Redbrook than to Wickersley?

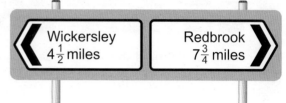

**3** Use your calculator to work these out. Give your answers as mixed numbers.

   **a** $4\frac{3}{4} + 1\frac{4}{5}$        **b** $3\frac{5}{6} + 4\frac{7}{10}$       **c** $2\frac{5}{8} + 3\frac{9}{16} + 5\frac{3}{5}$

   **d** $6\frac{9}{20} - 3\frac{1}{12}$      **e** $9\frac{7}{16} + 5\frac{3}{8} - 7\frac{1}{20}$    **f** $10\frac{3}{4} + 6\frac{2}{9} - 12\frac{3}{11}$

**(CM) 4** **a** Use your calculator to work out $\frac{18}{37} - \frac{23}{43}$.

   **b** Explain how your answer tells you that $\frac{23}{43}$ is greater than $\frac{18}{37}$.

**(PS) 5** A shape is rotated 90° clockwise and then a further 60° clockwise.

   What fraction of a turn will return it to its original position?

   Give both possible answers.

**6** Use your calculator to work these out. Give your answers as fractions in their simplest terms.

   **a** $\frac{3}{4} \times \frac{4}{5}$           **b** $\frac{5}{8} \times \frac{9}{16} \times \frac{3}{5}$       **c** $\frac{9}{20} \div \frac{1}{12}$

   **d** $\frac{3}{4} \div \frac{7}{48}$         **e** $\frac{7}{16} \times \frac{3}{8} \div \frac{1}{20}$     **f** $\frac{3}{4} \times \frac{2}{9} \div \frac{3}{11}$

**7** Use your calculator to work these out. Give your answers as mixed numbers.

   **a** $4\frac{3}{4} \times 1\frac{4}{5}$        **b** $7\frac{4}{5} \times 8\frac{9}{20}$       **c** $2\frac{5}{8} \times 3\frac{9}{16} \times 5\frac{3}{5}$

   **d** $6\frac{9}{20} \div 3\frac{1}{12}$      **e** $4\frac{3}{4} \div 2\frac{7}{48}$      **f** $9\frac{7}{16} \times 5\frac{3}{8} \div 7\frac{1}{20}$

**(EV) 8** The ribbon on a roll is $3\frac{1}{2}$ m long. Joe wants to cut pieces of ribbon that are each $\frac{1}{6}$ m long.

   He needs 50 pieces.

   How many rolls will he need?

**(PS) 9** An approximation for the number π is $\frac{22}{7}$. The formulae for the area, $A$, and circumference, $C$, of a circle with radius $r$ are $A = \pi r^2$ and $C = 2\pi r$.

   Work out the area of a circle with a circumference of 22 cm.

# Worked exemplars

 **1** The perimeter of a rectangle is $32\frac{1}{2}$ cm.

Use your calculator to work out one pair of possible values for the length and the width of the rectangle.

| This is a problem-solving question, so you need to make connections between different part of mathematics (in this case perimeters, substitution and fractions) and show your strategy clearly. | |
|---|---|
| Perimeter = 2 × length + 2 × width | Write down the correct formula for the perimeter of a rectangle. |
| 2 × length + 2 × width = $32\frac{1}{2}$<br>length + width = $32\frac{1}{2} \div 2$ | Put the value that you know into the formula. |
| length + width = $16\frac{1}{4}$ cm | Work out the total value of the length and width. |
| Possible length and width are:<br>Length = 10 cm<br>Width = $6\frac{1}{4}$ cm | Any two values with a sum of $16\frac{1}{4}$ would answer the question. Choosing 10 makes the calculation straightforward. |

 **2** Bob needs to travel 187 miles to a business meeting.

He sets off at 10:00 am and wants to arrive before the meeting starts at 3:00 pm.

He stops for one 20-minute break and his average speed when driving is $42\frac{1}{2}$ mph.

Will he make it on time?

| This is a mathematical reasoning question. There is a lot of information given, so make sure you show a clear chain of reasoning to get to your result. | |
|---|---|
| Time travelling = $187 \div 42\frac{1}{2} = 4\frac{2}{5}$ hours | Work out the time it takes to drive the distance. |
| $4\frac{2}{5}$ hours = 4 hours 24 minutes | Change the fraction into hours and minutes. $\frac{1}{5}$ of an hour is 12 minutes |
| Total time = 4 hours 24 minutes + 20 minutes<br>= 4 hours 44 minutes | Add on the time taken for a break. |
| 10:00 am + 4 hours 44 minutes = 2:44 pm, so he will arrive on time. | Add the total time to 10:00 am and write a conclusion. |

# Ready to progress?

I can convert fractions to decimals and terminating decimals to fractions.
I can write one quantity as a fraction of another.

I can add, subtract, multiply and divide decimals.
I can add, subtract, multiply and divide fractions.
I can calculate with mixed numbers.
I understand what reciprocals are and can work them out for numbers and fractions.

# Review questions

**1** Write down a decimal that is between $\frac{3}{10}$ and $\frac{1}{3}$.

**2** Plastic cups hold $\frac{2}{5}$ of a litre. How many plastic cups can be filled from seven 1-litre bottles of water?

**3** **a** Write $\frac{2}{5}$ as a decimal.      **b** Write 0.45 as a fraction.

**4** In a primary school, there are 144 boys and 156 girls. $\frac{1}{4}$ of the boys have blue eyes. $\frac{1}{3}$ of the girls have blue eyes. What fraction of the whole school has blue eyes?

**5** Write these fractions as recurring decimals.

**a** $\frac{5}{9}$      **b** $\frac{1}{3}$

**6** Zac got 28 out of 40 marks in an English test.

He got 33 out of 50 marks in a mathematics test.

Which of these scores is better?

**7** Write down the reciprocal of the reciprocal of $\frac{4}{5}$.

**8** In a month I drive 900 miles, on average. My car does 45 miles to one gallon of petrol. Petrol costs £1.40 a litre. There are 4.55 litres in a gallon. How much do I spend on petrol each year?

**9** Write down the reciprocal of each number.

**a** $\frac{1}{5}$      **b** $\frac{2}{3}$      **c** $\frac{10}{7}$      **d** 8

**10** Two shops sell the same type of cola.

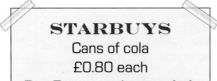

**STARBUYS**
Cans of cola
£0.80 each
**Buy 5 cans** and **get a sixth** one **FREE**

**Shoprite**
Cola sale
20% OFF
normal price
of £0.80 each

Lewis wanted to buy 12 cans. Which of these shops offers better value?

Show all your working.

**MR** **11**  **a** Use any method to work out $38 \times 44$.

     **b** Use your answer to part **a** to work out each of these.

     **i** $3.8 \times 44$        **ii** $38 \times 2.2$        **iii** $1.9 \times 8.8$

**MR** **12**  You are given that $57.6 \div 32 = 1.8$.

     Use this information to work out the answers to these divisions.

     **a** $57.6 \div 18$        **b** $57.6 \div 0.16$        **c** $57.6 \div 180$

**13**  **a** Work these out.

     **a** $\frac{1}{5} + \frac{3}{20}$        **b** $\frac{3}{4} \times \frac{8}{15}$        **c** $\frac{2}{3} \div \frac{4}{9}$

**14**  Work out $25.6 \times 1.6$.

**15**  Work out $3\frac{2}{3} - 1\frac{5}{8}$.

**16**  The formula for the perimeter, $P$, of a rectangle is:

     $P = 2l + 2w$

     where $l$ is the length and $w$ is the width.

     Work out the perimeter when $l = 5\frac{1}{8}$ cm and $w = 4\frac{1}{3}$ cm.

**17**  The formula for the area of a rectangle is area = length × width.

     Use this formula to work out the area of a rectangle of length $5\frac{2}{3}$ metres and width $3\frac{1}{4}$ metres.

**MR** **18**  This is a series of fractions where the numerators are the counting numbers starting at 1, that is, 1, 2, … 9, and the denominators are the counting numbers starting at 2, that is, 2, 3, … 10.

     $\frac{1}{2}, \frac{2}{3}, \frac{3}{4}, \frac{4}{5}, \ldots\ldots, \frac{8}{9}, \frac{9}{10}$

     **a** Work out each fraction as a decimal. Decide if it is recurring (R) or terminating (T).

     **b** Write all the denominators of the terminating decimals as a product of their prime factors.

     **c** Write down a rule for the denominator that determines whether or not the fraction will terminate.

**19**  A bag contains 400 coloured shapes. The shapes are either squares or triangles and are coloured either red or blue.

     $\frac{17}{50}$ of the shapes are blue squares. $\frac{19}{40}$ of the shapes are triangles.

     The number of red triangles is the same as the number of blue triangles.

     Copy and compete this table.

|           | Blue | Red | Total |
|-----------|------|-----|-------|
| Squares   |      |     |       |
| Triangles |      |     |       |
| Total     |      |     |       |

# 8 Algebra: Linear graphs

## This chapter is going to show you:

- how to plot negative coordinates
- how to work out the gradient of a line
- how to draw a straight-line graph from its equation
- how to work out the equation of a linear graph
- how to draw linear graphs parallel to other lines
- how to read information from a conversion graph
- how to use graphs to work out formulae and solve simultaneous linear equations.

## You should already know:

- how to substitute numbers into a formula
- how to read and estimate from scales
- how to plot a graph from a table of values.

## About this chapter

'A picture is worth a thousand words' is definitely true in mathematics: graphs can replace many lines of algebra as they show the relationship between two variables in a visual way.

A linear graph (often called a straight-line graph) shows two variables that increase at a constant rate. You will meet linear graphs and the relationships they represent in various situations in daily life.

When hiring a van to move house: $C = 25T + 20$ is the equation of the graph (right) used by a van hire company to calculate their fees, where $T$ is the time in hours and $C$ is the cost in pounds.

When buying tins at the supermarket: $C = 40T$ is the equation of the graph showing the cost in pence, $C$, of $T$ tin cans, where $C$ is given in pence. This sign from a supermarket, showing 40p cans sold as 5 for £2, shows that the supermarket chain understands how straight-line graphs work!

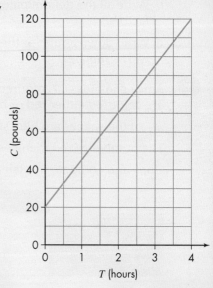

5 for £2

black eye beans in water
410 g (235 g)

40p

# 8.1 Graphs and equations

This section will show you how to:

- use flow diagrams to draw graphs
- work out the equations of horizontal and vertical lines
- work out the coordinates of the midpoint of a straight line.

## Flow diagrams

In its simplest form, a **flow diagram** consists of a single box containing a mathematical operation called a **function**. It shows the connection between **input** values and **output** values. For example, this flow diagram represents the function *multiplying by 3* for various inputs.

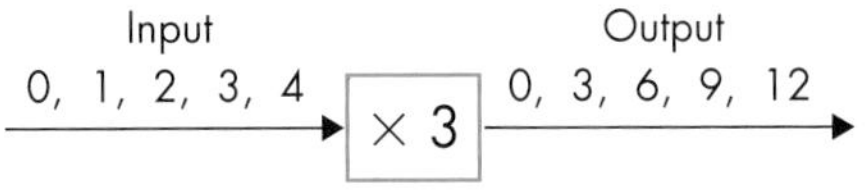

Input: 0, 1, 2, 3, 4 → $\times 3$ → Output: 0, 3, 6, 9, 12

You can use a **flow diagram** to create a set of coordinates.

Arrange the input and output values in a table.

| $x$ | 0 | 1 | 2 | 3 | 4 |
|---|---|---|---|---|---|
| $y$ | 0 | 3 | 6 | 9 | 12 |

The input values are the **x-values** and the output values are the **y-values**. These form a set of coordinates that you can plot on a graph. In this case, the coordinates are (0, 0), (1, 3), (2, 6), (3, 9) and (4, 12).

Some functions consist of more than one operation, so their flow diagrams consist of more than one box. In these cases, you need to match the input (first) values to the output (final) values. This means that 0 maps to 3, 1 maps to 5 and so on, with 4 mapping to 11. The middle values produced by the first operation(s) are just working numbers and you can miss them out.

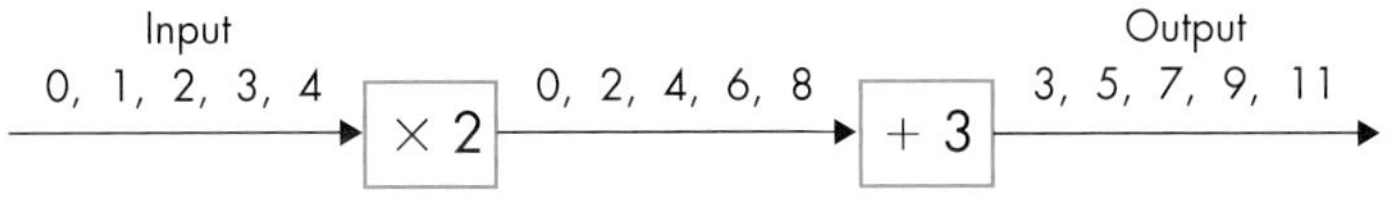

Input: 0, 1, 2, 3, 4 → $\times 2$ → 0, 2, 4, 6, 8 → $+ 3$ → Output: 3, 5, 7, 9, 11

So, the table for this two-box flow diagram looks like this.

| $x$ | 0 | 1 | 2 | 3 | 4 |
|---|---|---|---|---|---|
| $y$ | 3 | 5 | 7 | 9 | 11 |

This gives the coordinates (0, 3), (1, 5), (2, 7), (3, 9) and (4, 11).

The two flow diagrams above represent the equations $y = 3x$ and $y = 2x + 3$, as shown below.

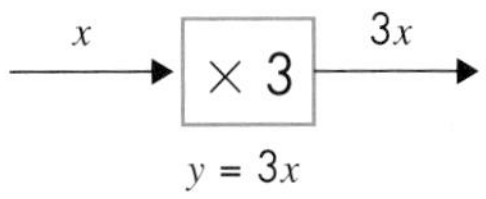

$x$ → $\times 3$ → $3x$

$y = 3x$

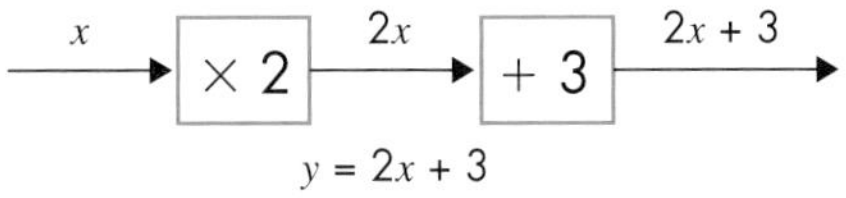

$x$ → $\times 2$ → $2x$ → $+ 3$ → $2x + 3$

$y = 2x + 3$

You can plot the coordinates for each equation on a set of axes to produce the graphs of $y = 3x$ and $y = 2x + 3$, as shown on the next page. The x-axis is horizontal and the y-axis is vertical.

**Remember:** Always label graphs.

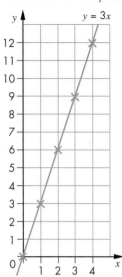

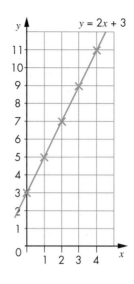

**Note:** The line drawn is a **line segment**, as it is only part of an infinitely long line. You should extend the line segment further than its endpoints. For example, you may need to know where the line crosses the *y*-axis.

In graph work you need to decide the range of values for the axes. To start with, diagrams like the one opposite will show you the range for your axes.

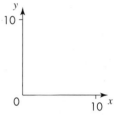

This diagram means draw the *x*-axis from 0 to 10 and the *y*-axis from 0 to 10.
**Note:** The *scale* on each axis does not always need to be the same.

## Plotting negative coordinates

Some graphs use **negative coordinates**.

The coordinates of the four points on this grid are:

A(2, 3)     B(−1, 2)     C(−3, −4)     D(1, −3)

The *x*-coordinate of all the points on line *X* is 3. So you can say the equation of line *X* is $x = 3$.

The *y*-coordinate of all the points on line *Y* is −2. So you can say the equation of line *Y* is $y = -2$.

**Note:** The equation of the *x*-axis is $y = 0$ and the equation of the *y*-axis is $x = 0$.

When you know the coordinates of two points, you can work out the coordinates of the **midpoint** of the line joining them.

Look at the points B(−1, 2) and C(−3, −4). If you lay your ruler so that it touches both points, you can see that the midpoint of the line BC will be the point (−2, −1).

You can work this out algebraically, by calculating the mean of the *x*-values and the mean of the *y*-values.

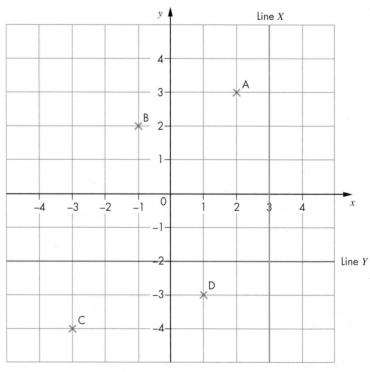

The midpoint of the line joining $(x_1, y_1)$ and $(x_2, y_2)$ is given by $(\frac{1}{2}(x_1 + x_2), \frac{1}{2}(y_1 + y_2))$.

For BC: the $x$-coo
the $y$-coo

the $x$-coordinate is $\frac{1}{2}(-1 + -3) = \frac{1}{2} \times -4$, which is $-2$

the $y$-coordinate is $(2 + -4) = \frac{1}{2} \times -2$, which is $-1$.

This gives the midpoint as $(-2, -1)$, as you have seen above.

Use the flow diagram to draw the graph of $y = 4x - 1$.

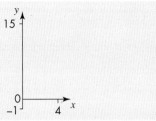

Work out the output for each input and enter the values in a table.

| $x$ | 0 | 1 | 2 | 3 | 4 |
|---|---|---|---|---|---|
| $y$ | −1 | 3 | 7 | 11 | 15 |

So, the coordinates are:

$(0, -1)$, $(1, 3)$, $(2, 7)$, $(3, 11)$, $(4, 15)$

Draw axes with the range shown above, plot these points and join them up.

Remember to label your graph.

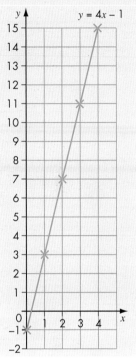

## Exercise 8A

**1**

**a** Write down the coordinates of each of the points A to J on the grid.

**b** Write down the coordinates of the midpoint of the line joining:

**i** A and B  **ii** H and I  **iii** D and J.

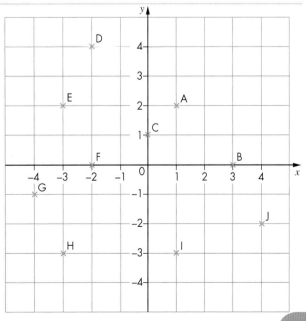

**2**  **a** Write down the equations of the lines labelled 1 to 4 on the grid.

  **b** Write down the equation of the line that is exactly halfway between:

  **i** line 1 and line 2      **ii** line 3 and line 4.

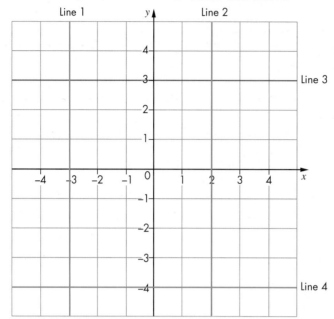

**3**  Draw the graph of $y = x + 2$.

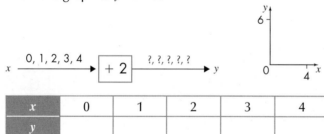

| $x$ | 0 | 1 | 2 | 3 | 4 |
|---|---|---|---|---|---|
| $y$ | | | | | |

**4**  Draw the graph of $y = 2x - 2$.

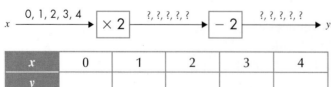
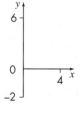

| $x$ | 0 | 1 | 2 | 3 | 4 |
|---|---|---|---|---|---|
| $y$ | | | | | |

**5**  Draw the graph of $y = \dfrac{x}{3} + 1$.

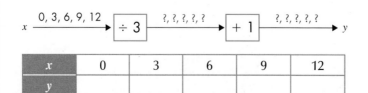

| $x$ | 0 | 3 | 6 | 9 | 12 |
|---|---|---|---|---|---|
| $y$ | | | | | |

**6** Draw the graph of $y = \dfrac{x}{2} - 4$. Use even $x$-values from 0 to 8.

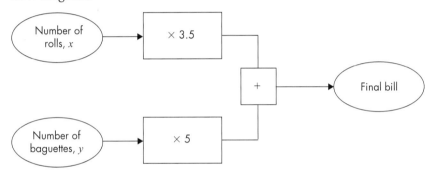

> **Hints and tips** If the $x$-value is divisible by a number, then choose multiples of that number as input values. It makes calculations and plotting points much easier.

**7** **a** Draw the graphs of $y = 2x$ and $y = x + 6$ on the same grid. Use 0 to 8 on the $x$-axis and 0 to 16 on the $y$-axis.

**b** At which point do the lines **intersect** (cross)?

**8** Draw the graph of $y = 5x - 1$. Choose your own inputs and axes.

**(MR)** **9** A tea shop sells two types of sandwiches. A roll costs £3.50 and a baguette costs £5.00.

To work out the cost of different combinations of sandwiches, they use a flow diagram.

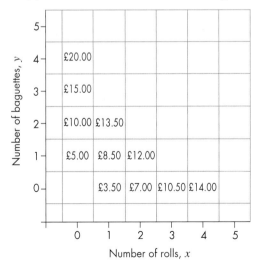

**a** Use the flow diagram to work out the cost of three rolls and two baguettes.

**b** The tea-shop owner is making a wall chart showing the costs of different orders.
Copy the chart and use the flow diagram to complete it.

**c** A group of friends paid £30.50. Can you be certain what their order was?

(MR) **10** A teacher reads out this problem:

'I am thinking of a number. I multiply it by 3 and add 1.'

**a** Represent this problem using a flow diagram.

**b** If the input is $x$ and the output is $y$, write down a relationship between $x$ and $y$.

**c** Draw a graph for $x$-values from 0 to 5.

**d** The final answer is 13. How can you use the graph to work out the number the teacher is thinking of?

(PS) **11** This flow diagram connects two variables, $Y$ and $X$.

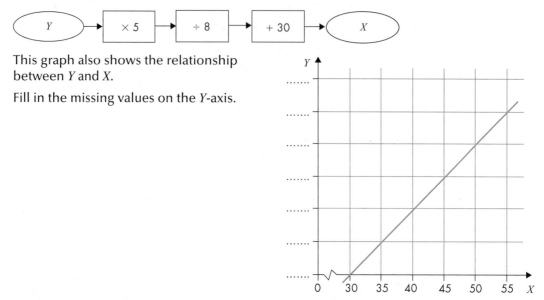

This graph also shows the relationship between $Y$ and $X$.

Fill in the missing values on the $Y$-axis.

**Hints and tips** If your graph does not start at zero, use a zigzag line (called a break) on the axis to show that some numbers have been left out.

# 8.2 Drawing linear graphs by finding points

## This section will show you how to:

• draw linear graphs without using flow diagrams.

This section is about drawing straight-line graphs. These are usually called **linear graphs**.

Here are some tips that will help you.

• You need to plot at least two points to draw a linear graph, but it is better to use three or more because that gives at least one point to act as a check.

• Use a sharp pencil and mark each point with an accurate cross.

• Position yourself so that your eyes are directly over the graph. If you look from the side, you will not be able to line up your ruler accurately.

This method can be quicker than using flow diagrams. However, if you prefer flow diagrams, then use them.

Example 2

Draw the graph of $y = 4x - 5$ for values of $x$ from 0 to 5.

**Note:** This is usually written as $0 \leqslant x \leqslant 5$.

Choose three values for $x$.

These should include the highest and lowest $x$-values and one in between. It is usually a good idea to choose 0 as one of the $x$-values.

Work out the $y$-values by substituting the $x$-values into the equation.

When $x = 0$, $y = 4(0) - 5 = -5$.   This gives the point $(0, -5)$.

When $x = 3$, $y = 4(3) - 5 = 7$.   This gives the point $(3, 7)$.

When $x = 5$, $y = 4(5) - 5 = 15$.   This gives the point $(5, 15)$.

Keep a record of your calculations in a table.

| $x$ | 0 | 3 | 5 |
|---|---|---|---|
| $y$ | −5 | 7 | 15 |

You are given the range of the $x$-axis, but you need to decide on the range for the $y$-axis. You can find this by looking at the coordinates that you have so far. The smallest $y$-value is −5; the largest is 15. Now draw the axes, plot the points and complete the graph.

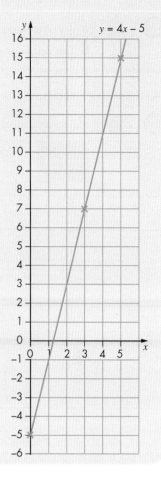

Read through these hints before drawing the linear graphs in Exercise **8B**.

• Use the highest and lowest values of $x$ given in the range.

• Don't choose $x$-values that are too close together, for example 1 and 2. Try to space them out so you can draw a more accurate graph.

• Always label your graph with its equation. This is particularly important when you are drawing two graphs on the same set of axes.

• Create a table of values.

## Exercise 8B

**Hints and tips**   Complete a table of values first; then you will know the range of the $y$-axis.

**1**   Draw each of these graphs for the given range of $x$.

   **a**  $y = x$ for $0 \leqslant x \leqslant 5$           **b**  $y = x - 1$ for $0 \leqslant x \leqslant 7$

   **c**  $y = 2x$ for $0 \leqslant x \leqslant 6$          **d**  $y = \frac{1}{2}x$ for $0 \leqslant x \leqslant 10$

**2**   Draw the graph of $y = 2x - 5$ for $0 \leqslant x \leqslant 5$.

**3**   Draw the graph of $y = \frac{x}{3} + 4$ for $-6 \leqslant x \leqslant 6$.

**4**
**a** On the same axes, draw the graphs of $y = 4x - 1$ and $y = 2x + 3$ for $0 \leqslant x \leqslant 5$.
**b** At which point do the two lines intersect?

CM **5**
**a** On the same axes, draw the graphs of $y = \frac{x}{3} - 1$ and $y = \frac{x}{2} - 2$ for $0 \leqslant x \leqslant 12$.
**b** On another set of axes, draw the graphs of $y = 3x + 1$ and $y = 3x - 2$ for $0 \leqslant x \leqslant 4$.
**c** Where possible, write down for parts **a** and **b** where the two lines intersect.
If it is not possible to write this down, give a reason why.

CM **6** Liam has completed a table for the equation $y = 2x + 1$ and drawn a graph.

| x | −3 | −2 | −1 | 0 | 1 | 2 | 3 |
|---|----|----|----|---|---|---|---|
| y | −7 | −5 | −3 | 1 | 3 | 5 | 7 |

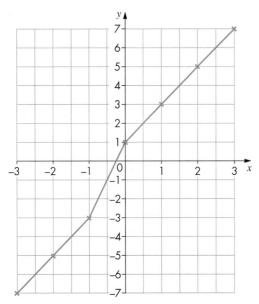

**a** How can you tell that he has made a mistake?
**b** Correct Liam's working.

EV **7** Ian the electrician used this formula to work out how much to charge for a job:
$C = 25 + 30H$, where $C$ is the charge (in £) and $H$ is how long the job takes (in hours).
Joan the electrician uses this formula: $C = 35 + 27.5H$.

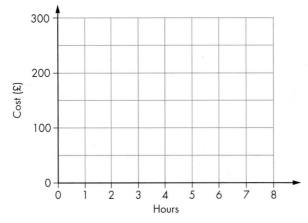

**a** On a copy of the grid, draw lines to show these formulae.
**b** Who would you hire for a job that takes 2 hours? Give a reason for your answer.

**(EV)** **8** Remi and Jada use different methods for finding points on the graph of $y = 4x + 2$.

Their methods are shown below.

Remi
$x = 0 \rightarrow y = 4 \times 0 + 2 = 2$
$x = 1 \rightarrow y = 4 \times 1 + 2 = 6$
$x = 2 \rightarrow y = 4 \times 2 + 2 = 10$
$x = 3 \rightarrow y = 4 \times 3 + 2 = 14$
$x = 4 \rightarrow y = 4 \times 4 + 2 = 18$

Jada
$x = 0 \rightarrow y = 2$ )+ 4
$x = 1 \rightarrow y = 6$ )+ 4
$10$ )+ 4
$14$ )+ 4
$18$ )+ 4

Whose method is more efficient?

**(PS)** **9** **a** Draw the graphs of $y = 4$, $y = x$ and $x = 1$ on a copy of this grid.

**b** What is the area of the triangle formed by the three lines?

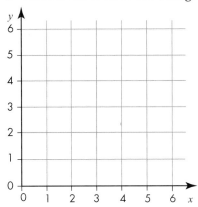

**(PS)** **10** The first two graphs show $y$ against $x$ and $y$ against $z$.

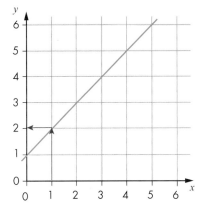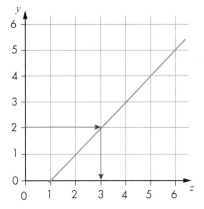

On a copy of the blank grid, show the graph of $x$ against $z$.

The first point has been done for you.

From the graph of $y$ against $x$, when $x = 1$, $y = 2$.

From the graph of $y$ against $z$, when $y = 2$, $z = 3$.

So when $x = 1$, $z = 3$.

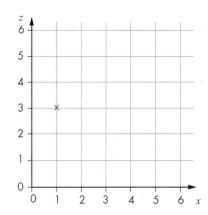

# 8.3 Gradient of a line

## This section will show you how to:

- work out the gradient of a straight line
- draw a line with a certain gradient.

The slope of a line is called its **gradient**. The steeper the slope of the line, the further the value of the gradient is from zero. So a gradient of 8 is steeper than a gradient of 3. Lines with gradients of 5 and –5 have the same steepness.

You can measure the gradient of the line by drawing a large right-angled triangle with part of the line as its hypotenuse (longest side). The gradient is then given by:

$$\text{gradient} = \frac{\text{distance measured up}}{\text{distance measured along}}$$

$$= \frac{\text{difference on } y\text{-axis}}{\text{difference on } x\text{-axis}}$$

For example, to measure the steepness of the line in the next diagram, you first draw a right-angled triangle where the hypotenuse is part of this line.

The gradient will be the same wherever you draw the triangle as you are calculating the ratio of two sides of the triangle. However, the calculations are much easier if you choose a sensible place, like where the line crosses the existing grid lines.

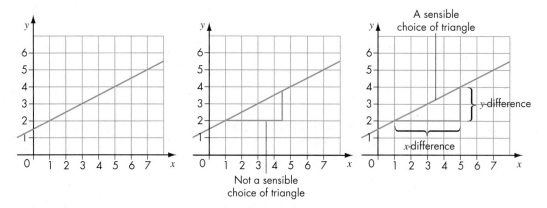

After you have drawn the triangle, measure (or count) how many squares there are on the vertical side. This is the difference between your $y$-coordinates. In the diagram above, this is 2.

You then measure (or count) how many squares there are on the horizontal side. This is the difference between your $x$-coordinates. In the diagram above, this is 4.

Then work out the gradient.

$$\text{gradient} = \frac{\text{difference of the } y\text{-coordinates}}{\text{difference of the } x\text{-coordinates}}$$

$$= \frac{2}{4}$$

$$= \frac{1}{2} \text{ or } 0.5$$

**Remember:** When a line slopes down from left to right, the gradient is negative, so there will be a minus sign in front of the fraction.

Example 3

Use the triangles drawn to work out the gradient of each line.

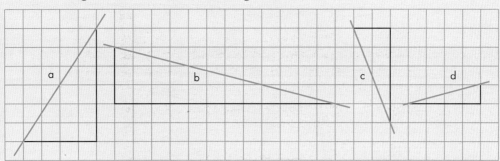

**a** $y$ difference = 6, $x$ difference = 4

Gradient is $6 \div 4 = \frac{3}{2}$ or 1.5.

**b** $y$ difference = 3, $x$ difference = 12

Line slopes down from left to right, so gradient is $-(3 \div 12) = -\frac{1}{4}$ or −0.25.

**c** $y$ difference = 5, $x$ difference = 2

Line slopes down from left to right, so gradient is $-(5 \div 2) = -\frac{5}{2}$ or −2.5.

**d** $y$ difference = 1, $x$ difference = 4

Gradient = $1 \div 4 = \frac{1}{4}$ or 0.25.

## Drawing a line with a given gradient

To draw a line with a given gradient, you reverse the process described above. Use the given gradient to draw the right-angled triangle. For example, take a gradient of 2.

Start at a suitable point (A in the diagrams below). A gradient of 2 means that for an $x$-step of 1 the $y$-step must be 2 (because 2 is the fraction $\frac{2}{1}$). So, move one square across and two squares up, and mark a dot.

Repeat this as many times as you like and draw the line. You can also move one square back and two squares down, which gives the same gradient, as the third diagram shows.

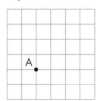
Stage 1

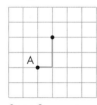
Stage 2

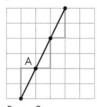
Stage 3

Example 4

Draw lines with these gradients.
**a** $\frac{1}{3}$
**b** −3
**c** $-\frac{1}{4}$

**a** This is a fractional gradient that has an $x$-step of 3 and a $y$-step of 1. Move three squares across and one square up every time.

**b** This is a negative gradient, so for every one square across, move three squares down.

**c** This is also a negative gradient and it is a fraction. So for every four squares across, move one square down.

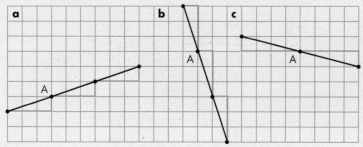

**1** Work out the gradient of lines **a** to **j**.

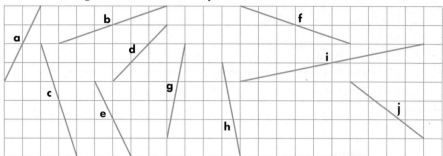

**2** Draw lines with these gradients.

**a** 4    **b** $\frac{1}{5}$    **c** $\frac{2}{5}$    **d** –3    **e** $-\frac{4}{5}$    **f** $\frac{3}{4}$

(EV) **3** Arianwen and Brianna are working out the gradient of the same line.

Arianwen

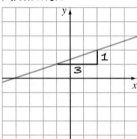

**Gradient** = $\frac{1}{3}$

Brianna

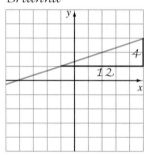

**Gradient** = $\frac{4}{12}$

**a** Whose calculation is correct?

**b** Whose method is more likely to give correct results in general?

(EV) **4** A playground supervisor suggests that a slide should be positioned with a gradient between $\frac{1}{2}$ and $\frac{1}{4}$.

**a** Why do you think the gradient has to be greater than $\frac{1}{4}$?

**b** Why do you think the gradient has to be less than $\frac{1}{2}$?

**c** Which of these slides satisfies the supervisor's safety suggestions?

A

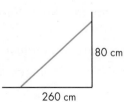

260 cm · 80 cm

B

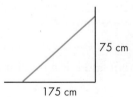

175 cm · 75 cm

C

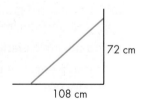

108 cm · 72 cm

D

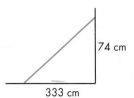

333 cm · 74 cm

E

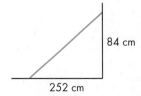

252 cm · 84 cm

F

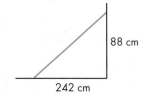

242 cm · 88 cm

**(MR) 5** The line on grid **e** is horizontal. The lines on grids **a** to **d** get nearer and nearer to the horizontal.

**a**  **b**  **c**  **d**  **e**

Work out the gradient of each line in grids **a** to **d**. Looking at the values you get, what do you think the gradient of a horizontal line is?

**(MR) 6** The line on grid **e** is vertical. The lines on grids **a** to **d** get nearer and nearer to the vertical.

**a**  **b**  **c**  **d**  **e**

Work out the gradient of each line in grids **a** to **d**. Looking at the values you get, what do you think the gradient of a vertical line is?

**(CM) 7** Raisa says the gradients of these two lines are the same.

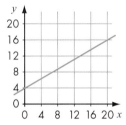

Why is Raisa wrong?

**(PS) 8** Put the following gradients in order of steepness, from the least steep to the steepest.

1 horizontal, 2 vertical    2 horizontal, 5 vertical    3 horizontal, 5 vertical

4 horizontal, 6 vertical    5 horizontal, 8 vertical    6 horizontal, 11 vertical

**9** Work out the gradient of each side of this pentagon.

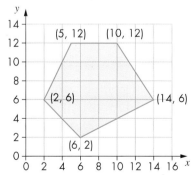

# 8.4 $y = mx + c$

This section will show you how to:
- draw graphs using the gradient-intercept method
- draw graphs using the cover-up method.

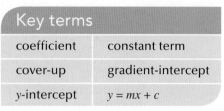

| Key terms | |
|---|---|
| coefficient | constant term |
| cover-up | gradient-intercept |
| $y$-intercept | $y = mx + c$ |

## Gradient-intercept method

When the equation of a straight line is written in the form $y = mx + c$, $m$ is the gradient and $c$ is where the line cuts the $y$-axis (called the **$y$-intercept**). $m$ is the **coefficient** of $x$ (the number in front of $x$) and $c$ is a **constant term**.

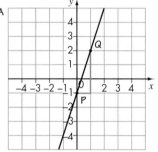

Here is the graph of $y = 3x - 1$.

The gradient is 3 and the $y$-intercept is $-1$.

Because the $y$-intercept, $c$, is $-1$, the graph goes through the $y$-axis at $-1$.

Because the gradient, $m$, is 3, for an $x$-step of one unit, there is a $y$-step of three units.

To draw this graph, start at $-1$ on the $y$-axis, move one square across and three squares up and mark the point with a dot or a cross (diagram **A**). Then repeat this from every new point (diagram **B**). You can also move one square back and three squares down.

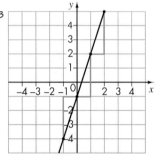

When you have enough points, join the dots (or crosses) to make the graph. Note that if the points are not in a straight line, you have made a mistake.

Drawing graphs this way is called the **gradient-intercept** method.

---

**Example 5**

Use the gradient-intercept method to draw the graph of $y = 2x - 5$.

$c = -5$      Mark this point with a dot.

$m = 2$      Move one square across and two squares up.

Repeat at least twice and plot the line.

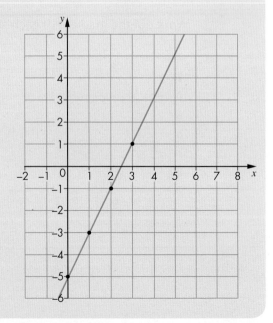

## Exercise 8D

**1** Draw these lines using the gradient-intercept method. Use the same grid, taking $x$ from –10 to 10 and $y$ from –10 to 10. If the grid gets too 'crowded', draw another one.

**a** $y = 2x + 6$      **b** $y = 3x - 4$      **c** $y = \frac{1}{2}x + 5$

**d** $y = x + 7$      **e** $y = 4x - 3$      **f** $y = 2x - 7$

**g** $y = \frac{1}{4}x - 3$      **h** $y = \frac{2}{3}x + 4$      **i** $y = 6x - 5$

**j** $y = x + 8$      **k** $y = \frac{4}{5}x - 2$      **l** $y = 3x - 9$

**2** **a** Using the gradient-intercept method, draw the following lines on the same grid. Use axes with ranges $-14 \leqslant x \leqslant 4$ and $-6 \leqslant y \leqslant 5$.

**i** $y = \frac{x}{2} + 3$      **ii** $y = \frac{x}{3} + 2$

**b** Where do the lines cross?

(MR) **3** Here are the equations of three lines.

A: $y = 3x - 1$     B: $2y = 6x - 4$     C: $y = 2x - 2$

**a** Change equation B to the form $y = mx + c$.

**b** State a mathematical property that lines B and C have in common.

**c** Which of the following points is the intersection of lines A and C?

(1, –4)     (–1, –4)     (1, 4)

**4** **a** What is the gradient of line A?

**b** What is the gradient of line B?

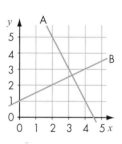

## Cover-up method

The $x$-axis has the equation $y = 0$. This means that all points on the $x$-axis have a $y$-value of 0.

The $y$-axis has the equation $x = 0$. This means that all points on the $y$-axis have an $x$-value of 0.

You can use these facts to draw any line that has an equation of the form $ax + by = c$.

Consider the graph of the line $4x + 5y = 20$.

Because the value of $x$ is 0 on the $y$-axis, you can solve the equation for $y$.

$$4(0) + 5y = 20$$
$$5y = 20$$
$$\Rightarrow y = 4$$

So the line passes through the point (0, 4) on the $y$-axis (diagram **A**).

Because the value of $y$ is 0 on the $x$-axis, you can also solve the equation for $x$.

$$4x + 5(0) = 20$$
$$4x = 20$$
$$\Rightarrow x = 5$$

So the line passes through the point (5, 0) on the $x$-axis (diagram **B**).

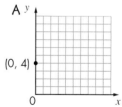

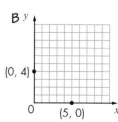

You only need two points to draw a line. (Normally, you would like a third point but, in this case, you can accept two.) Draw the graph by joining the points (0, 4) and (5, 0) (diagram **C**).

The graph of this type of equation can be drawn very easily, without much working at all, using the **cover-up method**.

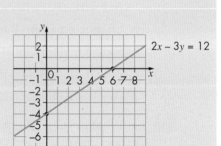

Start with the equation: $4x + 5y = 20$

Cover up the $x$-term.      $+ 5y = 20$

Solve the equation (when $x = 0$).     $y = 4$

Now cover up the $y$-term.     $4x + $  $= 20$

Solve the equation (when $y = 0$).     $x = 5$

This gives the points (0, 4) on the $y$-axis and (5, 0) on the $x$-axis.

---

**Example 6**

Draw the graph of $2x - 3y = 12$.

Solve the equation (when $x = 0$).     $-3y = 12$

                                 $y = -4$

Solve the equation (when $y = 0$).     $2x = 12$

                                 $x = 6$

This gives the points (0, −4) on the $y$-axis and (6, 0) on the $x$-axis.

## Exercise 8E

**1** Draw these lines using the cover-up method. Use the same grid, taking $x$ from −10 to 10 and $y$ from −10 to 10. If the grid gets too 'crowded', draw another.

   **a** $3x + 2y = 6$             **b** $4x + 3y = 12$         **c** $4x - 5y = 20$

   **d** $x + y = 10$               **e** $3x - 2y = 18$         **f** $x - y = 4$

   **g** $5x - 2y = 15$            **h** $2x - 3y = 15$         **i** $6x + 5y = 30$

   **j** $x + y = -5$               **k** $x + y = 3$            **l** $x - y = -4$

**2**  **a** Using the cover-up method, draw the following lines on the same grid.
Use axes with ranges $-2 \leqslant x \leqslant 6$ and $-3 \leqslant y \leqslant 6$.

     **i** $x + 2y = 6$           **ii** $2x - y = 2$

   **b** Where do the lines intersect (cross)?

(MR) **3** Here are the equations of three lines.

A: $2x + 6y = 12$    B: $x - 2y = 6$    C: $x + 3y = -9$

**a** State a mathematical property that lines A and B have in common.

**b** State a mathematical property that lines B and C have in common.

**c** State a mathematical property that lines A and C have in common.

**d** Line A crosses the $y$-axis at $(0, 2)$.

Line C crosses the $x$-axis at $(-9, 0)$.

For what values of $a$ and $b$ does the line $ax + by = 18$ pass through $(0, 2)$ and $(-9, 0)$?

(MR) **4** Match each equation with its graph.

**i** $y = 2x - 3$      **ii** $y = 1.5x$      **iii** $3y = x + 1$

**iv** $x + y = 2$      **v** $2x + y = 0$      **vi** $2x + 3y = -4$

**a** 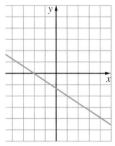    **b** 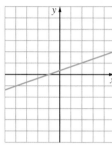    **c**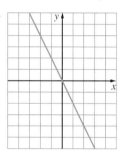

**d** 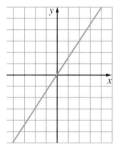    **e** 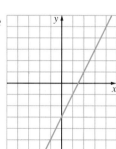    **f**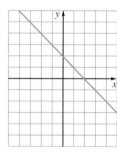

(PS) **5** The diagram shows an octagon ABCDEFGH.

The equation of the line through A and B is $y = 3$.

The equation of the line through B and C is $x + y = 4$.

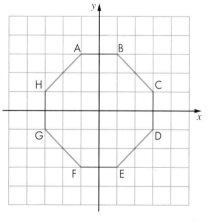

**a** Write down the equation of the lines through these vertices.

   **i** C and D      **ii** D and E      **iii** E and F

   **iv** F and G      **v** G and H      **vi** H and A

**b** The gradient of the line through F and B is 3.

Write down the gradient of the lines through these vertices.

   **i** A and E      **ii** G and C      **iii** H and D

(EV) **6** Elsa has been asked to plot the lines $2x + y = 10$ and $y = 10 - 2x$.

She has been taught the gradient-intercept method and the cover-up method.

Which method would you recommend for each line? Give reasons for your choices.

# 8.5 Finding the equation of a line from its graph

## This section will show you how to:

work out the equation of a line, using its gradient and $y$-intercept

- work out the equation of a line given two points on the line.

If you know the gradient, $m$, of a line and its $y$-intercept, $c$, on the $y$-axis, you can write down the equation of the line immediately.

For example, if $m = 3$ and $c = -5$, the equation of the line is $y = 3x - 5$.

All linear graphs can be written in the form $y = mx + c$. This gives a method to work out the equation of any line drawn on a pair of coordinate axes.

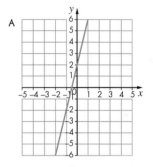

On the line in diagram A, the graph crosses the $y$-axis at $(0, 2)$ so $c = 2$.

Draw a triangle to measure the gradient of the line (diagram **B**).

$y$-step = 8

$x$-step = 2

gradient = $8 \div 2 = 4$

The line slopes up from left to right, so the gradient is positive.

So $m = 4$.

Then the equation of the line is $y = 4x + 2$.

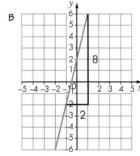

### Example 7

Work out the equation of the line that passes through the points $(2, 7)$ and $(10, 3)$.

Draw a sketch of the two points.

Calculate the $x$ and $y$ differences.     $y$ difference = 4

                                            $x$ difference = 8

Calculate the gradient.   $-(4 \div 8) = -\dfrac{1}{2}$

Remember, use a negative sign as the line slopes down from left to right.

Substitute $m = -\dfrac{1}{2}$ into $y = mx + c$.       $y = -\dfrac{1}{2}x + c$

Substitute values for $x$ and $y$ into $y = -\dfrac{1}{2}x + c$.

You could choose either point to substitute but $(2, 7)$ has smaller numbers.

$7 = -\dfrac{1}{2} \times 2 + c$

Take care with multiplying a negative number.

$7 = -1 + c$

$c = 8$

So, $y = -\dfrac{1}{2}x + 8$.

You can also work out the $x$ and $y$ differences without drawing a sketch, meaning you can work out the equation of a graph directly from just two points.

Example 8

Work out the equation of the line that passes through the points (2, –1) and (6, 11).

Calculate the $x$ and $y$ differences.

$y$ difference = $11 - -1 = 12$

$x$ difference = $6 - 2 = 4$

Calculate the gradient. $12 \div 4 = 3$

Substitute $m = 3$ into $y = mx + c$. $y = 3x + c$

Now choose one of the two points and substitute for $x$ and $y$. (6, 11) is simpler to use because both $x$ and $y$ are positive.

$11 = 3 \times 6 + c$

$11 = 18 + c$

$c = -7$

So, $y = 3x - 7$.

**Note:** Subtract the values in the same order each time, the coordinate of the first point minus the coordinate of the second point.

## Exercise 8F

**1** Give the equation of each of these lines. They all have positive gradients.

**a** 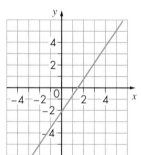 **b** 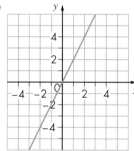 **c**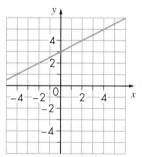

 **2** In each of these grids, there are two lines.

**a** 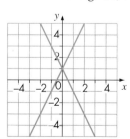 **b** 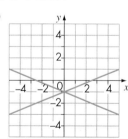 **c**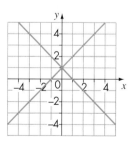

For each grid:

**i** work out the equation of each of the lines

**ii** describe any symmetries that you can see

**iii** describe any connection between the gradients of each pair of lines.

 **3** A straight line passes through the points (1, 3) and (2, 5).

**a** How can you tell that the line also passes through (0, 1)?

**b** How can you tell that the line has a gradient of 2?

**c** Work out the equation of the line that passes through (1, 5) and (2, 8).

**4** Give the equation of each of these lines. They all have negative gradients.

**a**

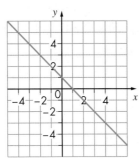

**b**

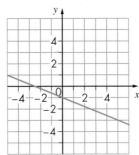

**c**

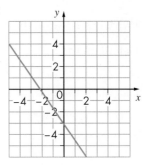

(PS) **5** In each of these grids, there are three lines. One of them is $y = x$.

**a**

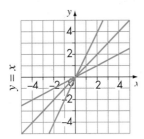

**b**

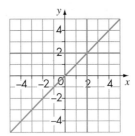

For each grid:

   **i** work out the equation of the other two lines

   **ii** describe any symmetries that you can see

   **iii** describe any connection between the gradients of each group of lines.

(MR) **6** On which of these lines does the point (12, 13) lie?

$$x + y = 25 \qquad y = \frac{1}{2}x + 7 \qquad y = 37 - 2x \qquad y = 19 - \frac{1}{2}x$$

$$y = 13 \qquad y = \frac{1}{4}x + 9 \qquad x = 12 \qquad y = \frac{2}{3}x + 5$$

**7** Use the facts to work out the equation of each line in the form $y = mx + c$.

  **a** gradient is –3; $y$-intercept is 5

  **b** gradient is 2; line passes through the point (5, 6)

  **c** $y$-intercept is –3; line passes through the point (2, 13)

  **d** line passes through the points (3, 19) and (9, 7)

  **e** line passes through the points (–9, –7) and (18, 11)

**8** Work out the equation of the line that passes through the points (6, 0) and (0, 5), giving the equation in the form $ax + by = c$.

(CM) **9** Chris has drawn a straight-line graph.

He says that the point (12, 8) will lie on the graph.

Helen says that the point (12, 8) will not lie on the graph.

Who is correct? Give a reason for your answer.

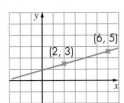

**10** For each set of four points:

    **i** work out the equation of the line passing through the points

    **ii** work out the value of $k$.

**a** $(75, 25), (17, 83), (50, 50), (k, 99)$      **b** $(46, 5), (46, 10), (46, -3), (k, 26)$

**c** $(13, 27), (48, 97), (32, 65), (k, 121)$      **d** $(-3, 16), (11, 30), (28, 47), (k, 2)$

# 8.6 The equation of a parallel line

This section will show you how to:

**Key term**

parallel

- work out the equation of a linear graph that is parallel to another line and passes through a specific point.

If two lines are **parallel**, then their gradients are equal.

Consider the line AB. Point A is at $(2, -1)$ and point B is at $(4, 5)$.

The gradient of AB is 3, so any parallel line can be written in the form $y = 3x + c$.

To work out the equation of the parallel line that passes through point C $(2, 8)$, substitute $x = 2$ and $y = 8$ into the equation $y = 3x + c$.

$$8 = 3 \times 2 + c$$

$$\Rightarrow c = 2$$

So the parallel line that passes through $(2, 8)$ is $y = 3x + 2$.

Copy the graph and add the line $y = 3x + 2$, to check.

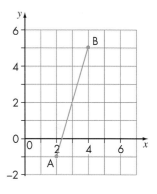

---

**Example 9**

Two points A and B are A $(0, 1)$ and B $(2, 4)$.

**a** Work out the equation of the line AB.

**b** Write down the equation of the line parallel to AB that passes through the point $(0, 5)$.

**c** Write down the equation of the line parallel to AB that passes through the point $(4, 10)$.

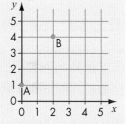

**a** The gradient of AB is $\frac{3}{2}$ and the $y$-intercept is $(0, 1)$, so the equation is $y = \frac{3}{2}x + 1$.

**b** The gradient is the same $\left(\frac{3}{2}\right)$ and the $y$-intercept is $(0, 5)$, so the equation is $y = \frac{3}{2}x + 5$.

**c** The gradient is $\left(\frac{3}{2}\right)$, so $y = \frac{3}{2}x + c$.

To work out the intercept, substitute $x = 4$ and $y = 10$ into the equation.

$$10 = \frac{3}{2} \times 4 + c$$

$$10 = 6 + c$$

$$c = 4$$

The equation is $y = \frac{3}{2}x + 4$.

## Exercise 8G

 **1** Here are the equations of three lines.

A: $y = 3x - 2$   B: $y = 3x + 1$   C: $y = 4x + 1$

Which two lines are parallel? Give a reason for your answer.

**2** Match the pairs of parallel lines.

| | | | |
|---|---|---|---|
| $x = 8$ | $y = 3 - x$ | $y = 3x - 5$ | $y = -\frac{1}{8}x$ |
| $2x + y = 9$ | $y = -\frac{1}{8}x + 6$ | $y = 10 - x$ | $y = 15 - 4x$ |
| $y = 7 - 4x$ | $x = -2$ | $y = 3x + 9$ | $2x + y = 13$ |

**3** Write down the equations of these lines.

**a** Parallel to $y = \frac{1}{2}x + 3$ and passes through $(0, -2)$

**b** Parallel to $y = -x + 2$ and passes through $(0, 3)$

**c** Parallel to $y = 3x + 2$ and passes through $(0, -8)$

**d** Parallel to $y = -\frac{1}{3}x - 2$ and passes through $(0, 10)$

 **4** Use the facts to work out the equation of each line in the form $y = mx + c$.

**a** Parallel to $y = 2x + 7$ and passes through $(3, 10)$

**b** Parallel to $y = 12 - 3x$ and passes through $(5, 2)$

**c** Passes through $(3, 8)$ and is parallel to the line that passes through the points $(5, 1)$ and $(7, 7)$

# 8.7 Real-life uses of graphs

This section will show you how to:

- convert from one unit to another unit by using a conversion graph
- use straight-line graphs to work out formulae.

You need to be able to read **conversion graphs** by finding a value on one axis and following it through to the other axis. Make sure you understand the scales on the axes to help you estimate the answers.

**Example 10**

This is a conversion graph between litres and gallons.

**a** How many litres are there in 5 gallons?

**b** How many gallons are there in 15 litres?

From the graph you can see that:

**a** 5 gallons are approximately equivalent to 23 litres.

**b** 15 litres are approximately equivalent to $3\frac{1}{4}$ gallons.

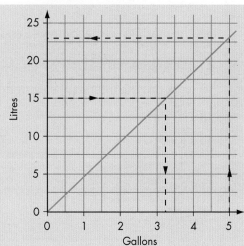

The graph opposite illustrates taxi fares in one part of England. It tells you that a fare will cost more, the further you go.

The taxi company charges a basic hire fee to start with of £2.00. This is shown on the graph as the point where the line cuts through the hire-charge axis (when distance travelled is 0).

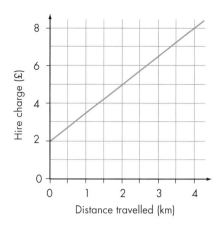

The gradient of the line is:

$\frac{8-2}{4} = \frac{6}{4}$ or 1.5

This represents the hire charge per kilometre travelled.

So the total hire charge is made up of two parts: a basic hire charge of £2.00 and an additional charge of £1.50 per kilometre travelled. This can be put in a formula as:

hire charge = £2.00 + £1.50 per kilometre

In this example, £2.00 is the constant term in the formula (the equation of the graph).

## Exercise 8H

**(MR)** **1** This is a conversion graph between kilograms (kg) and pounds (lb).

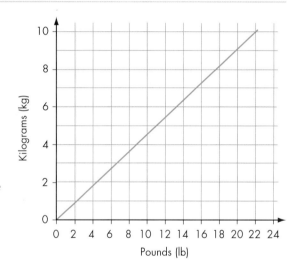

**a** Use the graph to make an approximate conversion of:

   **i** 18 lb to kilograms

   **ii** 5 lb to kilograms

   **iii** 4 kg to pounds

   **iv** 10 kg to pounds.

**b** Approximately how many pounds are equivalent to 1 kg?

**c** How could you use the graph to convert 48 lb to kilograms?

**d** Work out the gradient.

**e** What does the gradient represent?

**2** A conference centre displays this graph in their office to help staff calculate the approximate charge for a conference based on the number of people attending.

**a** Use the graph to calculate the approximate charge for:

   **i** 100 people      **ii** 550 people.

**b** Use the graph to estimate how many people can attend a conference at the centre for a charge of:

   **i** £300      **ii** £175.

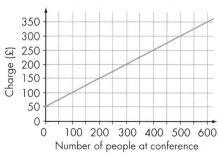

**3** Granny McAllister finds it hard to think in degrees Celsius so she always uses a conversion graph like the one shown to help her understand the weather forecast.

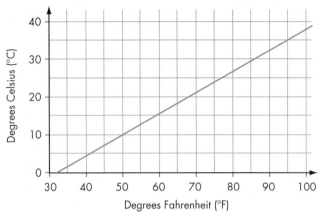

a Use the graph to work out the approximate conversions below.

  i 35 °C to Fahrenheit

  ii 20 °C to Fahrenheit

  iii 50 °F to Celsius

  iv 90 °F to Celsius

b Water freezes at 0 °C. What is the equivalent temperature in Fahrenheit degrees?

c Work out the gradient of the line. What does this represent?

**4** Tea is sold at a school fete between 1:00 pm and 3:30 pm. The number of cups of tea sold so far are recorded in a table every half-hour.

| Time | 1:00 | 1:30 | 2:00 | 2:30 | 3:00 | 3:30 |
|---|---|---|---|---|---|---|
| No. of cups of tea sold | 0 | 24 | 48 | 72 | 96 | 120 |

a Draw a graph to illustrate this information. Draw your horizontal axis from 1:00 pm to 4:00 pm and your vertical axis from 1 to 120.

b Use your graph to estimate the time when the sixtieth cup of tea was sold.

c Work out the gradient of the line. What does this represent?

**(PS)** **5** I used 700 units of fuel but lost my bill. My friends told me how much they were charged.

Bill was charged £57.50 for 850 units.

Wendy was charged £31 for 320 units.

Rhanni was charged £42 for 540 units.

Plot this information on a straight-line graph and work out my charge.

**(PS)** **6** Two taxi companies use these rules for calculating fares.

• CabCo:     £2.50 basic charge and £0.75 per kilometre

• YellaCabs:     £2.00 basic charge and £0.80 per kilometre

This map shows the distances, in kilometres, that three friends, Anya (A), Bettina (B) and Calista (C) live from a restaurant (R) and from each other.

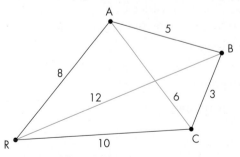

You may find a copy of the grid at the top of the next page useful to answer this question.

**a** If they each take an individual cab home from the restaurant, which company should they each choose?

**b** Work out the cheapest way they can travel home from the restaurant if two, or all three, share a cab.

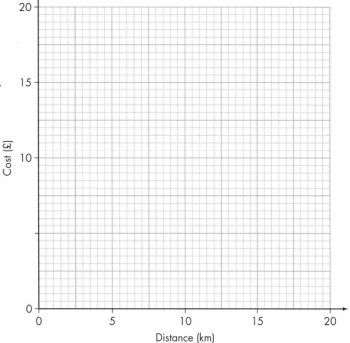

Cost (£)

Distance (km)

Hints and tips Draw a graph for both companies on the grid. Use this to work out the costs of the journeys.

**7** The graph shows the exchange rate between the American dollar and the British pound for three different months of a year.

**a** If George changed £1000 into dollars in March and another £1000 into dollars in December, approximately how much less did he get in December than in March?

**b** George went to America in March and stayed until July. In March, he changed £5000 into dollars. In July, he still had $2000 dollars left and he changed them back into pounds.

**i** How much, in dollars, did George spend between March and July?

**ii** How much, in pounds, did George spend between March and July?

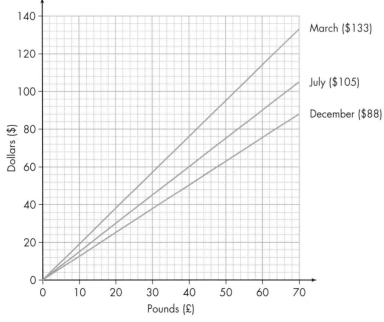

Dollars ($)

Pounds (£)

March ($133)

July ($105)

December ($88)

# 8.8 Solving simultaneous equations using graphs

This section will show you how to:

- solve simultaneous linear equations using graphs.

**Key term**

simultaneous equations

Two straight lines that are not parallel intersect at one point. This point is the solution of the equations of the lines. These are **simultaneous equations** represented by the lines.

You can solve simultaneous equations using algebra (as shown in Chapter 26) or graphs, as described below.

To work out the solution of the simultaneous equations $3x + y = 6$ and $y = 4x - 1$, follow these steps:

Draw the graph of $3x + y = 6$ using the cover-up method. It crosses the $x$-axis at $(2, 0)$ and the $y$-axis at $(0, 6)$.

Draw the graph of $y = 4x - 1$ by finding some points or by the gradient-intercept method. Using the gradient-intercept method, the graph crosses the $y$-axis at $-1$ and has a gradient of 4.

The point where the graphs intersect is $(1, 3)$. So the solution to the simultaneous equations is $x = 1$, $y = 3$.

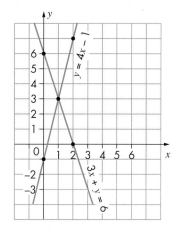

---

**Example 11**

Draw graphs to work out the solution of the simultaneous equations $y = x + 1$ and $x + y = 7$.

Plot both lines on the graph.

They intersect at $(3, 4)$, so $x = 3$ and $y = 4$.

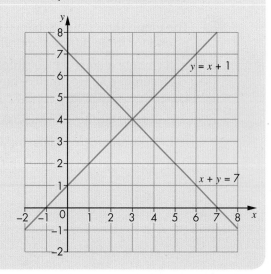

---

## Exercise 8I

In questions 1–8, draw the graphs to work out the solution of each pair of simultaneous equations.

**1**  $x + 4y = 8$
   $x - y = 3$

**2**  $y = 2x - 1$
   $3x + 2y = 12$

**3**  $y = 2x + 4$
   $y = x + 7$

**4**  $y = x + 8$
   $x + y = 4$

**5**  $y - 3x = 9$
   $y = x - 3$

**6**  $y = -x$
   $y = 4x - 5$

**7**  $3x + 2y = 18$
   $y = 3x$

**8**  $y = 3x + 2$
   $y + x = 10$

**9** One bun and two chocolate cakes cost £9.50.

Two buns and one chocolate cake cost £8.50.

Use $x$ to represent the cost of a bun and $y$ to represent the cost of a cake.

Use graphs to calculate the cost of a bun and the cost of a cake.

**10** The graph shows four lines.

P: $y = 4x + 1$      Q: $y = 2x + 2$      R: $y = x - 2$      S: $x + y + 1 = 0$

Which pairs of lines intersect at the following points?

**a** $(-1, -3)$      **b** $(\frac{1}{2}, -1\frac{1}{2})$

**c** $(\frac{1}{2}, 3)$      **d** $(-1, 0)$

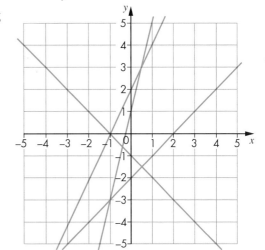

**11** Four lines have the following equations.

A: $y = x$      B: $y = 2$      C: $x = -3$      D: $y = -x$

These lines intersect at six different points.

Without drawing the lines accurately, write down the coordinates of the six intersection points.

| Hints and tips | Sketch the lines. |
| --- | --- |

**12** How many solutions does each pair of simultaneous equations have?

**a** $3x + 5y = 15$      **b** $2x + y = 7$      **c** $3x + 2y = 12$

     $3x + 5y = 10$          $4x = 14 - 2y$         $4x + y = 11$

| Hints and tips | Sketch the lines. |
| --- | --- |

**d** Which of these terms describes each pair of lines in **a** to **c**?

Not parallel          Parallel          Same line

# Worked exemplars

  This graph shows the conversion between two variables $x$ and $y$.

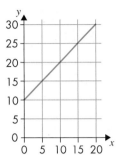

This graph shows the conversions between two variables $y$ and $z$.

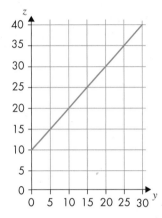

On the graph below, draw the conversion between $x$ and $z$.

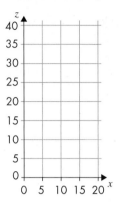

| This is a mathematical reasoning question where you have to construct a chain of reasoning using your mathematical skills and knowledge to achieve a given result. | |
| --- | --- |
| When $x = 0$, $y = 10$ and when $y = 10$, $z = 20$.<br>When $x = 20$, $y = 30$ and when $y = 30$, $z = 40$. | Connect at least two $x$-values to $y$-values. The most obvious choices are 0 and 20 as these are the ends of the line segment for $x$. |
| 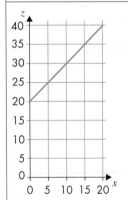 | Draw a line joining the pairs of $x$- and $z$-values, that is, (0, 20) to (20, 40). |
| When $x = 10$, $y = 20$.<br>When $y = 20$, $z = 30$. | It is a good idea to check your answer with a third point. |

**2** A shop on a cross-channel ferry accepts both pounds and euros.

The exchange rate is £1 = €1.25.

George pays £2.25 for five apples and three bananas.

Pierre pays €1.00 for one apple and two bananas.

**a** Represent this information in a graph.

**b** Using the graph, work out the cost of buying three apples and one banana in pounds.

| This is a problem-solving question. You need to plan a strategy to solve it and you should expect to use your knowledge of other areas of mathematics, in this case currency conversions. | |
|---|---|
| **a**   €1.00 ÷ 1.25 = £0.80 | Because the final answer will be in pounds, it makes sense to solve the whole question in pounds or pence. |
| The variables are:<br><br>$x$ = the price of an apple<br><br>$y$ = the price of a banana<br><br>The equations are:<br><br>$5x + 3y = 225$<br><br>$x + 2y = 80$ | |
| $5x + 3y = 225$<br><br>when $x = 0$, $y = 75$<br><br>when $y = 0$, $x = 45$<br><br>$x + 2y = 80$<br><br>when $x = 0$, $y = 40$<br><br>when $y = 0$, $x = 80$<br><br>*[graph showing two intersecting lines, intersection labelled $x = 30$, $y = 25$; axes labelled from 0 to 80]* | You can use any method for plotting the lines, but the cover-up method is simplest here. |
| **b**   The lines intersect when $x = 30$, $y = 25$.<br><br>An apple costs 30p.<br><br>A banana costs 25p. | |
| $3x + y = 3 \times 30 + 1 \times 25 = 115$<br><br>Three apples and one banana cost £1.15. | |

# Ready to progress?

I can read off values from a conversion graph.

I can draw a linear graph without being given a table of values.
I can work out the gradient of a line.
I can draw straight lines using the gradient-intercept method.

I can draw straight lines using the cover-up method.
I can work out the equation of a line from its graph.
I can work out the equation of a parallel line.
I can use graphs to work out formulae.
I can use graphs to solve simultaneous linear equations.

# Review questions

1  Draw the graph of $y = 3x + 4$ for $0 \leqslant x \leqslant 5$.

2  Draw the graph of $y = \dfrac{x}{2} - 3$ for $0 \leqslant x \leqslant 10$.

3  On a copy of the grid, draw the graph of $x + y = 6$.

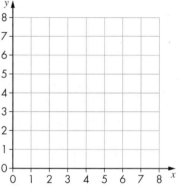

4  This graph shows the length of a spring when different weights are attached to it.

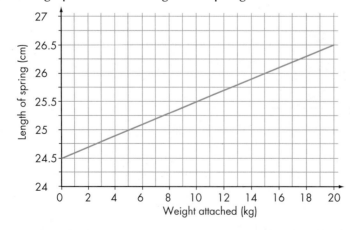

**a** Calculate the gradient of the line.

**b** How long is the spring when no weight is attached to it?

**c** By how much does the spring extend per kilogram?

**d** Write down the rule for calculating the length of the spring for different weights. Use $L$ for length and $W$ for weight.

(PS) **5** Match each line with a point it passes through.

Although some points lie on more than one line, there is only one way to pair them all up.

| Lines | Points |
|---|---|
| A: $y = x + 10$ | P: (6, 8) |
| B: $y = 7x - 8$ | Q: (2, 12) |
| C: $y = 14 - x$ | R: (8, 9) |
| D: $y = \frac{1}{2}x + 5$ | S: (3, 13) |
| E: $y = 4x + 1$ | T: (2, 6) |

**6** Draw the graphs to work out the solution of each pair of simultaneous equations.

**a**   $y = x$
  $x + y = 10$

**b**   $y = 2x + 3$
  $5x + y = 10$

**c**  $y = 5x + 1$
  $y = 2x + 10$

(PS) **7** Calculate the area of the rectangle bounded by these four lines.

$x = 12, y = 7, x = 3, y = -1$

**8** A has coordinates (0, 2). B has coordinates (2, 6).

**a** Work out the gradient of the line that passes through A and B.

**b** Work out the equation of the line that passes through A and B.

**c** Work out the equation of the line parallel to AB that passes through (3, 1).

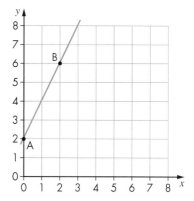

(MR) **9** A line $L_1$ passes through the points (23, 27), (40, 10) and (53, $k$).

**a** Work out the equation of the line.

**b** Work out the value of $k$.

**c** Work out the equation of the line $L_2$ which is parallel to $L_1$ and passes through the point (28, 12).

# 9 Algebra: Expressions and formulae

## This chapter is going to show you:

- how to use letters to represent numbers
- how to form simple algebraic expressions
- how to simplify expressions by collecting like terms
- how to substitute numbers into expressions and formulae
- how to expand and factorise expressions
- how to expand two pairs of brackets
- how to factorise quadratic expressions
- how to rearrange formulae.

## You should already know:

- the BIDMAS rule, which gives the order for the operations of arithmetic.

## About this chapter

If you circled one of these to describe mathematics, which would it be?

Art    Science    Sport    Language

In fact, you could circle them all.

But perhaps the most important description in the list above is mathematics as language. Mathematics is the only universal language. If you write the equation $3x = 9$, it will be understood by people in all countries. Algebra is the way to express the language of mathematics.

Algebra comes from the Arabic word *al-jab*r which means something similar to 'completion'. It was used in a book written in 820 AD by a Persian mathematician called al-Khwarizmi.

In the middle of the 17th century, René Descartes developed the basis of the algebra we use today. It can be understood by everyone, even if they do not speak the same language.

# 9.1 Basic algebra

This section will show you how to:

- write an algebraic expression
- recognise expressions, equations, formulae and identities.

| Key terms | |
|---|---|
| equation | expression |
| formula | identity |
| symbol | term |
| variable | |

Algebra is based on the idea that if something works with numbers, it will work with letters. The main difference is that working only with numbers gives a numerical answer. Working with letters gives an **expression** as the answer.

Algebra follows the same rules as arithmetic, and uses the same **symbols** ($+$, $-$, $\times$ and $\div$). Below are seven important algebraic rules.

- Write '4 more than $x$' as $4 + x$ or $x + 4$.
- Write '6 less than $p$' or '$p$ minus 6' as $p - 6$.
- Write '4 times $y$' as $4 \times y$ or $y \times 4$ or $4y$. $4y$ is the neatest way to write it.
- Write '$b$ divided by 2' as $b \div 2$ or $\frac{b}{2}$.
- When a number and a letter, or a letter and a letter, appear together, there is a hidden multiplication sign between them. So, $7x$ means $7 \times x$ and $ab$ means $a \times b$.
- Always write '$1 \times x$' as $x$.
- Write '$t$ times $t$' as $t \times t$ or $t^2$.

---

**Example 1**

What is the area of each of these rectangles?

**a** 4 cm by 6 cm      **b** 4 cm by $w$ cm      **c** $l$ cm by $w$ cm

The rule for working out the area of a rectangle is:

area = length × width

So, the area of rectangle **a** is $4 \times 6 = 24$ cm$^2$

The area of rectangle **b** is $4 \times w = 4w$ cm$^2$

The area of rectangle **c** is $l \times w = lw$ cm$^2$

---

If $A$ represents the area of rectangle **c** in the example above:

$$A = lw$$

This is an example of a rule expressed algebraically.

---

**Example 2**

What is the perimeter of each of these rectangles?

**a** 6 cm by 4 cm      **b** 4 cm by $w$ cm      **c** $l$ cm by $w$ cm

The rule for working out the perimeter of a rectangle is:

perimeter = twice the longer side + twice the shorter side

So, the perimeter of rectangle **a** is $2 \times 6 + 2 \times 4 = 20$ cm

The perimeter of rectangle **b** is $2 \times 4 + 2 \times w = 8 + 2w$ cm

The perimeter of rectangle **c** is $2 \times l + 2 \times w = 2l + 2w$ cm

---

If $P$ represents the perimeter of rectangle **c** in Example 2.

$P = 2l + 2w$

This is called a **formula**.

If $A$ represents the area then $A = lw$ is also a formula.

Here are some algebraic words that you need to know.

| | |
|---|---|
| **Variable** | Letters that are used to represent numbers are called variables. These letters can take on any value, so they are said to *vary*. |
| **Expression** | An expression is any combination of letters and numbers. |
| | For example, $2x + 4y$ and $\dfrac{p - 6}{5}$ are expressions. |
| **Equation** | An equation contains an 'equals' sign and at least one variable. The important fact is that a value can be found for the variable. This is called *solving the equation*. |
| **Formula** | You may already have seen lots of formulae (the plural of formula). These are like equations because they contain an 'equals' sign, but there is more than one variable. Formulae are rules for working out things such as area or the cost of taxi fares. |
| | For example, $V = x^3$, $A = \frac{1}{2}bh$ and $C = 3 + 4m$ are formulae. |
| **Identity** | These look like formulae, but the important fact about an identity is that it is true for all values, whether numbers or letters. |
| | For example, $5n \equiv 2n + 3n$, $2(2x + 5) \equiv 4x + 10$ and $(x + 1)^2 \equiv x^2 + 2x + 1$ are identities. Note that the special sign $\equiv$ is sometimes used to show an identity. |
| **Term** | Terms are the separate parts of expressions, equations, formulae and identities. |
| | For example, in $3x + 2y - 7$, there are three terms: $3x$, $+2y$ and $-7$. |

---

**Example 3**

State whether each of the following is an expression, equation, formula or identity.

A: $x - 5 = 7$    B: $P = 4x$    C: $2x - 3y$    D: $3n - n = 2n$

A is an equation as it can be solved to give $x = 12$.

B is a formula. This is the formula for the perimeter of a square with a side of $x$.

C is an expression with two terms.

D is an identity as it is true for all values of $n$.

---

## Exercise 9A

**1** Write down an algebraic expression for each of these.

**a** 2 more than $x$    **b** 6 less than $x$    **c** $h$ multiplied by $j$

**d** $y$ divided by $t$    **e** $a$ multiplied by $a$    **f** $g$ multiplied by itself

(PS) **2** **a** My dad is 72 and I am $T$ years old. How old will we each be in $x$ years' time?

**b** My mum is 64 years old. In two years' time, she will be twice as old as I will be. What age am I now?

(MR) **3** The answer to $3 \times 4m$ is $12m$.

Write down two *different* expressions where the answer is $12m$.

**(CM) 4** Three expressions for the perimeter of a rectangle with length $l$ and width $w$ are:

$P = l + w + l + w$

$P = 2l + 2w$

$P = 2(l + w)$

Show, using a numerical example, that they all give the same result.

> **Hints and tips** Pick some simple numbers for $l$ and $w$.

**(PS) 5** My sister is three years older than I am.

The sum of our ages is 29.

How old am I?

> **Hints and tips** If I am $x$ years old, work out how old my sister is in terms of $x$ and use this to set up an equation.

**(PS) 6** Ali has 65p and Heidi has 95p. How much should Heidi give to Ali so they both have the same amount?

**7** State whether each of the following is an expression, equation, formula or identity.

**a** $2x - 5$

**b** $s = \sqrt{A}$

**c** $2(x + 3) = 2x + 6$

**d** $2x - 3 = 1$

# 9.2 Substitution

## This section will show you how to:

- substitute into, simplify and use algebraic expressions.

One of the most important features of algebra is **substituting** real numbers into expressions and formulae.

The value of an expression, such as $3x + 2$, changes when you substitute different values of $x$ into it. For example, the expression $3x + 2$ has the value:

5 when $x = 1$     14 when $x = 4$

A formula gives the value of one variable as the other variables in the formula change. For example, the formula for the area, $A$, of a triangle of base $b$ and height $h$ is:

$$A = \frac{b \times h}{2}$$

When $b = 4$ cm and $h = 8$ cm:

$$A = \frac{4 \times 8}{2} = 16 \text{ cm}^2$$

**Example 4**

The formula for the area of a trapezium is:

$A = \dfrac{(a+b)h}{2}$

Work out the area of a trapezium when $a = 5$ m, $b = 9$ m and $h = 3$ m.

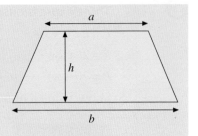

$A = \dfrac{(5+9) \times 3}{2} = \dfrac{14 \times 3}{2} = 21$ m$^2$

Whenever you substitute a number for a variable in an expression, it helps to put the value in brackets before working it out. This will help you to avoid calculation errors, especially with negative numbers.

To work out the value of $3x^2 - 5$ when $x = 3$

$$3(3)^2 - 5 = 3 \times 9 - 5$$
$$= 27 - 5$$
$$= 22$$

To work out the value when $x = -4$, write

$$3(-4)^2 - 5 = 3 \times 16 - 5$$
$$= 48 - 5$$
$$= 43$$

**Example 5**

Calculate the value of $L = a^2 - 8b^2$ when $a = -6$ and $b = 2$.

Substitute the values $a = -6$ and $b = 2$ for the letters $a$ and $b$.

$$L = a^2 - 8b^2$$
$$= (-6)^2 - 8(2)^2$$
$$= 36 - 8 \times 4$$
$$= 36 - 32$$
$$= 4$$

**Note:** If you do not use brackets and write $-6^2$, you might calculate it wrongly as $-36$.

**Example 6**

The formula for the electricity bill each quarter in a household is:

total bill = £7.50 + £0.07 × number of units.

A family uses 6720 units in a quarter.

How much is their total bill?

Substitute 6720 into the formula.

£7.50 + £0.07 × number of units = total bill

$7.5 + 0.07 \times 6720 = 477.9$

So, the total bill is £477.90.

# Exercise 9B

**1** Work out the value of $4k - 1$ for each of these values of $k$.

   **a** $k = 1$         **b** $k = 3$         **c** $k = 11$

**2** Evaluate $15 - 2f$ for each of these values of $f$.

   **a** $f = 3$         **b** $f = 5$         **c** $f = 8$

**(PS) 3** A taxi company uses the following rule to calculate their fares.

   Fare = £2.50 plus 50p per kilometre

   **a** How much is the fare for a 3-kilometre journey?

   **b** Farook pays £9.00 for his journey. How far did he travel?

   **c** Maisy's house is 5 kilometres from town. She has £5.50 left in her purse after a night out. Has she got enough to pay for a taxi home?

**4** $P = \dfrac{5w - 4y}{w + y}$ Calculate $P$ for each of the following values of $w$ and $y$.

   **a** $w = 3$ and $y = 2$   **b** $w = 6$ and $y = 4$   **c** $w = 2$ and $y = 3$

**5** $Z = \dfrac{y^2 + 4}{4 + y}$ Calculate $Z$ for each of these values of $y$.

   **a** $y = 4$         **b** $y = -6$       **c** $y = -1.5$

**(PS) 6** A holiday cottage costs £150 per day to rent.

   A group of friends decide to rent the cottage for seven days.

   **a** Which of the following formulae would represent the cost per person per day if there are $n$ people in the group and they share the cost equally?

   $\dfrac{150}{n}$         $\dfrac{150}{7n}$         $\dfrac{1050}{n}$         $\dfrac{150n}{7}$

   **b** Eventually 10 people go on the holiday.

      When they get the bill, they find that there is a discount for a seven-day rental. After the discount, they each find it costs them £12.50 less than they expected.

      How much does a seven-day rental cost?

> **Hints and tips** To check your choice in part **a**, make up some numbers and try them in the formulae. For example, take $n = 5$.

**(MR) 7** Kaz knows that $x$, $y$ and $z$ have the values 2, 8 and 11, but she does not know which variable has which value.

   **a** What is the maximum possible value of the expression $2x + 6y - 3z$?

   **b** What is the minimum possible value of the expression $5x - 2y + 3z$?

> **Hints and tips** You could just try all combinations but, in part **a**, the $6y$ term has to be the biggest and this will give you a clue to the other terms.

**8** $x$ and $y$ are different positive whole numbers.

Work out a possible pair of values for $x$ and $y$ so that the value of the expression $5x + 3y$ is:

**a** odd **b** prime.

Hints and tips You need to remember the prime numbers: 2, 3, 5, 7, 11, 13, 17, 19, …

 **9** **a** $p$ is an odd number and $q$ is an even number.

Say whether the following expressions are odd or even.

**i** $p + q$  **ii** $p^2 + q$  **iii** $2p + q$  **iv** $p^2 + q^2$

**b** $x$, $y$ and $z$ are all odd numbers.

Write an expression using $x$, $y$ and $z$ so that the value of the expression is always even.

 **10** A formula for the cost of delivery, in pounds, of orders from a do-it-yourself warehouse is:

$$D = 2M - \frac{C}{5}$$

where $D$ is the cost of the delivery, $M$ is the distance in miles from the warehouse and $C$ is the cost of the goods to be delivered.

**a** How much is the delivery cost when $M = 30$ and $C = 200$?

**b** Bob buys goods worth £300 and lives 10 miles from the warehouse.

The formula gives a negative value for the cost of delivery. What is this value?

**c** Martha buys goods worth £400. She calculates that her cost of delivery will be zero.

What is the greatest distance Martha could live from the warehouse?

 **11** Marvin hires a car for one day at a cost of £40. He wants to know how much it costs him for each mile he drives.

Petrol costs £1.20 per litre and he can drive 10 miles per litre of fuel.

Marvin works out the following formula for the cost per mile, $C$ in pounds, for $M$ miles driven:

$$C = 0.12 + \frac{40}{M}$$

**a** What does each term of the formula mean?

**b** How much will it cost per mile if Marvin drives 200 miles that day?

Hints and tips Use the information in the question in your explanation.

# 9.3 Expanding brackets

This section will show you how to:

- expand brackets such as $2(x - 3)$
- expand and simplify brackets.

| Key terms | |
|---|---|
| expand | like terms |
| multiply out | simplify |

In mathematics, to **'expand'** usually means **'multiply out'**. For example, expressions such as $3(y + 2)$ and $4y(2y + 3)$ can be expanded by multiplying them out.

Remember that there is an invisible multiplication sign between the outside number and the opening bracket. So $3(y + 2)$ is really $3 \times (y + 2)$ and $4y(2y + 3)$ is really $4y \times (2y + 3)$.

You expand by multiplying *everything inside* the brackets by what is outside the brackets.

So in the two examples above,

$3(y + 2) = 3 \times (y + 2) = 3 \times y + 3 \times 2 = 3y + 6$

$4y(2y + 3) = 4y \times (2y + 3) = 4y \times 2y + 4y \times 3 = 8y^2 + 12y$

Look at these next examples of expansion, where the term outside the brackets has been multiplied by the terms inside them.

$3(2t + 5) = 6t + 15$ $\qquad$ $m(p + 7) = mp + 7m$

$3t(2 + 5t - p) = 6t + 15t^2 - 3pt$ $\qquad$ $-2x(3 - 4x) = -6x + 8x^2$

$y(y^2 - 4x) = y^3 - 4xy$ $\qquad$ $4t(t^3 + 2) = 4t^4 + 8t$

Remember:

- the product of a negative and a positive is negative
- the product of a negative and a negative is positive.

This means that the signs change when a negative quantity is outside the brackets. For example,

$a(b + c) = ab + ac$ $\qquad$ $a(b - c) = ab - ac$

$-a(b + c) = -ab - ac$ $\qquad$ $-a(b - c) = -ab + ac$

$-(a - b) = -a + b$ $\qquad$ $-(a + b - c) = -a - b + c$

**Note:** A minus sign on its own in front of the brackets is actually $-1$, so:

$-(x - 3) = -1 \times (x - 3) = -1 \times x + -1 \times -3 = -x + 3$

## Exercise 9C

**1**  Expand these expressions.

**a** $2(3 + m)$ $\qquad$ **b** $5(2 + l)$ $\qquad$ **c** $3(4 - y)$ $\qquad$ **d** $4(5 + 2k)$

**e** $3(2 - 4f)$ $\qquad$ **f** $2(5 - 3w)$ $\qquad$ **g** $5(2k + 3m)$ $\qquad$ **h** $4(3d - 2n)$

**i** $t(t + 3)$ $\qquad$ **j** $k(k - 3)$ $\qquad$ **k** $4t(t - 1)$ $\qquad$ **l** $2k(4 - k)$

**m** $4g(2g + 5)$ $\qquad$ **n** $5h(3h - 2)$ $\qquad$ **o** $y(y^2 + 5)$ $\qquad$ **p** $h(h^3 + 7)$

**q** $k(k^2 - 5)$ $\qquad$ **r** $3t(t^2 + 4)$ $\qquad$ **s** $3d(5d^2 - d^3)$ $\qquad$ **t** $3w(2w^2 + t)$

**u** $5a(3a^2 - 2b)$ $\qquad$ **v** $3p(4p^3 - 5m)$ $\qquad$ **w** $4h^2(3h + 2g)$ $\qquad$ **x** $2m^2(4m + m^2)$

**2** A supermarket is offering £1 off a large tin of biscuits. Morris wants five tins.

    **a** If the normal price of one tin is £$t$, which expression below represents how much it will cost Morris to buy five tins?

        $5(t-1)$          $5t-1$          $t-5$          $5t-5$

    **b** Morris has £20 to spend. If each tin is £4.50, will he have enough money for five tins? Show working to justify your answer.

  **3** Dylan wrote the following.

$3(5x-4)=8x-4$

Dylan has made two mistakes.

    **a** What mistakes has Dylan has made?

    **b** What is the correct answer?

> **Hints and tips** It is not enough to give the right answer. You must try to explain, for example, why Dylan wrote 8 for $3 \times 5$ instead of 15.

**4** The expansion $2(x+3) = 2x+6$ can be shown by this diagram.

    **a** What expansion is shown in this diagram?

    **b** Write down one possible expansion that could be shown in this diagram.

| | |
|---|---|
| $12z$ | 8 |

## Collecting like terms

**Like terms** are terms that have the same letter(s) with the same power but can have different numerical coefficients (numbers in front). For example,

$m$, $3m$, $4m$, $-m$ and $76m$ are all like terms in $m$

$t^2$, $4t^2$, $7t^2$, $-t^2$, $-3t^2$ and $98t^2$ are all like terms in $t^2$

$pt$, $5tp$, $-2pt$, $7pt$, $-3tp$ and $103pt$ are all like terms in $pt$.

**Note:** All the $tp$ terms are also like terms to all the $pt$ terms.

When you have an expression with like terms, you can **simplify** the expression by combining the like terms together. For example,

| | | |
|---|---|---|
| $4h - h = 3h$ | $3y + 5y + 3 = 8y + 3$ | $2t^2 + 5t^2 = 7t^2$ |
| $2m + 6 + 3m = 5m + 6$ | $3ab + 2ba = 5ab$ | $10g - 4 - 3g = 7g - 4$ |

## Expand and simplify

When you expand two brackets, there are often like terms that you can collect together. Always simplify algebraic expressions as much as possible.

You can find a simplified expression for the perimeter of a rectangle with sides of $(5x - 8)$ cm and $(2x + 11)$ cm.

It has two of each side so: $2(5x - 8) + 2(2x + 11) = 10x - 16 + 4x + 22$

$$= (14x + 6) \text{ cm}$$

**Example 7**

Expand and simplify $3t(5t + 4) - 2t(3t - 5)$.

$3t(5t + 4) - 2t(3t - 5) = 15t^2 + 12t - 6t^2 + 10t$

$$= 9t^2 + 22t$$

## Exercise 9D

**1** Simplify these expressions.

**a** $4t + 3t$      **b** $2y + y$      **c** $3d + 2d + 4d$

**d** $5e - 2e$      **e** $4p - p$      **f** $3t - t$

**g** $2t^2 + 3t^2$      **h** $3ab + 2ab$      **i** $7a^2d - 4a^2d$

**2** Expand and simplify.

**a** $3(4 + t) + 2(5 + t)$      **b** $5(3 + 2k) + 3(2 + 3k)$

**c** $4(3 + 2f) + 2(5 - 3f)$      **d** $5(1 + 3g) + 3(3 - 4g)$

> **Hints and tips** Be careful with minus signs. For example, $-2(5e - 4) = -10e + 8$.

**3** Expand and simplify.

**a** $4(3 + 2h) - 2(5 + 3h)$      **b** $5(3g + 4) - 3(2g + 5)$

**c** $5(5k + 2) - 2(4k - 3)$      **d** $4(4e + 3) - 2(5e - 4)$

**4** Expand and simplify.

**a** $m(4 + p) + p(3 + m)$      **b** $k(3 + 2h) + h(4 + 3k)$

**c** $4r(3 + 4p) + 3p(8 - r)$      **d** $5k(3m + 4) - 2m(3 - 2k)$

**5** Expand and simplify.

**a** $t(3t + 4) + 3t(3 + 2t)$      **b** $2y(3 + 4y) + y(5y - 1)$

**c** $4e(3e - 5) - 2e(e - 7)$      **d** $3k(2k + p) - 2k(3p - 4k)$

**6** Expand and simplify.

**a** $4a(2b + 3c) + 3b(3a + 2c)$      **b** $3y(4w + 2t) + 2w(3y - 4t)$

**c** $5m(2n - 3p) - 2n(3p - 2m)$      **d** $2r(3r + r^2) - 3r^2(4 - 2r)$

**7** A two-carriage train has $f$ first-class seats and $2s$ standard-class seats.

A three-carriage train has $2f$ first-class seats and $3s$ standard-class seats.

On a weekday, 5 two-carriage trains and 2 three-carriage trains travel from Hull to Liverpool.

**a** Write down an expression for the total number of first-class and standard-class seats available on a weekday.

**b** On average in any day, half of the first-class seats are used. They cost £60 each.

On average in any day, three-quarters of the standard-class seats are used. They cost £40 each.

The rail company calculates the total amount of money earned on an average day on this route. Write an expression, in terms of $f$ and $s$, for the total amount earned from:

**i** first-class ticket sales    **ii** standard-class ticket sales.

**c** $f = 15$ and $s = 80$. It costs the rail company £30 000 per day to operate this route. How much profit do they make on an average day?

**8** Fill in whole-number values so that the following expansion is true.

$3(\ldots x + \ldots y) + 2(\ldots x + \ldots y) = 11x + 17$

> **Hints and tips** There is more than one answer. You need only give one.

**9** A rectangle with sides 5 and $3x + 2$ has a smaller rectangle with sides 3 and $2x - 1$ cut from it.

Work out the remaining area.

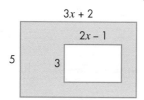

> **Hints and tips** Write out the expression for the difference between the two rectangles and then work it out.

# 9.4 Factorisation

## This section will show you how to:

- factorise an algebraic expression.

**Factorisation** is the opposite of expansion. It puts an expression back into the brackets it may have come from.

In factorisation, you have to look for the **common factors** in *every* term of the expression.

To factorise the expression $6t + 9m$, first look at the numerical coefficients 6 and 9. These have a common factor of 3.

Then look at the letters, $t$ and $m$. These do not have any common factors as they do not appear in both terms.

You can think of the expression as $3 \times 2t + 3 \times 3m$, which gives the factorisation:

$$6t + 9m = 3(2t + 3m)$$

**Note:** You can always check a factorisation by expanding the answer.

Factorise each expression.

**a** $6my + 4py$      **b** $5k^2 - 25k$      **c** $10a^2b - 15ab^2$

**a** First look at the numbers. These have a common factor of 2.

$m$ and $p$ do not occur in both terms but $y$ does. It is a common factor, so the factorisation is:

$6my + 4py = 2y(3m + 2p)$

**b** 5 is a common factor of 5 and 25 and $k$ is a common factor of $k^2$ and $k$.

$5k^2 - 25k = 5k(k - 5)$

**c** 5 is a common factor of 10 and 15, $a$ is a common factor of $a^2$ and $a$, $b$ is a common factor of $b$ and $b^2$.

$10a^2b - 15ab^2 = 5ab(2a - 3b)$

**Note:** If you multiply out each answer, you will get the expressions you started with.

## Exercise 9E

**1** Factorise the following expressions.

**a** $6m + 12t$      **b** $9t + 3p$      **c** $8m + 12k$

**d** $4r + 8t$      **e** $mn + 3m$      **f** $5g^2 + 3g$

**g** $4w - 6t$      **h** $3y^2 + 2y$      **i** $4t^2 - 3t$

**j** $3m^2 - 3mp$      **k** $6p^2 + 9pt$      **l** $8pt + 6mp$

**m** $8ab - 4bc$      **n** $5b^2c - 10bc$      **o** $8abc + 6bed$

**p** $4a^2 + 6a + 8$      **q** $6ab + 9bc + 3bd$      **r** $5t^2 + 4t + at$

**s** $6mt^2 - 3mt + 9m^2t$      **t** $8ab^2 + 2ab - 4a^2b$      **u** $10pt^2 + 15pt + 5p^2t$

**2** Three friends have a meal together. They each have a main meal costing £6.75 and a dessert costing £3.25.

Chris says that the bill will be $3 \times 6.75 + 3 \times 3.25$.

Mary says that she has an easier way to work out the bill as $3 \times (6.75 + 3.25)$.

**a** Why do Chris' and Mary's methods both give the correct answer?

**b** Why is Mary's method better?

**c** What is the total bill?

**3** Factorise the following expressions where possible. List those that do not factorise.

**a** $7m - 6t$      **b** $5m + 2mp$      **c** $t^2 - 7t$

**d** $8pt + 5ab$      **e** $4m^2 - 6mp$      **f** $a^2 + b$

**g** $4a^2 - 5ab$      **h** $3ab + 4cd$      **i** $5ab - 3b^2c$

**4** Three students are asked to factorise the expression $12m - 8$. These are their answers.

| Aidan | Bernice | Craig |
|---|---|---|
| $2(6m - 4)$ | $4(3m - 2)$ | $4m\left(3 - \frac{2}{m}\right)$ |

All the answers are accurately factorised, but only one is the normally accepted answer.

**a** Which student gave the correct answer?

**b** Why are the other two students' answers not acceptable as correct answers?

 **5** Why can't $5m + 6p$ be factorised?

 **6** Show that the perimeter of this shape can be written as $8(2x + 3)$.

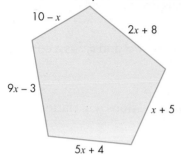

  **7** Alvin has correctly factorised the top and bottom of an algebraic fraction and cancelled the terms to give a final answer of $2x$. Unfortunately he spilt coffee on some of his work. What was the original fraction?

$$\frac{4x^2 - \blacksquare}{2(x \blacksquare} = \frac{4x \blacksquare}{\blacksquare - 3)} = 2x$$

# 9.5 Quadratic expansion

This section will show you how to:

- expand two linear brackets to obtain a quadratic expression.

> **Key terms**
>
> quadratic expansion
>
> quadratic expression

A **quadratic expression** is one where the highest power of the variables is 2. For example:

$$y^2 \qquad 3t^2 + 5t \qquad 5m^2 + 3m + 8$$

You can expand an expression such as $(3y + 2)(4y - 5)$ to give a quadratic expression.

Multiplying out pairs of these brackets is usually called **quadratic expansion**.

The rule for expanding expressions such as $(t + 5)(3t - 4)$ is similar to the rule for expanding single brackets:

multiply everything in one set of brackets by everything in the other set of brackets.

There are several methods for doing this. Examples 9 to 11 show the three main methods: expansion, FOIL and the box method.

**Expansion method**

Split the terms in the first set of brackets and make each of them multiply both terms in the second set of brackets. Then simplify the outcome.

Expand $(x + 3)(x + 4)$.

$$(x + 3)(x + 4) = x(x + 4) + 3(x + 4)$$
$$= x^2 + 4x + 3x + 12$$
$$= x^2 + 7x + 12$$

## Exercise 9F

**1**  Use the expansion method to expand each expression.

   **a**  $(x + 3)(x + 2)$     **b**  $(t + 4)(t + 3)$

   **c**  $(w + 1)(w + 3)$     **d**  $(m + 5)(m + 1)$

**2**  Use the expansion method to expand each expression.

   **a**  $(p + 10)(p - 7)$     **b**  $(u - 8)(u - 4)$

   **c**  $(k - 3)(k + 5)$     **d**  $(z - 9)(z - 3)$

 **3**  Find the mistake in each expansion.

   **a**  $(v + 5)(v + 7) = v^2 + 12v + 30$

   **b**  $(w - 8)(w + 10) = w^2 + 2w + 80$

   **c**  $(x - 6)(x - 4) = x^2 - 2x + 24$

   **d**  $(y + 11)(y + 1) = y^2 + 11y + 11$

   **e**  $(z - 7)(z - 2) = z^2 + 9z + 14$

**FOIL**

FOIL stands for First, Outer, Inner and Last. This is the order of multiplying the terms from each set of brackets.

Expand $(t + 5)(t - 2)$.

First terms give:    $t \times t = t^2$

Outer terms give: $t \times -2 = -2t$

Inner terms give:    $5 \times t = 5t$

Last terms give: $+5 \times -2 = -10$

$$(t + 5)(t - 2) = t^2 - 2t + 5t - 10$$
$$= t^2 + 3t - 10$$

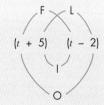

## Exercise 9G

**1** Use the FOIL method to expand each expression.

  **a** $(k + 3)(k + 5)$        **b** $(a + 4)(a + 1)$        **c** $(x + 4)(x - 2)$

  **d** $(t + 5)(t - 3)$        **e** $(w + 3)(w - 1)$        **f** $(f + 2)(f - 3)$

**2** Use the FOIL method to expand each expression.

  **a** $(r - 2)(r - 8)$        **b** $(s - 10)(s - 7)$        **c** $(d - 1)(d - 16)$

  **d** $(m - 6)(m - 3)$        **e** $(q - 9)(q - 11)$        **f** $(y - 5)(y - 8)$

**3** Complete each expansion.

  **a** $(a + 7)(a + 13) = a^2 + \ldots\ldots + 91$

  **b** $(b - 8)(b + 11) = b^2 + \ldots\ldots - 88$

  **c** $(c + 20)(c + 10) = c^2 + 30c + \ldots\ldots$

  **d** $(d - 7)(d - 4) = d^2 \ldots\ldots + 28$

  **e** $(e + 14)(e - 2) = e^2 + \ldots\ldots - \ldots\ldots$

---

**Example 11**

**Box method**

The box method is similar to the method for long multiplication.

Expand $(k - 3)(k - 2)$.

$(k - 3)(k - 2) = k^2 - 3k - 2k + 6$

$\qquad\qquad = k^2 - 5k + 6$

| × | $k$ | $-3$ |
|---|-----|------|
| $k$ | $k^2$ | $-3k$ |
| $-2$ | $-2k$ | $+6$ |

---

**Warning!** Be careful with the signs. This is the main place where mistakes are made in questions involving expanding brackets.

Note that, whichever method you use, it is important to show that you know there are four terms in the expansion before it is simplified.

## Exercise 9H

**1** Use the box method to expand each expression.

  **a** $(g + 1)(g - 4)$     **b** $(y + 4)(y - 3)$     **c** $(x - 3)(x + 4)$

  **d** $(p - 2)(p + 1)$     **e** $(k - 4)(k + 2)$     **f** $(y - 2)(y + 5)$

  **g** $(a - 1)(a + 3)$

> **Hints and tips**   A common error is to get minus signs wrong. $-2x - 3x = -5x$ and $-2 \times -3 = +6$

**2** The expansions in this question follow a pattern. Work out the first few and try to spot the pattern that will allow you to immediately write down the rest of the answers.

  **a** $(x + 3)(x - 3)$     **b** $(t + 5)(t - 5)$     **c** $(m + 4)(m - 4)$

  **d** $(t + 2)(t - 2)$     **e** $(y + 8)(y - 8)$     **f** $(p + 1)(p - 1)$

  **g** $(5 + x)(5 - x)$     **h** $(7 + g)(7 - g)$     **i** $(x - 6)(x + 6)$

**(PS)** **3** This rectangle is made up of four parts with areas of $x^2$, $2x$, $3x$ and $6$ square units.

Work out expressions for the sides of the rectangle, in terms of $x$.

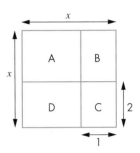

| $x^2$ | $2x$ |
|---|---|
| $3x$ | $6$ |

**(PS)** **4** This square has an area of $x^2$ square units.

It is split into four rectangles.

**a** Fill in the table below to show the dimensions and area of each rectangle.

| Rectangle | Length | Width | Area |
|---|---|---|---|
| A | $x - 1$ | $x - 2$ | $(x - 1)(x - 2)$ |
| B | | | |
| C | | | |
| D | | | |

**b** Add together the areas of rectangles B, C and D.

Expand any brackets and collect like terms together.

**c** Use the results to show why $(x - 1)(x - 2) = x^2 - 3x + 2$.

**(CM)** **5** Expand the expressions in parts **a** to **c**.

**a** $(x + 1)(x + 1)$    **b** $(x - 1)(x - 1)$    **c** $(x + 1)(x - 1)$

## Quadratic expansion with non-unit coefficients

All the terms in $x^2$ in Exercises 9F to 9H have a coefficient of 1 or –1.

The next two examples show you how to expand brackets containing terms in $x^2$ with coefficients that are not 1 or –1.

**Example 12**

Expand $(2t + 3)(3t + 1)$.

$(2t + 3)(3t + 1) = 6t^2 + 9t + 2t + 3$

$\qquad\qquad\qquad = 6t^2 + 11t + 3$

| × | $2t$ | $+3$ |
|---|---|---|
| $3t$ | $6t^2$ | $+9t$ |
| $+1$ | $+2t$ | $+3$ |

**Example 13**

Expand $(4x - 1)(3x - 5)$.

$(4x - 1)(3x - 5) = 4x(3x - 5) - (3x - 5)$     **Note:** $(3x - 5)$ is the same as $1(3x - 5)$.

$\qquad\qquad\qquad = 12x^2 - 20x - 3x + 5$

$\qquad\qquad\qquad = 12x^2 - 23x + 5$

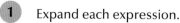

**1** Expand each expression.

**a** $(2x + 3)(3x + 1)$      **b** $(3y + 2)(4y + 3)$      **c** $(3t + 1)(2t + 5)$

**d** $(4t + 3)(2t - 1)$      **e** $(5m + 2)(2m - 3)$      **f** $(4k + 3)(3k - 5)$

**g** $(3p - 2)(2p + 5)$      **h** $(5w + 2)(2w + 3)$      **i** $(2a - 3)(3a + 1)$

**j** $(4r - 3)(2r - 1)$      **k** $(3g - 2)(5g - 2)$      **l** $(4d - 1)(3d + 2)$

**m** $(5 + 2p)(3 + 4p)$      **n** $(2 + 3t)(1 + 2t)$      **o** $(4 + 3p)(2p + 1)$

**p** $(6 + 5t)(1 - 2t)$      **q** $(4 + 3n)(3 - 2n)$      **r** $(2 + 3f)(2f - 3)$

**s** $(3 - 2q)(4 + 5q)$      **t** $(1 - 3p)(3 + 2p)$      **u** $(4 - 2t)(3t + 1)$

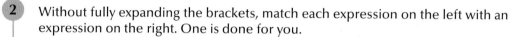

> **Hints and tips**   Give answers in the form $\pm ax^2 \pm bx \pm c$ even when the quadratic coefficient is negative.

**2** Without fully expanding the brackets, match each expression on the left with an expression on the right. One is done for you.

$(3x - 2)(2x + 1)$          $4x^2 - 4x + 1$

$(2x - 1)(2x - 1)$          $6x^2 - x - 2$

$(6x - 3)(x + 1)$          $6x^2 + 7x + 2$

$(4x + 1)(x - 1)$          $6x^2 + 3x - 3$

$(3x + 2)(2x + 1)$          $4x^2 - 3x - 1$

**3** Work out the pattern in each of the expressions in parts **a–o** so that you can write down the expansion without fully expanding the brackets.

**a** $(2x + 1)(2x - 1)$      **b** $(3t + 2)(3t - 2)$      **c** $(5y + 3)(5y - 3)$

**d** $(4m + 3)(4m - 3)$      **e** $(2k - 3)(2k + 3)$      **f** $(4h - 1)(4h + 1)$

**g** $(2 + 3x)(2 - 3x)$      **h** $(5 + 2t)(5 - 2t)$      **i** $(6 - 5y)(6 + 5y)$

**j** $(a + b)(a - b)$      **k** $(3t + k)(3t - k)$      **l** $(2m - 3p)(2m + 3p)$

**m** $(5k + g)(5k - g)$      **n** $(ab + cd)(ab - cd)$      **o** $(a^2 + b^2)(a^2 - b^2)$

(PS) **4** The diagram shows a square of side $a$ units with a square of side $b$ units cut from one corner.

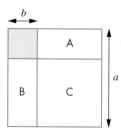

**a** Write an expression for the area remaining after the small square is cut away.

**b** The remaining area is cut into rectangles A, B and C, and rearranged as shown.

Write down the dimensions and area of the rectangle formed by A, B and C.

**c** Why does $a^2 - b^2 = (a + b)(a - b)$?

| B |
|---|
| A |
| C |

**5** Show that the areas that are shaded in these diagrams are the same.

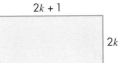

## Expanding squares

Whenever you see a linear bracketed term squared, such as $(x - 2)^2$, write the brackets down twice and then use whichever method you prefer to expand.

**Example 14**

Expand $(x + 3)^2$.

$(x + 3)^2 = (x + 3)(x + 3)$

$\qquad = x(x + 3) + 3(x + 3)$

$\qquad = x^2 + 3x + 3x + 9$

$\qquad = x^2 + 6x + 9$

**Example 15**

Expand $(3x - 2)^2$.

$(3x - 2)^2 = (3x - 2)(3x - 2)$

$\qquad = 9x^2 - 6x - 6x + 4$

$\qquad = 9x^2 - 12x + 4$

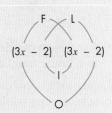

## Exercise 9J

**1** Expand the squares and simplify.

**a** $(x + 5)^2$    **b** $(m + 4)^2$    **c** $(6 + t)^2$    **d** $(3 + p)^2$

**e** $(m - 3)^2$    **f** $(t - 5)^2$    **g** $(4 - m)^2$    **h** $(7 - k)^2$

> **Hints and tips** Always write down the brackets twice. Do not try to take any short cuts.

**2** Expand the squares and simplify.

**a** $(3x + 1)^2$    **b** $(4t + 3)^2$    **c** $(2 + 5y)^2$    **d** $(3 + 2m)^2$

**e** $(4t - 3)^2$    **f** $(3x - 2)^2$    **g** $(2 - 5t)^2$    **h** $(6 - 5r)^2$

**i** $(x + y)^2$    **j** $(m - n)^2$    **k** $(2t + y)^2$    **l** $(m - 3n)^2$

**m** $(x + 2)^2 - 4$    **n** $(x - 5)^2 - 25$    **o** $(x + 6)^2 - 40$    **p** $(x - 2)^2 + 4$

**3** A teacher asks her class to expand $(3x + 1)^2$.

Bernice's answer is $9x^2 + 1$.

Pete's answer is $3x^2 + 6x + 1$.

  **a** What mistakes has Bernice made?

  **b** What mistakes has Pete made?

  **c** Work out the correct answer.

  **4** Use the diagram and algebra to show that $(2x - 1)^2 = 4x^2 - 4x + 1$.

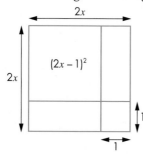

# 9.6 Quadratic factorisation

## This section will show you how to:

- factorise a quadratic expression of the form $x^2 + ax + b$ into two linear brackets.

Factorisation involves putting a quadratic expression back into its brackets (if possible).

You are going to factorise quadratic expressions of the type:

$x^2 + ax + b$

where $a$ and $b$ are integers.

There are some rules that will help you to factorise.

- The expression inside each set of brackets will start with an $x$.

- The signs in the quadratic expression show which signs to put after the $x$s.

  ○ When the second sign in the expression is positive, the signs in both sets of brackets are the same as the first sign.

    $x^2 + ax + b = (x + ?)(x + ?)$    since everything is positive

    $x^2 - ax + b = (x - ?)(x - ?)$    since $-\text{ve} \times -\text{ve} = +\text{ve}$

  ○ When the second sign is negative, the signs in the brackets are different.

    $x^2 + ax - b = (x + ?)(x - ?)$    since $+\text{ve} \times -\text{ve} = -\text{ve}$

    $x^2 - ax - b = (x + ?)(x - ?)$

  ○ When multiplied together, the two numbers in the brackets must give $b$.

  ○ The sum of the two numbers in the brackets give the coefficient of $x$, $a$.

**Example 16**

Factorise $x^2 - x - 6$.

The signs show that the brackets must be $(x + ?)(x - ?)$.

Two numbers that have a product of $-6$ and a sum of $-1$ are $-3$ and $+2$.

So, $x^2 - x - 6 = (x + 2)(x - 3)$.

**Example 17**

Factorise $x^2 - 9x + 20$.

The signs show that the brackets must be $(x - ?)(x - ?)$.

Two numbers that have a product of $+20$ and a sum of $-9$ are $-4$ and $-5$.

So, $x^2 - 9x + 20 = (x - 4)(x - 5)$.

## Exercise 9K

Factorise the expressions in questions **1–10**.

**1**　　**a** $x^2 + 5x + 6$　　**b** $t^2 + 5t + 4$　　**c** $m^2 + 7m + 10$　　**d** $k^2 + 10k + 24$

**2**　　**a** $p^2 + 14p + 24$　　**b** $r^2 + 9r + 18$　　**c** $w^2 + 11w + 18$　　**d** $x^2 + 7x + 12$

**3**　　**a** $a^2 + 8a + 12$　　**b** $k^2 + 10k + 21$　　**c** $f^2 + 22f + 21$　　**d** $b^2 + 20b + 96$

**4**　　**a** $t^2 - 5t + 6$　　**b** $d^2 - 5d + 4$　　**c** $g^2 - 7g + 10$　　**d** $x^2 - 15x + 36$

**5**　　**a** $c^2 - 18c + 32$　　**b** $t^2 - 13t + 36$　　**c** $y^2 - 16y + 48$　　**d** $j^2 - 14j + 48$

**6**　　**a** $p^2 - 8p + 15$　　**b** $y^2 + 5y - 6$　　**c** $t^2 + 2t - 8$　　**d** $x^2 + 3x - 10$

**7**　　**a** $m^2 - 4m - 12$　　**b** $r^2 - 6r - 7$　　**c** $n^2 - 3n - 18$　　**d** $m^2 - 7m - 44$

**8**　　**a** $w^2 - 2w - 24$　　**b** $t^2 - t - 90$　　**c** $h^2 - h - 72$　　**d** $t^2 - 2t - 63$

**9**　　**a** $d^2 + 2d + 1$　　**b** $y^2 + 20y + 100$　　**c** $t^2 - 8t + 16$　　**d** $m^2 - 18m + 81$

**10**　　**a** $x^2 - 24x + 144$　　**b** $d^2 - d - 12$　　**c** $t^2 - t - 20$　　**d** $q^2 - q - 56$

> **Hints and tips** First decide on the signs in the brackets, then look at the numbers.

**11** This rectangle is made up of four parts. Two of the parts have areas of $x^2$ square units and 6 square units.

| $x^2$ | |
|---|---|
| | 6 |

The sides of the rectangle are of the form $x + a$ and $x + b$.

There are two possible answers for $a$ and $b$.

Work out both answers and copy and complete the areas in the other parts of the rectangle.

**12**  **a** Expand $(x + a)(x + b)$.

**b** If $x^2 + 7x + 12 = (x + p)(x + q)$, use your answer to part **a** to write down the values of:

  **i** $p + q$          **ii** $pq$.

(CM) **c** How can you tell that $x^2 + 12x + 7$ will not factorise?

## Difference of two squares

In Exercise 9H, you multiplied out, for example, $(a + b)(a - b)$ and obtained $a^2 - b^2$. This type of quadratic expression, with only two terms which are both perfect squares separated by a minus sign, is called the **difference of two squares**.

Look at the expansion of $(x - 7)(x + 7)$.

$(x - 7)(x + 7) = x^2 - 7x + 7x - 49$

$\qquad\qquad\qquad = x^2 - 49$

Note that when you collect the $-7x$ and $+7x$ terms, they add together to make zero.

This process will only work when the terms in the brackets are different signs, which is why the difference of two squares has a minus sign.

You can use this idea to factorise expressions written as the difference of two squares.

$\qquad x^2 - 9 = (x - 3)(x + 3)$

$\qquad x^2 - 25 = (x - 5)(x + 5)$

Three conditions must be met for the difference of two squares to work.

• There must be two terms.

• They must separated by a negative sign.

• Each term must be a perfect square, say $x^2$ and $n^2$.

When these three conditions are met, the factorisation is:     $x^2 - n^2 = (x + n)(x - n)$

---

**Example 18**

Factorise $x^2 - 9$.

Recognise the difference of two squares, $x^2$ and 9.

So it factorises to $(x + 3)(x - 3)$.

Expanding the brackets shows that they do come from the original expression.

---

## Exercise 9L

 **1**  Each of the expressions is the difference of two squares. Factorise them.

  **a** $x^2 - 9$          **b** $t^2 - 25$          **c** $m^2 - 16$

  **d** $9 - x^2$          **e** $49 - t^2$          **f** $k^2 - 100$

  **g** $4 - y^2$          **h** $x^2 - 64$          **i** $t^2 - 81$

  > **Hints and tips**  Learn how to spot the difference of two squares as you will come across them a lot.

**2**

**a** A square has a side of $x$ units.

Write an expression for the area of the square.

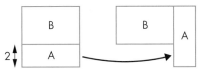

**b** A rectangle, A, 2 units wide, is cut from the square and placed at the side of the remaining rectangle, B.

A square, C, is then cut from the bottom of rectangle A to leave a final rectangle, D.

**i** What is the height of rectangle B?

**ii** What is the width of rectangle D?

**iii** What is the area of rectangle B plus rectangle A?

**iv** What is the area of square C?

**v** What is the area of rectangle D?

**c** Use your results to show why $x^2 - 4 = (x + 2)(x - 2)$.

 **3**

**a** Expand and simplify $(x + 2)^2 - (x + 1)^2$.

**b** Factorise $a^2 - b^2$.

**c** In your answer for part **b**, replace $a$ with $(x + 2)$ and $b$ with $(x + 1)$.

Expand and simplify the answer.

**d** What can you say about the answers to parts **a** and **c**?

**e** Simplify $(x + 1)^2 - (x - 1)^2$.

# 9.7 Changing the subject of a formula

**Key terms**

inverse operations

rearrange

subject

This section will show you how to:

• change the subject of a formula.

The **subject** is the variable (letter) in a formula or equation which stands on its own, usually on the left-hand side of the equals sign. For example, $x$ is the subject of each of the following formulae.

$$x = 5t + 4 \qquad x = 4(2y - 7) \qquad x = \frac{1}{t}$$

To change the existing subject to a different variable, you have to **rearrange** the formula to get that variable on the left-hand side. You do this by using **inverse operations**.

For example, to solve $2x + 7 = 13$:

Subtract 7 from both sides, which gives you $2x = 6$. Subtracting 7 is the inverse operation of adding 7.

Next, divide both sides by 2, which gives you $x = 3$. Dividing by 2 is the inverse operation of multiplying by 2.

Make $m$ the subject of this formula. $\qquad T = m - 3$

Add 3 to both sides so that the $m$ is on its own. $T + 3 = m$

Rewrite with $m$ on the left hand side. $\qquad\qquad m = T + 3$

Example 21

From the formula $P = 4t$, express $t$ in terms of $P$.

This is another way of asking you to make $t$ the subject.

$$P = 4t$$

Divide both sides by 4.    $\dfrac{P}{4} = \dfrac{4t}{4}$    $4 \div 4 = 1$

Rewrite with $t$ on the left-hand side.    $t = \dfrac{P}{4}$

---

Example 22

Make $m$ the subject of the formula $C = 2m^2 + 3$.

Subtract 3 from each side so that the $2m^2$ is on its own.  $C - 3 = 2m^2$

Divide both sides by 2.    $\dfrac{C - 3}{2} = \dfrac{2m^2}{2}$

Rewrite with $m^2$ on the left-hand side.    $m^2 = \dfrac{C - 3}{2}$

Take the square root of both sides.    $m = \sqrt{\dfrac{C - 3}{2}}$    Taking the square root is the inverse operation of squaring.

---

Example 23

The formula $v^2 = u^2 + 2as$ relates the final velocity ($v$) of an object to its initial velocity ($u$), its acceleration ($a$) and the distance moved ($s$).

**a**  Make $a$ the subject of the formula.

**b**  Work out the acceleration of a particle that has an initial velocity of 7 m/s, a final velocity of 11 m/s and a distance moved of 6 m.

**a**  $v^2 = u^2 + 2as$

Subtract $u^2$ from both sides so that the $2as$ is on its own.        $v^2 - u^2 = 2as$

Divide both sides by $2s$.    $a = \dfrac{v^2 - u^2}{2s}$

**b**  Substitute $u = 7$, $v = 11$ and $s = 6$.    $a = \dfrac{11^2 - 7^2}{2 \times 6}$

$$= \dfrac{121 - 49}{12}$$

$$= \dfrac{72}{12}$$

$$= 6 \text{ m/s}^2$$

## Exercise 9M

**1**  $T = 3k$        Make $k$ the subject.

**2**  $X = y - 1$        Express $y$ in terms of $X$.

**3**  $Q = \dfrac{p}{3}$        Express $p$ in terms of $Q$.

**4**  $A = 4r + 9$        Make $r$ the subject.

**5**  $p = m + t$        **a** Make $m$ the subject.        **b** Make $t$ the subject.

**6** $g = \dfrac{m}{v}$      Make $m$ the subject.

**7** $t = m^2$      Make $m$ the subject.

**8** $P = 2l + 2w$      Make $l$ the subject.

**9** $m = p^2 + 2$      Make $p$ the subject.

**10** The formula for converting Fahrenheit degrees to Celsius degrees is $C = \dfrac{5}{9}(F - 32)$.

     **a** Show that when $F = -40$, $C$ is also equal to $-40$.

     **b** Calculate the value of $C$ when $F = 68$.

     **c** Use this flow diagram to work out the formula for converting Celsius degrees to Fahrenheit degrees.

**(PS) 11** Distance, speed and time are connected by the formula:

distance = speed × time.

A delivery driver drove 126 km in 1 hour and 45 minutes. On the return journey, he was held up at some road works so his average speed decreased by 9 km per hour.

How long was he held up at the road works?

> **Hints and tips**   Work out the average speed for the first journey, then work out the average speed for the return journey.

**12** Kieran notices that the price of five buns is 75p more than the price of nine pies.

Let the price of a bun be $x$ pence and the price of a pie be $y$ pence.

     **a** Give the cost of one pie, $y$, in terms of the price of a bun, $x$.

     **b** The price of a bun is 60p and the price of a pie is 25p. Check that your formula is correct.

> **Hints and tips**   Set up a formula, using the information in the first sentence, then rearrange it.

**13** $v = u + at$      **a** Make $a$ the subject.      **b** Make $t$ the subject.

**14** $A = \dfrac{1}{4}\pi d^2$      Make $d$ the subject.

**15** $x = 5y - w$      **a** Make $y$ the subject.      **b** Express $w$ in terms of $x$ and $y$.

**16** $k = 2p^2$      Make $p$ the subject.

**17** $v = u^2 - t$      **a** Make $t$ the subject.      **b** Make $u$ the subject.

**18** $K = 5n^2 + w$      **a** Make $w$ the subject.      **b** Make $n$ the subject.

**(MR) 19**   **a** $U = K - \dfrac{P}{3D - Y}$   Make $D$ the subject.

     **b** Work out the value of $D$ when $U = -3$, $Y = 37$, $P = 77$ and $K = 4$.

# Worked exemplars

  Three students factorised $4x^2 - 8xy$.

Aliki wrote $4(x^2 - xy)$.

Bavleen wrote $4x(x - 2y)$.

Craig wrote $2x(2x - 4xy)$.

One student was given two marks, one was given one mark and one was given zero marks.

Which student scored which mark? Give reasons for your answers.

| | |
|---|---|
| This is a 'communicating mathematically' question. This means that you need to explain each step of your reasoning clearly. | |
| The correct factorisation is $4x(x - 2y)$. | First you need to factorise the expression to work out the correct answer. |
| Aliki scored zero marks. Her factorisation is incorrect because $4(x^2 - xy)$ equals $4x^2 - 4xy$ not $4x^2 - 8xy$. | Then you can compare each student's answer with the correct answer and assess what they got right. |
| Bavleen scored two marks. Her factorisation is correct and complete. There are no more factors in her brackets $(x - 2y)$. | |
| Craig scored one mark. His factorisation is correct: $2x(2x - 4xy)$ does equal $4x^2 - 8xy$ but it is only partly factorised. His brackets $(2x - 4xy)$ can be factorised further. | |

**2** A rectangle has a length of $2x + 4$ and width of $x + 2$.

**a** Show that the perimeter can be written as $6(x + 2)$.

**b** Mark says that the perimeter must always be an even number.

Prove that Mark is wrong.

---

Part **a** is a problem-solving question, which means that you need to find the connections between two topics, in this case collecting like terms and finding the perimeter. You need to show your method clearly.

| | |
|---|---|
| **a** Perimeter = $2 \times (2x + 4) + 2 \times (x + 2)$<br>$\quad\quad\quad\quad = 2 \times 2(x + 2) + 2(x + 2)$<br>$\quad\quad\quad\quad = 4(x + 2) + 2(x + 2)$<br>$\quad\quad\quad\quad = 6(x + 2)$<br>Alternatively,<br>Perimeter = $2x + 4 + 2x + 4 + x + 2 + x + 2$<br>$\quad\quad\quad\quad = 6x + 12$<br>$\quad\quad\quad 6x + 12 = 6(x + 2)$ | Write down a correct unsimplified expression for the perimeter.<br><br>You have to write your answer of $6(x + 2)$ by showing a complete accurate proof. |
| **b** Mark is wrong because although $6(x + 2)$ is always even when $(x + 2)$ is a whole number, there are values of $x$ for which $(x + 2)$ is not a whole number.<br><br>Example: $x = 2.5$, $6(x + 2) = 6 \times 4.5 = 27$<br>27 is not even.<br><br>[Alternatively, $6(x + 2) = 15$<br>$x + 2 = 2.5$<br>$x = 0.5$<br>It is possible for the perimeter to be a number which is not even.] | In part **b** you need to check if the statement is valid.<br><br>Either choose a value of $x$ for which the perimeter is not even (such as 2.5) or choose a perimeter which is not even (such as 15) and find the corresponding value of $x$.<br><br>Note that there are many answers. Any value that makes the value inside the brackets 'something and a half', such as $x = 1.5$ or $x = 2.5$ will work. |

9 Worked exemplars

227

# Ready to progress?

I can substitute numbers into expressions and formulae.

I can expand linear brackets and simplify expressions by collecting like terms.
I can factorise linear expressions.
I can expand two linear brackets.
I can rearrange simple formulae.

I can factorise quadratic expressions of the form $x^2 + ax + b$.
I can recognise the difference of two squares.

# Review questions

**1**  Maureen buys $x$ pencils each costing $y$p.

Write an expression for the amount of change she should get if she pays $t$p.

**(PS) 2**  **a**  A triangle has angles of $p°$ and $q°$. Write an expression for the size of the other angle.

**b**  Calculate the size of the other angle when $p = 26$ and $q = 95$.

**3**  Make $r$ the subject of these formulae.

**a**  $C = 2\pi r$

**b**  $A = \pi r^2$

**(PS) 4**  **a**  Write an expression for the sum of $(2x + 9)$, $(6x − 5)$, $(3x + 11)$ and $(9x + 1)$.

**b**  Work out the mean of the expressions $(2x + 9)$, $(6x − 5)$, $(3x + 11)$ and $(9x + 1)$.

**5**  $R = 5c − 9 + 3d + 2(c + 7)$

**a**  Make $c$ the subject of the formula.

**b**  Work out $c$ when $R = 20$ and $d = −9$.

**6**  Expand the brackets.        $(3x + 5y)(4x − 7y)$

**(PS) 7**  A formula for calculating a dose ($D$ ml) of medicine for a person of age $a$ is given by $D = \dfrac{4(a + 7)}{a + 10}$.

**a**  Calculate the dose for a person aged:

    **i** 14                   **ii** 30                   **iii** 65.

**b**  Work out the age of a person who would take a dose of:

    **i** 3.625 ml           **ii** 3.75 ml           **iii** 3.88 ml.

**(PS) 8**  Work out the area of a triangle with a base of $(5x − 11)$ metres and a height of $\dfrac{x^2 − 7}{x − 1}$ metres when $x = 4$.

**(PS) 9** The approximate surface area, $A$, of a cylinder with a radius, $r$, and a height, $h$ is given by the formula $A = 6r^2 + 6rh$.

  **a** Factorise $6r^2 + 6rh$.

  **b** Make $h$ the subject of the formula.

  **c** Calculate the height of a cylinder which has an area of 624 cm² and a radius of 8 cm.

**10** $y = 3(2x - 5)(x - 6)$

  **a** Show that when $x = 10$, $y = 180$.

  **b** Expand the expression.

**11** The formula $A = y + 0.01xy$ calculates the size of an amount $y$ after it has been increased by $x\%$.

  **a** Make $x$ the subject of the formula.

  **b** Factorise $y + 0.01xy$.

**(PS)**  **c** Use the formula to calculate the mass after 20 g has been increased by 10%.

**(PS) 12** A rectangle has a base of $(3x - 8)$ cm and a height of $(5x + 7)$ cm.

  **a** Work out simplified expressions for:     **i** its area

                                                 **ii** its perimeter.

  **b** If the perimeter of the rectangle is 70 cm, calculate its area.

**(MR) 13** **a** Expand $(2x + 1)^2$.

  **b** By substituting $x = 10$, work out the value of $21^2$.

**(PS) 14** The formula $v^2 = u^2 + 2as$ relates the final velocity ($v$) of an object to its initial velocity ($u$), its acceleration ($a$) and the distance moved ($s$).

  The formula $v = u + at$ relates the final velocity ($v$) to the initial velocity ($u$), the acceleration ($a$) and the time ($t$).

  A ball rolls for 2 seconds, increasing its velocity from 8 m/s to 12 m/s. Work out the distance moved by the ball.

**15** Factorise these expressions.

  **a** $2x - 16$        **b** $x^2 - 16x$

  **c** $x^2 - 16$       **d** $x^2 - 16x + 63$

**16** Factorise these expressions.

  **a** $x^2 - 100$        **b** $x^2 - 400$        **c** $x^2 - 40\,000$

**(CM) 17** Mark says that $x^2 - 20x - 21 = (x - 3)(x + 7)$.

  What is wrong with Mark's answer? Give the correct factorisation.

# 10 Ratio, proportion and rates of change: Ratio, speed and proportion

## This chapter is going to show you:

- what a ratio is
- how to divide an amount in a given ratio
- how to calculate speed
- how to solve problems involving direct proportion
- how to compare prices of products.

## You should already know:

- multiplication tables up to $12 \times 12$
- how to simplify fractions
- how to find a fraction of a quantity
- how to multiply and divide, with and without a calculator.

## About this chapter

This chapter is about comparing pieces of information. You use fractions, decimals, percentages, ratio and proportion in everyday life to help calculate quantities or to compare two or more pieces of information.

Shops and supermarkets always seem to have sales and offers. Every time you go shopping you see signs such as: '3 for 2', '2 for £3', 'Buy one, get one free', 'Save 25%' and 'A third off today only!' You need a good understanding of fractions, decimals, ratio and proportion to know if these offers are worthwhile. For example, which is the best bargain: '3 for the price of 2' or 'A third off the price of 1'? In fact they are the same.

# 10.1 Ratio

This section will show you how to:

- simplify a ratio
- express a ratio as a fraction
- divide amounts into given ratios
- complete calculations from a given ratio and partial information.

A ratio is a way of comparing the sizes of two or more quantities.

A ratio can be expressed in a number of ways. For example, the ratio of the quantities when you mix 5 centilitres of cordial with 20 centilitres of water is shown below.

```
cordial  :  water
   5     :    20
   1     :     4   (Divide both sides by 5 to simplify.)
```

> **Hints and tips** When you say a ratio, you do not say '5 colon 20' or '1 colon 4', you say '5 to 20' or '1 to 4'.

When you are comparing ratios, you may find it helpful to use a table. So if the ratio is 5 : 20 (as above), you can summarise the quantities of cordial and water in a table like this one.

| Cordial | 5 | 1 | 2 | 4 | 10 | 25 |
| --- | --- | --- | --- | --- | --- | --- |
| Water | 20 | 4 | 8 | 16 | 40 | 100 |

The value in each column is simply a multiplier or divisor of the value in a previous column.

This method is useful if you want to know how much cordial to mix with a litre (100 centilitres) of water. The last column shows that the answer is 25 centilitres.

How much water would you need if you only have 15 centilitres of cordial? You can find out by adding the numbers in the first and fifth columns, so the answer is 60 centilitres.

## Common units

When working with a ratio involving different units, always convert them all to a **common unit**. You can only simplify a ratio when the units of each quantity are the same, because the ratio itself has no units. You do this by dividing all the parts of the ratio by a common factor. This is the same idea you use when **cancelling** a fraction to its simplest form.

For example, you must convert the ratio of 125 g to 2 kg to the ratio of 125 g to 2000 g, so that you can simplify it.

```
125  :  2000
  5  :    80   (Divide both sides by 25.)
  1  :    16   (Divide both sides by 5.)
```

You can simplify the ratio 125 : 2000 to 1 : 16.

When a ratio has been simplified so that its parts do not have any common factors, it is in its **simplest form**.

Example 1

Express 25 minutes : 1 hour as a ratio in its simplest form.

The units must be the same, so change 1 hour into 60 minutes.

25 minutes : 1 hour = 25 minutes : 60 minutes

$\qquad\qquad$ = 25 : 60 $\qquad$ Cancel the units (minutes).

$\qquad\qquad$ = 5 : 12 $\qquad$ Divide both sides by 5.

So, 25 minutes : 1 hour simplifies to 5 : 12.

## Ratios as fractions

You can express ratios as fractions by using the total number of parts in the ratio as the denominator (bottom number) of each fraction. Then use the numbers in the ratio as the numerators. Always simplify a ratio before converting it to fractions. If the ratio is in its simplest form, the fractions will not cancel.

Example 2

A garden is divided into lawn and shrubs in the ratio 3 : 2.

What fraction of the garden is covered by:

**a** lawn $\qquad$ **b** shrubs?

The denominator (bottom number) of the fraction comes from adding the numbers in the ratio (so 2 + 3 = 5).

**a** $\quad$ The lawn covers $\frac{3}{5}$ of the garden.

**b** $\quad$ The shrubs cover $\frac{2}{5}$ of the garden.

## Exercise 10A

**1** $\quad$ Express each ratio in its simplest form.

$\qquad$ **a** $\;$ 6 : 18 $\qquad$ **b** $\;$ 16 : 24 $\qquad$ **c** $\;$ 20 to 50 $\qquad$ **d** $\;$ 25 to 40

$\qquad$ **e** $\;$ 15 : 10 $\qquad$ **f** $\;$ 28 to 12 $\qquad$ **g** $\;$ 0.5 to 3 $\qquad$ **h** $\;$ 2.5 to 1.5

**2** $\quad$ Write each ratio of quantities in its simplest form

$\qquad$ **a** $\;$ £5 to £15 $\qquad$ **b** $\;$ 125 g to 300 g $\qquad$ **c** $\;$ 34 kg to 30 kg $\qquad$ **d** $\;$ 3 kg to 750 g

$\qquad$ **e** $\;$ 1 hour to 1 day $\;$ **f** $\;$ 1.25 kg : 500 g $\qquad$ **g** $\;$ 4 weeks : 14 days $\quad$ **h** $\;$ 465 mm : 3 m

> **Hints and tips** $\quad$ Remember to express both parts in the same units before you simplify.

**3** $\quad$ A length of wood is cut into two pieces in the ratio 3 : 7. What fraction of the original length is the longer piece?

**4** $\quad$ Jack and Thomas find a bag of marbles. They share the marbles in the ratio of their ages. Jack is 10 years old and Thomas is 15 years old. What fraction of the marbles did Jack get?

**5** $\quad$ Dave and Sasha share a pizza in the ratio 2 : 3. They eat it all.

$\qquad$ **a** $\;$ What fraction of the pizza did Dave eat?

$\qquad$ **b** $\;$ What fraction of the pizza did Sasha eat?

**6** Amy gets $\frac{2}{3}$ of a packet of sweets. Her sister Susan gets the rest. Work out the ratio of sweets that each sister gets. Write it in the form Amy : Susan.

**7** **a** The recipe for a fruit punch is 1.25 litres of fruit crush to 6.25 litres of lemonade. What fraction of the punch is each ingredient?

**b** How much fruit crush will you need to mix with 2 litres of lemonade?

Hints and tips | Set up a table.

**c** You have half a litre of fruit crush. How much lemonade will you need?

**8** The recipe for a pudding is 125 g of sugar, 150 g of flour, 100 g of margarine and 175 g of fruit. What fraction of the pudding is made up by each ingredient?

(MR) **9** Andy plays 16 bowls matches. He wins $\frac{3}{4}$ of them.

He plays another $x$ matches and wins them all.

The ratio of wins : losses is now 4 : 1.

Work out the value of $x$.

(MR) **10** Three brothers share some money.

The ratio of Mark's share to David's share is 1 : 2.

The ratio of David's share to Paul's share is 1 : 2.

What is the ratio of Mark's share to Paul's share?

(PS) **11** Three brothers, Jarek, Jerzy and Justyn, share a block of chocolate in the ratio of their ages. Jarek gets half of the bar. Jerzy gets $\frac{3}{5}$ of the rest.

**a** What ratio, Jarek : Jerzy : Justyn, of the bar of chocolate does each brother get?

**b** Justyn is 8 years old. How old is Jarek?

(PS) **12** Three cows, Gertrude, Gladys and Henrietta, produced milk in the ratio 2 : 3 : 4. Henrietta produced $1\frac{1}{2}$ litres more than Gladys. How much milk did the three cows produce altogether?

## Dividing amounts in a given ratio

To divide an amount in a given ratio, you first look at the ratio to see how many parts there are altogether.

For example, 4 : 3 has 4 parts and 3 parts giving 7 parts altogether.

The whole amount is 7 parts.

You can work out 1 part by dividing the whole amount by 7.

Then you can work out 3 parts and 4 parts from 1 part.

**Example 3**

Divide £28 in the ratio 4 : 3.

There are two methods you can use.

**Method 1: Using a table**

Set up the first column then continue the columns as multiples.

| First part | 4 | 8 | 12 | 16 |
|------------|---|---|----|----|
| Second part | 3 | 6 | 9 | 12 |
| Total | 7 | 14 | 21 | 28 |

So £28 divided in the ratio 4 : 3 is £16 : £12.

*(continued)*

**Method 2: Dividing to find one part**

4 + 3 = 7 parts altogether.

So 7 parts = £28.

Divide by 7 to find 1 part.          1 part = £4

Multiply by 4 to find 4 parts.      4 × £4 = £16

Multiply by 3 to find 3 parts.      3 × £4 = £12

So £28 divided in the ratio 4 : 3 is £16 : £12.

You can also use fractions to divide an amount in a given ratio. First, express the whole numbers in the ratio as fractions with the same common denominator. Then multiply the amount by each fraction.

Example 4

Divide £40 between Peter and Hitan in the ratio 2 : 3.

**Method 1: Using a table**

Set up the first column then work out what you need to multiply 5 by, to get 40.

Do the same thing to the other rows.

| Peter | 2 | 2 × 8 = 16 |
|-------|---|------------|
| Hitan | 3 | 3 × 8 = 24 |
| Total | 5 | 5 × 8 = 40 |

So Peter receives 2 × 8 = £16 and Hitan receives 3 × 8 = £24.

**Method 2: Using fractions**

Change the ratio to fractions.

Peter's share = $\dfrac{2}{(2+3)} = \dfrac{2}{5}$          Hitan's share = $\dfrac{3}{(2+3)} = \dfrac{3}{5}$

So Peter receives £40 × $\dfrac{2}{5}$ = £16 and Hitan receives £40 × $\dfrac{3}{5}$ = £24.

## Exercise 10B

**1** Divide each amount in the given ratio.

  **a** 400 g in the ratio 2 : 3      **b** 280 kg in the ratio 2 : 5      **c** 5 hours in the ratio 7 : 5

  **d** £100 in the ratio 2 : 3 : 5    **e** £240 in the ratio 3 : 5 : 12    **f** 600 g in the ratio 1 : 5 : 6

**2** The ratio of women to men at Lakeside Gardening Club is 7 : 3. The total number in the group is 250.

  **a** How many members are women?

  **b** What percentage of the club members are men?

**3** A supermarket aims to stock branded goods and their own goods in the ratio 2 : 3. They stock 500 kg of breakfast cereal.

    **a** What percentage of the cereal stock is branded?

    **b** How much of the cereal stock is their own?

**4** Over a period of 12 years, the Illinois Department of Health tested a total of 357 horses for rabies. They reported that the ratio of horses with rabies to those without was 1 : 16.

    How many of the horses tested had rabies?

**5** Rewrite each scale as a ratio in its simplest form.

    **a** 1 cm to 4 km         **b** 4 cm to 5 km         **c** 2 cm to 5 km

    **d** 4 cm to 1 km         **e** 5 cm to 1 km         **f** 2.5 cm to 1 km

**6** Map A has a scale of 2 cm to 5 km. Map B, of the same area, has a scale of 1 cm to 10 km.

    **a** Rewrite these scales as ratios in their simplest form.

    **b** How long is a path that measures 0.8 cm on map A?

    **c** How long should a 12 km road be on map B?

    **d** A river is 1.2 cm long on map B. How long will it be on map A?

**7** You can simplify a ratio by changing it into the form 1 : $n$. For example, you can rewrite 5 : 7 like this.

$$\frac{5}{5} : \frac{7}{5} = 1 : 1.4$$

    Rewrite each ratio in the form 1 : $n$.

    **a** 5 : 8             **b** 4 : 13          **c** 8 : 9

    **d** 25 : 36         **e** 5 : 27         **f** 12 : 18

    **g** 5 hours : 1 day     **h** 4 hours : 1 week     **i** £4 : £5

**8** There are 150 cars in a car park. The ratio of diesel cars to petrol cars is 2 : 3.

    $\frac{1}{5}$ of the diesel cars are red. $\frac{4}{9}$ of the petrol cars are red.

    Are more than one-third of all the cars in the car park red? Show your working.

**9** Look at this number line.

    The ratio AB : BC is 2 : 3.

    Work out the number at B.

**10** Athos has 24 more marbles than Zena. The ratio of the numbers of marbles that they have is 4 : 1.

    How many marbles does Zena have?

**Example 5**

Alisha makes a fruit drink by mixing orange squash with water in the ratio 2 : 3.

How much water does she need to add to 5 litres of orange squash to make the drink?

**Method 1: Using a table**

| Squash | 2 | 1 | 5 |
|--------|---|---|---|
| Water | 3 | 1.5 | 7.5 |

**Method 2: Comparing ratio and quantity**

2 parts is 5 litres.

Divide by 2.    1 part is 2.5 litres.

Multiply by 3.    3 parts is 7.5 litres.

So, she needs 7.5 litres of water to make the drink.

**Example 6**

Two business partners, Lubna and Adama, divided their total profit in the ratio 3 : 5. Lubna received £2100. How much did Adama get?

**Method 1: Using a table**

| Lubna | 3 | $3 \div 3 = 1$ | $1 \times 2100 = 2100$ |
|-------|---|----------------|------------------------|
| Adama | 5 | $5 \div 3 = 1\frac{2}{3}$ | $1\frac{2}{3} \times 2100 = 3500$ |

**Method 2: Using fractions**

Lubna's £2100 was $\frac{3}{8}$ of the total profit. (Check that you know why.)

$\frac{1}{8}$ of the total profit = £2100 ÷ 3 = £700

So Adama's share, which was $\frac{5}{8}$, was £700 × 5 = £3500.

## Exercise 10C

  **1** Derek, aged 15, and Ricki, aged 10, shared all the conkers they found in the woods in the same ratio as their ages. Derek had 48 conkers.

 **a** Write down and simplify the ratio of their ages.

 **b** How many conkers did Ricki have?

 **c** How many conkers did they find altogether?

**2** A blend of tea is made by mixing Lapsang with Assam in the ratio 3 : 5. I have a lot of Assam tea but only 600 g of Lapsang. How much Assam do I need to make the blend, if I use all the Lapsang?

  **3** The ratio of female to male spectators at ice hockey games is 4 : 5. At the Giants' last match, 4500 men watched the match. What was the total attendance at the game?

**4** A teacher arranged the content of every lesson for Year 10 as 'teaching' and 'practising learnt skills' in the ratio 2 : 3.

    **a** If a lesson lasted 35 minutes, how much teaching would he do?

    **b** If he decided to teach for 30 minutes, how long would the lesson be?

**5** Three business partners, Kevin, John and Margaret, put money into a business in the ratio 3 : 4 : 5. They shared any profits in the same ratio. Last year, Margaret received £3400 out of the profits. How much did Kevin and John receive from the profits last year?

**6** **a** Iqra is making a drink from lemonade, orange juice and ginger ale in the ratio 40 : 9 : 1. If Iqra has only 4.5 litres of orange juice, how much of the other two ingredients does she need to make the drink?

    **b** Another drink made from lemonade, orange juice and ginger ale uses the ratio 10 : 2 : 1.

    Which drink has a larger proportion of ginger ale, Iqra's or this one? Show how you work out your answer.

**7** On an aeroplane, the ratio of business to premium to economy seats is 1 : 6 : 30.

A family of 8 book all the business seats. How many seats are there on the aeroplane altogether?

**8** Some boys and girls are waiting for school buses. 25 girls get on the first bus. The ratio of boys to girls at the stop is now 3 : 2.

15 boys get on the second bus. There are now the same number of boys as girls at the bus stop.

How many students were originally at the bus stop?

**9** A jar contains 100 ml of a mixture of oil and water in the ratio 1 : 4. Enough oil is added to make the ratio of oil to water 1 : 2.

How much water must be added to make the ratio of oil to water 1 : 3?

**10** A teacher asked her class to choose a number in the 10 times table then divide it into the ratio 1 : 3 : 5.

Zeke chose 10. Yoko chose 50. Will chose 90.

    **a** Who made the most sensible choice and why?

    **b** Zeke correctly worked out the values and wrote them down as 1.1 : 3.3 : 5.5.

    Yoko correctly worked out the values and wrote them down as 5.56 : 16.67 : 27.78.

    What mistake have they both made?

**11** In a brass band the ratio of boys to girls is 5 : 2.

The ratio of boys to adults is 4 : 3.

Work out the smallest number of people that could be in the band.

**12** A bag contains counters that are either red or blue or green.

The ratio of red counters to blue counters is 1 : 3.

The ratio of blue counters to green counters is 5 : 8.

Work out the fraction of green counters in the bag.

Give your answer in its simplest form.

**13** Students in a school only have blue, brown or green eyes.

$\frac{2}{7}$ of the students have green eyes.

The ratio of students with blue eyes to brown eyes is 3 : 7.

Work out the fraction of students that have brown eyes.

Give your answer in its simplest form.

# 10.2 Speed, distance and time

## This section will show you how to:

- recognise the relationship between speed, distance and time
- calculate average speed from distance and time
- calculate distance travelled from the speed and the time taken
- calculate the time taken on a journey from the speed and the distance.

The relationship between **speed**, **time** and **distance** can be expressed in three ways:

$$\text{speed} = \frac{\text{distance}}{\text{time}} \qquad \text{distance} = \text{speed} \times \text{time} \qquad \text{time} = \frac{\text{distance}}{\text{speed}}$$

This diagram will help you remember the relationships between distance ($D$), time ($T$) and speed ($S$).

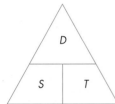

$$D = S \times T \qquad S = \frac{D}{T} \qquad T = \frac{D}{S}$$

In problems relating to speed, you generally mean average speed, as it would be unusual to maintain one exact speed for the whole of a journey.

---

**Example 7**

Paula drove a distance of 270 miles in 5 hours. What was her average speed?

Paula's average speed = $\dfrac{\text{total distance she drove}}{\text{total time she took}}$

$= \dfrac{270}{5}$

= 54 miles per hour (mph)

---

**Example 8**

Sarah drove from Sheffield to Peebles in $3\frac{1}{2}$ hours at an average speed of 60 mph. How far is it from Sheffield to Peebles?

distance = speed × time

Change the time to a decimal number.                     $3\frac{1}{2}$ hours = 3.5 hours

Use this to find the distance from Sheffield to Peebles.        60 × 3.5 = 210 miles

---

Sean is going to drive from Newcastle upon Tyne to Nottingham, a distance of 190 miles. He assumes he will drive at an average speed of 50 mph.

**a** How long will the journey take him, based on his assumption?

**b** Comment on your answer in part **a** and Sean's journey time if he undertakes the journey on a Friday afternoon or a Sunday morning.

**a** Sean's time for the journey = $\dfrac{\text{total distance he drove}}{\text{his average speed}}$

$$= \dfrac{190}{50}$$

$$= 3.8 \text{ hours}$$

Change 0.8 hours into minutes. $\qquad 0.8 \times 60 = 48$ minutes

So, Sean's journey will take 3 hours 48 minutes.

**b** Traffic will be heavier on Friday afternoon than on Sunday morning, so he is likely to take longer than 3 h 48 m on Friday afternoon and less than 3 h 48 m on Sunday morning.

**Remember** When you calculate a time and get a decimal answer, as in Example 9, do not mistake the decimal part for minutes. You must either:

• leave the time as a decimal number and give the unit as hours, or
• change the decimal part to minutes by multiplying it by 60 (1 hour = 60 minutes) and give the answer in hours and minutes.

## Exercise 10D 🖩

**1** A cyclist travels a distance of 90 miles in 5 hours. What was her average speed?

**2** How far along a motorway would you travel if you drove at 70 mph for 4 hours?

**3** I drive to Bude, in Cornwall, from Sheffield in about 6 hours. The distance from Sheffield to Bude is 315 miles. What is my average speed?

**4** The distance from Leeds to London is 210 miles. The train travels at an average speed of 90 mph. If I catch the 9:30 am train in London, at what time should I expect to arrive in Leeds?

**5** Copy and complete this table.

| | Distance travelled | Time taken | Average speed |
|---|---|---|---|
| a | 150 miles | 2 hours | |
| b | 260 miles | | 40 mph |
| c | | 5 hours | 35 mph |
| d | | 3 hours | 80 km/h |
| e | 544 km | 8 hours 30 minutes | |
| f | | 3 hours 15 minutes | 100 km/h |
| g | 215 km | | 50 km/h |

**Hints and tips** Remember to convert time to a decimal if you are using a calculator, for example 8 hours 30 minutes is 8.5 hours.

**Hints and tips** km/h means kilometres per hour, m/s means metres per second.

**6** The most common Ordnance Survey maps have a scale of 1 : 50 000.

    **a** How far is the actual distance represented by a distance of 1 cm on the map?

    **b** Ed plans a cycle ride. He estimates the distance on his Ordnance Survey map to be 78 cm. He plans to leave at 9 am and stop for about 30 minutes for a break. He assumes he will cycle at an average speed of 15 km/h. About what time will he be back? Show your working.

    **c** Comment on the effect that Ed's estimate of distance and assumed speed might have on the actual time of his ride.

> **Hints and tips** | 1 km = 1000 m = 100 000 cm

**7** Colin drives home from his son's house in 2 hours 15 minutes. He says that he drives at an average speed of 44 mph.

    **a** Change 2 hours 15 minutes to a decimal.

    **b** How far is it from Colin's home to his son's house?

**8** The distance between Paris and Le Mans is 200 km. The express train between Paris and Le Mans travels at an average speed of 160 km/h.

    **a** Calculate the time taken for the journey from Paris to Le Mans. Give your answer as a decimal number of hours.

    **b** Change your answer to part **a** to hours and minutes.

**9** This timetable shows a train journey from Sheffield to London by the Midland mainline. The distance travelled is 150 miles.

| Depart | | Arrive | | Travel by | Train company | Duration |
|---|---|---|---|---|---|---|
| 11:29 | Sheffield | 13:30 | London St Pancras Intl | Train | EAST MIDLANDS TRAINS | 02h 01 |

This timetable shows a train journey from Sheffield to London by the East Coast mainline. The distance from Sheffield to Doncaster is 20 miles and from Doncaster to London is 160 miles.

| Depart | | Arrive | | Travel by | Train company | Duration |
|---|---|---|---|---|---|---|
| 11:10 | Sheffield | 11:35 | Doncaster | Train | TRANSPENNINE EXPRESS | 00h 25 |
| 11:46 | Doncaster | 13:28 | London Kings Cross | Train | EAST COAST | 01h 42 |

    **a** Work out the average speed of each journey from Sheffield to London.

    **b** Work out the average speed of the train journey from Doncaster to London.

**10** A train travels at 50 km/h for 2 hours, then slows down to do the last 30 minutes of its journey at 40 km/h.

    **a** What is the total distance of this journey?

    **b** What is the average speed of the train over the whole journey?

**11** Change each speed to metres per second.

    **a** 36 km/h               **b** 12 km/h              **c** 60 km/h

> **Hints and tips** Remember that there are 3600 seconds in an hour and 1000 metres in a kilometre. So to change from km/h to m/s multiply by 1000 and divide by 3600.

**12** Change each speed to kilometres per hour.

    **a** 25 m/s               **b** 12 m/s               **c** 0.5 m/s

> **Hints and tips** To change from m/s to km/h multiply by 3600 and divide by 1000.

**13** A train travels at an average speed of 18 m/s.

The train set off at 7:30 am on a 40-km journey.

At approximately what time will it reach its destination?

> **Hints and tips** To convert a decimal fraction of an hour to minutes, multiply by 60.

**(PS) 14** At 9:00 am cyclist A sets off on a trail at an average speed of 16 km/h.

At 10:00 am cyclist B sets off from the same place, in the same direction at an average speed of 24 km/h.

At approximately what time will cyclist B catch up with cyclist A?

> **Hints and tips** Set up a table to show how far each cyclist has gone each 15 minutes after 10:00 am.

**(CM) 15** Rebecca says: 'If I travel for 10 minutes at 50 mph, then 10 minutes at 70 mph, then my average speed must be 60 mph.'

Nick says: 'If I travel for 10 miles at 40 mph, then 10 miles at 60 mph, then my average speed for the 20 miles will be 50 mph.'

Are they both correct? Show your working.

**(EV) 16** Josh and Nell need to travel from A to B.

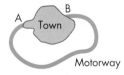

The distance across town is 20 miles. By motorway it is 50 miles.

The speed limit in town is 30 mph and on the motorway it is 70 mph.

They work out the time it will take them to go via the town and via the motorway if they travel at the speed limit.

Josh decides to go via the town. Nell decides to go via the motorway.

Who is most likely to get there first? Show your working and explain any assumptions that you make.

# 10.3 Direct proportion problems

This section will show you how to:

- recognise and solve problems that involve direct proportion.

Key terms

direct proportion

unitary method

unit cost

Suppose you buy 12 items that each cost the same. The total amount you spend is 12 times the cost of one item.

The total cost is in **direct proportion** to the number of items bought. The cost of a single item (the **unit cost**) is the constant factor that links the two quantities.

Direct proportion is not only concerned with costs. Any two related quantities can be in direct proportion to each other.

One way to solve any problem involving direct proportion is to start by finding the single unit value. This is called the **unitary method**. You can do this by using a table or just working out the single unit value. These methods are very similar, as you will see from Examples 10 and 11. The table can be useful if you have to do more complicated calculations.

> **Hints and tips** Before solving a direct proportion problem, think about it carefully to make sure that you know how to find the required single unit value.

---

**Example 10**

If eight pens cost £2.64, what is the cost of five pens?

**Method 1: Using a table**

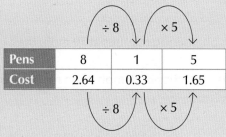

| Pens | 8 | 1 | 5 |
|------|------|------|------|
| Cost | 2.64 | 0.33 | 1.65 |

**Method 2: Unitary method**

First, find the cost of *one* pen.      £2.64 ÷ 8 = £0.33

Then multiply to find the cost of five pens.      £0.33 × 5 = £1.65

So, the cost of five pens is £1.65.

---

**Example 11**

Shop A sells a six pack of cola for £3.20. Shop B sells a 4 pack of cola for £2.10.

Which shop sells the cola more cheaply?

As we are comparing costs in two shops, the table method will not be much use, so use the unitary method. Work in pence.

Shop A: $\frac{320}{6}$ = 53.33... pence per can

Shop B: $\frac{210}{4}$ = 52.5 pence per can

So shop B is cheaper.

---

Example 12

Emma uses eight loaves of bread to make packed lunches for 18 people.

**a** How many packed lunches can she make from 20 loaves?

**b** How many loaves will she need to make packed lunches for 60 people?

## Method 1: Using a table

**a**

| | | ÷ 8 | × 20 | |
|---|---|---|---|---|
| Loaves | 8 | 1 | 20 | |
| Packed lunches | 18 | 2.25 | 45 | |

From 20 loaves she can make packed lunches for 45 people.

**b**

| | ÷ 18 | × 60 | |
|---|---|---|---|
| Loaves | 8 | $\frac{8}{18} = \frac{4}{9}$ | $\frac{4}{9} \times 60 = 26\frac{2}{3}$ |
| Packed lunches | 18 | 1 | 60 |

To make packed lunches for 60 people, she will require 27 loaves.

> **Hints and tips** Remember that she will need to round up to a whole number of loaves.

You can also use a table with known multiplication and division facts.

| | ÷ 3 | × 10 | |
|---|---|---|---|
| Loaves | 8 | $8 \div 3 = 2\frac{2}{3}$ | $26\frac{2}{3}$ |
| Packed lunches | 18 | $18 \div 3 = 6$ | 60 |

## Method 2: Unitary method

First, find how many lunches she can make from one loaf.

From one loaf she can make $18 \div 8 = 2.25$ lunches.

**a** So with 20 loaves she can make $2.25 \times 20 = 45$ lunches.

**b** Work out how many loaves she needs for one packed lunch. $8 \div 18 = \frac{4}{9}$ or 0.444...

So for 60 packed lunches she will need $\frac{4}{9}$ of $60 = 26\frac{2}{3}$ loaves, so she will need 27 loaves.

## Exercise 10E 🖩

**1** If 30 matches weigh 45 g, what would 40 matches weigh?

**2** Five bars of chocolate cost £2.90. Find the cost of nine bars.

**3** Eight men can chop down 18 trees in a day. How many trees can 20 men chop down in a day?

**4** Find the cost of 48 eggs when 15 eggs can be bought for £2.10.

**5** Seventy maths textbooks cost £875.

   **a** How much will 25 maths textbooks cost?

   **b** How many maths textbooks can you buy for £100?

> **Hints and tips** Remember to work out the value of one unit each time. Always check that answers are sensible.

**6** A lorry uses 80 litres of diesel fuel on a trip of 280 miles.

   **a** How much diesel would the same lorry use on a trip of 196 miles?

   **b** How far would the lorry get on a full tank of 100 litres of diesel?

**7** This is part of a recipe for 12 biscuits.

*Recipe*

200 g of margarine
400 g of sugar
500 g of flour
300 g of ground rice

   **a** What quantities are needed for:

   **i** 6 biscuits          **ii** 9 biscuits          **iii** 15 biscuits?

   (CM) **b** What is the maximum number of biscuits I could make if I had 1 kg of each ingredient?

**8** Peter the butcher sells sausages in packs of 6 for £2.30.

   Paul the butcher sells sausages in packs of 10 for £3.50.

   I have £10 to spend on sausages.

   (MR) **a** If I want to buy as many sausages as possible from one shop, which shop should I use? Show your working.

   (EV) **b** How can I get more sausages with my £10?

(PS) **9** 1.8 kg of flour make three loaves of bread. One loaf of bread makes 10 sandwiches. How much flour is needed to make enough bread for 400 sandwiches?

(EV) **10** Buns cost 40p each. Cakes cost 55p each. I spend exactly £4.35 on buns and cakes. How many of each did I buy?

# 10.4 Best buys

This section will show you how to:

- find the cost per unit mass
- find the mass per unit cost
- use the above to find which product is better value.

When you look around a supermarket and see all the different prices for the many different-sized packets, it is rarely obvious which are the **best buys**. However, with a calculator you can easily compare **value for money** by finding either the price of one gram (or kilogram) or the number of grams for £1 (or 1p).

To find:

- price of one gram (or kilogram), which is the cost per unit **mass**, divide cost by mass
- number of grams for £1 (or 1p), which is the mass per unit cost, divide mass by cost.

The next two examples show you how to do this.

Note that the amount something weighs is called its mass and is measured in grams and kilograms. In this section, you will use mass to represent the amount of a quantity.

---

**Example 13**

A 300-g tin of cocoa costs £1.20. Find:

**a** the cost of one gram of cocoa (the cost per unit mass)

**b** the number of grams per pound (the mass per unit cost).

First change £1.20 to 120p. Then use a calculator to divide.

**a** Cost per unit mass is $120 \div 300 = 0.4$p per gram

**b** Mass per unit cost is $300 \div 120 = 2.5$ g per penny

---

**Example 14**

A supermarket sells Whito soap powder in two different-sized packets. The medium size contains 800 g and costs £1.60 and the large size contains 2.5 kg and costs £4.75. Which is the better buy?

Find the mass per unit cost for both packets.

Medium packet      $800 \div 160 = 5$ g per penny

Large packet      $2500 \div 475 = 5.26$ g per penny

From these it is clear that the large size gives more 'mass per penny', which means that the large size is the better buy.

---

When you use a scaling method, you do not work out the cost of one unit. Instead you use common multiples to work out the cost of, say six items. For example, if you know that two items cost £4.50 in one shop and three of the same items cost £6.80 in another shop, you can work out the cost of six items in both shops. (£13.50 in the first shop and £13.60 in the second shop)

In some cases, it is easier to use a scaling method to compare prices and find **better value**.

Example 15

Which of these boxes of fish fingers is better value?

12 is a common factor of 24 and 36 so work out the cost of 12 fish fingers.

For the small box, 12 fish fingers cost £3.40 ÷ 2 = £1.70.

For the large box, 12 fish fingers cost £4.95 ÷ 3 = £1.65.

So the large box is better value.

In other cases, it is easier to use a table to compare the cost of one item in each case.

Example 16

Which of these packs of yoghurt is better value?

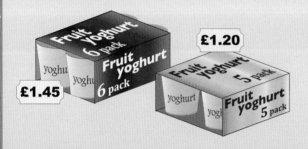

Set up a table for the six-pack.

| | ÷ 6 | × 5 |
|---|---|---|
| Pots | 6 | 1 | 5 |
| Price | £1.45 | £0.241 666… | £1.208 333… |

So five pots from the six-pack cost more than the five-pack.

You could also do this the other way.

Set up a table for the five-pack.

| | ÷ 5 | × 6 |
|---|---|---|
| Pots | 5 | 1 | 6 |
| Price | £1.20 | £0.24 | £1.44 |

So six pots from the five-pack cost less than the six-pack.

Either way, the five-pack is just better value!

**1** Compare the prices of the products in each pair. State which, if either, is the better buy.

  **a** Chocolate bars: £2.50 for a 5-pack, £4.50 for a 10-pack

  **b** Eggs: £1.08 for 6, £2.25 for 12

  **c** Car shampoo: £4.99 for 2 litres, £2.45 for 1 litre

  **d** Dishwasher tablets: £7.80 for 24, £3.90 for 12

**2** Compare the products in each pair. State which is the better buy. Explain your choice.

  **a** Coffee: a medium jar (140 g) for £1.10, a large jar (300 g) for £2.18

  **b** Toothpaste: a large tube (110 ml) for £1.79, a medium tube (75 ml) for £1.15

  **c** Frosted Flakes: a large box (750 g) for £1.64, a medium box (500 g) for £1.10

  **d** Hair shampoo: a medium bottle (400 ml) for £1.15, a large bottle (550 ml) for £1.60

**3** Julie wants to respray her car with yellow paint. In the local supermarket, she sees tins at these prices.

    Small tin        350 ml for £1.79

    Medium tin   500 ml for £2.40

    Large tin      1.5 litres for £6.70

  **a** What is the cost per litre of paint in the small tin?

  **b** Which tin is offered at the lowest price per litre?

**4** Tisco's sells bottled water in three sizes.

a Work out the cost per litre of the 40-cl bottle.

b Which bottle is the best value for money?

(MR) **5** Mary and Jane are arguing about which of them is better at mathematics.

Mary scored 49 out of 80 on a test.

Jane scored 60 out of 100 on a test of the same standard.

Based on these results, who is better at mathematics?

(PS) **6** Paula and Kelly are comparing their running times.

Paula completed a 10-mile run in 65 minutes.

Kelly completed a 10-kilometre run in 40 minutes.

Given that 8 kilometres are equal to 5 miles, which girl has the greater average speed?

(CM) **7** Sachets of cat food cost 35p each. Aldo's supermarket sells them in packs of 12 or 40.

A pack of 12 costs £3.60. A pack of 40 costs £11.50. Today they have an offer.

Which is the better value: the 12-pack, the 40-pack or the '5 for 4' offer? Show your working.

> **Buy any 5 items and pay for 4!**
>
> Cheapest of 5 items free.
>
> Offer only applies to individual items, not packs.

(CM) **8** Three people are comparing how much petrol their cars use.

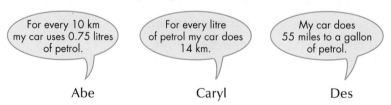

For every 10 km my car uses 0.75 litres of petrol.

For every litre of petrol my car does 14 km.

My car does 55 miles to a gallon of petrol.

Abe        Caryl        Des

4.55 litres = 1 gallon

a Whose car is the most economical?

b Why is Abe's statement not very helpful when comparing his car's petrol consumption to those of Caryl and Des?

Hints and tips Pick a distance and work out the amount of petrol they each use.

**9** Is 50% off the same as a 'Buy one get one free' offer. Give reasons to support your answer.

**10** In early 2015, the price of oil dropped and petrol became cheaper.

In the United States, the cost of petrol was $2.09 per US gallon.

In the UK the cost of petrol was £1.09 per litre.

The exchange rate was £1 = $1.51.

1 US gallon ≈ 3.8 litres.

Approximately how many times more expensive is petrol in the UK than the USA?

**11** VAT varies around Europe and is not always applied to the same goods. For example, the UK does not put VAT on most food, but in other countries VAT is put on food.

In Luxembourg VAT is not put on the price of concert tickets. In the UK the price of a concert ticket has VAT at 20% put on.

The tribute band Fink Ployd are on a European tour. Tickets are £42 in the UK before VAT is added and €65 in Luxembourg.

Work out the exchange rate between the pound (£) and the Euro (€) so that ticket prices in both countries are the same.

**12** A running club is ordering T-shirts for the runners who finish their race.

The race limit is 400 and the club expect all places to be taken.

Two companies are offering T-shirts

Shirts 4 U: £3.45 a shirt plus £75 set up fee.

Top Tees: Set up fee £100. First 200 shirts £3.99. All shirts after 200, £2.75.

**a** Which company is offering the better deal?

**b** The club treasurer says: 'Normally 5% of the runners who have entered do not turn up, so we only need 380 shirts'. Which company offers the better deal now?

**13** For a summer fayre at their school, form 11X decide to make and sell smoothies. They have two recipes.

| **Kiwi Krush** | **Banana Breeze** |
|---|---|
| Mix kiwi, banana and yoghurt in the ratio 2 : 1 : 1 by mass. | Mix banana, kiwi and yoghurt in the ratio 4 : 1 : 3 by mass. |

They find that each banana weighs an average of 150 grams after peeling.

A kiwi fruit weighs 90 grams after peeling.

The contents of a carton of yoghurt weigh 250 grams.

The students do not want to have to cut up and weigh fruit so they work out a recipe for each smoothie, using whole fruits and whole cartons of yoghurt.

**a** To make a batch of Kiwi Krush they use 3 cartons of yoghurt, 5 bananas and 17 kiwi fruits.

Give reasons why this is a sensible recipe.

**b** Work out a sensible recipe to make a batch of Banana Breeze, using whole fruits and whole cartons of yoghurt.

# Worked exemplars

 **1** Jonathan is comparing two ways to travel from his flat in London to his parents' house in Doncaster.

**Tube, train and taxi**

It takes 35 minutes to get to the railway station by tube in London.

The train journey from London to Doncaster takes 1 hour 40 minutes.

From Doncaster, it is 15 miles by taxi at an average speed of 20 mph.

**Car**

The car journey is 160 miles at an average speed of 50 mph.

Which journey takes longer: tube, train and taxi or car?

| | |
|---|---|
| This question assesses 'communicating mathematically', so you must display your methods clearly and include words to explain what your calculations show. Imagine that you will pass your answer to a friend or relative and ask yourself whether or not they could understand it. | |
| **For the taxi** time = distance ÷ speed $= \dfrac{15}{20}$ = 0.75 hour (or 45 minutes) | First, work out the time taken by the taxi. It is not essential to show the formula 'time = distance ÷ speed' but it is useful to draw the triangle that shows the relationship. Be careful with time as a decimal: 0.75 hours = 45 minutes. |
| Total time for the journey by tube, rail and taxi: 35 minutes + 1 hour 40 minutes + 45 minutes = 3 hours | Next, work out the total time for tube, train and taxi. |
| **For the car** time = distance ÷ speed $= 160 ÷ 50$ = 3.2 hours (or 3 hours 12 minutes) | Now work out the time taken by the car. 0.1 hours = 6 minutes so 0.2 hours = 12 minutes |
| Travelling by car takes 12 minutes longer. | Finally, compare the times taken. It is essential to write a final conclusion: do not assume it is obvious from your working. |

 **2** Which of the following ratios in not the same as 3 : 5?

$1 : 1\frac{2}{3}$        $1 : \frac{8}{5}$        $9 : 15$        $39 : 65$

| | |
|---|---|
| This is a reasoning question asking you to work out which ratio is not the same as 3 : 5. | |
| Either $1 : 1\frac{2}{3}$   or   $1 : \frac{8}{5}$ | Eliminate 9 : 15 and 39 : 65 as these will cancel down to 3 : 5. |
| The ratio 3 : 5 is the same as $1 : 1\frac{2}{3}$. | To get 1 as the first value divide by 3, 5 ÷ 3 = 1.666. |
| So $1 : \frac{8}{5}$ is not the same as 3 : 5. | |

(CM) 3  A, B, C and D are four points on a number line.

A       B   C      D

AB : BC = 7 : 3

BC : CD = 2 : 5

Work out the ratio AC : CD.

Give your answer in its simplest form.

| This question requires you to 'interpret and communicate information', so you will need to extend the information beyond what is stated explicitly. | |
|---|---|
| A    B   C <br> B   C    D | It may help to draw a diagram. |
| This table shows the given ratios.<br><br>| AB | 7 | |<br>| BC | 3 | 2 |<br>| CD | | 5 | | Set up a table and write in the information you are given. |
| BC has the same value in the first and second columns.<br><br>| AB | 7 | | 14 |<br>| BC | 3 | 2 | 6 |<br>| CD | | 5 | 15 | | Now extend and complete the third column.<br><br>Multiply AB and BC in the first column by 2 to give AB = 14 and BC = 6.<br><br>Multiply BC and CD in the second column by 3 to give BC = 6 and CD = 15. |
| Add AB and BC to find AC.<br><br>| AB | 7 | | 14 | 20 | 4 |<br>| BC | 3 | 2 | 6 | | |<br>| CD | | 5 | 15 | 15 | 3 | | Now extend and complete a fourth column by combining AB and BC from the third column to get AC = 20 (column 4) and cancel AC and CD by a factor of 5 (column 5). |
| The ratio of AC : CD is 4 : 3. | Give your answer. |

(PS) 4  A : B = 4 : 5. What fraction of B is A?

$\frac{4}{5}$       $\frac{4}{9}$       $\frac{5}{9}$       $\frac{5}{4}$

| This is a problem-solving question. The way the question is presented is designed to distract you. | |
|---|---|
| Let A be 4 and B be 5. | Choose the simplest possible values for A and B. |
| $\frac{4}{5}$ | The answer has nothing to do with ratios. A is $\frac{4}{5}$ of B. |

# Ready to progress?

I can simplify a ratio.

I can calculate average speeds from data.
I can calculate distance from speed and time.
I can calculate time from speed and distance.
I can compare prices of products to find the 'best buy'.

I can solve direct proportion problems.
I can use ratio to solve problems.

# Review questions

**1** The total cost of three pens is £1.20.

Work out the total cost of eight of these pens.

Give your answer in pounds.

**2** These are the ingredients for making apple pie for eight people.

**Apple pie for 8 people**

| 240 g flour | 5 eggs |
|---|---|
| 320 g apples | 210 ml milk |
| 105 g butter | |

**a** How much flour will Bill need to make an apple pie for five people?

**b** How much milk will Jenny need to make an apple pie for 18 people?

**3** Maura travelled 80 miles in 1 hour 40 minutes.

Work out Maura's average speed in miles per hour.

**4** A car travels for 5 hours.

Its average speed is 45 km/h.

Work out the total distance the car travels.

**5** Ron drives 270 km in 3 hours 45 minutes.

Work out Ron's average speed.

**(CM) 6** The ratio of the totals of the numbers in Box A and Box B is 2 : 3.

Box A                Box B

Swap two numbers, one from each box, so that the ratio of the totals of the numbers is now 9 : 11.

Show your working.

**7** A farmer has three fields. The area of field A is 1.73 hectares, the area of field B is 2.64 hectares and the area of field C is 0.95 hectares. Cattle need 0.065 hectares of space each, horses need 0.04 hectares of space each and sheep need 0.01 hectares of space each.

   **a** Show that if the farmer keeps horses in field A, cattle in field B and sheep in field C, he will be able to have a total of 178 animals.

   **b** Work out the combination of fields, cattle, horses and sheep that will allow the farmer to keep the maximum possible number of animals.

**8** Dawn has a box of chocolates.

24 of the chocolates have soft centres.

16 of the chocolates have hard centres.

Write down the ratio of the number of soft centres to the number of hard centres.

Give your ratio in its simplest form.

**9** There are some red counters and some blue counters in a bag.

The total number of counters is 78.

The ratio of the number of red counters to the number of blue counters is 1 : 5.

Work out the number of blue counters in the bag.

**10** A company mixes two types of lawn seed.

Britegreen grass seed costs £18 per kilogram.

Grassgreen grass seed costs £21 per kilogram.

The company mixes the seed in the ratio Britegreen : Grassgreen = 2 : 3.

   **a** How much Grassgreen will the company need to mix with 15 kg of Britegreen?

   **b** How much Britegreen will the company need to mix with 45 kg of Grassgreen?

   **c** How much will one kilogram of the mixture cost?

**11** Here is some information about the members of a club.

| | | Men | Women |
|---|---|---|---|
| Age of club members | 40 years or over | 60 | 42 |
| | Under 40 years | 35 | 63 |

   **a** Write the ratio of women aged 40 or over to women aged under 40. Give your answer in its simplest form.

   **b** What percentage of the whole club is men aged 40 years or older?

   **c** Fifty more members join the club. The ratio of women aged 40 or over to women aged under 40 is now 3 : 4. The percentage of men aged 40 years or over is now 28%. The number of women aged 40 years or over is now the same as the number of men aged under 40.

   Copy and complete the table to show the information after the 50 new members join.

| | | Men | Women |
|---|---|---|---|
| Age of club members | 40 years or over | | |
| | Under 40 years | | |

# 11 Geometry and measures: Perimeter and area

## This chapter is going to show you:

- how to work out the perimeters and areas of rectangles, triangles, parallelograms, trapeziums and compound shapes
- how to calculate the circumference and area of a circle and give your answers in terms of $\pi$.

## You should already know:

- what 'area' means
- the common units of length and area.

## About this chapter

Many centuries ago, the Greeks measured lengths based on the relative lengths of body parts, such as the foot (*pous*) and the finger (*dactylos*). An area called a *plethron* was the amount of land a pair of oxen could plough in one day.

Over the centuries, people realised that standard units would be easier and so adopted common measures. In the UK, the imperial system was used. However, more recently the UK and most other parts of the world, have moved to the metric system.

Perimeter and area both require an understanding of basic measurements. The word *perimeter* comes from two Greek words – *peri* meaning 'all around' and *meter* meaning 'measure'. So the perimeter of a shape is the length around its edge. Area is a measure of the amount of surface covered and is measured in square units.

In this chapter you will learn to calculate the perimeter and area of the common 2D shapes.

# 11.1 Rectangles

This section will show you how to:

- calculate the perimeter and area of a rectangle.

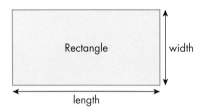

The perimeter of a rectangle is the sum of all its sides:

perimeter = 2 × (length + width)

The area of a rectangle is found by multiplying its length by its breadth:

area = length × width

---

**Example 1**

Work out the perimeter and area of this rectangle.

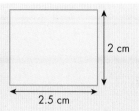

Perimeter = 2 × (2.5 + 2)

= 2 × 4.5

= 9 cm

Area = 2.5 × 2

= 5 cm²

---

**Example 2**

The area of a rectangle is 36 cm² and one side is 9 cm long.

What is the perimeter of the rectangle?

The area is: length × width = 36 cm²

9 × width = 36 cm²

So the breadth is 4 cm.

The perimeter is then 2 × (9 + 4) = 26 cm.

---

**Remember** that the diagrams in examples and exercises are not drawn accurately. You must calculate the angles, lengths and areas.

**1** Calculate the perimeter and the area of each rectangle.

a

7 cm
5 cm

b
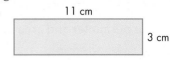
11 cm
3 cm

c

15 cm
3 cm

d
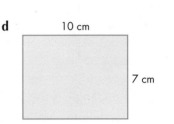
10 cm
7 cm

e
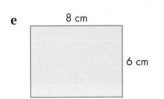
8 cm
6 cm

f
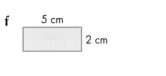
5 cm
2 cm

g
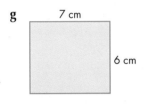
7 cm
6 cm

h
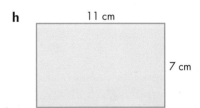
11 cm
7 cm

**2** The table shows some measurements for six rectangles. Copy and complete the table.

|   | Length | Breadth | Perimeter | Area |
|---|--------|---------|-----------|------|
| **a** | 6 cm | 4 cm | | |
| **b** | 10 cm | 8 cm | | |
| **c** | 10 cm | | | 30 cm² |
| **d** | | 3 cm | | 21 cm² |
| **e** | 6 cm | | 18 cm | |
| **f** | | 5 cm | 22 cm | |

**3** A rectangular field is 145 m long and 120 m wide.

**a** What length of fencing is needed to go all the way round the field?

**b** What is the area of the field?

**(MR) 4** A rugby pitch is 160 m long and 70 m wide. Show that, if 1 hectare = 10 000 m², the area of the rugby pitch is 1.12 hectares.

**(PS) 5** A farmer has a rectangular plot of land that is 95 m by 60 m. He buys bags of fertiliser to spread over all the land. If one bag covers 300 m², how many bags should he buy?

# 11.2 Compound shapes

This section will show you how to:

- calculate the perimeter and area of a compound shape made from rectangles.

A **compound shape** is made up of two or more shapes. Often you can split them into separate shapes, which makes it easier to calculate the area. To calculate the perimeter you have to work out the unknown lengths from the given lengths, as shown in the first example.

**Example 3**

Work out the perimeter of this compound shape.

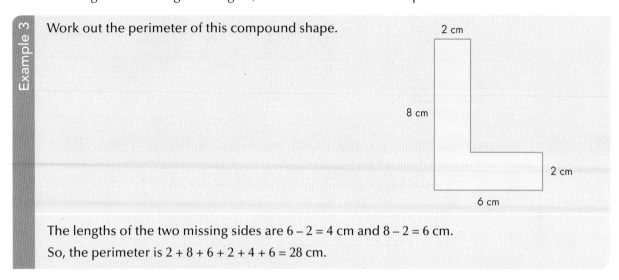

The lengths of the two missing sides are 6 − 2 = 4 cm and 8 − 2 = 6 cm.

So, the perimeter is 2 + 8 + 6 + 2 + 4 + 6 = 28 cm.

**Example 4**

Work out the area of this compound shape.

First, split the shape into two rectangles, A and B.

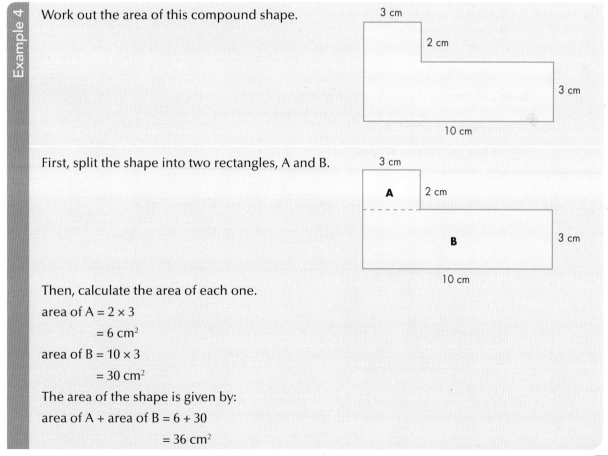

Then, calculate the area of each one.

area of A = 2 × 3

       = 6 cm$^2$

area of B = 10 × 3

       = 30 cm$^2$

The area of the shape is given by:

area of A + area of B = 6 + 30

       = 36 cm$^2$

1 Work out the perimeter of each of the following shapes.

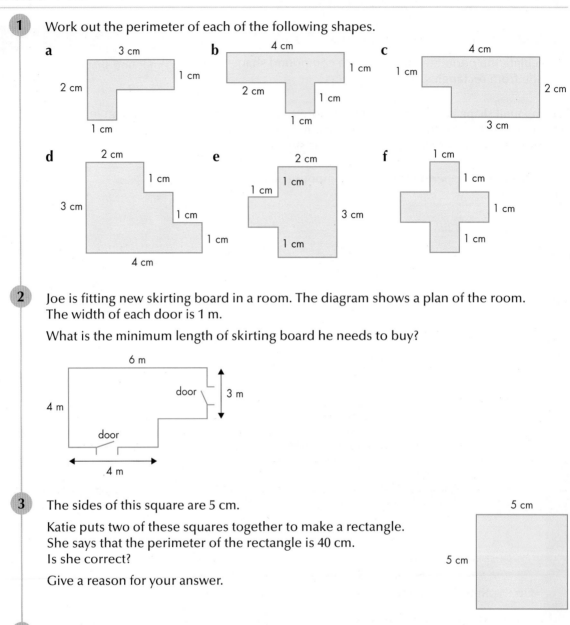

2 Joe is fitting new skirting board in a room. The diagram shows a plan of the room.
The width of each door is 1 m.

What is the minimum length of skirting board he needs to buy?

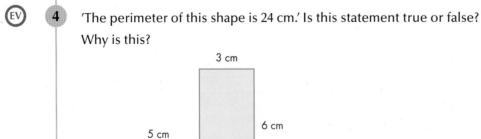

3 The sides of this square are 5 cm.

Katie puts two of these squares together to make a rectangle.
She says that the perimeter of the rectangle is 40 cm.
Is she correct?

Give a reason for your answer.

4 'The perimeter of this shape is 24 cm.' Is this statement true or false?

Why is this?

**5** Calculate the area of each of the compound shapes below.

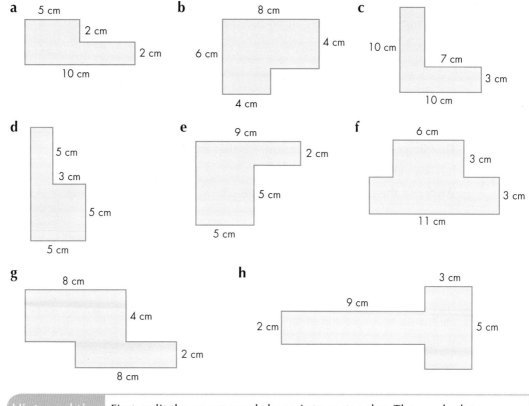

a    5 cm
     2 cm
     2 cm
     10 cm

b    8 cm
     6 cm    4 cm
     4 cm

c    10 cm
     7 cm
     3 cm
     10 cm

d    5 cm
     3 cm
     5 cm
     5 cm

e    9 cm
     2 cm
     5 cm
     5 cm

f    6 cm
     3 cm
     3 cm
     11 cm

g    8 cm
     4 cm
     2 cm
     8 cm

h    3 cm
     9 cm
     2 cm    5 cm

| Hints and tips | First, split the compound shape into rectangles. Then, calculate the area of each rectangle. Finally, add together the areas of the rectangles. |

| Hints and tips | Be careful to work out the length and width of each separate rectangle. You will usually have to add or subtract lengths to work out some of these. |

**PS** **6** A square lawn has side length 5 m. It has a rectangular path, 1 m wide, running all the way round the outside of it. Carlos is laying paving stones on the path. The area of each one is 1 m². How many paving stones does he need?

**EV** **7** Tom is working out the area of this shape.

This is what Tom wrote down.

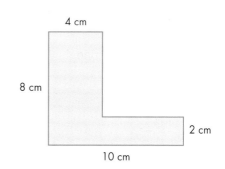

> 8 × 4 = 32
> 10 × 2 = 20
> So the area is 32 + 20 = 52 cm²

4 cm
8 cm
2 cm
10 cm

Explain why Tom is wrong.

(PS) **8** This compound shape is made from four rectangles that are all the same size.

Work out the area of the compound shape.

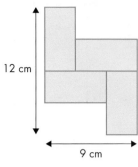

12 cm
9 cm

(PS) **9** This shape is made from five squares that are all the same size.

It has an area of 80 $cm^2$.

Work out the perimeter of the shape.

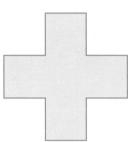

(MR) **10** Dave is painting this wall.

He buys a tin of paint that will cover 10 $m^2$.

Will he have enough paint?

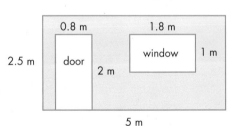

0.8 m
1.8 m
2.5 m
door
2 m
window
1 m
5 m

# 11.3 Area of a triangle

## This section will show you how to:

- calculate the area of a triangle
- use the formula for the area of a triangle.

**Key term**

perpendicular height

### Area of a right-angled triangle

You can see that the area of a right-angled triangle is half the area of the rectangle with the same base and height. So the formula is:

area of a right-angled triangle = $\frac{1}{2} \times$ base $\times$ height

Using algebra this is written as:

$A = \frac{1}{2}bh$

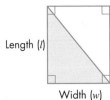

Length (l)
Width (w)
Height (h)
Base (b)

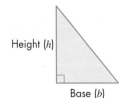

Work out the area of this right-angled triangle.

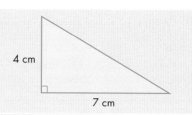

Area $= \frac{1}{2} \times 7$ cm $\times 4$ cm

$\quad\quad = \frac{1}{2} \times 28$ cm$^2$

$\quad\quad = 14$ cm$^2$

## Exercise 11C 🖩

**1** Work out the area and the perimeter of each triangle.

a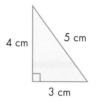

b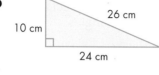

c

**2** Calculate the area of the shaded triangle RST.

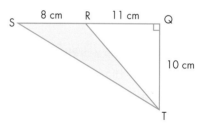

| Hints and tips | Work out the area of triangle QST and subtract the area of triangle QRT. |

**3** The diagram shows a garden. There is a square region round a tree where nothing can be planted.

What area can be planted?

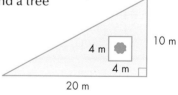

| Hints and tips | Work out the area of the triangle and subtract the area of the square. |

**4** Work out the area of the shaded part of each triangle.

a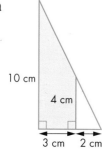

b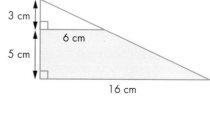

c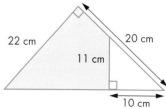

**5** Which of these three triangles has the largest area?

**a**
5 cm
20 cm

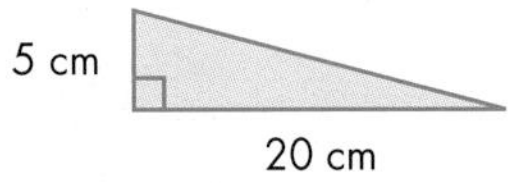

**b**
12 cm
8 cm

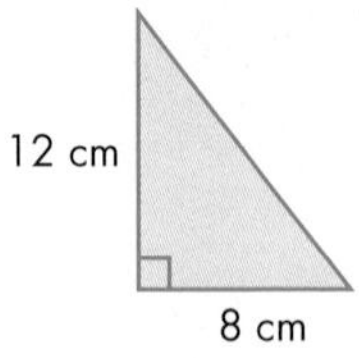

**c**
10 cm
15 cm

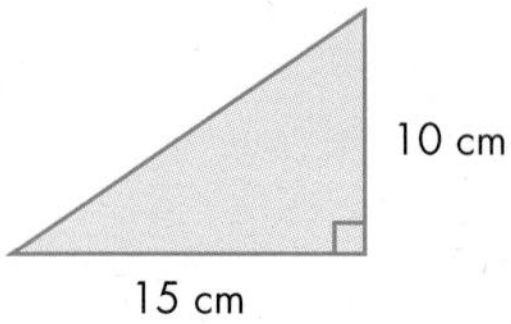

(EV) **6** This shape is made from two right-angled triangles.

Mai said the perimeter of the shape was 32 cm. Maggie said the area of the shape was 72 cm².

Comment on both girls' statements.

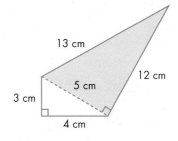

13 cm
12 cm
5 cm
3 cm
4 cm

(MR) **7** This compound shape is made from a rectangle and two identical right-angled triangles.

Show that the area of the shape is 108 cm².

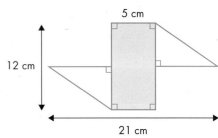

5 cm
12 cm
21 cm

(PS) **8** Chris is working out the area of one of the walls of his garden shed.

The diagram shows the dimensions.

**a** Calculate the area of the wall.

| Hints and tips | Work out the area of the rectangle and the right-angled triangle. |

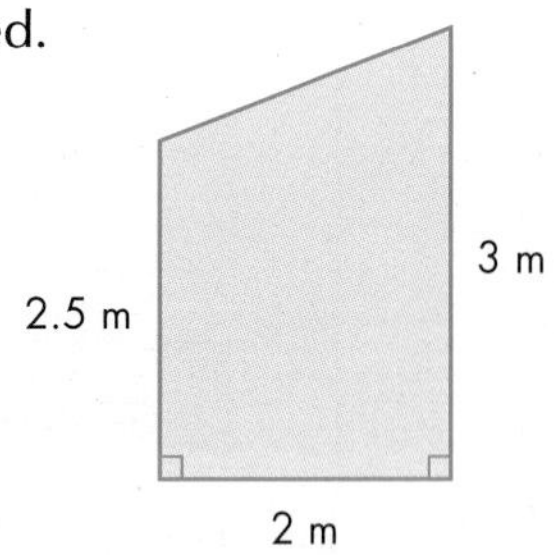

3 m
2.5 m
2 m

**b** Chris wants to repaint the outside of his shed.

The length of the shed is 4 m and Chris knows that a tin of paint covers 10 m².

How many tins of paint does he need?

## Area of any triangle

The area of any triangle is given by the formula:

area of a triangle = $\frac{1}{2}$ × base × **perpendicular height**

Using algebra this is written as:

$A = \frac{1}{2}bh$

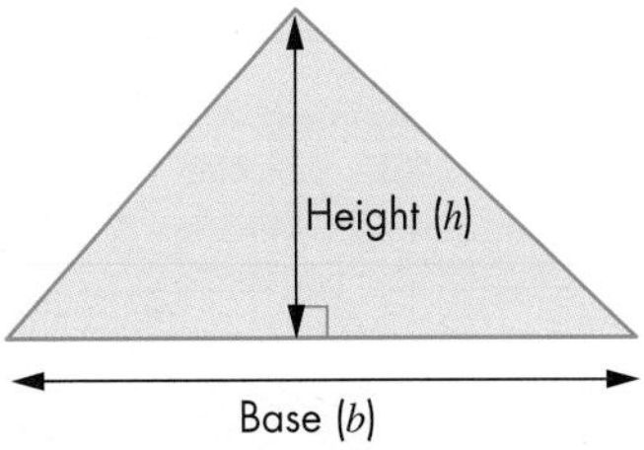

Height (*h*)

Base (*b*)

Calculate the area of this triangle.

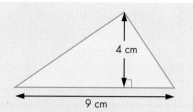

Area $= \frac{1}{2} \times 9 \times 4$

$= \frac{1}{2} \times 36$

$= 18 \text{ cm}^2$

Calculate the area of the shape shown below.

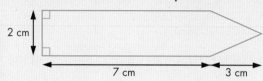

You can split this compound shape into a rectangle (R) and a triangle (T).

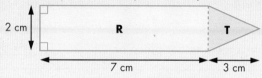

Area of the shape = area of R + area of T

$= 7 \times 2 + \frac{1}{2} \times 2 \times 3$

$= 14 + 3$

$= 17 \text{ cm}^2$

## Exercise 11D 🖩

**1** Calculate the area of each triangle

a
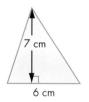

b
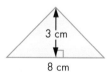

c

d
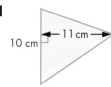

e
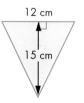

f
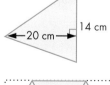

(PS) **2** This regular hexagon has an area of 48 cm².

What is the area of the rectangle that surrounds the hexagon?

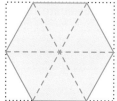

**3** Calculate the area of each of these shapes.

a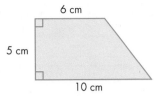

b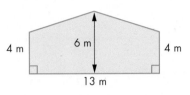

c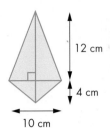

**4** Work out the area of each shaded shape.

a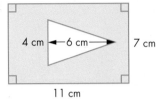

b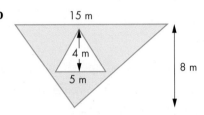

| Hints and tips | Work out the area of the outer shape and subtract the area of the inner shape. |

**(MR)** **5** Write down the dimensions of two different-sized triangles that have an area of 50 cm².

**(PS)** **6** Lee is making a kite. He cuts the kite from a rectangular piece of plastic measuring 60 cm by 40 cm.

a What is the area of the plastic left?

b Work out the area of material he will need to cover both sides of the kite.

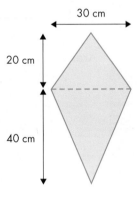

**(CM)** **7** Which triangle is the odd one out?

Give a reason for your answer.

a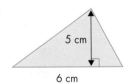

b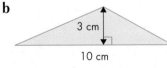

c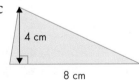

**(EV)** **8** Tim says it is impossible to draw a triangle with all of these properties:

- right-angled
- base 2 cm
- area 6 cm²
- perimeter 6 cm.

Evaluate Tim's statement.

# 11.4 Area of a parallelogram

This section will show you how to:

- calculate the area of a parallelogram
- use the formula for the area of a parallelogram.

The diagram shows that a **parallelogram** can be rearranged as a rectangle with the same base and height.

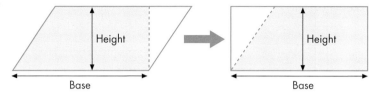

This means that the area of a parallelogram is the same as the area of a rectangle with the same base and perpendicular height. The formula is:

area of parallelogram = base × height

As an algebraic formula, this is:

$A = bh$

where $b$ is the length of the base and $h$ is the height of the parallelogram.

---

**Example 8**

Work out the area of this parallelogram.

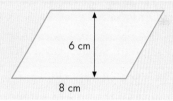

Area = 8 × 6

  = 48 cm²

---

## Exercise 11E

**1** Calculate the area of each parallelogram.

**a**

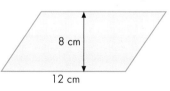

8 cm

12 cm

**b**

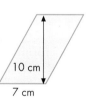

10 cm

7 cm

**c**

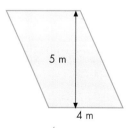

5 m

4 m

**d**

5 cm

25 cm

**e**

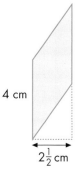

4 cm

$2\frac{1}{2}$ cm

**f**

14 m

8 m

**(EV)** **2** Sandeep says that the area of this parallelogram is 30 cm².

Is she correct? Give a reason for your answer.

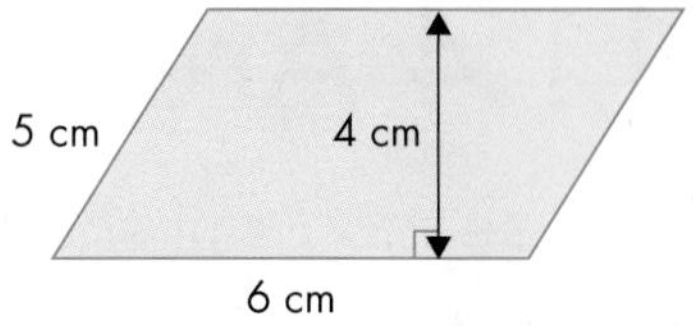

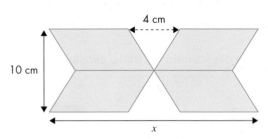

5 cm    4 cm

6 cm

**(CM)** **3** This shape is made from four identical parallelograms. The total area of the shape is 120 cm².

Freya says that the length marked $x$ on the diagram is 20 cm.

Show that Freya is incorrect.

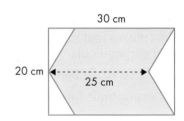

4 cm

10 cm

$x$

**(PS)** **4** This logo, made from two identical parallelograms, is cut from a sheet of card.

**a** Calculate the area of the logo.

**b** How many logos can be cut from a sheet of card that measures 1 m by 1 m?

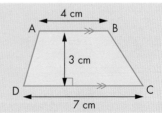

30 cm

20 cm

25 cm

# 11.5 Area of a trapezium

This section will show you how to:

- calculate the area of a trapezium
- use the formula for the area of a trapezium.

**Key term**

trapezium

You can calculate the area of a **trapezium** by working out the mean (average) of the lengths of its parallel sides and multiplying this by the perpendicular distance between them (the vertical height).

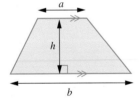

$a$

$h$

$b$

The area of a trapezium is given by this formula:

$A = \frac{1}{2}(a + b)h$

**Example 9**

Calculate the area of the trapezium ABCD.

4 cm

A        B

3 cm

D        C

7 cm

$A = \frac{1}{2}(4 + 7) \times 3$

$= \frac{1}{2} \times 11 \times 3$

$= 16.5 \text{ cm}^2$

**1** Copy and complete the table.

| Trapezium | Parallel side 1 | Parallel side 2 | Vertical height | Area |
|-----------|-----------------|-----------------|-----------------|------|
| a | 8 cm | 4 cm | 5 cm | |
| b | 10 cm | 12 cm | 7 cm | |
| c | 7 cm | 5 cm | 4 cm | |
| d | 5 cm | 9 cm | 6 cm | |
| e | 3 m | 13 m | 5 m | |
| f | 4 cm | 10 cm | | 42 cm² |
| g | 7 cm | 8 cm | | 22.5 cm² |
| h | 6 cm | | 5 cm | 40 cm² |

**2** Calculate the perimeter and area of each trapezium.

a
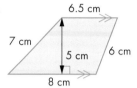

b
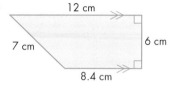

c
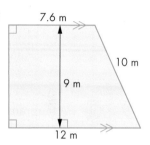

**Hints and tips** Make sure you use the correct measurement for the vertical height.

(MR) **3** A trapezium has an area of 25 cm². Its vertical height is 5 cm.

Work out a possible pair of lengths for the two parallel sides.

**4** Which of these shapes has the largest area?

a
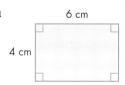

b
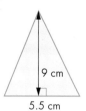

c
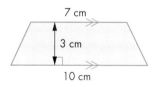

**5** Which of these shapes has the smallest area?

a
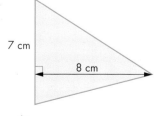

b
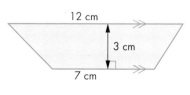

c

**(EV)** **6** Which of the following is the area of this trapezium?

**a** 45 cm² **b** 65 cm² **c** 70 cm²

Show your working.

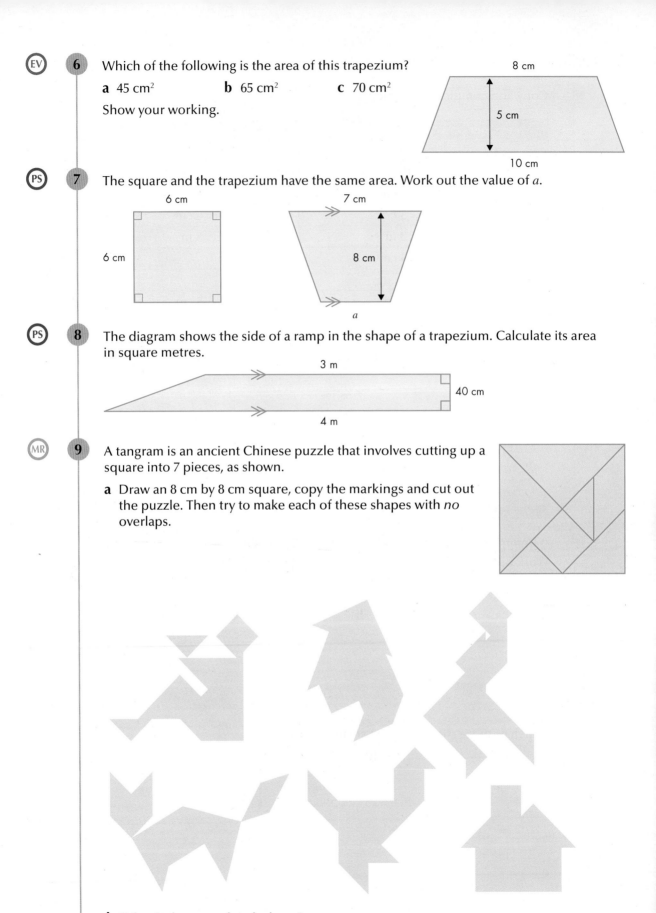

8 cm

5 cm

10 cm

**(PS)** **7** The square and the trapezium have the same area. Work out the value of *a*.

6 cm

6 cm

7 cm

8 cm

*a*

**(PS)** **8** The diagram shows the side of a ramp in the shape of a trapezium. Calculate its area in square metres.

3 m

40 cm

4 m

**(MR)** **9** A tangram is an ancient Chinese puzzle that involves cutting up a square into 7 pieces, as shown.

**a** Draw an 8 cm by 8 cm square, copy the markings and cut out the puzzle. Then try to make each of these shapes with *no* overlaps.

**b** What is the area of each shape?

# 11.6 Circles

This section will show you how to:

- recognise terms used for circle work
- calculate the circumference of a circle.

**Key terms**

| | |
|---|---|
| π (pi) | arc |
| chord | circumference |
| diameter | radius |
| sector | segment |
| tangent | |

## Circle terms

The following are common terms used for parts of circles.

| | |
|---|---|
| **O** | The **centre** of a circle. |
| **Diameter** | The 'width' of a circle. Any diameter passes through O. |
| **Radius** | The distance from O to the edge of a circle. The length of the diameter is twice the length of the radius. |
| **Circumference** | The perimeter of a circle. |
| **Chord** | A line joining two points on the circumference. |
| **Tangent** | A line that touches the circumference at one point only. |
| **Arc** | A part of the circumference of a circle. |
| **Sector** | A part of the area of a circle, lying between two radii and an arc. |
| **Segment** | A part of the area of a circle, lying between a chord and an arc. |

## Exercise 11G

1   Dave drew this face using different parts of a circle. Copy the face and label each part from the following selection of words.

sector     arc     segment     tangent     circle

**2** Hari got his circle definitions mixed up. Sort out his list and put each word with the correct definition.

| | |
|---|---|
| Radius | The width of a circle |
| Circumference | A part of a circle lying between two radii and an arc |
| Tangent | A part of the circumference of a circle |
| Sector | A part of the area of a circle, lying between a chord and an arc |
| Diameter | The distance from the centre to the edge of the circle |
| Chord | A line that touches the circumference of the circle at one point only |
| Arc | The perimeter of a circle |
| Segment | Line joining two points on a circumference |

 **3** Draw each of the following.

**a** A cat, using just circles and arcs

**b** A tent with an opening and a flag on the top, using just sectors

**c** An old-fashioned teacher's hat, using a segment and a tangent

**d** A boat, using a segment and sectors

 **4** See what different shapes you can draw just using parts of the circle.

## Circumference of a circle

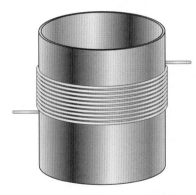

- Find a cylinder or tin and measure its diameter, $d$.
- Wrap a piece of string around the cylinder 10 times and measure the length, $L$, of the string.
- Divide $L$ by 10 to work out the circumference, $C$, of the cylinder.
- Divide the circumference, $C$, by diameter, $d$, and check you get the result 3.1.

If you were able to do this accurately you would get the result as 3.141 592…, which is the number $\pi$ that you will find on your calculator.

This result illustrates the relationship between the circumference, $C$, of a circle and its diameter, $d$.

$C = \pi d$

**Remember:** The diameter of a circle is twice the radius so, if you are given the radius, you will need to double it to get the diameter.

When you are working with $\pi$, you should use the value you are given or the value in your calculator.

**Example 10**

Calculate the circumference of the circle. Give your answer to 1 decimal place.

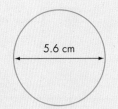

5.6 cm

$C = \pi d$

$\quad = \pi \times 5.6$

$\quad = 17.6$ cm (1 dp)

**Example 11**

Calculate the diameter of a circle that has a circumference of 40 cm.

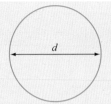

$d$

$C = \pi \times d$

$40 = \pi \times d$

$d = \dfrac{40}{\pi}$

$\quad = 12.7$ cm (1 dp)

## Exercise 11H 🖩

**1** Calculate the circumference of each circle. Give your answers to 1 decimal place.

a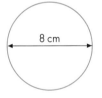
8 cm

b
5 cm

c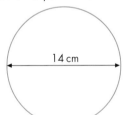
14 cm

d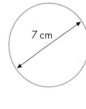
7 cm

e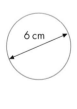
6 cm

f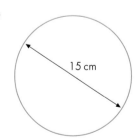
15 cm

g
9.2 cm

h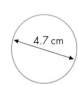
4.7 cm

**2** Work out the circumference of each of these coins. Give your answers to 1 decimal place.

**a** 1p coin (diameter 2 cm)

**b** 2p coin (diameter 2.6 cm)

**c** 5p coin (diameter 1.7 cm)

**d** 10p coin (diameter 2.4 cm)

3 Calculate the circumference of each circle. Give your answers to 1 decimal place.

a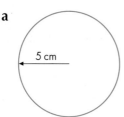
5 cm

b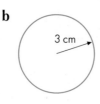
3 cm

c
1.5 cm

d
4 cm

e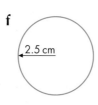
0.9 cm

f
2.5 cm

g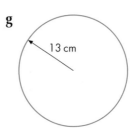
13 cm

h
6.3 cm

Hints and tips | Remember to double the radius to calculate the diameter, or use the formula $C = 2\pi r$.

 4 The radius of the wheels on Tim's bike is 31.5 cm.

a Calculate the circumference of one of the wheels. Give your answer to the nearest centimetre.

b Tim rides his bike for 1 km. How many complete revolutions does each wheel make?

 5 The diagram represents a race-track on a school playing field. The diameter of each circle is shown.

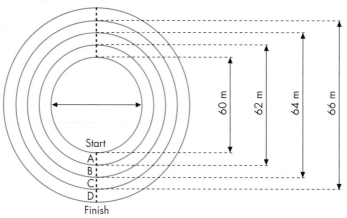

Start
A
B
C
D
Finish

60 m    62 m    64 m    66 m

In a race with four runners, each runner starts and finishes on the same inner circle of their lane after completing one circuit.

a Calculate the distance run by each of the runners.

b How much further than A does D have to run?

 6 A rope is wrapped eight times round a cylindrical post with a diameter of 35 cm. How long is the rope?

**7** A hamster has a treadmill with a diameter of 12 cm.

    **a** What is the circumference of the treadmill?

    **b** How many centimetres has the hamster run when the wheel has made 100 complete revolutions?

    **c** Give your answer to part **b** in metres.

    **d** One night, the hamster runs continuously and turns the wheel 100 000 times. How many kilometres has he run?

**8** A circle has a circumference of 314 cm. Calculate its diameter.

**9** What is the diameter of a circle with a circumference of 76 cm? Give your answer to 1 decimal place.

**10** What is the radius of a circle with a circumference of 100 cm? Give your answer to 1 decimal place.

 **11** A semicircular protractor has a diameter of 10 cm.

    Which of the following is the correct perimeter of the protractor?

    **a** 15.7 cm     **b** 25.7 cm     **c** 31.4 cm     **d** 41.4 cm

10 cm

 **12** **a** What are the distances round these trees? Assume that a tree trunk is circular.

     **i** Fir: diameter of 20 cm     **ii** Ash: diameter of 24 cm

     **iii** Elm: diameter of 22 cm     **iv** Oak: diameter of 26 cm

    **b** Compare differences between the tree circumferences. What connection do they have to $\pi$?

    **c** What would be the difference in length between a rope stretched tightly round Earth and another rope always held 1 m above it?

 **13** **a** Calculate the perimeter of shapes A and B.

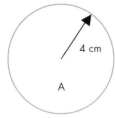 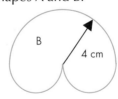 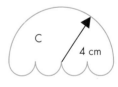

4 cm      B   4 cm      C   4 cm

A

    **b** Write down the perimeter of shape C.

 **14** A square has sides of length $a$ and a circle has radius $r$.

$a$

$a$

$r$

The perimeter of the square is equal to the circumference of the circle.

Show that $r = \dfrac{2a}{\pi}$.

**15** Ben wants to buy enough fencing to go round the curved part of a semicircular flowerbed with a diameter of 8 m.

The fencing is sold in 2-m lengths. How many lengths does Ben need?

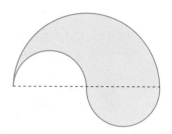

**16** The shape shown is made up of semicircles.

    **a** The radius of the small semicircle is 1 cm. Work out the circumference of the shaded shape.

    **b** Harry says that if the radius of the smaller semicircle is tripled, the perimeter of the whole shape will triple. Show that he is correct.

# 11.7 The area of a circle

This section will show you how to:

- calculate the area of a circle.

A circle can be divided into 32 sectors as shown.

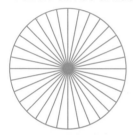

These 32 sectors are then cut out and rearranged together as the new shape shown.

This shape is close to a rectangle.

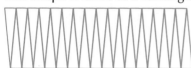

Notice that the length of the rectangle is half the circumference, that is $\frac{1}{2}\pi d$ or $\pi r$ (as $d = 2r$).

Notice that the width of the rectangle is the radius, $r$, of the circle.

So, the area of the circle = the area of rectangle

$$= \pi r \times r$$
$$= \pi r^2$$

This result illustrates the relationship between the area, $A$, and the radius, $r$, of a circle.

    $A = \pi r^2$

**Remember:** This formula uses the radius of a circle so, when you are given the diameter of a circle, you will need to halve it to get the radius.

Example 12

**Radius given**

Calculate the area of a circle with a radius of 7 cm.

7 cm

$A = \pi r^2$

$\quad = \pi \times 7^2$

$\quad = \pi \times 49$

$\quad = 153.9 \text{ cm}^2 \text{ (1 dp)}$

Example 13

**Diameter given**

Calculate the area of a circle with a diameter of 12 cm.

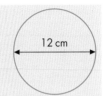
12 cm

First, halve the diameter to get the radius.　　　$12 \div 2 = 6 \text{ cm}$

Then, calculate the area.

Area $= \pi r^2$

$\quad = \pi \times 6^2$

$\quad = \pi \times 36$

$\quad = 113.1 \text{ cm}^2 \text{ (1 dp)}$

# Exercise 11I 🖩

**1**　Calculate the area of each circle. Give your answers to 1 decimal place.

**a**
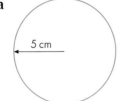
5 cm

**b**

3 cm

**c**

1.5 cm

**d**

4 cm

**e**

0.9 cm

**f**

2.5 cm

**g**

13 cm

**h**

6.3 cm

**2**　Work out the area of one face of each of these coins. Give your answers to 1 decimal place.

　**a** 1p coin (radius 1 cm)

　**b** 2p coin (radius 1.3 cm)

　**c** 5p coin (radius 0.85 cm)

　**d** 10p coin (radius 1.2 cm)

③ Calculate the area of each circle. Give your answers to 1 decimal place.

**a**
8 cm

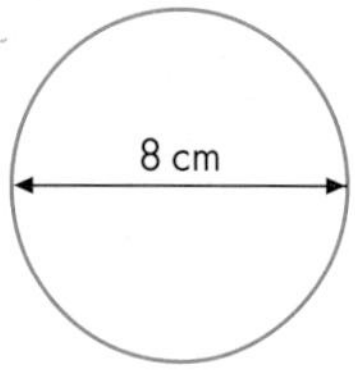

**b**
5 cm

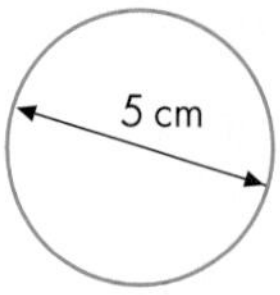

**c**
14 cm

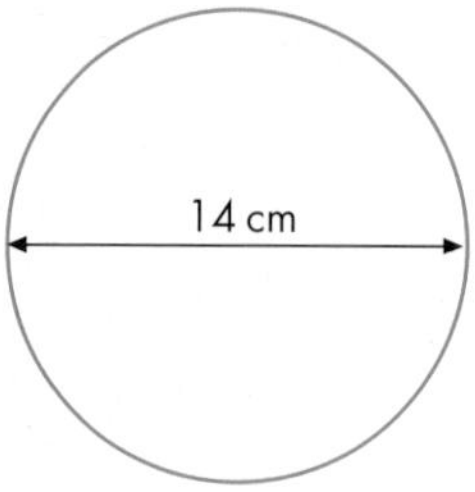

**d**
7 cm

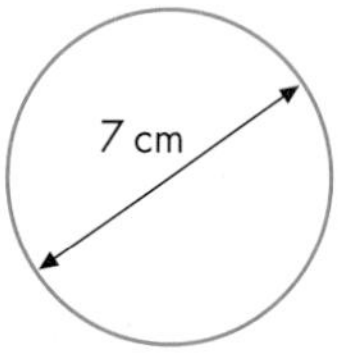

**e**
6 cm

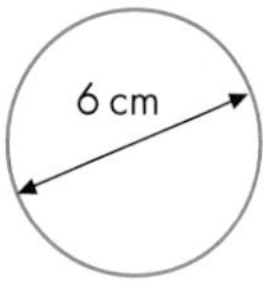

**f**
15 cm

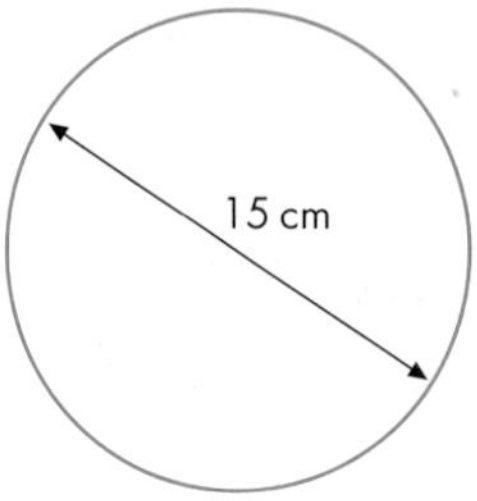

**g**
9.2 cm

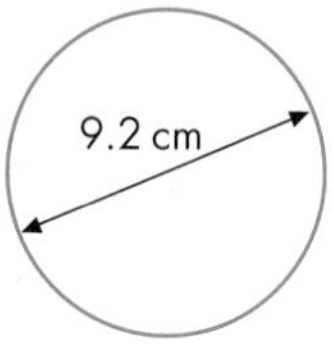

**h**
4.7 cm

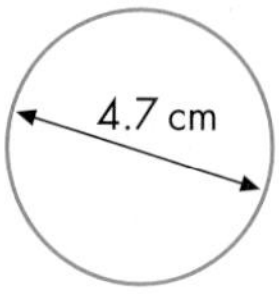

> **Hints and tips**  Remember to halve the diameter to calculate the radius.

④ Badges are stamped from rectangular strips as shown.

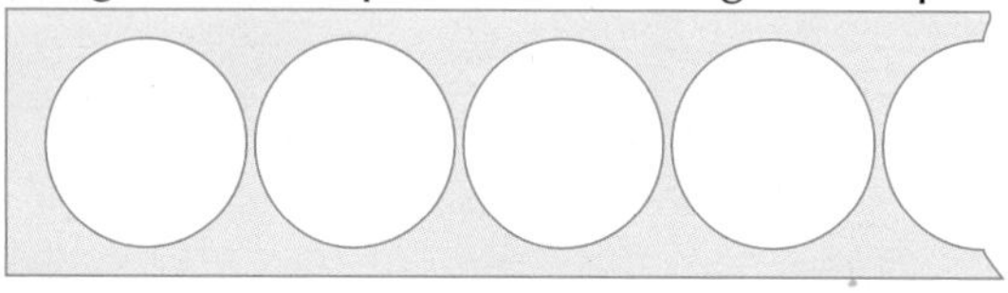

Each badge is made from a circle of radius 1.7 cm. Each rectangular strip measures 4 cm by 500 cm.

**a** What is the area of one surface of a badge?

**b** What is the area of the rectangular strip?

**c** How many badges can be stamped out of one 500-cm strip when there is a 0.2-cm gap between adjacent badges?

**d** What will be the total area of all the badges stamped out of one strip?

**e** How much waste is there for each strip?

⑤ A young athlete can throw the discus a distance of 35 m but is never sure about the direction in which he will throw it. What area of the field should be closed while he is throwing the discus?

⑥ Calculate **i** the circumference and **ii** the area of each of these circles. Give your answers to 1 decimal place.

**a**
9 cm

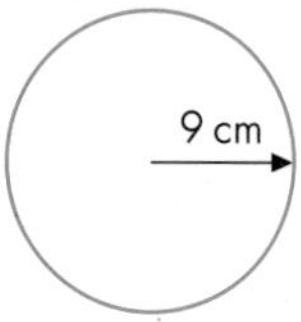

**b**
22 cm

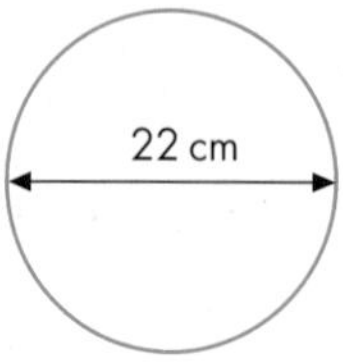

**c**
6.5 cm

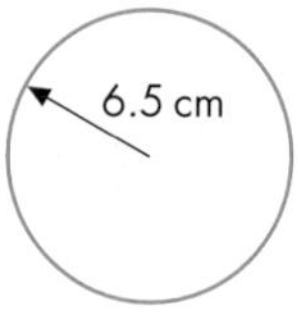

**d**
28 cm 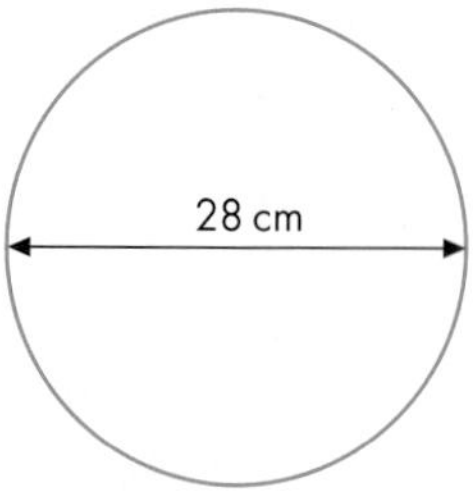

⑦ Calculate the area of a circle with a circumference of 110 cm.

**8** Calculate the area of these shapes. Give your answers to 1 decimal place.

a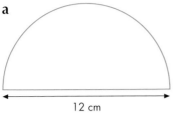

12 cm

Semicircle

b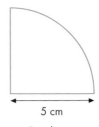

5 cm

Quadrant

**9** Calculate the area of the shaded part of each of these diagrams.

a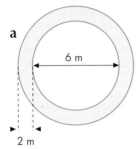

6 m

2 m

b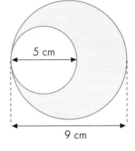

5 cm

9 cm

**Hints and tips** In each diagram, subtract the area of the small circle from the area of the large circle.

**CM** **10** A square has sides of length $a$ and a circle has radius $r$.

The area of the square is equal to the area of the circle.

Show that $r = \dfrac{a}{\sqrt{\pi}}$.

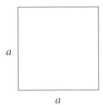

$a$

$a$

$r$

**EV** **11** A circle fits exactly inside a square with side length 10 cm.

Lewis said that 21% of the area of the shape is shaded.

Evaluate Lewis's statement.

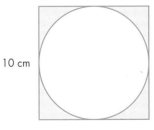

10 cm

10 cm

12    The shape shown is made up of semicircles.

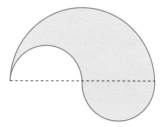

a   If the radius of the small semicircle is 1 cm, calculate the area of the shaded shape.

b   Repeat part **a** if the small semicircle has a radius of:

    **i**   2 cm                 **ii**   3 cm                 **iii**   4 cm.

c   What link is there between the radius, $r$, of the small semicircle and the area, $A$, of the shaded shape?

# 11.8 Answers in terms of $\pi$

This section will show you how to:

- give answers for circle calculations in terms of $\pi$.

Sometimes you do not want a numerical answer to a circle problem but need to give the answer in terms of $\pi$. (You could calculate the numerical answer later.)

---

**Example 14**

What are the circumference and area of this circle?

Leave your answers in terms of $\pi$.

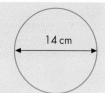

14 cm

Circumference = $\pi d$

           = $\pi \times 14$

           = $14\pi$ cm

      Area = $\pi r^2$

           = $\pi \times 7^2$

           = $\pi \times 49$

           = $49\pi$ cm$^2$

---

# Exercise 11J

In this exercise, give your answers in terms of π.

1. State the circumference of a circle with each of these measures.

   a diameter 4 cm

   b radius 10 cm

   c diameter 15 cm

   d radius 2 cm

2. State the area of a circle with each of these measures.

   a radius 4 cm

   b diameter 10 cm

   c radius 3 cm

   d diameter 18 cm

3. State the radius of the circle with a circumference of $50\pi$ cm.

4. State the radius of the circle with an area of $100\pi$ cm$^2$.

5. A circle has a circumference of 200 cm.

   Rory said the diameter is $\frac{200}{\pi}$ cm.

   Alice said the area is $\frac{1000}{\pi}$ cm$^2$.

   Comment on their statements.

6. Work out the area of each of these shapes.

   **a**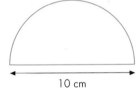
   10 cm

   **b**
   8 cm

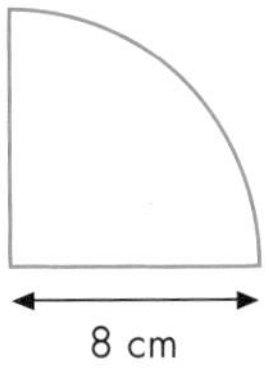

   **c**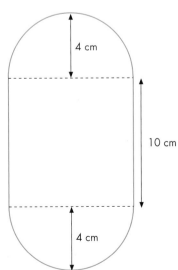

   **d**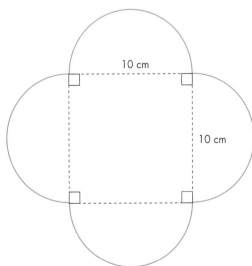

MR

**7**

**a** Work out the area of a semicircle with radius 8 cm.

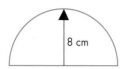

8 cm

**b** Work out the area of two semicircles with radii 4 cm.

4 cm

**c** Work out the area of four semicircles with radii 2 cm.

2 cm

**d** Use your answers to parts **a** to **c** to write down the area of eight semicircles with radii 1 cm.

PS

**8** A circle fits inside a semicircle of diameter 12 cm as shown.

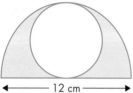

12 cm

Work out the area of the shaded region.

MR

**9** A shape is made from a rectangle and a quadrant of a circle.

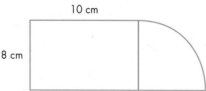

10 cm

8 cm

Which of these is the correct area of the shape?

**a** $(80 + 4\pi)$ cm$^2$

**b** $(80 + 8\pi)$ cm$^2$

**c** $(80 + 16\pi)$ cm$^2$

**d** $(80 + 32\pi)$ cm$^2$

# Worked exemplars

  **1**  A shape is made up of two squares and an equilateral triangle as shown. Show that the area of the shape is between 50 cm² and 75 cm².

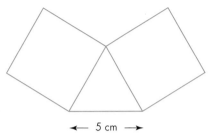

← 5 cm →

| This question assesses 'communicating mathematically', so you must display your methods clearly and include words to explain what your calculations show. | |
|---|---|
| Area of one square = 5 × 5 = 25 cm² <br><br> Area of two squares = 50 cm² <br><br> So, the area of the shape must be greater than 50 cm². | You need to recognise that you can calculate the accurate area of the squares but not the area of the triangle. |
| The area of the triangle must be less than the area of one of the squares. | You do not need to calculate the area of the triangle, only show that it must be less than the area of a square. |
| So, the area of the shape must be less than the area of three squares, which is 75 cm². | Write a conclusion for your working. |

  **2**  Four identical circles fit exactly in a square with side length $x$.

Work out the area of the shaded region.

Give your answer in terms of $\pi$.

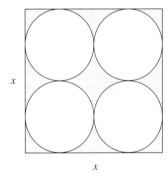

$x$

$x$

| This is a problem-solving question so you will need to show your strategy. | |
|---|---|
| Radius of each circle = $\frac{x}{4}$ | You need to work out the radius of one circle in order to calculate its area. |
| Area of each circle $= \pi \times \left(\frac{x}{4}\right)^2$ <br><br> $= \frac{\pi x^2}{16}$ <br><br> Area of four circles $= \frac{\pi x^2}{4}$ | Use this to calculate the area of the four circles. <br><br> Leave all your working in terms of $\pi$. |
| Area of square = $x^2$ <br><br> Area of shaded region = $x^2 - \frac{\pi x^2}{4}$ | Then subtract this from the area of the square to work out the area of the shaded part. |

# Ready to progress?

I can calculate the perimeter of a variety of shapes.

I can calculate the area of a triangle using the formula $A = \frac{1}{2}bh$.

I can calculate the area of a parallelogram using the formula $A = bh$.
I can calculate the area of a trapezium using the formula $A = \frac{1}{2}(a + b)h$.
I can calculate the area of compound shapes.
I can calculate the circumference of a circle.
I can calculate the area of a circle.
I can give answers to circle calculations in terms of $\pi$.

# Review questions

**1** What is the area of this shape?

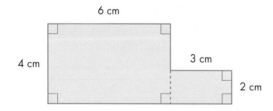

6 cm

4 cm

3 cm

2 cm

**(PS)** **2** The diagram shows a rectangular floor.
The length of the floor is 4 m.
The width of the floor is 3 m.

3 m

4 m

Tom is going to cover the floor with square tiles of side length 50 cm.
How many tiles does Tom need to completely cover the floor?

**(CM)** **3** Tilly says that the area of this triangle is 108 cm².
Show that she is incorrect.

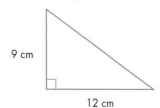

9 cm

12 cm

**4** This shape has an area of 44 cm². What is its vertical height?

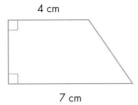

4 cm

7 cm

**(PS)** **5** The diagram shows the plan of a farmer's field that he will use for sheep to graze.

The farmer wants to have an area of at least 50 m² per sheep.

What is the maximum number of sheep the farmer should put in the field?

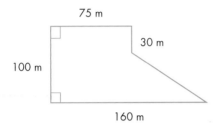

75 m

30 m

100 m

160 m

**(CM)** **6** The circumference of a circle is 30 cm.

Show that the area of the circle is 71.6 cm².

**(EV)** **7** A circle has a radius of 5 cm.

A parallelogram has a long side of 9 cm and vertical height of 8 cm.

Andy says: 'The difference between the area of the circle and the area of the parallelogram is 6.5 cm².'

Evaluate Andy's statement.

**(MR)** **8** The diagram shows two small circles inside a large circle.

The large circle has a radius of 10 cm.

Both of the small circles have a diameter of 5 cm.

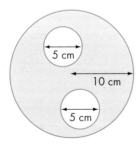

5 cm

10 cm

5 cm

**a** Write down the radius of each of the small circles.

**b** Show that the area of the shaded part is 87.5% of the area of the large circle.

**9** The circumference of a circle is 40 cm.

What is the radius of the circle? Give your answer in terms of π.

**(PS)** **10** Calculate the perimeter of a semicircle of diameter 10 cm.

Give your answer to 1 decimal place.

# 12 Geometry and measures: Transformations

## This chapter is going to show you:

- how to work out the order of rotational symmetry for a 2D shape
- how to translate, reflect, rotate and enlarge 2D shapes
- what is meant by a transformation
- how to add and subtract vectors.

## You should already know:

- how to find the lines of symmetry of a 2D shape
- how to draw lines with the equations $x = \pm a$, $y = \pm b$, $y = x$ and $y = -x$
- how to measure lines and angles.

## About this chapter

If you look around you, you will see two-dimensional designs everywhere: on clothes, curtains, furniture fabric, carpets, wallpaper and any flat surface that can be decorated. Many of these designs use symmetry to produce a repeated pattern. This makes it easy to manufacture a design on a fabric, for example, since a basic unit can be copied forever.

From a simple starting point, it is possible to use reflections, rotations and translations to build lots of different patterns. You might think that there are infinitely many repeating 'wallpaper designs' but, in fact, every repeating design you see can be classified as one of just 17 different types, depending on its symmetry.

Here are some examples of repeated designs from all over the world. After working through this chapter, you will be able to find examples of reflections, rotations and translations in them.

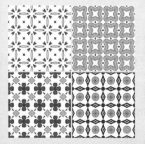  

# 12.1 Rotational symmetry

This section will show you how to:

- work out the order of rotational symmetry for a 2D shape
- recognise shapes with rotational symmetry.

**Key terms**

order of rotational symmetry

rotational symmetry

A 2D shape has **rotational symmetry** if it can be rotated (turned) about a point to look exactly the same in a new position. The **order of rotational symmetry** is the number of different positions where the shape looks the same when it is rotated about the point.

One way to work out the order of rotational symmetry for any shape is to trace it and count the number of times that the shape stays the same as you turn the tracing paper through one complete turn.

**Example 1**

Work out the order of rotational symmetry for this shape.

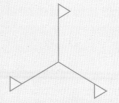

First, hold the tracing paper on top of the shape and trace the shape. Place the sharp end of a pencil on the centre of the shape and rotate the tracing paper, counting the number of times the tracing matches the original shape in one complete turn.

You will find three different positions. So the order of rotational symmetry for the shape is 3.

## Exercise 12A

**1** Work out the order of rotational symmetry for each of these shapes.

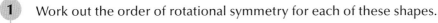

**2** Here are some Greek capital letters.

Write down the order of rotational symmetry for each one.

a   b   c   d   e

**PS 3** The upright capital letter A fits exactly onto itself only once. So, its order of rotational symmetry is 1. This means that it has no rotational symmetry.

Write down all the upright capital letters of the alphabet that have rotational symmetry of order 1.

**PS 4** Copy this grid. On your copy, shade in four squares so that the shape has rotational symmetry of order 2.

**5** Copy the table below. On your copy, write the letter for each shape in the correct box. The first one has been done for you.

| Order of rotational symmetry | | Number of lines of symmetry | | | |
|---|---|---|---|---|---|
| | | 0 | 1 | 2 | 3 |
| | 1 | | | | |
| | 2 | | A | | |
| | 3 | | | | |

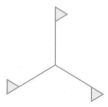

 **6** A shape with three flagpoles has rotational symmetry of order 3.

Sue says: 'The angle between each pole is 120°.'

Show that Sue is correct.

 **7** Rachel looked at a triangle and said: 'It has line symmetry but no rotational symmetry, so it must be isosceles.' Is Rachel correct? Draw diagrams to support your answer.

# 12.2 Translation

This section will show you how to:

- translate a 2D shape.

In this section, you will learn about **translation,** which is one for four basic transformations. A **transformation** changes the position or the size of a shape. In any transformation, the original shape is called the **object** and the transformed shape is called the **image**.

| Key terms | |
|---|---|
| image | object |
| transformation | translation |
| vector | |

A translation is the 'movement' of a shape from one place to another without reflecting it or rotating it. Every point in the shape moves in the same direction and through the same distance so the object and the image are always exactly the same size.

You use **vectors** to describe translations. A vector is the combination of a horizontal shift and a vertical shift. The next example shows how to do this.

Use vectors to describe the translations of the following triangles.

**a** A to B

**b** B to C

**c** C to D

**d** D to A

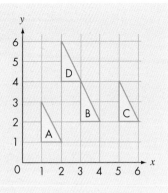

**a** The vector describing the translation from A to B is $\begin{pmatrix} 2 \\ 1 \end{pmatrix}$.

**b** The vector describing the translation from B to C is $\begin{pmatrix} 2 \\ 0 \end{pmatrix}$.

**c** The vector describing the translation from C to D is $\begin{pmatrix} -3 \\ 2 \end{pmatrix}$.

**d** The vector describing the translation from D to A is $\begin{pmatrix} -1 \\ -3 \end{pmatrix}$.

**Note:**

- The top number in the vector describes the horizontal movement. To the right +, to the left –.
- The bottom number in the vector describes the vertical movement. Upwards +, downwards –.
- These vectors are also called *direction vectors*.

## Exercise 12B

**1** Copy each shape onto squared paper and draw its image after the given translation.

**a** 3 squares right

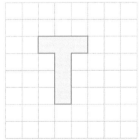

**b** 3 squares up

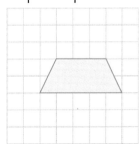

**c** 3 squares down

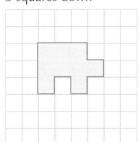

**d** 3 squares left

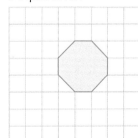

**2** Copy each shape onto squared paper and draw its image after the given translation.

**a** 4 squares right and 3 squares down

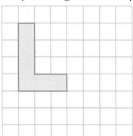

**b** 3 squares right and 3 squares up

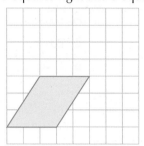

**c** 4 squares left and 3 squares down

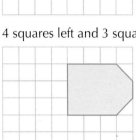

**d** 1 square left and 4 squares up

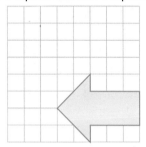

**3** Look at the triangles on this grid.

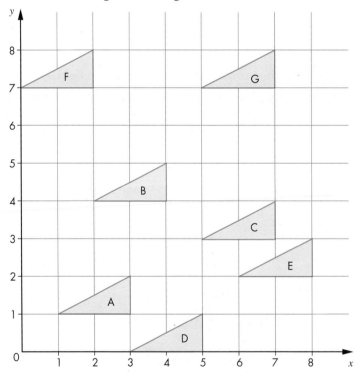

Use vectors to describe each translation.

**a  i** A to B     **ii** A to C     **iii** A to D     **iv** A to E

**b  i** B to A     **ii** B to C     **iii** B to F     **iv** B to G

**c  i** C to D     **ii** C to E     **iii** C to F     **iv** C to G

**d  i** D to E     **ii** E to B     **iii** F to C     **iv** G to D

**4** **a** Draw coordinate axes with the $x$-axis from –6 to 6 and the $y$-axis from –6 to 6. Draw a triangle with coordinates A(1, 1), B(2, 1) and C(1, 3).

**b** Draw the image of ABC after a translation with vector $\begin{pmatrix} 2 \\ 3 \end{pmatrix}$. Label this triangle P.

**c** Draw the image of ABC after a translation with vector $\begin{pmatrix} -1 \\ 2 \end{pmatrix}$. Label this triangle Q.

**d** Draw the image of ABC after a translation with vector $\begin{pmatrix} 3 \\ -2 \end{pmatrix}$. Label this triangle R.

**e** Draw the image of ABC after a translation with vector $\begin{pmatrix} -2 \\ -4 \end{pmatrix}$. Label this triangle S.

**5** Look at your diagram from question **4**. Use vectors to describe the translation that will move

| | | | |
|---|---|---|---|
| **a** P to Q | **b** Q to R | **c** R to S | **d** S to P |
| **e** R to P | **f** S to Q | **g** R to Q | **h** P to S. |

**6** Draw a set of coordinate axes with the $x$-axis from –5 to 5 and the $y$-axis from –5 to 5. Draw the triangle with coordinates A(0, 0), B(1, 0) and C(0, 1). How many different translations are there that use integer values only and move the triangle ABC to somewhere in the grid?

**7** In a game of *Snakes and ladders*, each of the snakes and ladders can be described by a translation.

Ladders $\quad \begin{pmatrix} 1 \\ 2 \end{pmatrix}, \begin{pmatrix} 2 \\ 5 \end{pmatrix}, \begin{pmatrix} -3 \\ 4 \end{pmatrix}, \begin{pmatrix} -2 \\ 3 \end{pmatrix}, \begin{pmatrix} 3 \\ 2 \end{pmatrix}$

Snakes $\quad \begin{pmatrix} 1 \\ -3 \end{pmatrix}, \begin{pmatrix} 3 \\ -4 \end{pmatrix}, \begin{pmatrix} -2 \\ -2 \end{pmatrix}, \begin{pmatrix} -1 \\ -3 \end{pmatrix}, \begin{pmatrix} 2 \\ -5 \end{pmatrix}$

**a** Put all five ladders and all five snakes described above onto a 10 by 10 coordinate grid to design a *Snakes and ladders* game board.

**b** Why are the bottom part of the vectors above:

**i** always positive for the ladders

**ii** always negative for the snakes?

**8** If a translation is given by $\begin{pmatrix} x \\ y \end{pmatrix}$, describe the translation that would take the image back to the original position of the object.

**9** A plane flies between three cities: A, B and C. It uses direction vectors, with distances in kilometres.

The direction vector for the flight from A to B is $\begin{pmatrix} 500 \\ 200 \end{pmatrix}$ and

the direction vector for the flight from B to C is $\begin{pmatrix} -200 \\ 300 \end{pmatrix}$.

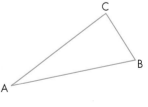

**a** Draw a diagram to show the three flights on centimetre-squared paper. Use a scale of 1 cm represents 100 km.

**b** Work out the direction vector for the flight from C to A.

**10** A boat travels between three jetties on a lake: X, Y and Z. It uses direction vectors, with distance in kilometres.

The direction vector from X to Y is $\begin{pmatrix} 3 \\ -1 \end{pmatrix}$ and the direction vector from Y to Z is $\begin{pmatrix} -2 \\ -3 \end{pmatrix}$.

**a** Draw a diagram to show journeys between X, Y and Z on centimetre-squared paper. Use a scale of 1 cm to 1 km.

**b** Work out the direction vector for the journey from Z to X.

(CM) **11** A triangle has been translated to a new position. How you work out the vector to take it back to its original position?

# 12.3 Reflections

This section will show you how to:

- reflect a 2D shape in a mirror line.

Key terms

mirror line    reflection

A **reflection** transforms a shape so that it becomes a mirror image of itself.

Notice that the reflection of each point in the object (original shape) is perpendicular to the **mirror line**. If you 'fold' the whole diagram along the mirror line, the object will coincide with its reflection or image.

The object and image will always be exactly the same size.

Object

Mirror line

Image

*Example 3*

**a** Reflect triangle ABC in the *x*-axis. Label the image P.

**b** Reflect triangle ABC in the *y*-axis. Label the image Q.

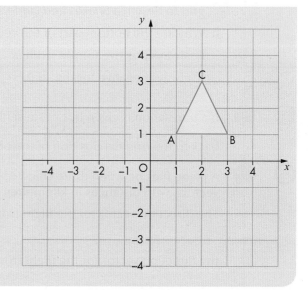

**a** The mirror line is the *x*-axis. So, each vertex on triangle P will be the same distance from the *x*-axis as the corresponding vertex on the object.

**b** The mirror line is the *y*-axis. So, each vertex on triangle Q will be the same distance from the *y*-axis as the corresponding vertex on the object.

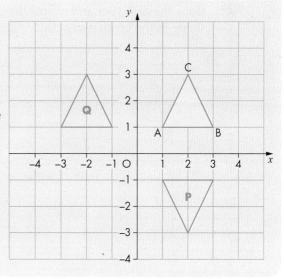

## Exercise 12C

**1** Copy these shapes and mirror lines onto squared paper. Draw the reflection of each shape in the given mirror line.

**a**

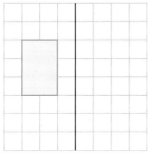

**b**

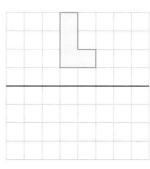

**c**

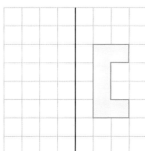

**d**

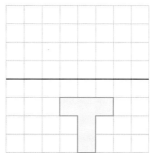

**2** Copy these shapes and mirror lines onto squared paper. Draw the reflection of each shape in the given mirror line.

**a** 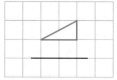 **b**  **c**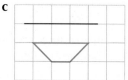

**d** 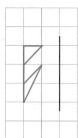 **e** 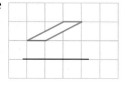 **f** 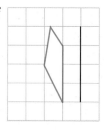 **g**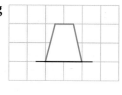

**3** Copy these shapes and mirror lines onto squared paper. Draw the reflection of each shape in the given mirror line.

**a**  **b** 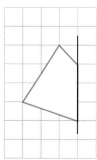 **c**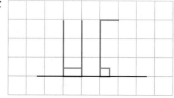

**4** Copy this diagram onto squared paper.

   **a** Reflect rectangle ABCD in the $x$-axis. Label the image R.

   **b** Reflect rectangle ABCD in the $y$-axis. Label the image S.

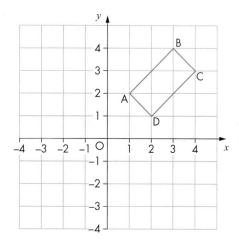

**5** **a** Draw a pair of axes. Label the $x$-axis from –5 to 5 and the $y$-axis from –5 to 5.

   **b** Draw the triangle with coordinates A(1, 1), B(3, 1) and C(4, 5).

   **c** Reflect the triangle ABC in the $x$-axis. Label the image P.

   **d** Reflect triangle P in the $y$-axis. Label the image Q.

   **e** Reflect triangle Q in the $x$-axis. Label the image R.

   **f** Describe the reflection that will move triangle ABC to triangle R.

**CM** **6**  **a** Draw a pair of axes. Label the x-axis from –5 to 5 and the y-axis from –5 to 5.

**b** Reflect the points A(2, 1), B(5, 0), C(–3, 3) and D(3, 2) in the x-axis.

**c** What do you notice about the values of the coordinates of the reflected points?

**d** If the point (a, b) was reflected in the x-axis, what would be the coordinates of the reflected point?

**CM** **7**  **a** Draw a pair of axes. Label the x-axis from –5 to 5 and the y-axis from –5 to 5.

**b** Reflect the points A(2, 1), B(0, 5), C(3, –2) and D(4, –3) in the y-axis.

**c** What do you notice about the values of the coordinates of the reflected points?

**d** If the point (a, b) was reflected in the y-axis, what would be the coordinates of the reflected point?

**PS** **8** By using the middle square as the starting square ABCD, describe how to keep reflecting the square to obtain the final shape in the diagram.

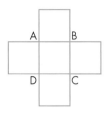

**MR** **9** Triangle A is drawn on a grid and reflected to give Triangle B.

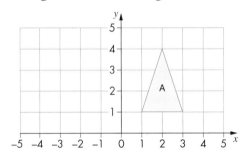

The coordinates of B are (–4, 4), (–3, 1) and (–5, 1).

What is the equation of the mirror line?

**10** Copy these triangles and mirror lines onto squared paper, leaving plenty of space on the opposite side of the mirror lines. Draw the reflection of each triangle in its mirror line.

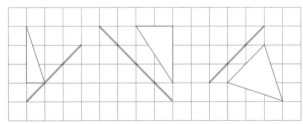

Hints and tips Turn the page around so that the mirror lines are vertical or horizontal.

**11** **a** Draw a pair of axes and the lines $y = x$ and
$y = -x$, as shown.

**b** Draw the triangle with coordinates A(2, 1), B(5, 1)
and C(5, 3).

**c** Draw the reflection of triangle ABC in the $x$-axis
and label the image P.

**d** Draw the reflection of triangle P in the line $y = -x$
and label the image Q.

**e** Draw the reflection of triangle Q in the $y$-axis and
label the image R.

**f** Draw the reflection of triangle R in the line $y = x$ and
label the image S.

**g** Draw the reflection of triangle S in the $x$-axis and label the image T.

**h** Draw the reflection of triangle T in the line $y = -x$ and label the image U.

**i** Draw the reflection of triangle U in the $y$-axis and label the image W.

**j** What single reflection will move triangle W to triangle ABC?

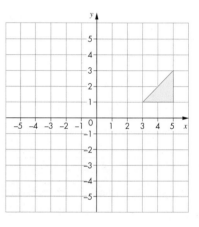

**12** Copy the diagram and reflect the triangle in these lines.

**a** $y = x$        **b** $x = 1$

**c** $y = -x$      **d** $y = -1$

**13** **a** Draw a pair of axes. Label the $x$-axis from −5 to 5 and the $y$-axis from −5 to 5.

**b** Draw the line $y = x$.

**c** Reflect the points A(2, 1), B(5, 0), C(−3, 2) and D(−2, −4) in the line $y = x$.

**d** What do you notice about the values of the coordinates of the reflected points?

**e** If the point $(a, b)$ was reflected in the line $y = x$, what would be the coordinates of
the reflected point?

**14** **a** Draw a pair of axes. Label the $x$-axis from −5 to 5 and the $y$-axis from −5 to 5.

**b** Draw the line $y = -x$.

**c** Reflect the points A(2, 1), B(0, 5), C(3, −2) and D(−4, −3) in the line $y = -x$.

**d** What do you notice about the values of the coordinates of the reflected points?

**e** If the point $(a, b)$ was reflected in the line $y = -x$, what would be the coordinates of
the reflected point?

**15** Triangle A has been reflected in a straight line. What reflection will return the image
to its original position?

# 12.4 Rotations

This section will show you how to:

- rotate a 2D shape about a point.

**Key terms**

angle of rotation

centre of rotation

rotation

A **rotation** transforms a shape to a new position by turning it about a fixed point called the **centre of rotation**.

See how the shapes below are rotated around the points shown.

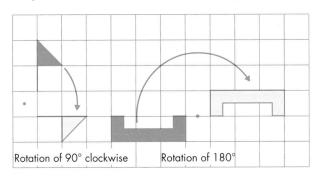

Rotation of 90° clockwise | Rotation of 180°

**Note**

- The direction of turn or the **angle of rotation** is expressed as clockwise or anticlockwise.
- The position of the centre of rotation is always given.
- The rotations 180° clockwise and 180° anticlockwise are the same.
- When a shape is rotated, the rotated shape is exactly the same size as the original shape.

You can use tracing paper to help you with rotation questions.

**Example 4**

Draw the image of this shape after it has been rotated through 90° clockwise about the point X.

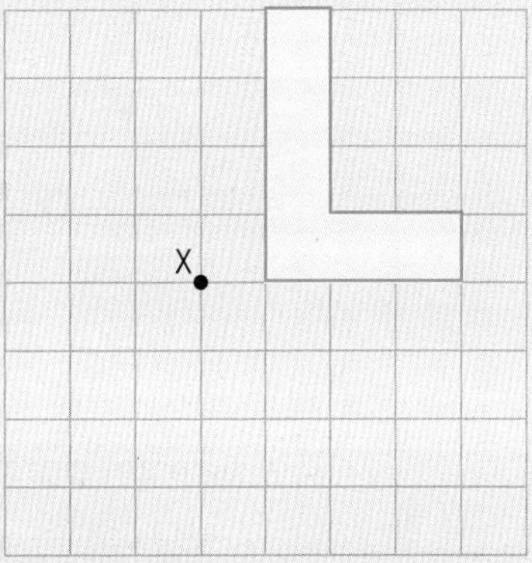

First trace the object shape and fix the centre of rotation with a pencil point. Then rotate the tracing paper through 90° clockwise.

The tracing now shows the position of the image.

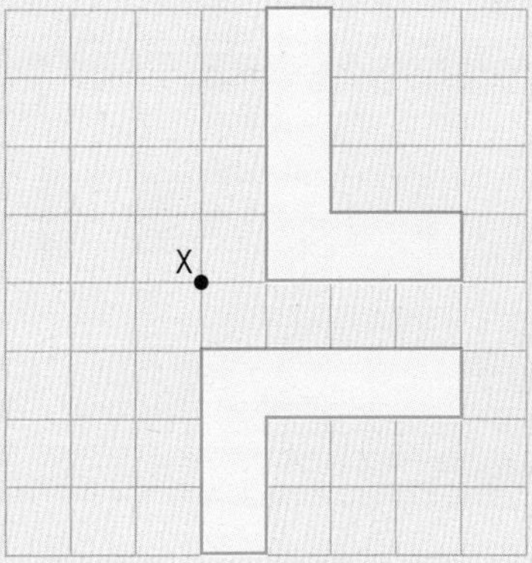

**1** Copy each diagram onto squared paper. Draw the image of each triangle after the given rotation about the centre, X.

**a** $\frac{1}{2}$-turn  **b** $\frac{1}{4}$-turn clockwise

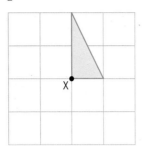 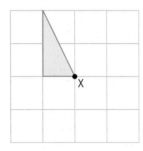

**c** $\frac{1}{4}$-turn anticlockwise  **d** $\frac{3}{4}$-turn clockwise

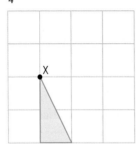 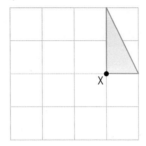

**2** Copy each diagram onto squared paper. Draw the image of each shape after the given rotation about the centre, X.

**a** $\frac{1}{2}$-turn  **b** $\frac{1}{4}$-turn clockwise

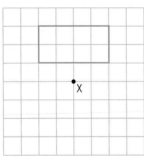 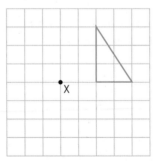

**c** $\frac{1}{4}$-turn anticlockwise  **d** $\frac{3}{4}$-turn clockwise

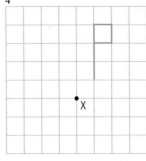 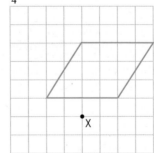

**3** Copy each shape and its centre of rotation onto squared paper, leaving plenty of space around each one.

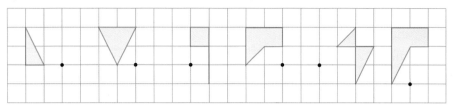

    **a** Rotate each shape about its centre of rotation:

      **i** first by 90° clockwise (call the image A)

      **ii** then by 180° anticlockwise (call the image B).

    **b** Describe, in each case, the rotation that would take:

      **i** A back to its original position    **ii** A to B.

  **4**   **a** Copy the diagram.

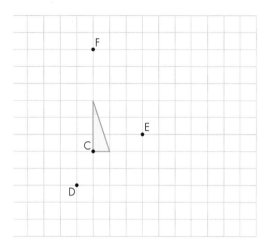

    **b** Rotate the shaded triangle 90° anticlockwise so that point C moves to point D.

    **c** Rotate the shaded triangle shown 90° clockwise so that point C moves to point E.

    **d** Rotate the shaded triangle 180° so that point C moves to point F.

> **Hints and tips** The centre of rotation might not be on the shape.

**5** A graphic designer made this routine for creating a design.

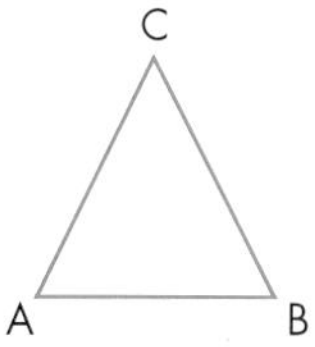

- Start with a triangle ABC.
- Reflect the triangle in the line AB.
- Rotate the whole shape about point C 90° clockwise, then a further 90° clockwise, then a further 90° clockwise.

    **a** From any triangle of your choice, create a design using the above routine.

     **b** Describe in detail how you could use triangle ABC in a series of rotations to create a regular hexagon.

 **6** By using the middle square as a starting square ABCD, describe how to keep rotating the square to obtain the final shape in the diagram.

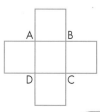

**7** Copy the diagram and rotate the given triangle as described.

   **a** 90° clockwise about (0, 0)

   **b** 180° about (3, 3)

   **c** 90° anticlockwise about (0, 2)

   **d** 180° about (–1, 0)

   **e** 90° clockwise about (–1, –1)

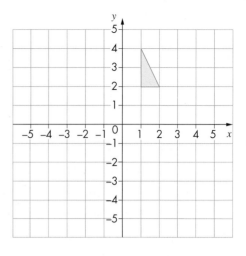

**8** **a** Why are the rotations 90° clockwise and 270° anticlockwise equivalent?

   **b** What other rotations are equivalent to these rotations?

      **i** 270° clockwise       **ii** 90° clockwise

      **iii** 60° anticlockwise     **iv** 100° anticlockwise

**CM** **9** Show that a reflection in the $x$-axis followed by a reflection in the $y$-axis is equivalent to a rotation of 180° about the origin.

**EV** **10** Raith said that a reflection in the line $y = x$ followed by a reflection in the line $y = -x$ is equivalent to a rotation of 180° about the origin.

Comment on Raith's statement.

**11** **a** Draw a regular hexagon ABCDEF with centre O.

   **b** Describe a transformation, with centre of rotation O, that will give the following movements.

      **i** Triangle AOB to triangle BOC       **ii** Triangle AOB to triangle COD

      **iii** Triangle AOB to triangle DOE      **iv** Triangle AOB to triangle EOF

   **c** Describe the transformations that will move the rhombus ABCO to these positions.

      **i** Rhombus BCDO      **ii** Rhombus DEFO

**12** Triangle A, as shown on the grid, is rotated to form triangle B.

The coordinates of the vertices of B are (0, –2), (–3, –2) and (–3, –4).

Describe fully the rotation that maps triangle A onto triangle B.

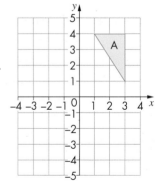

**CM** **13** Triangle B is rotated anticlockwise to form triangle C. Describe the transformation that will move triangle C back to triangle B.

# 12.5 Enlargements

## This section will show you how to:

* enlarge a 2D shape by a scale factor.

**Key terms**

centre of enlargement

enlargement

An **enlargement** changes the size of a shape to give a similar image.
It always has a **centre of enlargement** and a scale factor.

Every length of the enlarged shape will be:

original length × scale factor.

The distance of each image point on the enlargement from the centre of enlargement will be:

distance of original point from centre of enlargement × scale factor.

The diagram shows the enlargement of triangle ABC by scale factor 3 from the centre of enlargement X.

**Note:**

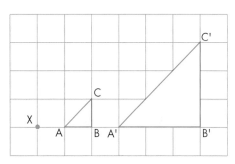

* Each length on the enlargement A′B′C′ is three times the corresponding length on the original shape. This means that the corresponding sides are in the same ratio:
AB : A′B′ = AC : A′C′ = BC : B′C′ = 1 : 3

* The distance of any point on the enlargement from the centre of enlargement is three times the distance from the corresponding point on the original shape to the centre of enlargement.

There are two ways to enlarge a shape: the ray method and the coordinate, or counting squares, method.

## Ray method

This is the *only* way to construct an enlargement when the diagram is not on a grid.

See how the triangle ABC has been enlarged by scale factor 3 from the centre of enlargement X.

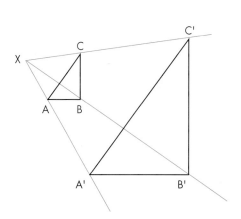

Notice that the rays have been drawn from the centre of enlargement to each vertex and beyond.

The distance from X to each vertex on triangle ABC is measured and multiplied by 3 to give the distance from X to each vertex A′, B′ and C′ for the enlarged triangle A′B′C′.

Once each image vertex has been found, the whole enlarged shape can then be drawn.

Check the measurements and work through the calculations yourself.

Notice again that the length of each side on the enlarged triangle is three times the length of the corresponding side on the original triangle.

## Counting squares method

In this method, you use the coordinates of the vertices to 'count squares'.

Triangle ABC has been enlarged by scale factor 3 from the centre of enlargement (1, 2).

To work out the coordinates of each image vertex, first work out the horizontal and vertical distances from each original vertex to the centre of enlargement. Then multiply each of these distances by 3 to find the position of each image vertex.

For example, to work out the coordinates of C', work out the distance from the centre of enlargement (1, 2) to the point C(3, 5).

horizontal distance = 2          vertical distance = 3

Multiply each distance by 3.

new horizontal distance = 6      new vertical distance = 9

So, the coordinates of C' are (1 + 6, 2 + 9) = (7, 11).

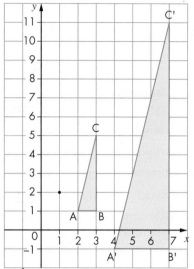

Notice again that the length of each side is three times as long in the enlargement.

## Fractional enlargement

You can have an enlargement in mathematics that is actually smaller than the original shape!

Triangle ABC has been enlarged by a scale factor of $\frac{1}{2}$ from the centre of enlargement X to give triangle A'B'C'.

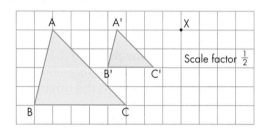

## Exercise 12E

**1** Copy each of these shapes with its centre of enlargement. Use the ray method to enlarge it by the given scale factor.

a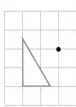

b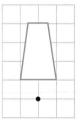

c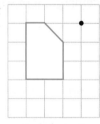

d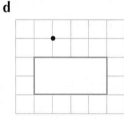

Scale factor 2      Scale factor 3      Scale factor 2      Scale factor 3

**2** Copy each of these shapes and grids onto squared paper. Enlarge each shape by scale factor 2 from the given centre of enlargement.

a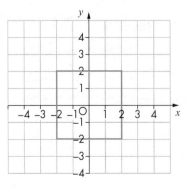

Centre of enlargement (–1, 1)

b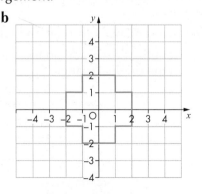

Centre of enlargement (–2, –3)

**3** Copy each shape and centre of enlargement. Then enlarge each shape by a scale factor of $\frac{1}{2}$ from the given centre of enlargement.

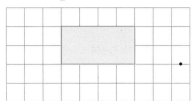 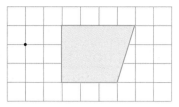

**4** Copy this diagram onto squared paper.

**a** Enlarge rectangle A by scale factor $\frac{1}{3}$ from the origin. Label the image B.

**b** Write down the ratio of the lengths of the sides of rectangle A to the lengths of the sides of rectangle B.

**c** Work out the ratio of the perimeter of rectangle A to the perimeter of rectangle B.

**d** Work out the ratio of the area of rectangle A to the area of rectangle B.

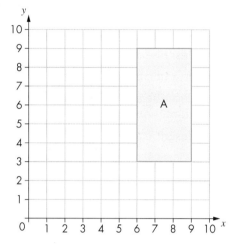

(MR) **5** A triangle ABC has vertices A(1, 3), B(2, –1) and C(–2, –2). It is enlarged by scale factor 2 from the centre of enlargement (1, 1).

What are the coordinates of the vertices of the enlarged shape?

(CM) **6** Copy this diagram onto squared paper.

**a** Enlarge A by a scale factor of 2 from a centre (5, 5). Label It C.

**b** What transformation will map shape C onto shape B?

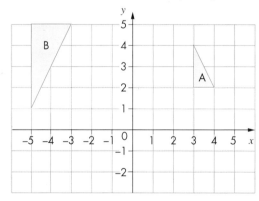

(PS) **7** Triangle A has vertices with coordinates (2, 1), (4, 1) and (4, 4).

Triangle B has vertices with coordinates (–5, 1), (–5, 7) and (–1, 7).

**a** Describe fully the single transformation that will take triangle A to triangle B.

**b** Describe fully the single transformation that will take triangle B to triangle A.

# 12.6 Using more than one transformation

This section will show you how to:

- use more than one transformation.

You often have to use more than one transformation. In this exercise, you will practise combining the transformations you have met so far.

Remember:

- to describe a translation fully, you need to use a vector
- to describe a reflection fully, you need to use a mirror line
- to describe a rotation fully, you need a centre of rotation, an angle of rotation and the direction of turn
- to describe an enlargement fully, you need a centre of enlargement and a scale factor.

## Exercise 12F

**1** AB is a line where A is (1, 2) and B is (3, 4).

Draw a grid showing

  **a** AB reflected in the $x$-axis. Draw and label the reflection CD.

  **b** CD is rotated by 90° clockwise about the origin. Draw and label the rotation EF.

  **c** What are the coordinates of the points E and F?

(MR) **2** A point Q(5, 2) is rotated by 180° about the point (0, 2) to the point R.

  **a** What are the coordinates of point R?

  **b** Point R is reflected in the $x$-axis to point T. What are the coordinates of point T?

  **c** What single transformation would have taken Q directly to the point T?

(CM) **3** Describe fully the transformations that will map the shaded triangle onto each of the triangles A–F.

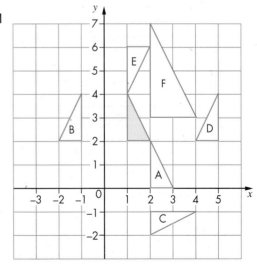

**4** A designer uses the following routine.

- Start with a rectangle ABCD.
- Reflect ABCD in the line AC.
- Rotate the whole new shape about C through 180°.
- Enlarge the whole shape by scale factor 2, centre of enlargement point A.

Start with any rectangle of your choice and create the design above.

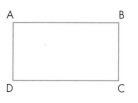

**(CM) 5** Describe fully the transformations that will give the following movements.

**a** $T_1$ to $T_2$

**b** $T_1$ to $T_6$

**c** $T_2$ to $T_3$

**d** $T_6$ to $T_2$

**e** $T_6$ to $T_5$

**f** $T_5$ to $T_4$

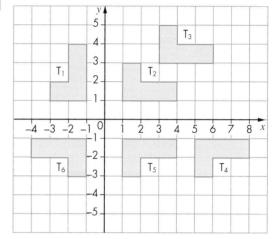

**(CM) 6** **a** Plot a triangle T with vertices (1, 1), (2, 1) and (1, 3).

**b** Reflect triangle T in the $y$-axis and label the image $T_b$.

**c** Rotate triangle $T_b$ 90° anticlockwise about the origin and label the image $T_c$.

**d** Reflect triangle $T_c$ in the $y$-axis and label the image $T_d$.

**e** Describe fully the transformation that will move triangle $T_d$ back to triangle T.

**7** Work out the coordinates of the image of the point (3, 5) after a clockwise rotation of 90° about the point (1, 3).

**(CM) 8** Describe fully at least three different transformations that could move the square labelled S to the square labelled T.

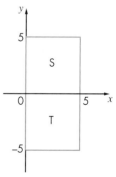

**(PS) 9** The point A(4, 4) has been transformed to the point B(4, –4). Describe as many different transformations as you can that could transform point A to point B.

(CM) **10** Copy the diagram onto squared paper.

**a** Triangle A is translated by the vector $\begin{pmatrix} -2 \\ -6 \end{pmatrix}$ to give triangle B.

Triangle B is then enlarged by a scale factor $\frac{1}{2}$ from (2, 4) to give triangle C.

Draw triangles B and C on the diagram.

**b** Describe fully the single transformation that will take triangle C to triangle A.

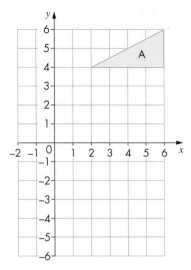

(EV) **11** Helen said that if you reflect a shape in line $y = A$ then reflect that image in a line $y = B$ (where $A$ and $B$ are any numbers), the final image is always a rotation of the original shape.

Is this is true?

# 12.7 Vectors

This section will show you how to:

- represent vectors
- add and subtract vectors.

**Key terms**

direction

magnitude

resultant vector

scalar

A **scalar** is a quantity that has a **magnitude** (length) but no **direction**. The mass of a bus (10 tonnes) and the length of a line (25.4 mm) are scalars.

A vector is a quantity that has both magnitude and direction. It may be represented by a straight line, with an arrow to show its direction. The length of the line represents the magnitude of the vector.

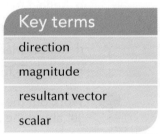

In the same way that you used vectors to describe a translation in terms of horizontal and vertical shifts, this vector **a** represents the translation or movement from A to B.

You can write vectors in other ways.

- In textbooks, a vector may be shown as a small letter, such as **a**, printed in bold type. When you write the vector down, you underline it, like this: <u>a</u>.

- You may also write the vector **a** as $\overrightarrow{AB}$. The arrow above the letters shows the direction: here it is from A to B.

**Example 5**

Write each vector as a column vector.

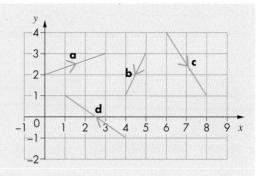

$$a = \begin{pmatrix} 3 \\ 1 \end{pmatrix} \qquad b = \begin{pmatrix} -1 \\ -2 \end{pmatrix} \qquad c = \begin{pmatrix} 2 \\ -3 \end{pmatrix} \qquad d = \begin{pmatrix} -3 \\ 2 \end{pmatrix}$$

Multiplying a vector by a number (scalar) alters its magnitude (length) but not its direction. For example, the vector 2**a** is twice as long as the vector **a**, but acts in the same direction.

A negative vector, for example –**b**, has the same magnitude as the vector **b**, but acts in the opposite direction.

## Exercise 12G

**1** Draw lines to represent these vectors. (Remember to include the arrow.)

$$a = \begin{pmatrix} 4 \\ 2 \end{pmatrix} \qquad b = \begin{pmatrix} -3 \\ -5 \end{pmatrix} \qquad c = \begin{pmatrix} -2 \\ 3 \end{pmatrix} \qquad d = \begin{pmatrix} 1 \\ -4 \end{pmatrix} \qquad e = \begin{pmatrix} 3 \\ 0 \end{pmatrix}$$

**2**  **a** Draw the points A(3, 2), B(6, 8) and C(4, –1) on a coordinate grid.

**b** Write these as column vectors.

   **i** $\overrightarrow{AB}$              **ii** $\overrightarrow{BC}$            **iii** $\overrightarrow{AC}$

**3** A is the point (–3, 2). $\overrightarrow{AB} = \begin{pmatrix} 3 \\ -2 \end{pmatrix}$ and $\overrightarrow{AC} = \begin{pmatrix} 6 \\ 3 \end{pmatrix}$.

**a** Write down the coordinates of B and C.     **b** What is the column vector for $\overrightarrow{BC}$?

**4** A triangle has vertices at A(–3, 1), B(1, 0) and C(1, 3).

Give the coordinates of triangle ABC after:

**a** a translation of $\begin{pmatrix} 5 \\ 1 \end{pmatrix}$             **b** a translation of $\begin{pmatrix} -3 \\ -2 \end{pmatrix}$.

**5** B is the point (4, 2). $\overrightarrow{AB} = \begin{pmatrix} 3 \\ -3 \end{pmatrix}$

Give the coordinates of point A.

**6** A translation, T, transforms the point A(3, –2) to A′(5, 3).

**a** What will be the coordinates of point B(1, 3) after the same translation?

**b** C′(8, –1) has been transformed by T. What were the original coordinates of point C?

## Adding and subtracting of vectors

Here are two non-parallel vectors, **a** and **b**.

Think about what **a** + **b** might represent.

**a** + **b** is the translation by **a** followed by the translation by **b**.

You can see this on a vector diagram.

The vector **a** + **b** is called the **resultant vector**.

Similarly, **a** – **b** is the translation by **a** followed by the translation by –**b** as shown below.

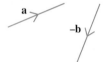 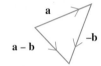

This grid is made from parallelograms that are all the same size.

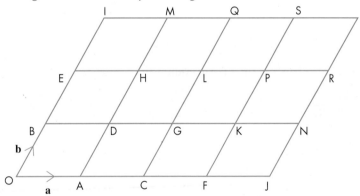

Each small parallelogram represents two independent vectors, **a** and **b**.

You can define the position of any point on this grid, with reference to O, as a vector written in terms of **a** and **b**.

For example:

$\overrightarrow{OK} = 3\mathbf{a} + \mathbf{b}$

$\overrightarrow{OE} = 2\mathbf{b}$.

You can also define the vector linking any two points in terms of **a** and **b**.

$\overrightarrow{HT} = 3\mathbf{a} + \mathbf{b}$    $\overrightarrow{PN} = \mathbf{a} - \mathbf{b}$    $\overrightarrow{MK} = 2\mathbf{a} - 2\mathbf{b}$    $\overrightarrow{TP} = -\mathbf{a} - \mathbf{b}$

**Note:**

- $\overrightarrow{OK}$ and $\overrightarrow{HT}$ are equal vectors because they have exactly the same length and the same direction
- $\overrightarrow{MK}$ and $\overrightarrow{PN}$ are parallel vectors but $\overrightarrow{MK}$ has twice the magnitude of $\overrightarrow{PN}$.

Example 6

**a** Refer to the grid above. Write down each vector in terms of **a** and **b**.

    **i** $\overrightarrow{BH}$            **ii** $\overrightarrow{HP}$            **iii** $\overrightarrow{GT}$

    **iv** $\overrightarrow{TI}$            **v** $\overrightarrow{FH}$            **vi** $\overrightarrow{BQ}$

**b** What is the relationship between the vectors in each pair?

    **i** $\overrightarrow{BH}$ and $\overrightarrow{GT}$        **ii** $\overrightarrow{BQ}$ and $\overrightarrow{GT}$        **iii** $\overrightarrow{HP}$ and $\overrightarrow{TI}$

**a**  **i** $\mathbf{a} + \mathbf{b}$           **ii** $2\mathbf{a}$           **iii** $2\mathbf{a} + 2\mathbf{b}$

    **iv** $-4\mathbf{a}$         **v** $-2\mathbf{a} + 2\mathbf{b}$        **vi** $2\mathbf{a} + 2\mathbf{b}$

**b**  **i** $\overrightarrow{BH}$ and $\overrightarrow{GT}$ are parallel and $\overrightarrow{GT}$ has twice the magnitude of $\overrightarrow{BH}$ because it is twice its length.

    **ii** $\overrightarrow{BQ}$ and $\overrightarrow{GT}$ are equal.

    **iii** $\overrightarrow{HP}$ and $\overrightarrow{TI}$ are opposite directions and $\overrightarrow{TI}$ has twice the magnitude of $\overrightarrow{HP}$ because it is twice its length.

## Exercise 12H

**1** On this grid, $\overrightarrow{OA}$ is **a** and $\overrightarrow{OB}$ is **b**.

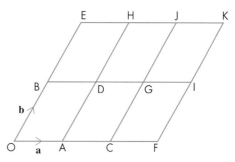

  **a** Name three other vectors that are equivalent to **a**.

  **b** Name three other vectors that are equivalent to **b**.

  **c** Name three vectors that are equivalent to –**a**.

  **d** Name three vectors that are equivalent to –**b**.

**2** Look again at the grid in question **1**. Write each of these vectors in terms of **a** and **b**.

  **a** $\overrightarrow{OC}$      **b** $\overrightarrow{OE}$      **c** $\overrightarrow{OD}$      **d** $\overrightarrow{OG}$      **e** $\overrightarrow{OJ}$      **f** $\overrightarrow{OH}$

  **g** $\overrightarrow{AG}$      **h** $\overrightarrow{AK}$      **i** $\overrightarrow{BK}$      **j** $\overrightarrow{DI}$      **k** $\overrightarrow{GJ}$      **l** $\overrightarrow{DK}$

**MR** **3**  **a** What do the answers to parts **2c** and **2g** tell you about the vectors $\overrightarrow{OD}$ and $\overrightarrow{AG}$?

  **b** On the grid in question **1**, there are three vectors equivalent to $\overrightarrow{OG}$. Name all three.

**MR** **4**  **a** What do the answers to parts **2c** and **2e** tell you about vectors $\overrightarrow{OD}$ and $\overrightarrow{OJ}$?

  **b** On the grid in question **1**, there is one other vector that is twice the size of $\overrightarrow{OD}$. Which is it?

  **c** On the grid in question **1**, there are three vectors that are three times the size of $\overrightarrow{OA}$. Name all three.

**5** Copy this grid. Use the information below to mark the points C to K on your grid.

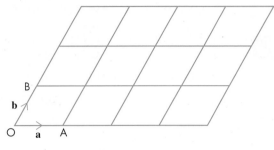

a $\overrightarrow{OC} = 2\mathbf{a} + 3\mathbf{b}$      b $\overrightarrow{OD} = 2\mathbf{a} + \mathbf{b}$

c $\overrightarrow{OE} = \mathbf{a} + 2\mathbf{b}$      d $\overrightarrow{OF} = 3\mathbf{a} + 3\mathbf{b}$

e $\overrightarrow{OG} = \mathbf{a} + \mathbf{b}$      f $\overrightarrow{OH} = 2\mathbf{a} + 2\mathbf{b}$

g $\overrightarrow{OI} = 3\mathbf{a} + 2\mathbf{b}$      h $\overrightarrow{OJ} = 3\mathbf{b}$

i $\overrightarrow{OK} = 4\mathbf{a} + \mathbf{b}$

**(CM)** **6** a Look at your grid from question **5**. What can you say about the points O, F, G and H?

b How could you tell this by looking at the vectors in question **5** parts **d**, **e** and **f**?

c L is a point on the edge of the grid, on the extended line OD. Write the vector $\overrightarrow{OL}$ in terms of **a** and **b**.

**7** On this grid, $\overrightarrow{OA}$ is **a** and $\overrightarrow{OB}$ is **b**.

Write these vectors in terms of **a** and **b**.

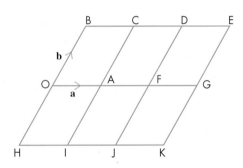

a $\overrightarrow{OH}$      b $\overrightarrow{OK}$      c $\overrightarrow{OJ}$      d $\overrightarrow{OI}$

e $\overrightarrow{OC}$      f $\overrightarrow{CO}$      g $\overrightarrow{AK}$      h $\overrightarrow{DI}$

i $\overrightarrow{JE}$      j $\overrightarrow{AB}$      k $\overrightarrow{CK}$      l $\overrightarrow{DK}$

**(CM)** **8** a What do your answers to parts **7e** and **7f** tell you about the vectors $\overrightarrow{OC}$ and $\overrightarrow{CO}$?

b On the grid in question **7**, there are five other vectors in the opposite direction to $\overrightarrow{OC}$. Name at least three.

**(CM)** **9** a What do your answers to parts **7j** and **7k** tell you about vectors $\overrightarrow{AB}$ and $\overrightarrow{CK}$?

b On the grid in question **7**, there are two vectors that are twice the size of $\overrightarrow{AB}$ and are in the opposite direction. Name both of them.

c On the grid in question **7**, there are three vectors that are three times the size of $\overrightarrow{OA}$ and in the opposite direction. Name all three.

**(PS)** **10** The diagram shows two sets of parallel lines.

$\overrightarrow{OA} = \mathbf{a}$ and $\overrightarrow{OB} = \mathbf{b}$
$\overrightarrow{OC} = 3\overrightarrow{OA}$ and $\overrightarrow{OD} = 2\overrightarrow{OB}$

a Write these vectors in terms of **a** and **b**.
i $\overrightarrow{OF}$      ii $\overrightarrow{OC}$      iii $\overrightarrow{EG}$      iv $\overrightarrow{CE}$

b Write down two vectors that can be written as $3\mathbf{a} - \mathbf{b}$.

# Worked exemplars

**1** The grid shows several transformations of the shaded triangle.

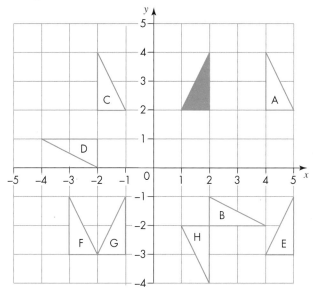

**a** Describe fully the transformation that takes:

**i** the shaded triangle to shape A

**ii** the shaded triangle to shape E.

**b** There are two reflections shown in the diagram that do not involve the shaded triangle. Identify the triangles involved and fully describe each reflection.

**c** Pete says that to transform shape C to shape D you need a translation and then a reflection. Kathy says you need a rotation and then a reflection.

Comment on what each person says.

| This first part of this question assesses communicating mathematics so you need to show that you know how to identify and fully describe the transformations using correct mathematical terms. The final part of the question assesses evaluation skills so you need to assess the accuracy of each person's comments. | |
|---|---|
| **a** **i** Reflection in the line $x = 3$ <br><br> **ii** Translation of $\begin{pmatrix} 3 \\ -5 \end{pmatrix}$ | It is not enough to simply identify, for example, a reflection. You must also give the line of the reflection. |
| **b** Triangle F to G, reflection in line $x = -2$ <br><br> Triangle H to B, reflection in the line $y = x$, between $(0, 0)$ and $(4, -4)$ | You could describe F to G as G to F, and H to B as B to H; the order doesn't matter as long as the objects and images are correct. |
| **c** For Pete's suggestion, translate the shape by $\begin{pmatrix} 0 \\ -1 \end{pmatrix}$ and then reflect in the line between $(-4, 3)$ and $(-1, 0)$. For Kathy's suggestion, rotate shape C 90° anticlockwise about $(-2, 2)$ and then reflect in the line $y = 1.5$. <br> So, they are both correct. | You need to work through both statements and show why they are correct or incorrect. |

Work out the single transformation that is equivalent to a rotation of 90° clockwise about the origin, followed by a reflection in the line $y = x$.

| This is a problem-solving question so you will need to show your strategy. | |
| --- | --- |
| 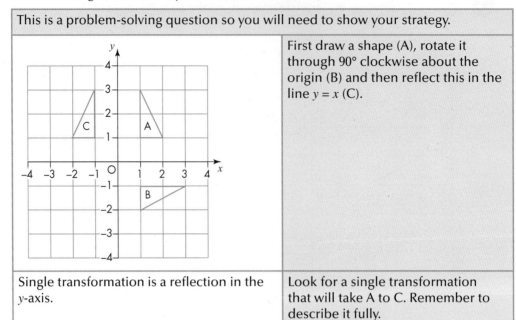 | First draw a shape (A), rotate it through 90° clockwise about the origin (B) and then reflect this in the line $y = x$ (C). |
| Single transformation is a reflection in the $y$-axis. | Look for a single transformation that will take A to C. Remember to describe it fully. |

# Ready to progress?

I can recognise and describe rotational symmetry.

I can translate a 2D shape by a given vector.
I can reflect a 2D shape in a given mirror line.
I can rotate a 2D shape around a given point through a quarter turn or a half turn, either clockwise or anticlockwise.
I can describe a rotation of a given 2D shape around a given point through any angle.
I can enlarge a 2D shape about any point using a positive fractional scale factor.

I can add and subtract vectors.

# Review questions

**1** This shape is made from three equilateral triangles and a regular hexagon.

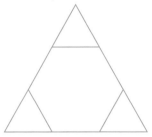

**a** Write down the order of rotational symmetry of the shape.

**b** Copy the shape and draw all its lines of symmetry.

**2** Copy this diagram onto squared paper.

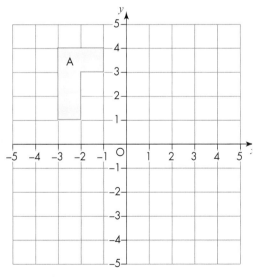

**a** Reflect shape A in the x-axis. Label this B.

**b** Reflect shape A in the line $y = x$. Label this C.

**c** Describe the transformation that will take shape B to shape C.

(CM) **3** The diagram shows two identical shapes, A and B.

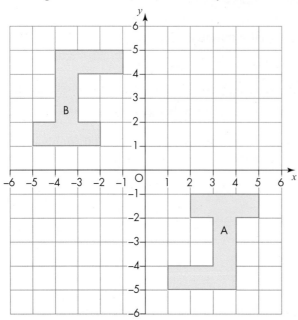

Describe fully the *single* transformation that takes shape A to shape B.

**4** Triangle ABC has coordinates A(1, 1), B(2, 1) and C(1, 3).

Triangle DEF is an enlargement of triangle ABC by scale factor 2, with A as the centre of enlargement. Triangle GHI is an enlargement of triangle ABC by scale factor 2, with (5, 2) as the centre of enlargement.

Describe fully the transformation that will map triangle GHI onto triangle DEF.

**5** **a** Copy the diagram and rotate triangle P 90° clockwise about the point (–1, 1). Label this triangle A.

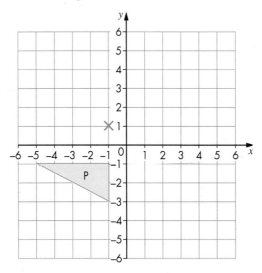

**b** Rotate triangle P 90° anticlockwise about the point (–1, 1). Label this triangle B.

**c** Describe the transformation that takes triangle A to the triangle B.

(CM) **6** Draw a coordinate grid with $x$- and $y$-axes from –5 to 5.

Draw the triangle with coordinates A(2, 1), B(4, 1) and C(4, 2).

**a** Reflect triangle ABC in the line $y = x$. Label the image T.

**b** Reflect triangle ABC in the line $y = -x$. Label the image Q.

**c** Describe fully the transformation that would take triangle Q to triangle T.

**7** Triangle ABC has coordinates A(3, 1), B(5, –2) and C(–4, 0).

It is translated by the vector $\begin{pmatrix} 6 \\ -1 \end{pmatrix}$.

What are the new coordinates of the translated triangle?

**8** PQRS is a rectangle.

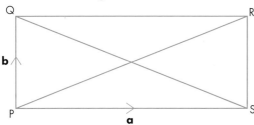

State each vector in terms of **a** and **b**.

**i** $\overrightarrow{QP}$      **ii** $\overrightarrow{SP}$      **iii** $\overrightarrow{SR}$      **iv** $\overrightarrow{QR}$      **v** $\overrightarrow{PR}$      **vi** $\overrightarrow{QS}$

(EV) **9** Kieron said that if you rotate a shape by 90° and then reflect it in the $y$-axis, the final image is always a reflection of the original shape.

Draw a diagram showing that this is not always true.

(PS) **10** A single transformation, T, transforms the line $y = 1 + 2x$ to the line $y = 1 - 2x$.

Describe two possible options for T.

# 13 Probability: Probability and events

## This chapter is going to show you:

- how to use the probability scale and the language of probability
- how to work out the probability of an outcome of an event happening
- how to work out the probability of an outcome of an event not happening
- how to recognise mutually exclusive and exhaustive outcomes
- how to work out experimental probabilities and relative frequencies from experiments
- how to predict the likely number of successful outcomes, given the number of trials and the probability of any one outcome
- how to use systematic strategies to list and count outcomes.

## You should already know:

- how to cancel, add and subtract fractions
- that an event or trial has outcomes
- that outcomes cannot always be predicted and that the laws of chance apply to everyday events
- the meaning of the term 'bias'.

## About this chapter

Chance is a part of everyday life. Decisions are frequently made based on probability.

When they discuss probabilities, different people may give different answers because of their different views. For example, one person may not agree that there is an 80% chance of United winning the game. They may well say that there is only a 70% chance of United winning tomorrow. A lot depends on what that person believes or has experienced.

Probability is a branch of mathematics that describes the chance of different outcomes of events. It may be taken into account for lots of different events, from playing games to tomorrow's weather.

Probability was first thought about in connection with games of chance, such as throwing a dice or spinning a roulette wheel. In the sixteenth and seventeenth centuries, mathematicians started to think about the mathematics of chance in games. Probability theory developed as a branch of mathematics in the seventeenth century when French roulette players asked mathematicians Blaise Pascal and Pierre de Fermat for help in their gambling.

In the twenty-first century, probability theory has a wide range of everyday applications from controlling the flow of traffic through road systems to running telephone exchanges and looking at patterns of the spread of infections. As you work through this chapter you will see how frequently the language of probability is used.

# 13.1 Calculating probabilities

This section will show you how to:

- use the probability scale and the language of probability
- calculate the probability of an outcome of an event.

**Key terms**

| |
|---|
| equally likely |
| event |
| outcome |
| probability |
| probability fraction |
| probability scale |
| random |
| trial |

You hear somebody talking about the probability of something happening almost daily. They usually use words such as 'chance', 'likelihood' or 'risk' rather than 'probability'. For example:

- 'What is the likelihood of rain tomorrow?'
- 'What chance does she have of winning the 100 m sprint?'
- 'Is there a risk that his company will go bankrupt?'

You can give a value to the chance of any of these things happening. This value is called the **probability**.

The probability of something happening can be anywhere between impossible and certain. This is represented on the **probability scale**.

**Note:** All probabilities lie somewhere in the range of 0 to 1.

Something that is impossible has a probability of 0. For example, the probability that pigs will fly is 0.

Something that is certain to happen has a probability of 1. For example, the probability that the sun will rise tomorrow is 1.

In probability, an **event** is an activity, such as throwing a dice, that may have several possible **outcomes**. A **trial** is one attempt at performing an event.

---

**Example 1**

Draw arrows on the probability scale to show the probability of each event.

a You will get a head when throwing a coin.

b You will get a 6 when throwing a dice.

c You will have maths homework this week.

a Each outcome of this event has an even chance. (This is commonly described as a fifty-fifty chance.)

b This event is fairly unlikely.

c This event is likely.

The arrows show the approximate probabilities on the probability scale.

---

Look carefully at the probability scale in Example 1. How can you work out the value, from 0 to 1, for each probability?

You can start by writing down all the possible outcomes for a particular event. For example, when you throw a coin there are two **equally likely** outcomes: heads or tails. If you want to score a head, only one outcome is *successful*. So, you can say that there is a 1 in 2, or 1 out of 2, probability of getting a head. This is usually given as a **probability fraction**.

You can write the probability of any outcome as P(outcome). So the probability of getting a head is:

$$P(head) = \frac{1}{2} \text{ or } 50\% \text{ or } 0.5$$

You may also see probabilities given as percentages or decimals, for example, in the weather forecasts on TV.

$$P(rain) = \frac{1}{2} \text{ or } 50\% \text{ or } 0.5$$

The probability of any outcome of an event is defined as:

$$P(outcome) = \frac{\text{number of ways the outcome can happen}}{\text{total number of possible outcomes}}$$

This calculation always leads to a fraction, which you should write in its simplest form.

Another important probability term is at **random**. This means that the outcome cannot be predicted or affected by anyone.

---

**Example 2**

A bag contains five red balls and three blue balls. A ball is taken out at random.

What is the probability that it is:

**a** red **b** blue **c** green?

Use the formula:

$$P(outcome) = \frac{\text{number of ways the outcome can happen}}{\text{total number of possible outcomes}}$$

to work out these probabilities.

There are eight equally likely outcomes.

**a** There are five red balls out of a total of eight, so $P(red) = \frac{5}{8}$.

**b** There are three blue balls out of a total of eight, so $P(blue) = \frac{3}{8}$.

**c** There are no green balls, so this outcome is impossible, which means that P(green) = 0.

---

**Example 3**

Charlotte spins this spinner and records the number on the side on which it lands.

What is the probability that the score is:

**a** 2 **b** odd **c** less than 5?

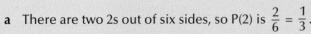

There are six equally likely outcomes.

**a** There are two 2s out of six sides, so P(2) is $\frac{2}{6} = \frac{1}{3}$.

**b** There are four odd numbers, so P(odd) is $\frac{4}{6} = \frac{2}{3}$.

**c** All of the numbers are less than 5, so this is a certainty, which means that P(less than 5) = 1.

# Exercise 13A

**1** State whether each of these is impossible, very unlikely, unlikely, even chance, likely, very likely or certain.

   **a** Taking a heart from an ordinary pack of cards

   **b** Christmas Day being on 25 December

   **c** Someone in your class being left-handed

   **d** You living to be 100

   **e** Obtaining a score of 7 when throwing a fair, six-sided dice

   **f** You watching some TV this evening

   **g** A newborn baby being a girl

**2** Work out the probability of each outcome. Remember to write the probability fractions in their lowest terms.

> **Hints and tips** If an outcome is impossible, you write the probability as 0. If it is certain, you write the probability as 1.

   **a** Throwing a 2 with a fair, six-sided dice

   **b** Throwing a 6 with a fair, six-sided dice

   **c** Throwing a fair coin and getting a tail

   **d** Taking a queen from an ordinary pack of cards

   **e** Taking a heart from an ordinary pack of cards

   **f** Taking a black card from an ordinary pack of cards

   **g** Throwing a 6 or less with a fair, six-sided dice

   **h** Taking a black queen from an ordinary pack of cards

   **i** Taking an ace from an ordinary pack of cards

   **j** Throwing a 7 with a fair, six-sided dice

**3** A bag contains only blue balls. If one ball is taken out at random, what is the probability that it is:

   **a** black       **b** blue?

**4** Ten cards numbered from 1 to 10 (inclusive) are placed in a hat. Bob takes a card out of the bag without looking. What is the probability that he takes out:

   **a** the number 7       **b** an even number       **c** a number greater than 6

   **d** a number less than 3       **e** a number from 3 to 8 (inclusive)?

**5** A bag contains three blue beads, two pink beads and one black bead. Craig takes a bead from the bag without looking. What is the probability that he takes out:

   **a** a blue bead       **b** a pink bead       **c** the black bead?

**6** In a Christmas raffle, 2500 tickets are sold. Mr Weller has 50 tickets. What is the probability that he wins the first prize?

**7** A card is taken at random from an ordinary pack of cards. What is:

   **a** P(jack)       **b** P(10)       **c** P(red card)       **d** P(red jack)

**(EV) 8** A bag contains 25 coloured balls. 12 are red, 7 are blue and the rest are green. Mahmood takes a ball at random from the bag.

   **a** Write down:

      **i** P(he takes a red)       **ii** P(he takes a blue)       **iii** P(he takes a green).

   **b** Add together the three probabilities. What do you notice?

   **c** Give a reason for your answer to part **b**.

(PS) **9** Adam, Bianca, Charles, Debbie and Elizabeth are in the same class. Their teacher asks two of these students, at random, to do a special job.

**a** Write down all the possible ways of choosing two students.

> Hints and tips There are 10 ways altogether.

**b** How many pairs give two boys?

**c** What is the probability of choosing two boys?

**d** How many pairs give a boy and a girl?

**e** What is the probability of choosing a boy and a girl?

**f** What is the probability of choosing two girls?

(MR) **10** A teacher chooses, at random, a student in her class to take a message.

Tom says: 'The probability that the teacher chooses a boy is 50%.'

Explain why Tom may not be correct.

(PS) **11** This table shows the numbers of boys and girls in 12 classes at Bradway School.

| Year | Y1 | | Y2 | | Y3 | | Y4 | | Y5 | | Y6 | |
|------|----|----|----|----|----|----|----|----|----|----|----|----|
| Class | P | Q | R | S | T | U | W | X | Y | Z | K | L |
| Girls | 7 | 8 | 8 | 10 | 10 | 10 | 9 | 11 | 8 | 12 | 14 | 15 |
| Boys | 9 | 10 | 9 | 10 | 12 | 13 | 11 | 12 | 10 | 8 | 16 | 17 |

One student is chosen at random from each class.

Which class has the highest probability of having a boy chosen?

# 13.2 Probability that an outcome will not happen

This section will show you how to:

- calculate the probability of an outcome not happening when you know the probability of that outcome happening.

The probability of throwing a 6 on a fair, six-sided dice is $P(6) = \frac{1}{6}$.

There are five outcomes that are not sixes: 1, 2, 3, 4 and 5.

So, the probability of *not* throwing a six on a dice is $P(\text{not a 6}) = \frac{5}{6}$.

Notice that:

$P(6) = \frac{1}{6}$ and $P(\text{not a 6}) = \frac{5}{6}$

so:

$P(6) + P(\text{not a 6}) = 1$.

If you know that $P(6) = \frac{1}{6}$, then $P(\text{not a 6})$ is:

$1 - \frac{1}{6} = \frac{5}{6}$.

So if you know P(outcome happening), then:

P(outcome not happening) = 1 – P(outcome happening).

Example 4

What is the probability of *not* taking an ace from an ordinary pack of cards?

First, find the probability of taking an ace.

P(taking an ace) is $\frac{4}{52} = \frac{1}{13}$

Therefore:

P(not taking an ace) is $1 - \frac{1}{13} = \frac{12}{13}$

Example 5

Zara is always early, on time or late for work.

The probability that she is early is 0.1; the probability that she is on time is 0.5.

What is the probability that she is late?

As all of the possibilities are covered, since she must be 'early', 'on time' or 'late', the total probability is 1. So:

P(early) + P(on time) = 0.1 + 0.5 = 0.6

So, the probability of Zara being late is $1 - 0.6 = 0.4$.

## Exercise 13B

1. **a** The probability of winning a prize in a raffle is $\frac{1}{20}$. What is the probability of not winning a prize in the raffle?

   **b** The probability that snow will fall during the Christmas holidays is 45%. What is the probability that it will not snow during the Christmas holidays?

   **c** The probability that Paddy wins a game of chess is 0.7 and the probability that he draws a game is 0.1. What is the probability that he loses a game?

2. Mia takes a card from an ordinary pack of cards.

   Find the probability that the card she takes is:

   **a i** a picture card (jack, queen or king)　　**ii** not a picture card

   **b i** a club　　**ii** not a club

   **c i** an ace or a king　　**ii** neither an ace nor a king.

3. This is the starting section from a board game.

   | Start | Owens Park | Take a chance | Curry Mile | Oxford Road | Pay £500 in tax | Salford Quays | Exchange Quays | Station | Old Trafford | Rest area | The Lowry | Trafford Park |
   |---|---|---|---|---|---|---|---|---|---|---|---|---|

   You throw a single dice and move, from the start, the number of places shown by the dice.

   What is the probability of *not* landing on:

   **a** a brown square　　　**b** the station　　　**c** a coloured square?

4. A pencil case contains six red pens and five blue pens. Aaron takes out a pen at random. What is the probability that he takes out:

   **a** a red pen　　　**b** a blue pen　　　**c** a pen that is not blue?

**5** A bag contains 50 balls. 10 are green, 15 are red and the rest are white. Chun takes a ball from the bag at random. What is the probability that she takes:

    **a** a white ball                      **b** a green ball

    **c** a ball that is not green           **d** a ball that is not red?

**6** A box contains seven bags of cheese and onion crisps, two bags of beef crisps and six bags of plain crisps. Iklil takes out a bag of crisps at random. What is the probability that he takes:

    **a** a bag of cheese and onion crisps         **b** a bag of beef crisps

    **c** a bag of crisps that are not beef          **d** a bag of crisps that is either plain or beef

    **e** a bag of crisps that is neither plain nor beef     **f** a bag of prawn cocktail crisps?

**(MR) 7** Noah and Phoebe are playing a board game. On the next turn, Noah will lose if he throws an even number and Phoebe will lose if she throws a 5 or a 6.

Who has the better chance of *not* losing on the next turn?

Explain your answer.

**(CM) 8** Hamzah is told: 'The chance that you win this game is 0.3.'

Hamzah says: 'So I have a 0.7 chance of losing.'

Explain why Hamzah may be wrong.

**9** These letter cards are put into a bag.

| M | A | T | H | E | M | A | T | I | C | A | L |

    **a** Zach takes a letter card at random.

        **i** What is the probability that he takes a letter A?

        **ii** What is the probability that he does not take a letter A?

    **b** Zach takes an M from the original set of cards and keeps it. Rosie takes a letter from those remaining.

        **i** What is P(A) now?

        **ii** What is P(not A) now?

**10** The weather tomorrow will be sunny, cloudy or raining.

If P(sunny) = 40% and P(cloudy) = 25%, what is P(raining)?

**11** At morning break, Priya has a choice of coffee, tea or hot chocolate.

If P(she chooses coffee) = 0.3 and P(she chooses hot chocolate) = 0.2, what is P(she chooses tea)?

# 13.3 Mutually exclusive and exhaustive outcomes

## This section will show you how to:

- recognise mutually exclusive and exhaustive outcomes.

Key terms

exhaustive

mutually exclusive

Outcomes that are **mutually exclusive** cannot happen at the same time, such as 'throwing an odd number' and 'throwing an even number' with a dice.

When two outcomes are mutually exclusive, you can work out the probability of them occurring by adding up their separate probabilities.

---

**Example 6**

A bag contains 12 red balls, eight green balls, five blue balls and 15 black balls. A ball is drawn at random. What is the probability that the ball is:

**a**   red          **b**   black          **c**   red or black          **d**   not green?

**a**   P(red) is $\frac{12}{40} = \frac{3}{10}$.

**b**   P(black) is $\frac{15}{40} = \frac{3}{8}$.

**c**   P(red or black) is P(red) + P(black) = $\frac{12}{40} + \frac{15}{40}$

$$= \frac{27}{40}$$

**d**   P(not green) is 1 – P(green) = $1 - \frac{8}{40}$

$$= \frac{32}{40}$$

$$= \frac{4}{5}$$

---

**Example 7**

Trevor throws an ordinary six-sided dice.

**a**   What is the probability that he throws:          **i**   an even number          **ii**   an odd number?

**b**   What is the total of the answers to part **a**?

**a**   **i**   P(even) = $\frac{1}{2}$          **ii**   P(odd) = $\frac{1}{2}$

**b**   $\frac{1}{2} + \frac{1}{2} = 1$

---

Outcomes such as those in the example above are mutually exclusive because they can never happen at the same time. They are also **exhaustive** outcomes because there are no other possibilities. The probabilities of an exhaustive set of mutually exclusive outcomes add up to 1.

**1** Iqbal throws an ordinary dice. What is the probability that he throws:

**a** a 2       **b** a 5       **c** a 2 or a 5?

**2** A card is taken at random from an ordinary pack of cards. What is the probability that the card is:

**a** a heart       **b** a club       **c** a heart or a club?

**3** A letter is taken at random from this set of letter cards.

| P | R | O | B | A | B | I | L | I | T | Y |
|---|---|---|---|---|---|---|---|---|---|---|

What is the probability of each outcome?

**a** P(B)       **b** P(a vowel)       **c** P(B or a vowel)

**4** A bag contains 10 white balls, 12 black balls and 8 red balls. A ball is drawn at random from the bag. What is the probability that it will be:

**a** white       **b** black       **c** black or white

**d** not red       **e** neither red nor black?

**5** John needs his calculator for his mathematics lesson. It is always in his pocket, his school bag or his locker. The probability that it is in his pocket is 0.35 and the probability that it is in his school bag is 0.45. What is the probability that:

**a** he will have the calculator for the lesson (pocket or school bag)

**b** his calculator is in his locker?

**6** Aneesa downloads 20 music tracks of which 12 are rock, 5 are pop and 3 are classical. She picks a track at random. What is the probability that it will be:

**a** rock or pop       **b** pop or classical       **c** not pop?

(MR) **7** The probability that it will rain on Monday is 0.5, the probability that it will rain on Tuesday is 0.5 and the probability that it will rain on Wednesday is 0.5.

Kelly argues that it is certain to rain on Monday, Tuesday or Wednesday because 0.5 + 0.5 + 0.5 = 1.5, which is bigger than 1 so it is a certain event. Explain why she is wrong.

(PS) **8** In a TV game show, contestants throw a dart at a dartboard resembling the one shown.

The angle at the centre of each black sector is 15°.

If the dart lands in a black sector, the contestant loses.

What is the probability that a contestant who throws the dart at random, and hits the board, wins?

**9** The probability of it snowing on any one day in February is $\frac{1}{4}$.

One year, there was no snow for the first 14 days.

Ciara said: 'The chance of it snowing on any day in the rest of February must now be $\frac{1}{2}$.'

Explain why Ciara is wrong.

 **10** During morning break, Saskia has a choice of coffee, tea or hot chocolate. She also has a choice of a ginger biscuit, a rich tea biscuit or a doughnut.

The table shows the probabilities that she chooses each drink and snack.

| Drink | Coffee (C) | Tea (T) | Hot chocolate (H) |
|---|---|---|---|
| Probability | 0.2 | 0.5 | 0.3 |

| Snack | Ginger biscuit (G) | Rich tea biscuit (R) | Doughnut (D) |
|---|---|---|---|
| Probability | 0.3 | 0.1 | 0.6 |

**a** Leon says that the probability that Saskia has coffee and a ginger biscuit is 0.5. Explain why Leon is wrong.

**b** There are nine possible combinations of drink and snack. Two of these are coffee and a ginger biscuit (C, G) and coffee and rich tea biscuit (C, R).

Write down the other seven combinations.

**c** The probability that Saskia chooses coffee and a ginger biscuit is calculated by working out $0.2 \times 0.3 = 0.06$, and the probability that she chooses coffee and a rich tea biscuit is calculated by $0.2 \times 0.1 = 0.02$.

**i** Work out the other seven probabilities.

**ii** Add up all nine probabilities. What do you notice about your answer?

# 13.4 Experimental probability

This section will show you how to:

- calculate experimental probabilities and relative frequencies from experiments
- recognise different methods for estimating probabilities.

These are the results when Tebor threw a fair coin 10 times.

In this experiment, he threw six heads.

Tebor asked some friends to help him repeat the experiment.

The table shows the results of the different experiments.

| Number of throws | Number of heads | Number of heads / Number of throws |
|---|---|---|
| 10 | 6 | 0.6 |
| 50 | 24 | 0.48 |
| 100 | 47 | 0.47 |
| 200 | 92 | 0.46 |
| 500 | 237 | 0.474 |
| 1000 | 488 | 0.488 |
| 2000 | 960 | 0.48 |
| 5000 | 2482 | 0.4964 |

He drew a graph of the results, plotting the first column against the last column.

As the total number of throws increases, the value of $\dfrac{\text{number of heads}}{\text{number of throws}}$ gets closer and closer to 0.5.

The value of $\dfrac{\text{number of heads}}{\text{number of throws}}$ is called the **experimental probability**. As the number of throws or trials increases, the value of the experimental probability gets closer to the true or **theoretical probability**.

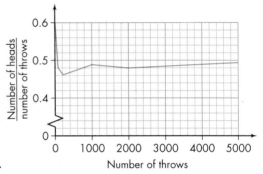

The experimental probability is also known as the **relative frequency** of an outcome. The relative frequency is an estimate for the theoretical probability. It is given by:

$$\text{relative frequency of an outcome} = \frac{\text{number of times the outcome occurs}}{\text{total number of trials}}.$$

**Example 8**

The frequency table shows the speeds of 160 vehicles that pass a radar speed check on a dual carriageway.

| Speed (mph) | 20–29 | 30–39 | 40–49 | 50–59 | 60–69 | 70+ |
|---|---|---|---|---|---|---|
| Frequency | 14 | 23 | 28 | 35 | 52 | 8 |

What is the experimental probability that a car picked at random is travelling faster than 70 mph?

8 cars travel faster than 70 mph, so the experimental probability or relative frequency is $\dfrac{8}{160} = \dfrac{1}{20}$.

## Finding probabilities

You can find the probability of an outcome in one of three ways.

1   If you can work out the theoretical probability of an outcome – for example, taking a king from an ordinary pack of cards – you can use equally likely outcomes.

2   Some probabilities, such as people buying a certain brand of dog food, cannot be calculated by using equally likely outcomes. To find the probabilities for such trials, you can perform an

experiment or conduct a survey. This is called 'collecting **experimental data**'. The more data you collect, the better the estimate is.

3   The probabilities involved in some events, such as an earthquake occurring in Japan, cannot be calculated by either of the above methods. However, you can look at data collected over a long period of time and make an estimate (sometimes called a 'best guess') at the chance of something happening. This is called 'looking at historical data'.

Which method – A, B or C – would you use to estimate the probabilities for events **a** to **e**?

   A: Use equally likely outcomes

   B: Conduct a survey or collect data

   C: Look at historical data

**a**   Someone in your class will go abroad for a holiday this year.

**b**   You will score a tail when you throw a coin.

**c**   Your bus home will be late.

**d**   It will snow on Christmas Day.

**e**   You will take a red seven from an ordinary pack of cards.

**a**   You would have to ask all the members of your class what they intended to do for their holidays this year. You would therefore conduct a survey, method B.

**b**   There are two possibilities – head or tail – so P(tail) = $\frac{1}{2}$. This is an equally likely outcome, so method A.

**c**   If you catch the bus every day, you can collect data over several weeks. This would be method C.

**d**   If you check whether it snowed on Christmas Day for the last few years, you would be able to make a good estimate of the probability. This would be method C.

**e**   There are 2 red sevens out of 52 cards, so you can calculate the probability of taking one of them: P(red seven) is $\frac{2}{52} = \frac{1}{26}$. This is method A.

## Exercise 13D

**1**   Which method – A, B or C – would you use to estimate or state the probabilities for events **a** to **h**?

   A: Use equally likely outcomes

   B: Conduct a survey or experiment

   C: Look at historical data

**a**   How people will vote in the next election.

**b**   A drawing pin dropped on a desk will land point up.

**c**   A Premier League team will win the FA Cup.

**d**   You will win a school raffle.

**e**   The next car to drive down the road will be red.

**f**   You will throw a 'double 6' with two fair, six-sided dice.

**g**   Someone in your class likes classical music.

**h**   A person picked at random from your school will be a vegetarian.

**2** Naseer throws a fair, six-sided dice and records the number of 6s that he gets after various numbers of throws. The table shows his results.

| Number of throws | 10 | 50 | 100 | 200 | 500 | 1000 | 2000 |
|---|---|---|---|---|---|---|---|
| Number of 6s | 2 | 4 | 10 | 21 | 74 | 163 | 329 |

  **a** Calculate the experimental probability of scoring a 6 at each stage that Naseer recorded his results.

  **b** How many ways can a normal dice land?

  **c** How many of these ways give a 6?

  **d** What is the theoretical probability of throwing a 6 with a dice?

**3** Marie made a five-sided spinner, like this one.

  She used it to play a board game with her friend Sarah. The girls thought that the spinner was not very fair as it seemed to land on some numbers more than others. They spun the spinner 200 times and recorded the results, as shown in the table.

| Side spinner lands on | 1 | 2 | 3 | 4 | 5 |
|---|---|---|---|---|---|
| Number of times | 19 | 27 | 32 | 53 | 69 |

  **a** Work out the relative frequency of the spinner landing on each number.

  **b** How many times would you expect each number to occur if the spinner is fair?

  **c** Do you think that the spinner is fair? Give a reason for your answer.

(MR) **4** A sampling bottle is a sealed bottle with a clear plastic tube at one end. When the bottle is tipped up, one of the balls inside will fall into the tube.

Kenny's sampling bottle contains 200 balls that are either blue or white. Kenny conducts an experiment to see how many blue balls there are in the bottle. He takes various numbers of samples and records how many of them revealed a blue ball. His results are shown in the table.

| Number of samples | Number of blue balls | Experimental probability |
|---|---|---|
| 10 | 2 | |
| 100 | 25 | |
| 200 | 76 | |
| 500 | 210 | |
| 1000 | 385 | |
| 5000 | 1987 | |

  **a** Copy the table and complete it by calculating the experimental probability of getting a blue ball at each stage.

  **b** Using this information, how many blue balls do you think there are in the bottle?

**MR** **5** A four-sided dice has faces numbered 1, 2, 3 and 4. The score is the face on which it lands. Five students decide to throw the dice to test whether it is biased. They each throw it a different number of times. Their results are shown in the table.

| Student | Total number of throws | Score | | | |
|---------|------------------------|---|---|---|---|
| | | 1 | 2 | 3 | 4 |
| Ayesha | 20 | 7 | 6 | 3 | 4 |
| Brian | 50 | 19 | 16 | 8 | 7 |
| Caryl | 250 | 102 | 76 | 42 | 30 |
| Deema | 80 | 25 | 25 | 12 | 18 |
| Evan | 150 | 61 | 46 | 26 | 17 |

**a** Which student will have the most reliable set of results? Why?

**b** Add up all the score columns and work out the relative frequency of each score. Give your answers correct to 2 decimal places.

**c** Is the dice biased? Explain your answer.

**CM** **6** At a computer factory, tests were carried out to see how many faulty computer chips were produced in one week.

| | Monday | Tuesday | Wednesday | Thursday | Friday |
|---|--------|---------|-----------|----------|--------|
| **Sample** | 850 | 630 | 1055 | 896 | 450 |
| **Number faulty** | 10 | 7 | 12 | 11 | 4 |

On which day was the number of faulty chips produced the highest? Use what you know about probability to explain your answer.

**PS** **7** Andrew made an eight-sided spinner. He tested it out to see if it was fair.

He spun the spinner and recorded the results.

Unfortunately, he spilt coffee over his results table, so he could not see the middle part.

| Number spinner lands on | 1 | 2 | 3 | | 6 | 7 | 8 |
|-------------------------|---|---|---|---|---|---|---|
| **Frequency** | 18 | 19 | 22 | | 19 | 20 | 22 |

Copy and complete the table for Andrew. Assume the spinner was fair.

**EV** **8** Stefan threw a coin 1000 times to see how many heads he scored.

He said: 'If this is a fair coin, I should get exactly 500 heads.'

Explain why he is wrong.

**MR** **9** Roza has an eight-sided spinner, marked like this.

She tests it by spinning it 100 times and records the results, as shown in the table.

| Colour | Red | Blue | Black | Green |
|--------|-----|------|-------|-------|
| **Frequency** | 48 | 13 | 28 | 11 |

Roza says the spinner is fair as the frequencies are close to what may be expected.

Sylwester says the spinner is unfair as there are far more reds than any other colour.

Who is correct? Give reasons for your answer.

# 13.5 Expectation

This section will show you how to:

- predict the likely number of successful outcomes, given the number of trials and the probability of any one outcome.

When you know the probability of an outcome, you can predict how many times you would expect that outcome to happen in a certain number of trials. This is called **expectation**.

**Note:** This is what you *expect*. It is not necessarily what is going to happen.

---

**Example 10**

A bag contains 20 balls, nine of which are black, six are white and five are yellow. Pieter takes a ball at random from the bag, notes its colour and then puts it back in the bag. He does this 500 times.

**a** How many times would you expect him to take a black ball?

**b** How many times would you expect him to take a yellow ball?

**c** How many times would you expect him to take a black ball or a yellow ball?

**a** P(black ball) = $\frac{9}{20}$

So, expected number of black balls is $\frac{9}{20} \times 500 = 225$.

**b** P(yellow ball) is $\frac{5}{20} = \frac{1}{4}$

So, expected number of yellow balls is $\frac{1}{4} \times 500 = 125$.

**c** Expected number of black or yellow balls is 225 + 125 = 350.

---

**Example 11**

Four in ten cars sold in Britain are made by Japanese companies.

**a** What is the probability that the next car to be driven down your road will be Japanese?

**b** If there are 2000 cars in a multi-storey car park, how many of them would you expect to be Japanese?

**a** P(Japanese car) is $\frac{4}{10}$ which cancels to $\frac{2}{5}$ (= 0.4).

**b** Expected number of Japanese cars in 2000 cars is 0·4 × 2000 = 800 cars.

---

## Exercise 13E

**1 a** What is the probability of throwing a 6 with an ordinary fair dice?

**b** I throw an ordinary fair dice 150 times. How many times can I expect to get a score of 6?

**2 a** What is the probability of throwing a head with a coin?

**b** I throw a coin 2000 times. How many times can I expect to get a head?

**3 a** A card is taken at random from an ordinary pack of cards. What is the probability that it is:

**i** a black card   **ii** a king   **iii** a heart   **iv** the king of hearts?

**b** I take a card from a pack of cards and replace it. I do this 520 times. How many times would I expect to get:

**i** a black card   **ii** a king   **iii** a heart   **iv** the king of hearts?

**4** The ball in a roulette wheel can land in one of 37 spaces that are marked with numbers from 0 to 36 inclusive. I always choose the same number, 13.

 **a** What is the probability of the ball landing in 13?

 **b** If I play all evening and there are exactly 185 spins of the wheel in that time, how many times could I expect the ball to land on the number 13?

**5** In a bag there are 30 balls, 15 of which are red, five are yellow, five are green and five are blue. A ball is taken out at random, its colour noted and the ball replaced. If this is done 300 times, how many times would you expect:

 **a** a red ball                  **b** a yellow or blue ball

 **c** a ball that is not blue     **d** a pink ball?

**6** A class does the experiment described in question **5** 1000 times. Approximately how many times would they expect to get:

 **a** a green ball    **b** a ball that is not blue?

**7** At the local school fayre, the tombola stall gives out a prize if you take a numbered ticket from the drum and it ends in 0 or 5. There are 300 tickets in the drum altogether and the probability of getting a winning ticket is 0.4.

 **a** What is the probability of getting a losing ticket?

 **b** How many winning tickets are there in the drum?

**(MR)** **8** Josie said: 'When I throw a dice, I expect to get a score of 3.5.'

 'Impossible,' said Paul. 'You can't score 3.5 with a dice.'

 'Do this and I'll prove it,' said Josie.

 **a** Throw an ordinary dice 60 times. Copy and complete the table for the expected number of times each score will occur.

| Score | | | | | | |
|---|---|---|---|---|---|---|
| Expected occurrences | | | | | | |

 **b** Now work out the average score that is expected over 60 throws.

 **c** There is an easy way to get an answer of 3.5 for the expected average score. Explain what it is.

**9** The table shows the probabilities of seeing some cloud types on any day.

| Cumulus | 0.3 |
|---|---|
| Stratocumulus | 0.25 |
| Stratus | 0.15 |
| Altocumulus | 0.11 |
| Cirrus | 0.05 |
| Cirrocumulus | 0.02 |
| Nimbostratus | 0.005 |
| Cumulonimbus | 0.004 |

 **a** What is the probability of not seeing one of the above clouds in the sky?

 **b** On how many days of the year would you expect to see altocumulus clouds in the sky?

(PS) **10** Every evening Tamara and Chris take a card from an ordinary pack to see who washes up.

If they take a king or a jack, Chris washes up.

If they take a queen, Tamara washes up.

Otherwise, they wash up together.

In a year of 365 days, how many days would you expect them to wash up together?

(CM) **11** A market gardener is supplied with tomato plant seedlings.
She knows that the probability that any plant will develop a disease is 0.003.

How will she calculate the number of tomato plants that are likely not to develop a disease?

(PS) **12** I have 20 tickets for a raffle and I know that the probability of my winning the prize is 0.05.
How many tickets were sold altogether in the raffle?

# 13.6 Choices and outcomes

This section will show you how to:

- apply systematic listing and counting strategies to identify all outcomes for a variety of problems.

**Key terms**

combination

factorial

systematic counting

When you need to work out how many different ways a given situation may occur you need to apply a **systematic counting** or listing strategy. There are some mathematical techniques you can use to do this but, most of the time, you just have to show a logical approach.

---

**Example 12**

Here are five letter cards.

How many different ways can these be arranged in a line of five? For example, one way is ATHMS.

When you take the first card, you have a choice of five.
You can take any one of the five, so you can choose in five ways.

When you take the second card, there are only four left.
You can choose in four ways.

Similarly, you can choose the third card in three ways, and so on.

This gives a total of $5 \times 4 \times 3 \times 2 \times 1 = 120$ possible arrangements of the five cards.

Note that you can write the calculation $5 \times 4 \times 3 \times 2 \times 1$ as $5!$, which you say as 'five **factorial**'. Most calculators have a factorial button, which is often a 'SHIFT' function, and looks like this.

$1! = 1$ and, surprisingly, $0! = 1$ as well. Try these on your calculator.

---

Example 13

Mary has a blue hat (BH), a green hat (GH) and a red hat (RH).
She also has a blue scarf (BS), a red scarf (RS), a white scarf (WS) and a purple scarf (PS).
One day she chooses a hat and scarf at random.

a List all the possible pairs of hat and scarf she could choose.

b Work out the probability that she wears a hat and scarf of the same colour.

a Start with a blue hat and write down all the scarf combinations.
  Then repeat with the green hat, and so on.

  (BH, BS), (BH, RS), (BH, WS), (BH, PS),

  (GH, BS), (GH, RS), (GH, WS), (GH, PS),

  (RH, BS), (RH, RS), (RH, WS), (RH, PS)

b There are 12 different combinations but only two of these (BH, BS) and (RH, RS) give the
  same colour hat and scarf.

  So the probability of her choosing the same colour hat and scarf is $\frac{2}{12} = \frac{1}{6}$.

Example 14

a How many three-digit numbers can you make using just 1, 2, 3 and 4?

b How many of these will be between 200 and 400, with no repeated digits?

a Examples include 123, 224 and 444 because this time you can use the same digit more
  than once.

  Therefore, you have four choices for the first digit, then four choices for the second digit
  and four choices for the third digit.

  This gives a total of $4 \times 4 \times 4 = 4^3$ or 64 different three-digit numbers.

b To work out how many of these are between 200 and 400, with no repeated digits, you could
  try writing them all out but this will take a long time and you may miss one. It is better
  to be logical.

  The numbers will start with 2 or 3, so they will be of the form 2■■ or 3■■ where ■ stands
  for another digit.

  So in 2■■, the first digit must be 2, the second could be 1, 3 or 4, so there are three choices.

  When you have chosen this digit, because the question says 'no repeated digits', you only
  have two choices for the third digit.

  This makes $3 \times 2 = 6$ possible numbers that start 2■■.

  You can use the same logic for numbers starting 3■■, so there are six three-digit numbers
  that start 3■■.

  This gives you 12 possible numbers between 200 and 400 with no repeated digits.

  Check for yourself.
  The numbers are 213, 214, 231, 234, 241, 243, 312, 314, 321, 324, 341 and 342.

Example 15

In Mathsland, vehicle registration plates comprise 1, 2 or 3 letters and 1, 2 or 3 numbers. Given that the number cannot start with zero, how many possible vehicles can be registered?

You need to apply a systematic counting strategy.

Start with a single letter. There are 26 letters. Choosing two gives $26 \times 26 = 676$ options. Choosing three gives $26 \times 26 \times 26 = 17\,576$ options.

Now consider the numbers.

For the first number there are nine options (1–9).

The second number may be zero, so there are 10 choices (0–9) and, therefore, for two numbers, there are $9 \times 10 = 90$ options.

Again, for the third number there are 10 choices, so for three numbers there are $9 \times 10 \times 10 = 900$.

Now combine all of the possible arrangements of letters and numbers: one letter and one number, one letter and two numbers, … until you get to three letters and three numbers. You can write this as a table.

| Numbers \ Letters | 1 (26) | 2 (676) | 3 (17 576) |
|---|---|---|---|
| 1 (9) | 234 | 6 084 | 158 184 |
| 2 (90) | 2 340 | 60 840 | 1 581 840 |
| 3 (900) | 23 400 | 608 400 | 15 818 400 |

Adding up all of the options gives 18 259 722 different **combinations**.

There are about 35 million vehicles on the road in Britain today!

## Exercise 13F

**1** A café offers a breakfast deal.

   **a** How many different breakfast deals are possible?

   **b** Naz is a vegetarian. How many breakfast deals could he have from this café?

> **Breakfast Deal £2.99**
> Choose 1 item from each list.
>
> | Breakfast sandwich | Extras | Drink |
> |---|---|---|
> | Ham and Egg | Beans | Tea |
> | Egg and Tomato | Mushrooms | Coffee |
> | Ham and Tomato | Pork sausage | Juice |
> | Egg and Cheese | | Beef extract |

**2** Salim has five different shirts and eight different ties. How many different combinations of shirt and tie are possible?

**3** Don has a blue shirt (BS), a white shirt (WS) and a green shirt (GS). He also has a blue tie (BT), a red tie (RT) and a green tie (GT). Don picks a shirt and tie combination at random.

   **a** List all the different combinations of shirt and tie he could pick.

   **b** Work out the probability that the shirt and tie are different colours.

**4** Mobile phone numbers always start with 07. The other nine digits use the numbers from 0 to 9. How many different mobile phone numbers are possible?

**5** When one coin is thrown, there are two outcomes: head (H) and tail (T).

   **a** When two coins are thrown there are four possible outcomes: HH, HT, TH and TT. What is the probability that, when two coins are thrown, the outcome is a head and a tail, in any order?

   **b** When three coins are thrown there are eight possible outcomes. List them and work out the probability of getting two heads and a tail, in any order.

**6** How many times would you write the digit 9 if you wrote down all the numbers from 0 to 100?

**7** These are five light switches in a row.

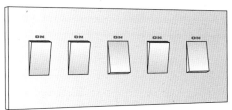

How many on/off combinations are there?

**8** A pizza parlour offers a choice of any three toppings, from a list of eight. Assuming that a customer cannot choose the same topping more than once, how many different combinations of toppings are there?

**9** A code consists of two letters followed by two digits. The letters and digits can be repeated and the digits can include zero.

Are there more than 50 000 different possible codes?

**10 a** All citizens of Futureworld are given an ID number. This is X for women or Y for men, followed by one letter, followed by a six-digit number (000 000 is not allowed), followed by one of the letters A, B, C or D. How many different ID numbers are possible?

**b** Write down the probability that the ID number ends in A.

**11 a** How many different arrangements of the four letters in the word TIME are there?

**b** There are four proper words that can be made from the letters T, I, M and E. These are TIME, ITEM, EMIT and MITE. If the four letters are arranged at random, what is the probability that they spell a proper word?

**12** Salim takes two of these coins at random.

Work out the probability that the total value of the two coins will be greater than £1.

**13 a** How many numbers between 1 and 200 have 6 as *at least* one of their digits?

**b** If a number between 1 and 200 is chosen at random, what is the probability that it contains at least one 6?

**c** If a number between 1 and 200 is chosen at random, what is the probability that the first digit is 6?

**14** These keypads each have a four-key code.

**Keypad A**

**Keypad B**

**a** How many different codes are possible for each keypad?

**b** Keypad **B** is reprogrammed so that the four-key code must start with a letter and be followed by three numbers. How many codes are possible now?

**c** Omar knows that the four-key code on keypad **A** uses the digits 2, 4, 6 and 8. He tries a four-digit code using these numbers at random. What is the probability that he gets it right?

**15** **a** Use your calculator to work out:       **i** 7!     **ii** 20!

**b** 60! ≈ $8.32 \times 10^{81}$ which is about the number of atoms in the whole universe.

What is the largest factorial that you can work out with your calculator?

# Worked exemplars

**1**  400 tickets are sold in a raffle. There is only one prize.

Mr Raza buys five tickets for himself and sells another 40.

Mrs Raza buys 10 tickets for herself and sells another 50.

Mrs Hewes just sells 52 tickets.

**a** What is the probability of:

   **i** Mr Raza winning the raffle

   **ii** Mr Raza selling the winning ticket to someone else?

**b** What is the probability of either Mr or Mrs Raza selling the winning ticket to someone else?

**c** What is the probability of Mrs Hewes not selling a winning ticket?

**d** Who has the greatest chance, Mr or Mrs Raza or Mrs Hughes, of either winning the raffle or selling the winning ticket? Give a reason for your answer.

   Give your answers as fractions in their simplest form.

| | |
|---|---|
| This is a mathematical reasoning question and so you must communicate your method clearly. Do not just write down probabilities without some explanation. | |
| **a** **i** $\dfrac{5}{400} = \dfrac{1}{80}$ <br><br> **ii** $\dfrac{40}{400} = \dfrac{1}{10}$ | Remember to cancel the fractions and make sure you read the information given in the question. |
| **b** $\dfrac{40}{400} + \dfrac{50}{400} = \dfrac{90}{400}$ <br><br> $= \dfrac{9}{40}$ | Remember that 'or' means add the two separate events. <br><br> When adding the two fractions, they must have the same denominator. |
| **c** $1 - \dfrac{52}{400} = \dfrac{348}{400}$ <br><br> $= \dfrac{87}{100}$ | This is a question where you are working out the probability of an event not happening. |
| **d** P(Mr Raza either winning the raffle or selling the winning ticket) $= \dfrac{45}{400}$ <br><br> P(Mrs Raza either winning the raffle or selling the winning ticket) $= \dfrac{60}{400}$ <br><br> P(Mrs Hewes either winning the raffle or selling the winning ticket) $= \dfrac{52}{400}$ <br><br> Mrs Raza has the greatest chance as $\dfrac{60}{400}$ is the largest fraction. | As you will need to compare fractions to solve the problem, there is no need to cancel the three probability fractions. <br><br> Make sure you state your conclusion clearly and give a reason. |

**2**  **a** Megan takes a card at random from an ordinary pack of cards. Write down the probability of each outcome.

　**i** P(ace)　　　　　　**ii** P(picture card)　　　**iii** P(not a club)

**b** Megan takes a card from an ordinary pack of cards at random. Then she returns the card to the pack and takes another card. She repeats the experiment until she has taken a card 100 times.

Approximately many hearts can she expect to take? Give a reason for your answer.

| In this question you have to interpret and evaluate results and assumptions. | |
|---|---|
| **a**　**i**　$\dfrac{4}{52} = \dfrac{1}{13}$ | There are 4 aces in an ordinary pack of 52 cards.<br><br>Remember to cancel probability fractions. |
| **ii**　P(picture card) = P(jack) +<br>　　P(queen) + P(king)<br><br>　　$= \dfrac{12}{52}$<br><br>　　$= \dfrac{3}{13}$ | There are 4 jacks, 4 queens and 4 kings in an ordinary pack of cards, making 12 cards in total. |
| **iii**　P(not a club) = 1 – P(club)<br>　　　　$= 1 - \dfrac{13}{52}$<br>　　　　$= 1 - \dfrac{1}{4}$<br>　　　　$= \dfrac{3}{4}$ | This is a question where you are working out the probability of an event not happening. |
| **b**　P(taking a heart) = $\dfrac{13}{52}$<br>　　　　　　$= \dfrac{1}{4}$<br>So she can expect to take<br>approximately $\dfrac{1}{4} \times 100 = 25$ hearts. | Show that you understand the difference between theoretical probability and expectation. |

# Ready to progress?

I can use the probability scale and understand the language of probability.
I can use fractions, decimals or percentages to work out a probability of an outcome.
I can work out the probability of an outcome not happening.
I can recognise mutually exclusive and exhaustive outcomes.

I know the different methods for estimating probabilities.
I can calculate experimental probabilities and relative frequencies from experiments.
I can predict the likely number of successful outcomes, given the number of trials and the probability of any one outcome.

I can use systematic counting strategies to identify arrangements.

# Review questions

1. Bill planted some bulbs in October. The ticks in the table show the months in which each type of bulb flowers.

|  |  | Month | | | | | |
|---|---|---|---|---|---|---|---|
|  |  | Jan | Feb | March | April | May | June |
| Type of bulb | Alium |  |  |  |  | ✓ | ✓ |
|  | Crocus | ✓ | ✓ |  |  |  |  |
|  | Daffodil |  | ✓ | ✓ | ✓ |  |  |
|  | Iris | ✓ | ✓ |  |  |  |  |
|  | Tulip |  |  |  | ✓ | ✓ |  |

   a In which months do tulips flower?

   b Which type of bulb flowers in March?

   c In which month do most types of bulb flower?

   d Which type of bulb flowers in the same months as the iris?

   Ben puts one of each of these bulbs in a bag. He takes a bulb from the bag without looking.

   e i Write down the probability that he will take a crocus bulb.

   ii Copy the probability scale and mark with a cross (×) the probability that he will take a bulb that flowers in February.

   |———————————|———————————|
   0                                               1

2. There are six red crayons, eight blue crayons and ten green crayons in a box.

   Maya takes a crayon, at random, from the box.

   Write down the probability that she takes a green crayon.

**3** A bag contains some beads that are red or green or blue or yellow.

The table shows the number of beads of each colour.

| Colour | Red | Green | Blue | Yellow |
|---|---|---|---|---|
| Number of beads | 13 | 12 | 15 | 10 |

Harvey takes a bead at random from the bag. Work out these probabilities.

**a** P(he takes a red bead)

**b** P(he takes a yellow bead)

**c** P(he takes a red or green bead)

**d** P(he does not take a blue bead)

CM **4** **a** Sienna has a spinner that has five equal sections. The numbers 1, 2 and 3 are written on the spinner.

She spins the spinner once. On what number is the spinner least likely to land?

**b** Sienna thinks that the chance of getting a 2 is $\frac{2}{3}$. Explain why she is wrong.

MR **5** Nathan has a four-sided dice.

The sides of the dice are numbered 1, 2, 3 and 4.

The table shows the probability of the dice landing on each of the numbers 1, 2 and 3.

| Number | 1 | 2 | 3 | 4 |
|---|---|---|---|---|
| Probability | 0.2 | 0.1 | 0.2 | |

**a** Work out the probability that the dice will land on the number 4.

**b** Is the dice biased? Give a reason for your answer.

PS **6** Ellie has a bag in which there are nine counters, all green. Maria has a bag in which there are 15 counters, all red. Jade has a bag in which there are only blue counters.

**a** Ellie and Maria put all their counters into a box.

What is the probability of taking a green counter from the box?

**b** Jade now adds some of her blue counters to the box.

The probability of taking a blue counter from the box is $\frac{1}{3}$.

How many blue counters does Jade put in the box?

**CM** **7** Sam and Tomas are playing a game with dice.

a Sam uses a biased dice. The probability that he throws a 6 is 0.2.

  i Write down the probability that he does not throw a 6.

  ii Sam throws this biased dice 60 times. Work out an estimate for the number of times he will throw a 6.

b Tomas uses a fair dice. He throws this fair dice 60 times. Is he likely to throw more 6s than Sam?

  Explain your answer.

**MR** **8** Khalid has a triangular spinner with sections coloured white (W), green (G) and blue (B).

Khalid spins his spinner 20 times and records the colour it lands on each time.

| W | W | B | G | G | W | B | G | G | W |
| G | B | G | B | G | W | G | B | G | B |

a Copy and complete the relative frequency table.

| Colour | White (W) | Green (G) | Blue (B) |
|---|---|---|---|
| **Relative frequency** | | | |

b The table below shows the relative frequencies after this spinner has been spun 100 times.

| Colour | White (W) | Green (G) | Blue (B) |
|---|---|---|---|
| **Relative frequency** | $\frac{21}{100}$ | $\frac{52}{100}$ | $\frac{27}{100}$ |

Which of the two relative frequency tables gives the better estimate of the probability of the spinner landing on white? Give a reason for your answer.

**PS** **9** Asher throws a coin and a dice together.

a List all the ways that they could land.

b If the coin lands on heads, the score on the dice is doubled.

  If the coin lands on tails, the score on the dice stays the same.

  Work out the probability that the combination results in a score of 6.

**PS** **10** Two numbers are taken at random from the numbers 2, 11, 14 and 22.

Work out the probability that the total of the two numbers is a square number.

**11** a A combination lock has three wheels. Each wheel has the digits 0 to 9 on it.

  How many different combinations are there?

b A combination is chosen at random. Work out the probability that all three digits are the same.

**MR** c A different combination lock has $n$ dials. Each dial has the digits 1 to 9. There are nearly 60 000 different combinations. Work out the value of $n$. Show your working.

# 14 Geometry and measures: Volumes and surface areas of prisms

## This chapter is going to show you:

- how to calculate the volume of a composite shape made from cuboids
- how to calculate the volume and surface area of a prism
- how to calculate the volume and surface area of a cylinder.

## You should already know:

- the formula for the circumference of a circle
  (circumference $C = \pi \times$ diameter or $C = \pi d$)
- the formula for the area of a circle (area $= \pi \times$ radius$^2$ or $A = \pi r^2$)
- the formula for the volume of a cuboid
  (volume = length × width × height or $V = lwh$)
- the common metric units to measure area, volume and capacity
  shown below.

| Area | Volume | Volume to capacity |
|---|---|---|
| 100 mm$^2$ = 1 cm$^2$ | 1000 mm$^3$ = 1 cm$^3$ | 1000 cm$^3$ = 1 litre |
| 10 000 cm$^2$ = 1 m$^2$ | 1 000 000 cm$^3$ = 1 m$^3$ | 1 m$^3$ = 1000 litres |

## About this chapter

Volumes tell us how much space there is inside any structure. Whether it is a house, barn, aeroplane, car or office, the volume is important. Did you know, for example, that in England there is a regulation that states the number of people who can use an office, based on the volume of the room?

Surface areas are important too. If you want to paint a room or garden fence, calculating the surface area will help you work out how many tins of paint you need.

So how do you measure areas and volumes? Some shapes and objects are easier to measure than others. In this chapter, you will learn formulae that can be used to calculate areas and volumes of different shapes, based on a few measurements. Many of these formulae were first worked out thousands of years ago. The fact that these formulae are still in use today shows how important they are.

# 14.1 3D shapes

This section will show you how to:

- use the correct terms when working with 3D shapes.

3D shapes have **faces**, **vertices** and **edges**.

**Note:** Vertices is the plural of vertex.

You can see from the diagram that 3D shapes have three dimensions: a length, a width and a height. They also have a **volume**. This is the amount of space inside a 3D shape. It can be measured in cubic millimetres (mm³), cubic centimetres (cm³) or cubic metres (m³).

A cube with an edge of 1 cm has a volume of 1 cm³ and each face has an area of 1 cm².

A cube with an edge of 2 cm has a volume of 8 cm³ and each face has an area of 4 cm².

A cube with an edge of 3 cm has a volume of 27 cm³ and each face has an area of 9 cm².

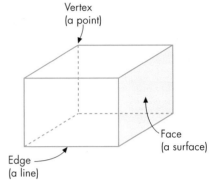

1 cm

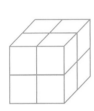
2 cm

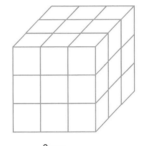
3 cm

---

**Example 1**

The diagram shows a set of steps.

**a** How many cubes, each 1 cm by 1 cm by 1 cm, are there in these steps?

**b** What is the volume of the steps?

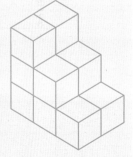

**a** Count the cubes. Remember to include those hidden at the back.

6 + 4 + 2 = 12

**b** The volume of each cube is 1 cm³.

So the volume of the steps is 12 × 1 = 12 cm³.

# Exercise 14A

 **1**  **a** Copy and complete the table for the 3D shapes shown.

The hidden faces, vertices and edges are shown with dashed lines.

| Shape | Name | Number of faces ($F$) | Number of vertices ($V$) | Number of edges ($E$) |
|---|---|---|---|---|
| | Cuboid | | | |
| | Square-based pyramid | | | |
| | Triangular-based pyramid (or tetrahedron) | | | |
| | Octahedron | | | |
| | Triangular prism | | | |
| | Hexagonal prism | | | |
| | Hexagonal-based pyramid | | | |

**b** Look at the numbers in the completed table.

What is the connection between the number of faces, $F$, the number of vertices, $V$, and the number of edges, $E$ for each shape?

**2** **a** For each shape, work out:

    **i** the area of the front face        **ii** the volume.

       Shape A                                   Shape B

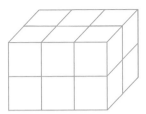
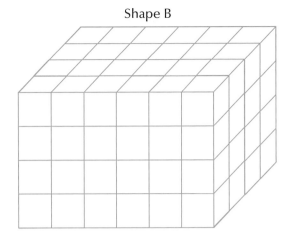

**b** Copy and complete these sentences.

    **i** The length, width and height of shape B is … times as big as the length, width and height of shape A.

    **ii** The area of the front face of shape B is … times as big as the front face of shape A.

    **iii** The volume of shape B is …. times as big as the volume of shape A.

**c** For each shape, work out:

    **i** the area of the front face        **ii** the volume.

       Shape C                                   Shape D

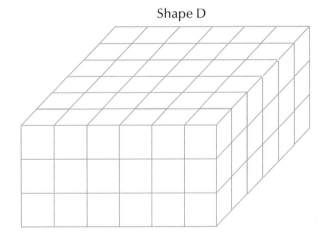

**d** Copy and complete these sentences.

    **i** The length, width and height of shape D is … times as big as the length, width and height of shape C.

    **ii** The area of the front face of shape D is … times as big as the front face of shape C.

    **iii** The volume of shape D is …. times as big as the volume of shape C.

**(CM)** **3** Joy says: 'There are 100 cubes in the shape.' How might she have calculated this?

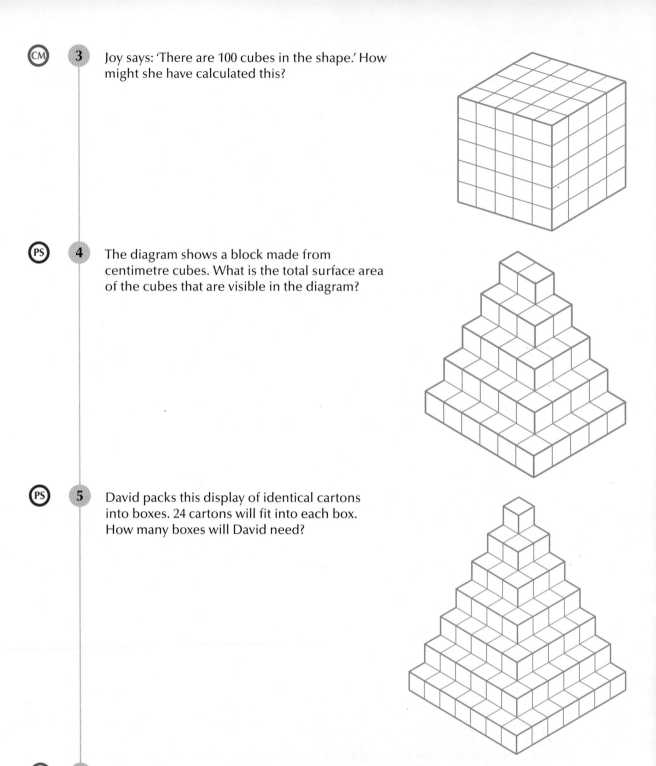

**(PS)** **4** The diagram shows a block made from centimetre cubes. What is the total surface area of the cubes that are visible in the diagram?

**(PS)** **5** David packs this display of identical cartons into boxes. 24 cartons will fit into each box. How many boxes will David need?

**(CM)** **6** David and Amber were looking at a pentagonal prism. David said: 'If you add the number of edges to the number of faces, it comes to 22.' Is he correct? Give reasons for your answer.

# 14.2 Volume and surface area of a cuboid

## This section will show you how to:

- calculate the surface area and volume of a cuboid.

**Key terms**

capacity

surface area

You will come across examples of cuboids every day, such as breakfast cereal packets, shoe boxes, mp3 players – and even this book.

### Volume

A cuboid has six rectangular faces.

You can use this formula to calculate the volume of a cuboid.

volume = length × width × height

This can be written using algebra as:

$V = l \times w \times h$ or $V = lwh$

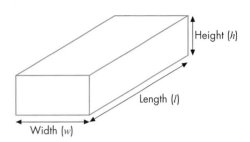

### Surface area

To calculate the **surface area** of a cuboid, you need to work out the total area of its six rectangular faces. Notice that each pair of opposite rectangles has the same area. So:

area of top and bottom rectangles = 2 × length × width or $2lw$

area of front and back rectangles = 2 × height × width or $2hw$

area of two side rectangles = 2 × height × length or $2hl$

This gives the formula for the surface area of a cuboid.

surface area $(A) = 2lw + 2hw + 2hl$

---

**Example 2**

Calculate the volume and surface area of this cuboid.

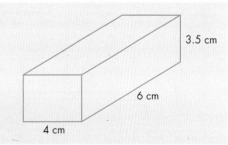

Use the formula for the volume.

$V = lwh$

$= 6 \times 4 \times 3.5$

$= 84 \text{ cm}^3$

Use the formula for the surface area.

$A = 2lw + 2hw + 2hl$

$= (2 \times 6 \times 4) + (2 \times 3.5 \times 4) + (2 \times 3.5 \times 6)$

$= 48 + 28 + 42$

$= 118 \text{ cm}^2$

The word **capacity** is used to describe the amount of liquid or gas that a container can hold. This is often measured in litres (*l*).

$$1000 \text{ millilitres (ml)} = 1 \text{ litre}$$
$$100 \text{ centilitres (cl)} = 1 \text{ litre}$$
$$1000 \text{ cm}^3 = 1 \text{ litre}$$
$$1 \text{ m}^3 = 1000 \text{ litres}$$

**Remember:** $1 \text{ cm}^3 = 1000 \text{ mm}^3$ and $1 \text{ m}^3 = 1\,000\,000 \text{ cm}^3$

**Example 3**

A petrol tank measures 50 cm by 60 cm by 25 cm. How many litres can it hold?

Start by working out the volume of the tank.    $V = lwh$
$$= 50 \times 60 \times 25$$
$$= 75\,000 \text{ cm}^3$$

Now work out how many litres a tank with this volume can hold.

$$1000 \text{ cm}^3 = 1 \text{ litre}$$

So $75\,000 \text{ cm}^3 = 75 \text{ litres}$

The petrol tank can hold 75 litres.

## Exercise 14B 🖩

**1** For each cuboid calculate:

   **i** the volume       **ii** the surface area.

   **a**

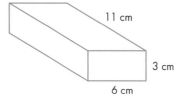

11 cm, 3 cm, 6 cm

   **b**

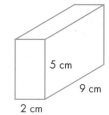

5 cm, 9 cm, 2 cm

   **c**

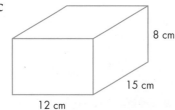

8 cm, 15 cm, 12 cm

   **d**

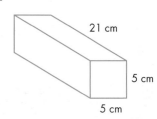

21 cm, 5 cm, 5 cm

**2** Calculate the capacity of a fish tank with length 40 cm, width 30 cm and height 20 cm. Give your answer in litres.

**3** Calculate the volume of cuboids with these dimensions.

   **a** base area 40 cm², height 4 cm

   **b** height 4 cm, base with one side 10 cm and the other side 2 cm longer

   **c** area of top face 25 cm², depth 6 cm

**4** Calculate:

   **i** the volume

   **ii** the surface area of cubes
      with each of these edge lengths.

   **a** 4 cm        **b** 7 cm        **c** 10 mm        **d** 5 m        **e** 12 m

**(PS) 5** Safety regulations state that, in a room where people sleep, there should be a minimum volume of 12 m$^3$ for each person. A dormitory is 20 m long, 13 m wide and 4 m high.

   What is the greatest number of people who can safely sleep in the dormitory?

**6** Copy and complete the table for cuboids **a** to **e**.

|   | Length | Width | Height | Volume |
|---|--------|-------|--------|--------|
| **a** | 8 cm | 5 cm | 4.5 cm | |
| **b** | 12 cm | 8 cm | | 480 cm$^3$ |
| **c** | 9 cm | | 5 cm | 270 cm$^3$ |
| **d** | | 7 cm | 3.5 cm | 245 cm$^3$ |
| **e** | 7.5 cm | 5.4 cm | 2 cm | |

**(MR) 7** A tank contains 320 litres of water. The base of the tank measures 65 cm by 31 cm.

   Show that the depth of water is 159 cm.

**(EV) 8** Peter booked storage space advertised as 168 m$^3$ with a height of 3.5 m. He calculated the area of the floor to be 50 m$^2$. Is Peter's answer sensible? Comment on your answer.

**9** What is the side length of a cube with each of these volumes?

   **a** 27 cm$^3$        **b** 125 m$^3$        **c** 8 mm$^3$        **d** 1.728 m$^3$

**10** Calculate the volume of these shapes.

   **a**                        **b**

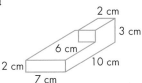 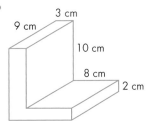

**(MR) 11** A cuboid has a volume of 125 cm$^3$ and a total surface area of 160 cm$^2$.

   Could it be a cube? Give a reason for your answer.

**(PS) 12** The volume of a cuboid is 1000 cm$^3$. What is the smallest surface area of the cuboid?

# 14.3 Volume and surface area of a prism

This section will show you how to:

- calculate the volume and surface area of a prism.

**Key terms**

cross-section

prism

A **prism** is a 3D shape that has the same **cross-section** running all the way through it.

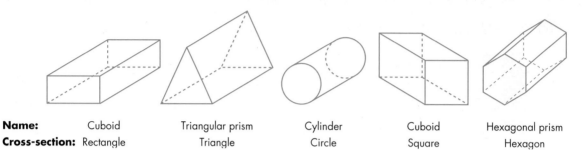

| **Name:** | Cuboid | Triangular prism | Cylinder | Cuboid | Hexagonal prism |
|---|---|---|---|---|---|
| **Cross-section:** | Rectangle | Triangle | Circle | Square | Hexagon |

## Volume

You can calculate the volume of a prism by multiplying the area of its cross-section by the length of the prism (or height, if the prism is standing on end).

That is, volume of prism = area of cross-section × length or $V = Al$.

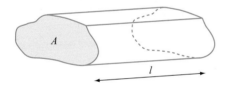

---

**Example 4**

Calculate the volume of the triangular prism.

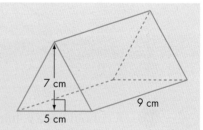

7 cm

5 cm

9 cm

First work out the area of the triangular cross-section.

$$\text{Area} = \frac{\text{base} \times \text{height}}{2}$$

$$= \frac{5 \times 7}{2}$$

$$= 17.5 \text{ cm}^2$$

Use this to work out the volume.

Volume = area of cross-section × length

$$= 17.5 \times 9$$

$$= 157.5 \text{ cm}^3$$

## Surface area

You calculate the surface area of a prism by working out the area of each face and adding them together, as you did for the cuboid.

Work out the surface area of the prism in Example 4. Its sloping side is 7.4 cm.

Calculate the area of each different face.

| | |
|---|---|
| triangular end face | 17.5 cm² (from Example 4) |
| sloping face | 7.4 × 9 = 66.6 cm² |
| base | 5 × 9 = 45 cm² |

Add together the area of each face.

  2 × triangular end face + 2 × sloping face + base

= 2 × 17.5 + 2 × 66.6 + 45

= 213.2 cm²

## Exercise 14C 🖩

**1** For each prism calculate:

  **i** the area of the cross-section        **ii** the volume.

**a**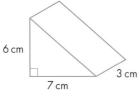

6 cm
7 cm
3 cm

**b**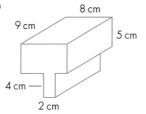

8 cm
9 cm
5 cm
4 cm
2 cm

**c**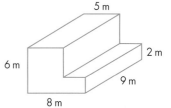

5 m
6 m
2 m
9 m
8 m

**2** Calculate the volume of each prism.

**a**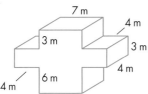

7 m
4 m
3 m
3 m
4 m
6 m
4 m

**b**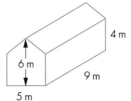

4 m
6 m
9 m
5 m

**c**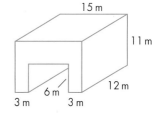

15 m
11 m
12 m
6 m
3 m
3 m

**(CM) 3** A swimming pool is 10 m wide and 25 m long.

It is 1.2 m deep at one end and 2.2 m deep at the other end. The floor slopes uniformly from one end to the other.

**a** Sketch the regular cross-section that shows this pool is a prism.

**b** The pool is filled with water at a rate of 2 m³ per minute. How long will it take to fill the pool?

**(PS) 4** A conservatory is in the shape of a prism. Calculate the volume of air inside the conservatory in cubic metres.

3 m
1.5 m
2 m
1.7 m

**(MR) 5** A girl builds 27 cubes, each of edge 2 cm, into a single large cube. Show that with 37 more cubes, she could build a larger cube with each edge 2 cm longer than the first one.

**6** These prisms have a uniform cross-section in the shape of a right-angled triangle. For each prism calculate:

**i** the volume **ii** the total surface area.

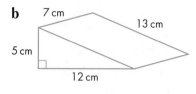

**a**

3.5 cm
5 cm
4 cm
3 cm

**b**

7 cm
13 cm
5 cm
12 cm

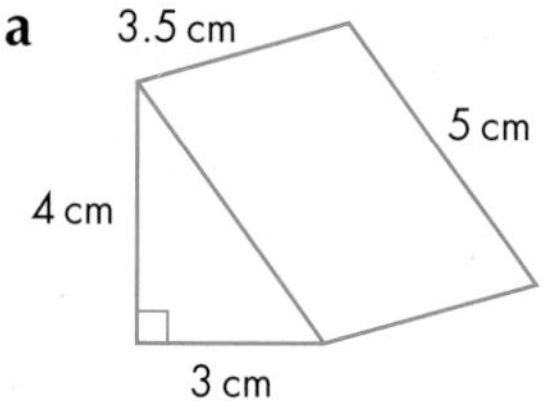

PS **7** The diagram shows the cross-section of a girder (in the shape of a prism). The girder is 2 m in length and made of iron. 1 $cm^3$ of iron has a mass of 79 g. What is the mass of the girder?

8 cm
8 cm
8 cm
12 cm
8 cm
8 cm
8 cm
30 cm

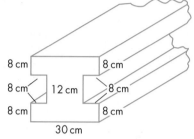

EV **8** Zuzanna and her daughter, Maisy, were trying to work out the volume of this prism. Zuzanna says it has a volume of 26 880 $cm^3$. Maisy says: 'Don't be silly mum!' Why does Maisy say this? Calculate the actual volume of the prism.

7 cm
20 cm
12 cm
16 cm
7 cm

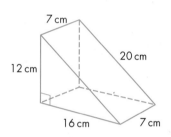

# 14.4 Volume and surface area of cylinders

## This section will show you how to:

- calculate the volume and surface area of a cylinder.

| Key term |
|---|
| cylinder |

### Volume

A **cylinder** is an example of a prism, so you can calculate its volume by multiplying the area of one of its circular ends by the height. That is

volume = $\pi r^2 h$

where $r$ is the radius of the cylinder and $h$ is its height or length.

r
h

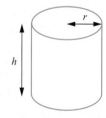

**Example 6**

What is the volume of a cylinder with a radius of 5 cm and a height of 12 cm?

Volume = area of circular base × height

$= \pi r^2 h$

$= \pi \times 5^2 \times 12$

$= 942$ $cm^3$ (nearest whole number)

## Surface area

The total surface area of a cylinder is made up of the area of its curved surface plus the area of its two circular ends.

The curved surface area, when opened out, is a rectangle with length equal to the circumference of the circular end.

curved surface area = circumference of end × height of cylinder

$$= 2\pi rh \text{ or } \pi dh$$

area of one end $= \pi r^2$

This gives total surface area $= 2\pi rh + 2\pi r^2$ or $\pi dh + 2\pi r^2$.

What is the total surface area of a cylinder with a radius of 15 cm and a height of 2.5 m?

First, you must change the dimensions to a *common unit*. Use centimetres in this case.

2.5 m = 2.5 × 100 cm

= 250 cm

Total surface area $= 2\pi rh + 2\pi r^2$

$$= 2 \times \pi \times 15 \times 250 + 2 \times \pi \times 15^2$$

$$= 23\,562 + 1414$$

$$= 24\,976 \text{ cm}^2$$

## Exercise 14D ▦

1 For each cylinder calculate:

   **i** the volume   **ii** the total surface area.

   Round your answers to a suitable degree of accuracy.

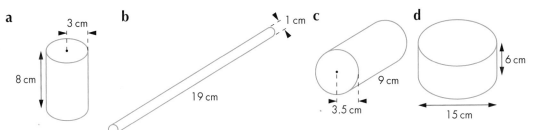

2 Work out the volume of a cylinder with each set of dimensions. Round your answers to a suitable degree of accuracy.

   **a** base radius 4 cm, height 5 cm

   **b** base diameter 9 cm, height 7 cm

   **c** base diameter 13.5 cm, height 15 cm

   **d** base radius 1.2 m, length 5.5 cm

 3 The diameter of a cylindrical marble column is 60 cm and its height is 4.2 m. The cost of making this column is quoted as £67.50 per cubic metre. Show that the total cost is approximately £80.

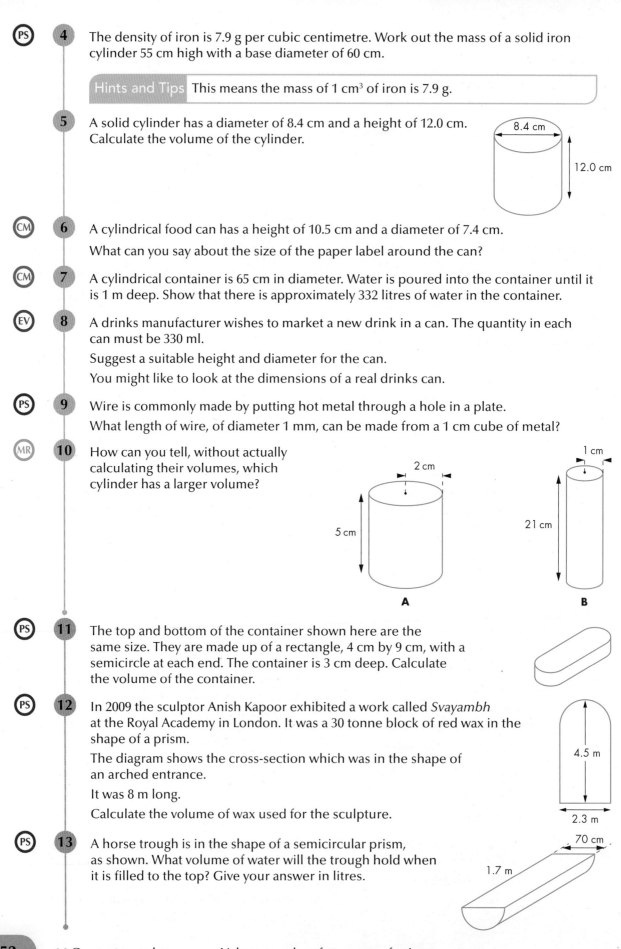

(PS) **4** The density of iron is 7.9 g per cubic centimetre. Work out the mass of a solid iron cylinder 55 cm high with a base diameter of 60 cm.

Hints and Tips This means the mass of 1 cm³ of iron is 7.9 g.

**5** A solid cylinder has a diameter of 8.4 cm and a height of 12.0 cm. Calculate the volume of the cylinder.

8.4 cm

12.0 cm

(CM) **6** A cylindrical food can has a height of 10.5 cm and a diameter of 7.4 cm.

What can you say about the size of the paper label around the can?

(CM) **7** A cylindrical container is 65 cm in diameter. Water is poured into the container until it is 1 m deep. Show that there is approximately 332 litres of water in the container.

(EV) **8** A drinks manufacturer wishes to market a new drink in a can. The quantity in each can must be 330 ml.

Suggest a suitable height and diameter for the can.

You might like to look at the dimensions of a real drinks can.

(PS) **9** Wire is commonly made by putting hot metal through a hole in a plate.

What length of wire, of diameter 1 mm, can be made from a 1 cm cube of metal?

(MR) **10** How can you tell, without actually calculating their volumes, which cylinder has a larger volume?

2 cm

5 cm

**A**

1 cm

21 cm

**B**

(PS) **11** The top and bottom of the container shown here are the same size. They are made up of a rectangle, 4 cm by 9 cm, with a semicircle at each end. The container is 3 cm deep. Calculate the volume of the container.

(PS) **12** In 2009 the sculptor Anish Kapoor exhibited a work called *Svayambh* at the Royal Academy in London. It was a 30 tonne block of red wax in the shape of a prism.

The diagram shows the cross-section which was in the shape of an arched entrance.

It was 8 m long.

Calculate the volume of wax used for the sculpture.

4.5 m

2.3 m

(PS) **13** A horse trough is in the shape of a semicircular prism, as shown. What volume of water will the trough hold when it is filled to the top? Give your answer in litres.

70 cm

1.7 m

# Worked exemplars

**1** A baker uses square and circular tins to make his cakes.

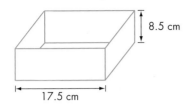

8.5 cm

17.5 cm

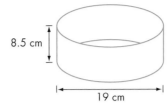

8.5 cm

19 cm

He thinks that the square tin will hold more cake mix. Show that he is correct.

| This question assesses 'communicating mathematically', so you must display your methods clearly and include words to explain what your calculations show and why this means the baker is correct. | |
|---|---|
| Volume of the square tin $= lwh$<br>$= 17.5 \times 17.5 \times 8.5$<br>$= 2603.125$ cm$^3$<br>Diameter of circular tin is 19 cm, so radius = 9.5 cm.<br>Volume of the circular tin $= \pi r^2 h$<br>$= \pi \times 9.5^2 \times 8.5$<br>$= 2409.9943$ cm$^3$ | You need to show clearly how you calculate the volume of each shape.<br><br>There is no need to round the answers it is only important to know which one is larger. |
| 2603.125 > 2409.9943, so the baker is correct that the square tin will hold more cake mix than the circular tin. | Finish by stating that the baker is correct and how your working shows this. |

**2** A firm makes small cylindrical rods with radius 5 mm and length 10 cm.

They are made from metal blocks, measuring 60 cm by 60 cm by 2 m, that are melted and reformed.

How many rods can be made from one block of metal?

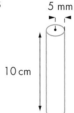

5 mm

10 cm

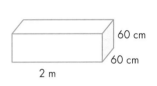

60 cm

60 cm

2 m

| This is a problem-solving question so you need to decide on a strategy to solve the problem and show all the stages of your thinking. | |
|---|---|
| Dimensions of block<br>length 200 cm, width 60 cm, height 60 cm<br>Dimensions of cylinder<br>radius 0.5 cm, height 10 cm | You need to convert all the dimensions to common units to solve this problem. You can either do this at the beginning or as you go. In this case, changing to centimetres requires the least conversions. |
| Volume of block $= lwh$<br>$= 200$ cm $\times 60$ cm $\times 60$ cm<br>$= 720\,000$ cm$^3$<br>Volume of a rod $= \pi r^2 h$<br>$= \pi \times 0.5^2 \times 10$<br>$= 2.5\pi$ cm$^3$ | Calculate the volume of the block and the volume of a rod *without rounding*. |
| Number of rods $= \dfrac{720\,000}{2.5\pi}$<br>$= 91\,673.247\,22$<br>$= 91\,673$ | Now divide the volume of the block by the volume of 1 rod.<br><br>If you round before the end, your result may be inaccurate. If you had rounded the volume of the rod to 7.85 cm$^3$, you would get the result 91 719. |

# Ready to progress?

I can calculate the volume and surface area of cuboids.

I can calculate the volume and surface area of prisms.
I can calculate the volume and surface area of cylinders.
I can calculate the volume of composite shapes.

# Review questions

**1** A box is a cuboid with length 6 cm, width 5 cm and height 2 cm.

   **a** Calculate the volume of the box.    **b** Calculate the surface area of the box.

**2** The diagram shows a solid prism made from centimetre cubes.

What is the total surface area of the prism?

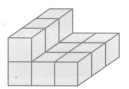

**MR** **3** Candles are packed into small boxes measuring 8 cm by 8 cm by 7 cm.

The small boxes are packed into a large box measuring 35 cm by 32 cm by 32 cm.

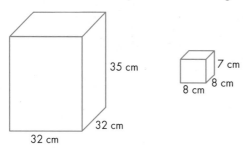

Show that 80 small candle boxes will completely fill one large box.

**MR** **4** The diagram shows a cylinder.

Show that the total surface area of the cylinder is 721 cm².

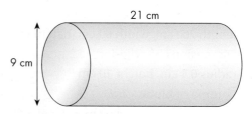

**PS** **5** A solid cube of side 20 cm has a circular hole cut through.

The hole has a diameter of 12 cm.

Calculate the volume remaining.

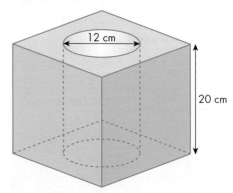

(PS) **6** A solid wooden cylinder has a radius of 5 cm and a height of 12 cm.

1 cm³ of the wood has a mass of 0.65 g.

Calculate the total mass of the cylinder. Give your answer correct to a suitable degree of accuracy.

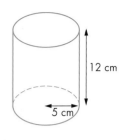

12 cm

5 cm

(PS) **7** Calculate the volume of this right-angled triangular prism.

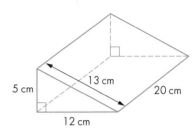

5 cm    13 cm    20 cm

12 cm

(PS) **8** The diagram shows a solid prism.

The prism is made from plastic.

1 cm³ of this plastic has a mass of 0.45 g.

Calculate the mass of the prism.

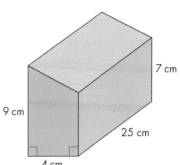

7 cm

9 cm

25 cm

4 cm

# 15 Algebra: Linear equations

## This chapter is going to show you:

- how to solve linear equations with the variable on one side
- how to solve linear equations involving brackets
- how to solve linear equations with fractions
- how to solve linear equations with the variable on both sides
- how to set up linear equations from real-life problems.

## You should already know:

- the basic language of algebra
- how to expand brackets and collect together like terms
- that addition and subtraction are opposite (inverse) operations
- that multiplication and division are opposite (inverse) operations.

## About this chapter

The first equation ever written using a modern equals sign, was:

$$I\ 4.\overline{z}e. \underline{\quad+\quad} .I\ 5.\mathcal{S}===== 7\ I.\mathcal{S}.$$

It was written by Robert Recorde in 1557. In today's notation it would be written as $14\sqrt{x} + 15 = 71$ and the solution as $x = 16$.

### Why don't planes fall out of the sky?

Bernoulli's principle states that as the speed of a fluid increases, its pressure decreases.

In its simplest form, the equation can be written as:

$p + q = p_0$

where $p$ = static pressure, $q$ = dynamic pressure and $p_0$ is the total pressure.

### How can a couple of kilograms of plutonium have enough energy to wipe out a city?

Einstein's theory of special relativity states that the speed of light is the same for all observers, even if one of them is moving at half the speed of light. It also connects mass and energy in the equation:

$E = mc^2$

where $E$ is the energy, $m$ is the mass and $c$ is the speed of light.

# 15.1 Solving linear equations

This section will show you how to:

- solve linear equations such as $3x - 1 = 11$ where the variable only appears on one side
- use inverse operations and inverse flow diagrams
- solve equations by balancing
- solve equations in which the variable (the letter) appears in the numerator of a fraction.

Key terms

balancing

inverse flow diagrams

solution

A teacher gave these instructions to her class.

> Think of a number.
>
> Double it.
>
> Add 3.

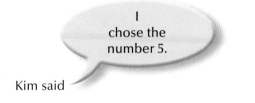

Kim said

> I chose the number 5.

Freda said

> My final answer was 11.

What algebraic expression represents the teacher's statement?

Can you work out Kim's answer and the number that Freda started with?

Kim's answer will be $2 \times 5 + 3 = 13$.

Freda's answer can be found using an equation.

You form an equation by putting one expression equal to a number or another expression.

The **solution** to an equation is the value of the variable that makes the equation true. For example, the equation for Freda's answer is:

$$2x + 3 = 11$$

where $x$ represents Freda's number.

The value of $x$ that makes this true is $x = 4$.

To solve an equation, you have to 'undo' it. That is, you have to reverse the processes that set up the equation in the first place.

Freda did two things. First she multiplied by 2 and then she added 3. The reverse process is *first* to subtract 3 and *then* to divide by 2.

So, to solve:

$$2x + 3 = 11$$

subtract 3 from both sides $\quad 2x + 3 - 3 = 11 - 3$

$$2x = 8$$

divide both sides by 2. $\quad \dfrac{2x}{2} = \dfrac{8}{2}$

$$x = 4$$

You need to know how to set up an equation so that you can undo it in the right order.

You are going to learn three ways to solve equations.

The methods are very similar. You will have to decide which method you prefer, although you should know how to use all three.

Whenever you solve an equation, you should always check that your answer works in the original equation.

For example, to check the answer to Freda's equation, put $x = 4$ into the equation. This gives:

$$2 \times 4 + 3 = 8 + 3$$
$$= 11$$

which is correct.

## Inverse operations

One method of solving equations is to use inverse operations. The opposite or inverse operation to addition is subtraction (and vice versa) and the opposite or inverse operation to multiplication is division (and vice versa).

That means you can 'undo' the four basic operations by using the inverse operation.

Solve these equations.

**a** $w + 7 = 9$      **b** $x - 8 = 10$      **c** $2y = 8$      **d** $\frac{z}{5} = 3$

**a** The opposite operation to $+ 7$ is $- 7$, so the solution is $9 - 7 = 2$.

Check: $2 + 7 = 9$

**b** The opposite operation to $- 8$ is $+ 8$, so the solution is $10 + 8 = 18$.

Check: $18 - 8 = 10$

**c** $2y$ means $2 \times y$. The opposite operation to $\times 2$ is $\div 2$, so the solution is $8 \div 2 = 4$.

Check: $2 \times 4 = 8$

**d** $\frac{z}{5}$ means $z \div 5$. The opposite operation to $\div 5$ is $\times 5$, so the solution is $3 \times 5 = 15$.

Check: $15 \div 5 = 3$

## Exercise 15A

**1** Solve the following equations by applying the inverse of the operation on the left-hand side to the right-hand side.

     **a** $x + 6 = 10$          **b** $w - 5 = 9$          **c** $2x = 10$

     **d** $3x = 18$           **e** $\frac{z}{3} = 8$           **f** $4x = 10$

 **2** The solution to the equation $5x = 20$ is $x = 4$.

Write down two different equations for which the solution is 4.

**3** These nine equations have answers from 1 to 9, once each. Solve the equations.

     **a** $a + 6 = 12$        **b** $4b = 36$         **c** $\frac{c}{7} = 1$

     **d** $d - 2 = 1$         **e** $5e = 5$          **f** $6f = 30$

     **g** $g + 5 = 9$         **h** $h + 11 = 13$     **i** $\frac{i}{4} = 2$

**Hints and tips** Solve the equations to give you a clue and look for similarities.

## Inverse flow diagrams

Another way to solve linear equations is to use **inverse flow diagrams**.

This flow diagram represents the instructions that Kim and Freda were given by their teacher.

The **inverse flow diagram** looks like this.

Running Freda's answer through this gives:

So, Freda started with 4 to get an answer of 11.

**Example 2**

Use an inverse flow diagram to solve the following equation.

$3x - 4 = 11$

Flow diagram

Inverse flow diagram

Put the value on the right-hand side of the equals sign through the inverse flow diagram.

So, the answer is $x = 5$.

Checking the answer gives:

$5 \times 3 - 4 = 11$

which is correct.

## Exercise 15B

 **1**    Use inverse flow diagrams to solve each equation. Remember to check that each answer works for its original equation.

   **a** $3x + 5 = 11$     **b** $3x - 13 = 26$     **c** $3x - 7 = 32$     **d** $4y - 19 = 5$

   **e** $2x - 10 = 8$     **f** $\frac{x}{5} + 2 = 3$     **g** $\frac{t}{3} - 4 = 2$     **h** $\frac{y}{4} + 1 = 7$

   **i** $\frac{k}{2} - 6 = 3$     **j** $\frac{h}{8} - 4 = 1$

**Hints and tips**   Remember the rules of BIDMAS. $3x + 5$ means do $x \times 3$ first then $+ 5$ in the flow diagram. Do the opposite (inverse) operations in the inverse flow diagram.

**2** The diagram shows a two-step number machine.

Work out a value for the input that gives the same value for the output.

 **3** A man buys two apples and gets 46p change from £1.

He wants to know the cost of each apple.

By setting up a flow diagram (or otherwise) work out the cost of one apple.

## Balancing

You can also solve equations by **balancing** or performing the same operation on both sides of the equals sign.

Mary had two bags of marbles, each containing the same number of marbles. She also had five spare marbles.

She put them on scales and balanced them with 17 single marbles.

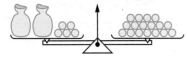

How many marbles were there in each bag?

If $x$ is the number of marbles in each bag, then the equation representing Mary's balanced scales is:

$2x + 5 = 17$

Take five marbles from each pan.

$2x + 5 - 5 = 17 - 5$

$\phantom{2x + 5 - 5 = }2x = 12$

Now halve the number of marbles on each pan.

That is, divide both sides by 2.

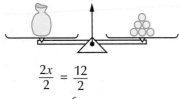

$\frac{2x}{2} = \frac{12}{2}$

$\phantom{\frac{2x}{2} = }x = 6$

Checking the answer gives $2 \times 6 + 5 = 17$, which is correct.

Example 3

Solve these equations by balancing.

**a** $3x - 5 = 16$

**b** $\frac{x}{2} + 2 = 10$

**a** Add 5 to both sides.
$$3x - 5 + 5 = 16 + 5$$
$$3x = 21$$
Divide both sides by 3.
$$\frac{3x}{3} = \frac{21}{3}$$
$$x = 7$$
Checking the answer gives:
$$3 \times 7 - 5 = 16$$
which is correct.

**b** Subtract 2 from both sides.
$$\frac{x}{2} + 2 - 2 = 10 - 2$$
$$\frac{x}{2} = 8$$
Multiply both sides by 2.
$$\frac{x}{2} \times 2 = 8 \times 2$$
$$x = 16$$
Checking the answer gives:
$$16 \div 2 + 2 = 10$$
which is correct.

The solution to an equation may be a negative number. Remember that when you multiply or divide a negative number by a positive number, then the answer is also a negative number. For example:

$$-3 \times 4 = -12 \qquad \text{and} \qquad -10 \div 5 = -2$$

## Exercise 15C

**1** Solve each of the following equations by balancing. Remember to check that each answer works for its original equation.

**a** $x + 4 = 60$    **b** $3y - 2 = 4$    **c** $3x - 7 = 11$

**d** $5y + 3 = 18$    **e** $\frac{w}{3} - 5 = 2$    **f** $\frac{x}{8} + 3 = 12$

**g** $\frac{m}{7} - 3 = 5$    **h** $\frac{x}{5} + 3 = 3$    **i** $\frac{h}{7} + 2 = 1$    **j** $\frac{f}{6} - 2 = 8$

| Hints and tips | Be careful with negative numbers. |

**2** A teacher gave these instructions to her class.

'Think of a number. Divide it by 3 and then subtract 6.'

Mandy says: 'My answer is –1.'

Andy says: 'My starting number is 6.'

**a** What answer did Andy get?

**b** What number did Mandy start with?

**3** The solution of the equation $4x + 17 = 9$ is $x = -2$.

Make up two different equations of the form $ax + b = c$ where $a$, $b$ and $c$ are positive whole numbers, for which the answer is also –2.

**4** A teacher asked her class to solve the equation $2x - 1 = 7$.

Amanda wrote:        Betsy wrote:

$$2x - 1 = 7 \qquad\qquad 2x - 1 = 7$$
$$2x - 1 - 1 = 7 - 1 \qquad 2x - 1 + 1 = 7 + 1$$
$$2x = 6 \qquad\qquad 2x = 8$$
$$2x - 2 = 6 - 2 \qquad 2x \div 2 = 8 \div 2$$
$$x = 4 \qquad\qquad x = 4$$

When the teacher read out the correct answer of 4, both students ticked their work as correct.

**a** Which student used a correct method?

**b** What mistakes did the other student make?

## Fractional equations

To solve equations with fractions you will need to multiply both sides of the equation by the denominator at some stage. It is important to do the inverse operations in the right order.

Sometimes you need to simplify the constant term before multiplying by the denominator of the fraction (as in Example 4). However, if all of the left-hand side is part of the fraction, you need to multiply both sides by the denominator first (as in Example 5).

It is essential to check your answer in the original equation.

**Example 4**

Solve this equation and check your answer.    $\dfrac{x}{3} + 1 = 5$

Subtract 1 from both sides.          $\dfrac{x}{3} = 4$

Now multiply both sides by 3.       $x = 12$

Check in the original equation.      $\dfrac{12}{3} + 1 = 5$

**Example 5**

Adam opened a packet of biscuits and ate two of them before sharing the rest with his four friends. As a result, each of his friends received three biscuits.

How many biscuits were in the packet originally?

Set up the equation.

If there were $x$ biscuits, he took away 2 and then shared $(x - 2)$ biscuits between 5 people.

So                            $\dfrac{x - 2}{5} = 3$

Multiply both sides by 5.        $x - 2 = 15$

Now add 2 to both sides.         $x = 17$

Check in the original equation.    $\dfrac{17 - 2}{5} = \dfrac{15}{5}$
$$= 3$$

There were 17 biscuits in the packet originally.

## Exercise 15D

**1** Solve each of these equations. Remember to check that each answer works for its original equation.

  **a** $\dfrac{k+1}{2} = 3$    **b** $\dfrac{h-4}{8} = 3$    **c** $\dfrac{w+1}{6} = 1$

  **d** $\dfrac{x+5}{4} = 10$    **e** $\dfrac{y-3}{6} = 5$    **f** $\dfrac{f+2}{5} = 5$

**2** A teacher read out the text below to her class.

'I am thinking of a number. I subtract 5 from it and then divide the result by 4.

The answer is 7. What number did I think of to start with?'

  **a** What was the number the teacher thought of?

  **b** Bryn misunderstood the instructions and got the operations the wrong way round. What number did Bryn think the teacher started with?

**3** Solve these equations.

  **a** $\dfrac{3x+10}{2} = 8$    **b** $\dfrac{2x+1}{3} = 5$    **c** $\dfrac{5y-2}{4} = 3$

  **d** $\dfrac{6y+3}{9} = 1$    **e** $\dfrac{2x-3}{5} = 4$    **f** $\dfrac{5t+3}{4} = 1$

**4** The solution to the equation $\dfrac{2x-3}{5} = 3$ is $x = 9$.

Make up two more *different* equations of the form $\dfrac{ax-b}{c} = 3$ for which $x$ is also 9, where $a$, $b$ and $c$ are positive whole numbers.

**5** A teacher asked her class to solve the equation $\dfrac{2x+4}{5} = 6$.

  Anwar wrote:

  $2x + 4 = 6 \times 5$

  $2x + 4 - 4 = 30 - 4$

  $2x = 26$

  $2x \div 2 = 26 \div 2$

  $x = 13$

  Brody wrote:

  $\dfrac{2x}{5} = 6 + 4$

  $2x = 6 + 4 + 5$

  $2x = 15$

  $2x - 2 = 15 - 2$

  $x = 13$

When the teacher read out the correct answer of 13, both students ticked their work as correct.

  **a** Which student used the correct method?

  **b** What mistakes did the other student make?

**6** Five friends went for a meal in a cafe. The bill was £$x$. They added a £10 tip and shared the total cost equally between them.

Each person paid £9.50.

  **a** Set this problem up as an equation.

  **b** Solve the equation and find the cost of the bill before the tip was added.

**7** The mean value of the expressions $(3x + 7)$, $(x - 9)$, $(5x + 11)$ and $(6x - 5)$ is 11.

  **a** Work out the value of $x$.

  **b** Check that your answer is correct. Show all your working.

# 15.2 Solving equations with brackets

When you have an equation that contains brackets, first multiply out the brackets and then solve the equation (as before).

Solve $5(x + 3) = 25$ and check your answer.

$$5(x + 3) = 25$$

First multiply out the brackets.   $5x + 15 = 25$

Subtract 15 from both sides.   $5x = 25 - 15$

$$= 10$$

Divide by 5.   $\dfrac{5x}{5} = \dfrac{10}{5}$

$$x = 2$$

Check your answer.   $5(2 + 3) = 5 \times 5$

$$= 25$$

A trapezium has parallel sides of $(x + 1)$ cm and $(2x - 9)$ cm and a perpendicular height of 6 cm. Its area is 21 cm².

Work out the value of $x$.

Write down the formula for the area of a trapezium.   $A = \dfrac{1}{2}(a + b)h$

Substitute the information.   $21 = \dfrac{1}{2}(x + 1 + 2x - 9) \times 6$

$$21 = \dfrac{1}{2}(3x - 8) \times 6$$

Simplify.   $21 = 3(3x - 8)$

Multiply out the brackets.   $21 = 9x - 24$

Add 24 to both sides.   $45 = 9x$

Divide both sides by 9.   $x = 5$

Check in the original equation.   $\dfrac{1}{2}(5 + 1 + 10 - 9) \times 6 = 21$

## Exercise 15E

**1** Solve each of these equations. Some of the answers may be fractions or negative numbers. Remember to check that each answer works for its original equation. Use your calculator if necessary.

**a** $6(3k + 5) = 39$   **b** $5(2x + 3) = 27$   **c** $9(3x - 5) = 9$

**d** $2(x + 5) = 6$   **e** $5(x - 4) = -25$   **f** $3(t + 7) = 15$

**g** $2(3x + 11) = 10$   **h** $4(5t + 8) = 12$

> **Hints and tips**  When you expand the brackets, remember to multiply *everything* inside the brackets by what is outside.

(MR) **2** Fill in values for $a$, $b$ and $c$ so that the answer to this equation is $x = 4$.

$a(bx + 3) = c$

(PS) **3** The diagram shows a square.

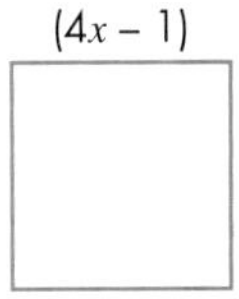

$(4x - 1)$

Work out $x$ if the perimeter is 44 cm.

(PS) **4** Max thought of a number. He then multiplied his number by 3. He added 4 to the answer. He then doubled that answer to get a final value of 38. What number did he start with?

(CM) **5** Show that the answer to this equation is $x = 6$.

$8(x - 7) - 5(x + 4) - (19 - x) + 71 = 0$

(PS) **6** A heptagon has two angles of $(3x - 17)°$ and one angle of $(4x - 36)°$. The remaining angles are all $(2x + 13)°$. Work out the size of the largest angle in the heptagon.

| Hints and tips | A heptagon has seven sides. |

# 15.3 Solving equations with the variable on both sides

## This section will show you how to:

- solve equations where the variable appears on both sides of the equals sign.

When a letter (or variable) appears on both sides of an equation, you can use the balancing method to collect all the terms containing the variable on the left-hand side of the equation.

**Example 8**

Solve the equation $5x + 4 = 3x + 10$.

| Solve the equation | $5x + 4 = 3x + 10$. |
|---|---|
| Subtract $3x$ from both sides. | $2x + 4 = 10$ |
| Subtract 4 from both sides. | $2x = 6$ |
| Divide both sides by 2. | $x = 3$ |

When an equation also contains brackets, you must multiply them out first as shown in the next example.

**Example 9**

Solve the equation $3(2x + 5) + x = 2(2 - x) + 2$.

$$3(2x + 5) + x = 2(2 - x) + 2$$

Multiply out both brackets. $\qquad 6x + 15 + x = 4 - 2x + 2$

Simplify both sides. $\qquad\qquad 7x + 15 = 6 - 2x$

Add $2x$ to both sides. $\qquad\qquad 9x + 15 = 6$

Subtract 15 from both sides. $\qquad\qquad 9x = -9$

Divide both sides by 9. $\qquad\qquad\qquad x = -1$

When there are more of the letters on the right-hand side of the equation, it is easier to turn the equation round first.

**Example 10**

Solve this equation. $\qquad 2x + 3 = 6x - 13$

There are more $x$s on the right-hand side, so turn the equation round. $\quad 6x - 13 = 2x + 3$

Subtract $2x$ from both sides. $\qquad\qquad\qquad\qquad\qquad 4x - 13 = 3$

Add 13 to both sides. $\qquad\qquad\qquad\qquad\qquad\qquad 4x = 16$

Divide both sides by 4. $\qquad\qquad\qquad\qquad\qquad\qquad x = 4$

## Exercise 15F

**1** Solve each of these equations.

   **a** $2x + 3 = x + 5$      **b** $5y + 4 = 3y + 6$      **c** $4a - 3 = 3a + 4$

   **d** $5t + 3 = 2t + 15$     **e** $7p - 5 = 3p + 3$      **f** $6k + 5 = 2k + 1$

   **g** $4m + 1 = m + 10$     **h** $8s - 1 = 6s - 5$

> **Hints and tips** Do the same to both sides. Show all your working. Rearrange *before* you simplify. If you try to rearrange and simplify at the same time, you are more likely to make a mistake.

**2** Terry says: 'I am thinking of a number. I multiply it by 3 and subtract 2.'

June says: 'I am thinking of a number. I multiply it by 2 and add 5.'

Terry and June find that they both thought of the same number and both got the same final answer.

What number did they think of?

> **Hints and tips** Set up expressions; make them equal and solve.

**3** Solve each of these equations.

   **a** $2(d + 3) = d + 12$     **b** $5(x - 2) = 3(x + 4)$     **c** $3(2y + 3) = 5(2y + 1)$

   **d** $3(h - 6) = 2(5 - 2h)$    **e** $4(3b - 1) + 6 = 5(2b + 4)$    **f** $2(5c + 2) - 2c = 3(2c + 3) + 7$

**CM** **4** **a** Why can't the equation $3(2x + 1) = 2(3x + 5)$ be solved?

**b** Why are there an infinite number of solutions to the equation $2(6x + 9) = 3(4x + 6)$?

**PS** **5** Wilson has eight silver coins of the same value and seven pennies.

Chloe has eleven silver coins of the same value as those that Wilson has. She also has five pennies.

Wilson says: 'If you give me one of your silver coins and four pennies, we will have the same amount of money.'

What is the total value of all the silver coins that Wilson and Chloe have?

> **Hints and tips** Call the value of the coin $x$ and set up the expressions, for example, Wilson has $8x + 7$, and then take one $x$ and 4 from Chloe and add one $x$ and 4 to Wilson. Then put the expressions equal and solve.

**PS** **6** Work out the area of the rectangle.

> **Hints and tips** Use the fact that the opposite sides of a rectangle are equal to obtain the equation $5x + 29 = x + 17$.

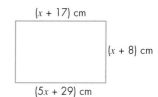

$(x + 17)$ cm

$(x + 8)$ cm

$(5x + 29)$ cm

**MR** **7** The diagram shows two number machines that perform the same operations.

**a** Starting with an input value of 7, work through the left-hand machine to get the output.

**b** Work out an input value that gives the same value for the output.

**c** Write down the algebraic expressions in the right-hand machine for an input of $n$. (The first operation has been filled in for you.)

**d** Set up an equation for the same input and output and show each step in solving the equation to get the answer in part **b**.

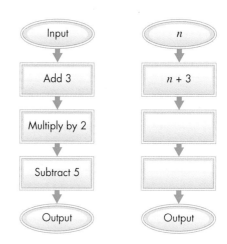

**PS** **8** Mary has a large bottle and a small bottle of cola. The large bottle holds 50 cl more than the small bottle.

From the large bottle she fills four cups and has 18 cl left over.

From the small bottle she fills three cups and has 1 cl left over.

How much cola does each bottle hold?

> **Hints and tips** Set up expressions for both, using $x$ as the amount of cola in a cup. Make them equal but remember to add 50 to the small bottle expression to allow for the difference. Solve for $x$, then work out how much is in each bottle.

# Worked exemplars

  In the table, the letters $a$, $b$, $c$ and $d$ represent different numbers. The total of each row is given at the side of the table.

Work out the values of $a$, $b$, $c$ and $d$.

| $a$ | $a$ | $a$ | $a$ | 32 |
|---|---|---|---|---|
| $a$ | $a$ | $b$ | $b$ | 36 |
| $a$ | $a$ | $b$ | $c$ | 33 |
| $a$ | $b$ | $c$ | $d$ | 31 |

| | |
|---|---|
| This is a problem-solving question so you need to make connections between different parts of mathematics (in this case equations and substitution) and show your strategy clearly. | |
| $4a = 32$ <br> $2a + 2b = 36$ <br> $2a + b + c = 33$ <br> $a + b + c + d = 31$ | You could start by writing out the four equations. |
| $4a = 32$ <br> $a = 8$ | Only one of these equations can be solved immediately ($4a = 32$) because the others all include two or more variables. <br><br> <table><tr><td>8</td><td>8</td><td>8</td><td>8</td><td>32</td></tr><tr><td>8</td><td>8</td><td>$b$</td><td>$b$</td><td>36</td></tr><tr><td>8</td><td>8</td><td>$b$</td><td>$c$</td><td>33</td></tr><tr><td>8</td><td>$b$</td><td>$c$</td><td>$d$</td><td>31</td></tr></table> |
| $2a = 2 \times 8 = 16$ <br> So, $16 + 2b = 36$ <br> $\Rightarrow 2b = 20$ | Now you know that $a = 8$, you can substitute it into the other equations and simplify them. <br><br> It will help to work out $2a$ first. <br><br> You can use the sign $\Rightarrow$ which means 'so' or 'this means that'. |
| $2b = 20$ <br> $b = 10$ | Now you can solve $2b = 20$ and use the answer to simplify the other equations. |
| $16 + b + c = 33$ <br> $b + c = 17$ <br> $b$ is 10, so $10 + c = 17$ <br> $c = 7$ | |
| $8 + b + c + d = 31$ <br> $b$ is 10 and $c$ is 7, so $8 + 10 + 7 + d = 31$ <br> and $d = 6$ | |
| $a = 8$, $b = 10$, $c = 7$, $d = 6$ | Make sure you give a full answer to the question at the end. |

**2** The solution of the equation $5x + 3 = 3x + 7$ is $x = 2$.

Write down a different equation of the form $ax + b = cx + d$ for which the solution is also $x = 2$.

| This is a mathematical reasoning question so you need to demonstrate your mathematical skills to work out one of the many possible solutions. | |
|---|---|
| **Method One**<br><br>$6x + b = 3x + d$ | Start by choosing any values for $a$ and $c$, such as 6 and 3. It is sensible to choose positive integers, although it is acceptable to use negative numbers or fractions. |
| $6 \times 2 + b = 3 \times 2 + d$<br>$12 + b = 6 + d$<br>$6 + b = d$ | Since $x$ must equal 2, work out $6 \times 2$ on the left and $3 \times 2$ on the right and simplify the answer. |
| $6x + 7 = 3x + 13$ | You can then choose *any* values for $b$ and $d$ so that $d$ is 6 more than $b$. Again, it is sensible to choose positive integers, such as 13 and 7, although you could choose negative numbers or fractions. |
| $6x + 7 = 3x + 13$<br>$\quad\quad 6x = 3x + 6$ (Subtract 7 from<br>$\quad\quad\quad\quad\quad\quad\quad$ both sides.)<br>$\quad\quad 3x = 6$ (Subtract $3x$ from<br>$\quad\quad\quad\quad\quad\quad\quad$ both sides.)<br>$\quad\quad x = 2$ (Divide both<br>$\quad\quad\quad\quad\quad\quad\quad$ sides by 3.) | Check your answer works by solving the equation and ensuring that $x = 2$. |
| $6x + 7 = 3x + 13$ | |
| **Method Two**<br><br>If $x = 2$ is a solution:<br>$2a + b = 2c + d$ | Start by substituting $x = 2$ into the equation to find a relationship between $a$, $b$, $c$ and $d$. |
| Let $a = 1$, $b = 2$ and $c = 3$:<br>$2 \times 1 + 2 = 2 \times 3 + d$<br>$\quad\quad 2 + 2 = 6 + d$<br>$\quad\quad\quad\quad 4 = 6 + d$<br>$\quad\quad\quad\quad d = -2$ | Choose values for $a$, $b$ and $c$.<br><br>Solve the equation to find the value of $d$. |
| $x + 2 = 3x - 2$ | Substitute the values into the equation $ax + b = cx + d$. |

# Ready to progress?

I can solve linear equations by using inverse operations, inverse flow diagrams and balancing.

I can solve linear equations containing brackets and fractions.
I can solve linear equations where the variable appears on both sides.
I can set up and solve linear equations from practical and real-life situations.

# Review questions

**1** Solve these equations

  **a** $4x + 9 = 33$        **b** $5x - 13 = 32$        **c** $12x + 11 = 19$

  **d** $3x + 22 = 7$        **e** $\frac{x}{6} - 7 = 5$

**(PS) 2** A boy is $Y$ years old. His father is 25 years older than he is. The sum of their ages is 31. How old is the boy?

**3** Choose the correct answer for the equation $5(x + 2) = 2(x - 4)$.

  $x = -2$      $x = -6$      $x = 2$      $x = 6$

**4** Solve these equations.

  **a** $7x - 4 = 2x - 19$        **b** $6(x + 5) = 12$        **c** $\frac{3x + 17}{4} = 2$

**5** For which of these equations is the answer $x = 9$?

  **a** $2(x - 6) = 26$        **b** $\frac{2x - 1}{3} = 7$        **c** $x + 13 = 3x - 5$

**(PS) 6** A carpet costs £12.75 per square metre.

The shop charges £35 for fitting. The final bill was £137.

How many square metres of carpet were fitted?

**(PS) 7** A rectangular room is 3 m longer than it is wide. The perimeter is 16 m.

Carpet costs £9.00 per square metre.

Show that it will cost just under £125 to carpet the room.

**(PS) 8** A square has sides of $(5x - 6)$ m and $(14 - 3x)$ m.

  **a** Show why $5x - 6 = 14 - 3x$.

  **b** Calculate the area of the square.

**9** Solve the equation $\frac{3(x - 2)}{4} = 2(x - 7)$.

**(PS) 10** A semicircle has a perimeter of 35 cm.

Calculate the diameter of the semicircle.

You may use $\pi = 3$.

> **Hints and tips**   Use the formula for the circumference of a circle, $C = \pi d$.

**11** This diagram shows the traffic flow through a one-way system in a town centre.

Cars enter at A and at each junction the fractions show the proportion of cars that take each route.

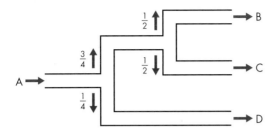

**a** $x$ cars enter at A. How many come out of each of the exits, B, C and D?

**b** If 300 cars exit at B, set up and solve an equation to work out how many cars entered at A.

**c** If 500 cars exit at D, set up and solve an equation to work out how many exit at B.

  A teacher asked her class to work out three angles of a triangle that were consecutive even numbers.

Tammy wrote:
$$x + x + 2 + x + 4 = 180$$
$$3x + 6 = 180$$
$$3x = 174$$
$$x = 58$$

So the angles are 58°, 60° and 62°.

The teacher then asked the class to work out four angles of a quadrilateral that are consecutive even numbers.

Can this be done? Give reasons for your answer.

# 16 Ratio and proportion and rates of change: Percentages and compound measures

## This chapter is going to show you:

- how to convert between fractions, decimals and percentages
- how to use a percentage multiplier
- how to work out percentage increase and decrease
- how to work out one quantity as a percentage of another
- how to calculate compound measures (rates of pay, density, pressure).

## You should already know:

- the multiplication tables up to $12 \times 12$
- how to simplify fractions
- how to multiply and divide, with and without a calculator
- how to substitute values into expressions.

## About this chapter

Percentages are used a lot in everyday life. A test mark may be given as a percentage, even when the test is marked out of 80! Banks lend money, but will charge you a percentage of the loan when you pay back what you borrowed. They also give you a percentage of any money you save with them – but not as much as they charge for a loan! Governments use percentages to explain what is happening in the country. For example, 'unemployment has gone down by 5%' or 'the cost of living has increased by 3% over the last year'. Understanding percentages will help you make the right choices about loans, buying houses and many other aspects of your life in the future.

# 16.1 Equivalent percentages, fractions and decimals

This section will show you how to:

- convert percentages to fractions and decimals and vice versa.

**Key terms**

equivalent

percentage

per cent

**Per cent** means 'out of 100' and so a **percentage** is the same as a fraction with 100 as the denominator. 100% means the *whole* of something.

You can change a percentage to a fraction by writing it as a fraction with denominator 100.

For example:

$32\% = \frac{32}{100}$ and this simplifies (or cancels) to $\frac{8}{25}$.

You can change any percentage to an **equivalent** decimal by dividing the percentage number by 100. To do this, you move the digits two places to the right.

For example:

$65\% = 65 \div 100$

$\qquad = 0.65$

You can convert any decimal to a percentage by multiplying it by 100% (moving the digits two places to the left).

For example:

$0.43 = 0.43 \times 100\%$

$\qquad = 43\%$

To convert a fraction to a percentage, divide the numerator by the denominator and multiply by 100%.

For example:

$\frac{2}{5} = 2 \div 5 \times 100\%$

$\quad = 40\%$

You can convert some fractions to percentages by finding an equivalent fraction with the denominator 100.

Then the numerator is the percentage.

For example:

$\frac{2}{5} = \frac{40}{100}$

$\quad = 40\%$

You may find it useful to learn the percentage and decimal equivalents of some common fractions.

$\frac{1}{2} = 0.5 = 50\%$

$\frac{1}{4} = 0.25 = 25\% \qquad \frac{3}{4} = 0.75 = 75\% \quad \frac{1}{8} = 0.125 = 12.5\%$

$\frac{1}{10} = 0.1 = 10\% \qquad \frac{1}{5} = 0.2 = 20\%$

$\frac{1}{3} = 0.33 = 33\frac{1}{3}\% \qquad \frac{2}{3} = 0.67 = 66\frac{2}{3}\%$

This table shows a summary of the conversions.

| Convert from percentage to: | |
|---|---|
| Decimal | Fraction |
| Divide the percentage by 100; for example:<br><br>$52\% = 52 \div 100$<br><br>$\quad = 0.52$ | Make the percentage into a fraction with a denominator of 100 and simplify if possible; for example:<br><br>$52\% = \frac{52}{100}$<br><br>$\quad\quad = \frac{13}{25}$ |

| Convert from decimal to: | |
|---|---|
| Percentage | Fraction |
| Multiply the decimal by 100%; for example:<br><br>$0.65 = 0.65 \times 100\%$<br><br>$\quad\quad = 65\%$ | If the decimal has 1 decimal place then put it over the denominator 10; if it has 2 decimal places put it over the denominator 100, and so on. Then simplify if possible; for example:<br><br>$0.65 = \frac{65}{100}$<br><br>$\quad\quad = \frac{13}{20}$ |

| Convert from fraction to: | |
|---|---|
| Percentage | Decimal |
| Divide the numerator by the denominator and multiply by 100%; for example:<br><br>$\frac{7}{8} = 7 \div 8$<br><br>$\quad = 0.875$<br><br>$\quad = 87.5\%$<br><br>If the denominator is a factor of 100, multiply the numerator and denominator by a number that will make the denominator 100. Then the numerator is the percentage; for example:<br><br>$\frac{3}{20} = \frac{15}{100}$<br><br>$\quad\quad = 15\%$ | Divide the numerator by the denominator; for example:<br><br>$\frac{9}{40} = 9 \div 40$<br><br>$\quad\quad = 0.225$ |

**Example 1**

Convert each number to a decimal.

**a** 78%  **b** 35%  **c** $\frac{3}{25}$  **d** $\frac{7}{40}$

**a** $78\% = 78 \div 100$  **b** $35\% = 35 \div 100$
$\quad\quad = 0.78$  $\quad\quad\quad = 0.35$

**c** $\frac{3}{25} = 3 \div 25$  **d** $\frac{7}{40} = 7 \div 40$
$\quad\quad = 0.12$  $\quad\quad = 0.175$

Convert each number to a percentage.

**a** 0.85      **b** 0.125      **c** $\dfrac{7}{20}$      **d** $\dfrac{3}{8}$

**a** $0.85 = 0.85 \times 100\%$
$\quad\quad = 85\%$

**b** $0.125 = 0.125 \times 100\%$
$\quad\quad\quad = 12.5\%$

**c** $\dfrac{7}{20} = \dfrac{35}{100}$
$\quad\quad = 35\%$

**d** $\dfrac{3}{8} = 3 \div 8 \times 100\%$
$\quad\quad = 0.375 \times 100\%$
$\quad\quad = 37.5\%$

Convert each number to a fraction in its simplest form.

**a** 0.45      **b** 0.4      **c** 32%      **d** 15%

**a** $0.45 = \dfrac{45}{100}$
$\quad\quad = \dfrac{9}{20}$

**b** $0.4 = \dfrac{4}{10}$
$\quad\quad = \dfrac{2}{5}$

**c** $32\% = \dfrac{32}{100}$
$\quad\quad = \dfrac{8}{25}$

**d** $15\% = \dfrac{15}{100}$
$\quad\quad = \dfrac{3}{20}$

# Exercise 16A

**1** Write each percentage as a fraction in its simplest form.

    **a** 8%      **b** 35%      **c** 90%

**2** Write each percentage as a decimal.

    **a** 27%      **b** 85%      **c** 13%

**3** Write each decimal as a fraction in its simplest form.

    **a** 0.12      **b** 0.4      **c** 0.45

**4** Write each decimal as a percentage.

    **a** 0.29      **b** 0.6      **c** 1.25

**5** Write each fraction as a percentage.

    **a** $\dfrac{7}{25}$      **b** $\dfrac{3}{10}$      **c** $\dfrac{19}{20}$

    **d** $\dfrac{17}{50}$      **e** $\dfrac{11}{40}$      **f** $\dfrac{7}{8}$

**6** Write each fraction as a decimal.

    **a** $\dfrac{9}{15}$      **b** $\dfrac{3}{40}$      **c** $\dfrac{19}{25}$

    **d** $\dfrac{5}{16}$      **e** $\dfrac{1}{20}$      **f** $\dfrac{1}{8}$

**(MR)** **7** Gillian told her dad that she got 100% of her spellings correct. She told her mum that there were 25 spellings to learn.

How many spellings did Gillian get wrong?

**8** **a** If 23% of students go home for lunch, what percentage do not go home for lunch?

**b** If 61% of the population buy tickets for the National Lottery, what percentage do not buy tickets for the National Lottery?

**c** If 37% of the adults who use a gym are men, what percentage are women?

**9** I calculated that I spend 28% of my time sleeping and 45% working.

What percentage of my time do I have left for everything else?

**(MR)** **10** Approximately what percentage of each bottle is filled with liquid?

a   b   c

**(PS)** **11** Helen made a cake. James ate it. The diagram shows the amount of cake left at the end of each day.

Monday    Tuesday    Wednesday    Thursday    Friday

**a** What percentage of the cake is left at the end of each day?

**b** What percentage did he eat on Wednesday?

**(CM)** **12** In a mathematics test, Chris got 24 marks out of a possible 40.

**a** Write this as a fraction in its simplest form.

**b** Write this as a decimal.

**c** Write this as a percentage.

Chris's dad asked him how he got on in his test.

**d** Which of the values in **a**, **b** and **c** would his dad find the easiest to understand?

**13** Evie's end-of-year test scores are given in the table below.

**a** Copy and complete the table.

Give the percentages to the nearest whole number.

| Subject | Result | Percentage |
|---|---|---|
| Mathematics | 38 out of 60 | |
| English | 29 out of 35 | |
| Science | 27 out of 70 | |
| History | 56 out of 90 | |
| Technology | 58 out of 75 | |

**b** All the tests were of the same academic standard.

Which was Evie's best result?

**14** On Friday, two students were absent from my class of 20. What percentage of my class was this?

**15** Copy and complete the table to show the equivalent values.

| Percentage | Decimal | Fraction |
|---|---|---|
| 34% | | |
| | 0.85 | |
| | | $\dfrac{3}{40}$ |
| 45% | | |
| | 0.3 | |
| | | $\dfrac{2}{3}$ |
| 84% | | |
| | 0.45 | |
| | | $\dfrac{3}{8}$ |

**16** The manager of a garage wants to order 27 000 litres of fuel. A full fuel tanker holds 30 000 litres. What fraction or percentage of a full tanker load should he order?

# 16.2 Calculating a percentage of a quantity

This section will show you how to:

• calculate a percentage of a quantity.

To calculate a percentage of a **quantity**, first express the percentage as an equivalent fraction or a decimal and then multiply this by the quantity.

The next examples show you how to calculate a percentage without using a calculator by first finding the value of 10% or 1%.

Example 4

Calculate these quantities.

**a** 10% of 54 kg      **b** 15% of 54 kg

**a** 10% = $\dfrac{1}{10}$, so:     10% of 54 kg = $\dfrac{1}{10}$ of 54 kg

$$= 54 \text{ kg} \div 10$$

$$= 5.4 \text{ kg}$$

**b** 15% = 10% + 5%

From part **a**,      10% = 5.4 kg

so          5% = 2.7 kg

and        15% = 5.4 + 2.7

$$= 8.1 \text{ kg}$$

**Calculate 12% of £80.**

10% of £80 is £8 and 1% of £80 is £0.80.

12% = 10% + 1% + 1%

$\quad$ = £8 + £0.80 + £0.80

$\quad$ = £9.60

## Using a percentage multiplier

To work out percentage calculations on a calculator, express the percentage as a decimal. Then use this to complete the calculation.

For example:

- 13% is the same as 0.13
- 20% is the same as 0.2 (or 0.20).

These are called percentage multipliers.

**Calculate these quantities.**　　**a** 45% of 160 cm　　**b** 52% of £460

**a** 45% = 0.45

To calculate 45% of 160, key in:

The display shows 72, so 45% of 160 cm is 72 cm.

**b** 52% = 0.52

To calculate 52% of 460, key in:

The display shows 239.2, so 52% of £460 is £239.20.

Remember to write amounts of money with 2 decimal places.

## Exercise 16B

**1** Write down the multiplier for each percentage.

$\quad$ **a** 88%　　　　**b** 30%　　　　**c** 25%

$\quad$ **d** 8%　　　　　**e** 115%

**2** Write down the percentage that is equivalent to each multiplier.

$\quad$ **a** 0.78　　　　**b** 0.4　　　　**c** 0.75

$\quad$ **d** 0.05　　　　**e** 1.1

**3** Calculate these quantities.

$\quad$ **a** 15% of £300　　　　**b** 6% of £105　　　　**c** 23% of 560 kg

$\quad$ **d** 45% of 2.5 kg　　　**e** 12% of 9 hours　　　**f** 21% of 180 cm

$\quad$ **g** 4% of £3　　　　　**h** 35% of 8.4 m　　　　**i** 95% of £8

$\quad$ **j** 11% of 308 minutes　**k** 20% of 680 kg　　　**l** 45% of £360

(MR) **4** There are 1200 students at Bradbury School. The canteen manager estimates that 40% of students will buy school lunches. Her estimate is accurate to within 2%.

    **a** How many lunches should she cook each day to be sure that she has enough?

    **b** What is the greatest possible number of lunches she will have left over?

**5** An estate agent charges 2% commission when he sells a house. He sells a house for £120 500. How much commission will he earn?

**6** A department store has 250 employees. During a flu epidemic, 14% of the store's employees were absent.

    **a** What percentage of the employees were at work?

    **b** How many employees were at work?

**7** There were 42 600 fans at a rugby match at Twickenham. 20% of them were women. How many of the fans were women?

**8** 350 trains arrived at St Pancras railway station on one day. Of these trains, 5% arrived early and 13% arrived late. How many trains arrived on time?

(PS) **9** 90 000 tickets for an FA Cup final were allocated as follows.

The two teams playing were each given 30% of the tickets.

10% of the tickets were shared among other football teams.

The Football Association was given 20% of the tickets.

The Referees' Association was given 1% of the tickets.

The rest of the tickets were given to celebrities.

Approximately how many celebrities were given tickets for the FA Cup final?

(PS) **10** There are 1500 students at a school. The headteacher estimates that 60% of them will come to a special event. The caretaker is told to put out one seat for each student expected to attend, plus an extra 10% in case more attend. How many seats should she put out?

> Hints and tips    It is not 70% of the number of students in the school.

**11** The attendance record at a school in the week before Christmas was:

Monday 96%    Tuesday 98%    Wednesday 100%    Thursday 94%    Friday 88%

The school has 850 students. Calculate the number of students at school on each of these days.

**12** Calculate each quantity.

    **a** 12.5% of £26    **b** 6.5% of 34 kg    **c** 26.8% of £2100

    **d** 7.75% of £84    **e** 16.2% of 265 m    **f** 0.8% of £3000

**13** Air consists of 80% nitrogen and 20% oxygen (by volume). A man's lungs have a capacity of 600 cm$^3$. How much of each gas will he have in his lungs when he has just taken a deep breath?

**14** A factory estimates that 1.5% of all the garments it produces will have faults. The factory produces 850 garments each week. How many of these are likely to have a fault?

**15** The annual premium for Marie's house insurance is 0.3% of the value of her house. Her house is valued at £90 000. What will be the annual premium?

 **16** Average prices in a shop increased by 3% in 2013 and a further 3% in 2014. Was the actual average price increase in 2014 more, less or the same as in 2013?

Explain how you decided.

# 16.3 Increasing and decreasing quantities by a percentage

This section will show you how to:

- increase and decrease quantities by a percentage.

During a sale, shops often use percentages to describe how much prices have fallen. When newspapers describe increases in travel fares or wages, they often use percentages to describe how much prices or salaries have risen. This allows you to compare the new amount with the old.

### Increasing by a percentage

There are two methods for increasing a quantity by a percentage. The first is to work out the increase and add it to the original amount, as shown in the next example.

| Example 7 | | |
|---|---|---|
| Increase £6 by 5%. | | |
| Work out 5% of £6. | $6 \times 5 \div 100 = £0.30$ | |
| Add £0.30 to the original amount. | $£6 + £0.30 = £6.30$ | |

The second is to use a multiplier. For example, an increase of 6% is equivalent to the original 100% plus the extra 6%. This is a total of 106%, or $\frac{106}{100}$, and is equivalent to the multiplier 1.06.

| Example 8 | |
|---|---|
| Increase £6.80 by 5%. | |
| Find the multiplier. | A 5% increase gives a multiplier of 1.05. |
| Multiply. | $£6.80 \times 1.05 = £7.14$ |

### Decreasing by a percentage

There are also two methods for decreasing by a percentage. The first is to work out the decrease and subtract it from the original amount.

| Example 9 | |
|---|---|
| Decrease £8 by 4%. | |
| Work out 4% of £8. | $(4 \div 100) \times 8 = £0.32$ |
| Subtract it from the original amount. | $£8 - £0.32 = £7.68$ |

The second method is to use a multiplier. For example, a 7% decrease is equivalent to 7% less than the original 100%, so it represents 100% – 7% = 93% of the original. This is a multiplier of 0.93.

Decrease £8.60 by 5%.

Find the multiplier.    A decrease of 5% gives a multiplier of 0.95.

Multiply.    £8.60 × 0.95 = £8.17

## Exercise 16C 🖩

**1** Work out the multiplier you would use to increase a quantity by each percentage.

   **a** 10%           **b** 3%          **c** 20%

   **d** 7%           **e** 12%

**2** Work out the multiplier you would use to decrease a quantity by each percentage.

   **a** 8%           **b** 15%        **c** 25%

   **d** 9%           **e** 12%

**3** Increase each amount by the given percentage.

   **a** 340 kg by 15%      **b** 670 cm by 23%      **c** 130 g by 95%

   **d** £82 by 75%        **e** 640 m by 15%       **f** £28 by 8%

**4** Decrease each amount by the given percentage.

   **a** 860 m by 15%       **b** 96 g by 13%        **c** 480 cm by 25%

   **d** 180 minutes by 35%  **e** 86 kg by 5%       **f** £65 by 42%

**5** A large factory employed 640 workers. Then it reduced the number of workers by 30%. How many workers were employed after the reduction?

**CM** **6** Kerry wants to buy a sweatshirt (£19), a tracksuit (£26) and some running shoes (£56) from the same shop. It costs £25 to join the shop's premium club. If she joins the club, she can get 20% off the cost of these goods.

   Should she join the club? Show your working to support your answer.

**CM** **7** Kevin earns £27 500 each year. He is offered a pay rise of 7% or an extra £150 per month. Which should he accept? Give calculations to support your answer.

**8** VAT was 17.5% in 2010. It was increased to 20% in January 2011.

   **a** A TV cost £245 before VAT was added. How much more expensive was the TV after the VAT increase?

**EV**    **b** An item costs £$x$ before VAT is added. Which one of these calculations gives the increase in the total cost of the item as a result of the VAT increase?

     $x \times 1.175 \div 1.2$    $x \times 1.2 \div 1.175$    $x \times 0.025$    $x \div 0.025$

**EV**    **c** An item costs £$y$ including 17.5% VAT. Which of these calculations would give the cost of the item after the rate of VAT increases to 20%?

     $y \div 1.175 \times 1.2$    $y \div 1.2 \times 1.175$    $y \times 0.025$    $y \div 0.025$

**9** A cereal packet normally contains 300 g of cereal and costs £1.40.

There are two special offers.

Offer A: 20% more for the same price

Offer B: Same amount for 20% off the normal price

Which is the better offer?

**a** Offer A      **b** Offer B      **c** Both the same      **d** Cannot tell

Give reasons for your choice.

(CM) **10** BookWorms increased its prices by 5% and then increased them by 3%. Books Galore increased its prices by 3% and then increased them by 5%.

Which shop's prices increased by the greater percentage?

**a** BookWorms      **b** Books Galore      **c** Both the same      **d** Cannot tell

Show your working.

(CM) **11** Shop A increased its prices by 4% and then by another 4%. Shop B increased its prices by 8%.

Which shop's prices increased by the greater percentage?

**a** Shop A      **b** Shop B      **c** Both the same      **d** Cannot tell

Give reasons for your answer.

**12** A computer cost £450 at the start of 2013. At the start of 2014 the price was increased by 5%. At the start of 2015 the price was decreased by 10%. How much did the computer cost at the start of 2015?

(CM) **13** Show that a 10% decrease followed by a 10% increase gives the same result as a 1% decrease.

> **Hints and tips**   Choose a starting amount and work through the calculations.

# 16.4 Expressing one quantity as a percentage of another

This section will show you how to:

- express one quantity as a percentage of another
- work out percentage change.

Follow these steps to express one quantity as a percentage of another.

**Step 1:** Make sure that both quantities are written in the same units.

**Step 2:** Set up the first quantity as a fraction of the second.

**Step 3:** Convert the fraction into a percentage by multiplying by 100.

| Key terms |
|-----------|
| percentage change |
| percentage decrease |
| percentage increase |
| percentage loss |
| percentage profit |

Express £6 as a percentage of £40.

Set up the fraction and multiply by 100. $\frac{6}{40} \times 100 = 15\%$

Express 75 cm as a percentage of 2.5 m.

First, change both quantities to the same units. 2.5 m = 250 cm

Now express 75 cm as a percentage of 250 cm.

Set up the fraction and multiply by 100. $\frac{75}{250} \times 100 = 30\%$

## Percentage change

A **percentage change** may be a **percentage increase** or a **percentage decrease**.

$$\text{Percentage change} = \frac{\text{change}}{\text{original amount}} \times 100$$

You can use this to calculate **percentage profit** or **percentage loss** in a financial transaction.

Jake buys a car for £1500 and sells it for £1800. What is Jake's percentage profit?

Jake's profit is £300.

His percentage profit is given by: $\frac{\text{profit}}{\text{original amount}} \times 100$

$= \frac{300}{1500} \times 100$

$= 20\%$

## Using a multiplier (or decimal)

To use a multiplier, divide the increase by the original quantity. Then change the resulting decimal to a percentage.

Express 5 as a percentage of 40.

Set up the fraction or decimal.       $5 \div 40 = 0.125$

Convert the decimal to a percentage.   $0.125 = 12.5\%$

# Exercise 16D 🖩

**1** Express each fraction as a percentage. Round your answers if necessary.

   **a** £5 of £20          **b** £4 of £6.60         **c** 241 kg of 520 kg

   **d** 3 hours of 1 day      **e** 25 minutes of 1 hour    **f** 12 m of 20 m

   **g** 125 g of 600 g        **h** 12 minutes of 2 hours   **i** 1 week of a year

   **j** 1 month of 1 year       **k** 25 cm of 55 cm        **l** 105 g of 1 kg

**2** Liam gets £2.50 pocket money. He spends 80p at the tuck shop. What percentage of his pocket money was this?

**3** In Greece, there are 3 654 000 acres of farm land. Olives are grown on 237 000 acres of this land. What percentage of the farm land is used for olives?

**4** In 1981 it rained in Manchester on 123 days of the year. What percentage of the days of the year was this?

**5** Find the percentage profit on each item. Give your answers to one decimal place.

| | Item | Wholesale price paid by the shop (£) | Retail or selling price (£) |
|---|---|---|---|
| **a** | CD player | 60 | 89.50 |
| **b** | TV set | 210 | 345.50 |
| **c** | Computer | 750 | 829.50 |

**6** In 2012 Melchester County Council raised £14 870 000 in council tax. In 2013 it raised £15 597 000 in council tax. What was the percentage increase?

**7** When Blackburn Rovers won the championship in 1995, they lost only 4 of their 42 league games. What percentage of games did they *not* lose?

**8** These are the results from two tests taken by Calum and Stacey. Both tests are out of the same mark.

| | Test A | Test B |
|---|---|---|
| **Calum** | 12 | 17 |
| **Stacey** | 14 | 20 |

Whose result has the greater percentage increase from test A to test B?

Show your working.

**(CM) 9** This is how a supermarket advertises its own brand of cat food.

Trading standards investigate the claim.

They observe that, over one hour, 46 people buy cat food and 38 of these buy the store's own brand.

Does this data support the store's claim?

> 8 out of 10 cat owners choose our cat food.

**(CM) 10** $x$, $y$ and $z$ are three quantities. $x$ is 60% of $y$ and $y$ is 75% of $z$. What percentage is $x$ of $z$?

> Hint and tips    Choose a value for $z$.

**11**   $x$, $y$ and $z$ are three quantities. $x$ is 75% of $y$ and $x$ is 60% of $z$. What percentage is $y$ of $z$?

**12**   In 2000 the population of a town was 4800 and 30% of this population owned a mobile phone. In 2015 the number of people in the town had increased by 20% and 70% of them owned a mobile phone. By what percentage has the number of people in the town owning a mobile phone increased?

 **13**   This letter appeared in a newspaper. Comment on the figures in the letter.

> Dear Sir,
> In your last edition you said that the vote for Amir Patel was 5% greater than the vote for John Smith. The relevant percentages of the total vote were 31% and 26% respectively. A 31% share of the vote is in fact 19% greater than a 26% share. Did you mean that Amir Patel's vote was 5% more than John Smith's?

# 16.5 Compound measures

This section will show you how to:

- recognise and solve problems involving the compound measures of rates of pay, density and pressure.

**Key terms**

compound measure

density

pressure

**Compound measures** always involve three variables. The three variables can be connected by a 'triangle' that shows the relationship between them. You have already met the triangle that shows the relationship between speed, distance and time.

## Rates of pay

Many jobs are paid on an hourly rate. The amount earned is calculated by the rule:

pay ($P$) = hours worked ($H$) × hourly rate ($R$)

These three variables are connected by this triangle.

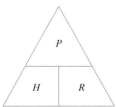

It shows the three relationships between the three variables.

$$P = H \times R \qquad H = \frac{P}{R} \qquad R = \frac{P}{H}$$

**Example 15**

Vikki works 42 hours one week. Her hourly rate of pay is £11.20.

From her weekly earnings, 20% is deducted as tax, 8% is deducted as National Insurance (NI) and £3.50 is deducted for union fees.

How much does Vikki take home after deductions?

| | |
|---|---|
| Calculate her pay. | $42 \times 11.20 = £470.40$ |
| Calculate her tax. | $0.2 \times 470.40 = £94.08$ |
| Calculate her NI. | $0.08 \times 470.40 = £37.63$ |
| Find the total of her deductions. | $£94.08 + £37.63 + £3.50 = £135.21$ |
| Subtract to find her take-home pay. | $£470.40 - £135.21 = £335.19$ |

## Exercise 16E

**1** Work out the total pay for each person.

   **a** 40 hours at £6.50 per hour       **b** $37\frac{1}{2}$ hours at £8.20 per hour

   **c** 35 hours at £9.25 per hour       **d** $42\frac{1}{2}$ hours at £6.80 per hour

**2** Work out the hourly rate for each payment.

   **a** £300 for 40 hours' work         **b** £380.10 for 42 hours' work

   **c** £217.50 for $37\frac{1}{2}$ hours' work     **d** £268.75 for 25 hours' work

**3** Work out the number of hours worked for each job.

   **a** £321.10 at £8.45 per hour      **b** £390.10 at £9.40 per hour

   **c** £211.75 at £6.05 per hour      **d** £502 at £12.55 per hour

**(CM) 4** Mary sees two job adverts for a cook.

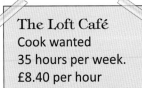

The Loft Café
Cook wanted
35 hours per week.
£8.40 per hour

Café Creme
Cook wanted
38 hours per week.
£8.15 per hour

   **a** Give a reason why Mary may think the job at The Loft Café sounds better.

   **b** Give a reason why Mary may think the job at Café Creme sounds better.

**5** Adele is a joiner. Her normal working week is from 8 until 5 Monday to Friday with a 1-hour lunch break. Her hourly rate is £13.50.

She is paid 'time and a half' if she works at weekends. This means she gets one and a half times her normal hourly rate.

   **a** How much does Adele earn for a normal working week?

   **b** One week she works for 8 hours at the weekend. How much does she earn that week?

**6** Sasha works for 35 hours at her normal hourly rate and 6 hours at 'time and a half'. She earns a total of £303.60. What is her hourly rate of pay?

**7** One week Bernice works her normal hours and 8 hours at 'time and a half'. She is paid £375. The next week she works her normal hours and 4 hours at 'time and a half'. She is paid £330. How many hours is her normal week? Show your working.

(PS) **8** Bill works 40 hours a week at an hourly rate of £$x$. Ben works 32 hours a week. They both get exactly the same weekly pay. What is Ben's hourly rate? Give your answer in terms of $x$.

(PS) **9** Steve works for $37\frac{1}{2}$ hours at an hourly rate of £11.80. He pays 20% of his pay in income tax. He also pays National Insurance at a rate of $x$%. This leaves him with £327.45. What is the value of $x$? Show your working.

(PS) **10** Antoine works for a whole number of hours and is paid a whole number of pounds for each hour. He earns £407. He works more than one hour a week and the number of hours is higher than the hourly rate. Work out how many hours he works and his hourly rate.

## Density

**Density** is a compound measure that combines mass and volume. It is usually written in grams per cubic centimetre (g/cm³). For example, if a metal has a density of 7.5 g/cm³ then 1 cm³ of the metal weighs 7.5 g. The relationship between the three quantities is:

$$\text{density} = \frac{\text{mass}}{\text{volume}}$$

This triangle shows the relationships between mass, density and volume.

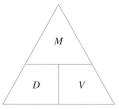

$M = DV$    mass = density × volume

$D = \dfrac{M}{V}$    density = mass ÷ volume

$V = \dfrac{M}{D}$    volume = mass ÷ density

**Note:** Density is defined in terms of mass and volume. The common metric units for mass are grams and kilograms. Common units for volume are cubic metres (m³) and cubic centimetres (cm³).

**Example 16**

A piece of metal has a mass of 30 g and a volume of 4 cm³. What is the density of the metal?

$$\text{density} = \frac{\text{mass}}{\text{volume}}$$

$$= \frac{30}{4}$$

$$= 7.5 \text{ g/cm}^3$$

**Example 17**

What is the mass of a piece of rock that has a volume of 34 cm³ and a density of 2.25 g/cm³?

mass = density × volume

$$= 2.25 \times 34$$

$$= 76.5 \text{ g}$$

## Pressure

**Pressure** is the force per unit area. It is expressed in newtons (N) per square metre ($m^2$). A force of 1 N applied to 1 $m^2$ is 1 pascal (Pa). Other units to measure pressure are pounds per square inch and bars.

The relationship between pressure ($P$), force ($F$) and area ($A$) is pressure $= \dfrac{\text{force}}{\text{area}}$.

This triangle shows the relationships between pressure, force and area.

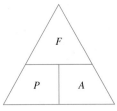

$P = \dfrac{F}{A}$    pressure = force ÷ area

$F = PA$    force = pressure × area

$A = \dfrac{F}{P}$    area = force ÷ pressure

As an example of pressure, think about pushing your thumb onto a piece of wood. Not much happens. Now use your thumb to push a drawing pin onto the wood. The drawing pin will go into the wood. The force you are applying is the same but the area it is applied to is much smaller. Therefore, the pressure is much greater.

The downward force exerted by an object is the result of its mass and the effect of gravity. On Earth, an object with a mass of $x$ kg exerts a downward force of $xg$ newtons, where $g$ is the acceleration due to gravity. $g = 9.81$ $m/s^2$, so a mass of 1 kg exerts a force of 9.81 N. To make calculations easier, $g$ is often rounded to 10 $m/s^2$.

**Example 18**

**a**  When does a woman exert the greater pressure on the floor: when she is wearing walking boots or high-heeled shoes? Explain your answer.

**b**  A woman has a mass of 55 kg. She is wearing a pair of high-heeled shoes. Each shoe has an area of 40 $cm^2$ for the sole and 1 $cm^2$ for the heel. Take $g = 10$ $m/s^2$.

    **i**  When she is standing on both shoes with the heel down, what is the average pressure exerted on the ground?

    **ii**  She swivels around on the heel of one shoe only. How much pressure, in pascals, is exerted on the ground?

Remember that 1 $cm^2$ = 0.0001 $m^2$.

**a**  A woman exerts more pressure on the floor when she is wearing high-heeled shoes, as they have a much smaller contact area with the floor than walking boots do.

**b**  **i**  Force is 55 × 10 = 550 N

    Area is 82 × 0.0001 $m^2$ = 0.0082 $m^2$  This is is the area of two soles and two heels, 2 × (40 + 1) $cm^2$.

    Pressure = force ÷ area

    So average pressure is 550 ÷ 0.0082 ≈ 67 000 Pa.

    **ii**  On one heel, the pressure is 550 ÷ 0.0001 = 5 500 000 Pa.

# Exercise 16F

**1** A piece of wood has a mass of 6 g and a volume of 8 cm³. Calculate its density.

**2** A force of 20 N acts over an area of 5 m². What is the pressure?

**3** A piece of metal 12 cm³ has a mass of 100 g. Calculate its density.

**4** A pressure of 5 Pa acts on an area of $\frac{1}{2}$ m². What force is exerted?

**5** A piece of plastic, 20 cm³ in volume, has a density of 1.6 g/cm³. Calculate its mass.

**6** A crate weighs 200 N and exerts a pressure of 40 Pa on the ground. Calculate the area of the base of the crate.

**7** Calculate the volume of a piece of wood with a mass of 102 g and a density of 0.85 g/cm³.

**8** The density of marble is 2.8 g/cm³. Find the mass of a marble model with a volume of 56 cm³.

**(CM)** **9** A steel block is in the shape of a cuboid, 30 cm by 20 cm by 10 cm. On which face should it be stood to exert least pressure?

**10** Which of these will exert a greater pressure: carrying a full shopping bag by the handles or carrying the same bag in your arms? Explain your answer.

**(PS)** **11** Why do camels have large, wide feet?

**12** A gold bar is in the shape of a cuboid with dimensions 5 cm by 10 cm by 20 cm. The density of gold is 19.3 g/cm³. The bar is placed on a weighing scale on its largest face.

    **a** What figure will show on the scale?

**(CM)**     **b** The bar is now placed on the scale on its smallest face. What figure will now show on the scale? Explain your answer.

    **c** Work out the pressure exerted on the scale in **a** and **b**. Take $g = 10$ m/s².

**(MR)** **13** Two statues look identical and both appear to be made out of gold but one of them is a fake.

Each statue has a volume of approximately 200 cm³.

The density of gold is 19.3 g/cm³.

The first statue has a mass of 5.2 kg.

The second statue has a mass of 3.8 kg.

Which one is the fake?

**14** A piece of metal has a mass of 345 g and a volume of 15 cm³.

A different piece of metal has a mass of 400 g and a density of 25 g/cm³.

Which piece of metal has the bigger volume and by how much?

**(MR)** **15** When does a man exert more pressure, when he is wearing shoes or when he is wearing slippers? Give a reason for your answer.

**16** Give two reasons why skis help people move quickly over snow.

 **17** If you are in the middle of an iced-up pond and the ice starts to crack, which is the safer thing to do: walk quickly to the side of the pond or lay down and slide to the side of the pond? Give a reason for your answer.

 **18** Large trucks in the US are called 'eighteen-wheelers'. Why do they have 18 wheels?

 **19** These containers are standing on a table. They all have the same mass.

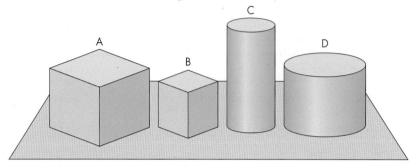

Write down the letters, in order, starting with the container that exerts least pressure to the one that exerts most pressure.

**20** A cylinder and a cone have the same mass, same height and the same base radius.

   **a** Which exerts more pressure on the ground when they stand on their bases? Give a reason for your answer.

   **b** What would the answer be if they were turned upside down? Give a reason for your answer.

 **21** Two pieces of scrap metal are melted down to make a single piece of metal.

The first piece has a mass of 1.5 tonnes and a density of 7000 kg/m³.

The second piece has a mass of 1 tonne and a density of 8000 kg/m³.

Work out the total volume of the new piece of metal.

**22** A cuboid has a mass of 20 kg and a volume of 0.4 m³.

When placed on each of its three different faces in turn, the pressures exerted are 400 Pa, 250 Pa and 500 Pa.

Take $g = 10$ m/s². Work out the dimensions of the cuboid.

**23** The acceleration due to gravity on the Moon is about $\frac{1}{6}$ of what it is on Earth.
Say whether each of these statements is true or false.

   **a** Buzz's mass is the same on Earth and the Moon.

   **b** Buzz's weight is the same on Earth and the Moon.

   **c** The pressure exerted by Buzz's feet when standing in his spacesuit is the same on Earth as on the Moon.

# Worked exemplars

 **1** Laura buys a house for £240 000.

She pays a deposit of 20%.

She takes out a 25-year mortgage for the rest of the cost.

She will have to pay 0.525% of the mortgage each month.

How much more than £240 000 will the house eventually cost?

| This is a problem-solving question where you need to translate a real-life problem into a series of mathematical steps. Show your working clearly and use words to explain what you are calculating. | |
|---|---|
| Deposit = 0.2 × £240 000 <br> = £48 000 <br> Mortgage = £240 000 – £48 000 <br> = £192 000 | Work out how much the mortgage will be. |
| Monthly cost = £192 000 × 0.525 ÷ 100 <br> = £1008 | Work out the monthly cost of the mortgage. |
| Total mortgage repayments <br> = 25 × 12 × £1008 <br> = £302 400 | Work out the total of the mortgage repayments. |
| Total amount paid = £302 400 + £48 000 <br> = £350 400 | Work out the total amount she pays. |
| Extra money paid = £350 400 – £240 000 <br> = £110 400 | Subtract the cost of the house from the total amount she paid. <br><br> You could reduce the steps in your calculation by subtracting the initial loan amount from the total mortgage repayments: 302 400 – 192 000 = £110 400. |

  **2** Two cubes have the same mass.

Cube A has a side of 2 cm and is made from a material with a density of 34 g/cm$^3$.

Cube B has a side of 3 cm.

Work out the density of the material in cube B.

Give your answer to the nearest gram per cubic centimetre.

| In this question you need to interpret and communicate information accurately. | |
|---|---|
| Mass of cube A: $2^3$ × 34 = 272 g | Work out the mass of cube A. <br><br> Remember, the volume of a cube is the cube of the length of the side, which is $2^3$ in this case. |
| Density of cube B: 272 ÷ $3^3$ = 10.074 g/cm$^3$ | Work out the density of Cube B. <br><br> Divide the mass by the volume. |
| Density of cube B is 10 g/cm$^3$. | Round to the nearest g/cm$^3$. |

# Ready to progress?

I can recognise and calculate equivalent fractions, decimals and percentages.
I can calculate a percentage of a quantity.

I can calculate percentage increases and decreases.
I can use percentage multipliers to carry out percentage calculations.
I can solve simple problems involving compound measures.

I can solve more complex problems involving compound measures, such as density and pressure.

# Review questions

**1** A bill is £120 plus VAT. VAT is charged at 20%. What is the total bill?

**2** The price of a fridge is reduced by 15% in a sale. The original cost of the fridge was £245. What is the sale price?

**3** Bonny works $7\frac{1}{2}$ hours each day from Monday to Friday. She is paid £7.30 per hour. On Saturday she works for 4 hours and gets paid 'time and a half'. How much does she earn that week?

**4** Frank works 42 hours a week and is paid £373.80. What is his hourly rate of pay?

(PS) **5** The speed limit on UK motorways is 70 mph. The speed limit on French motorways is 120 km/h. 5 miles = 8 kilometres. Which country has the higher speed limit?

(CM) **6** A farmer has three fields. The area of field A is 1.73 hectares, the area of field B is 2.64 hectares and the area of field C is 0.95 hectares. Cattle need 0.065 hectares of space each, horses need 0.04 hectares of space each and sheep need 0.01 hectares of space each.

   **a** Show that, if the farmer keeps horses in field A, cattle in field B and sheep in field C, she will be able to have a total of 178 animals.

   **b** Work out the combination of fields, cattle, horses and sheep that will allow the farmer to keep the maximum possible number of animals.

(MR) **7** In a sale, a shoe shop reduces its prices by 50%. On the last day of the sale it reduces the sale prices by 50%. At what percentage of the original price are the shoes now being sold?

(PS) **8** Zena works for 40 hours at normal time, 6 hours at 'time and a half' and 3 hours at 'double time'. She is paid £494. Work out her hourly rate of pay.

**9** A metal cuboid with dimensions 10 cm by 5 cm by 4 cm has a mass of 1.6 kg. Work out the density of the metal. Give your answer in g/cm³.

**10** Which of these multipliers would you use to work out a reduction of 12%?

     0.12           1.12          0.88          0.12

**11** What pressure is applied by a force of 120 N acting over an area of $\frac{1}{2}$ m²?

16 Ratio and proportion and rates of change: Percentages and compound measures

**12** Two shops sell the same model of TV.

Price in shop A: £450 plus 20% VAT

Price in shop B: £630 (includes VAT) with 15% off

In which shop is the TV cheaper?

**13** A tin of dog food costs 45p.

A shop has a special offer on the dog food.

Julie wants 30 tins of dog food.

**a** Work out how much she pays.

The normal price of a dog collar and lead is £15.

In a sale, the price of the collar and lead is reduced by 12%.

**b** Work out the sale price of the collar and lead.

**14** A washing machine normally costs £350. It is reduced by 8% in a sale.

How much is the sale price of the washing machine?

**15** At the start of 2014, 1600 in Town A did not have a job. The local council set a target to reduce this number by at least 25%. At the end of the year, there were 1152 without a job. Did the council meet its target?

**16** A shop promises to pay the VAT on all the items its sells and so reduces the prices by 20%. Comment on the shop's offer.

**17** A mixed golf club has 250 members. There are 50% more men than women.

**a** Work out:

**i** the number of men          **ii** the number of women

that are members of the golf club.

10% of the men and 15% of the women are left-handed.

**b** What percentage of the total number of members is left-handed?

**18** The cost of coffee increased by 15% one week but fell back to the original price the next week.

By what percentage did the cost of coffee fall in the second week? Round your answer to the nearest whole number.

# 17 Ratio and proportion and rates of change: Percentages and variation

## This chapter is going to show you:

- how to calculate compound interest and repeated percentage change
- how to calculate a reverse percentage
- how to solve problems where two variables are in direct proportion
- how to solve problems where two variables are in inverse proportion
- how to recognise graphs that show direct and inverse proportion
- how to work out problems about growth and decay
- how to work out problems about original values.

## You should already know:

- the multiplication tables up to $12 \times 12$
- how to simplify fractions
- how to multiply and divide, with and without a calculator
- how to substitute values into expressions
- how to solve simple algebraic equations.

## About this chapter

In many real-life situations, variables are connected by a rule or relationship. Sometimes, as one variable increases the other increases. Sometimes, as one variable increases the other decreases. For example, as plants get older they become taller, as a car goes faster the length of time the journey takes decreases and as more songs are downloaded there is less money left on a gift voucher. Sometimes there is no pattern to how things change. The plant is unlikely to grow at a constant rate, as it will depend on the amount of sunshine and water, but the speed and time of a journey are connected by the rule: time = distance ÷ speed. This chapter will show you how to work out problems involving variation and also how to do some more complex percentage calculations.

# 17.1 Compound interest and repeated percentage change

This section will show you how to:

- calculate simple interest
- calculate compound interest
- solve problems involving repeated percentage change.

Banks and building societies pay interest on savings accounts. **Simple interest** is when the same percentage of the original amount (**principal**) is paid each year. This is rarely used now and most banks pay **compound interest** on savings accounts.

For compound interest, the interest earned each year is added to the original amount. The new total then earns interest at the **annual rate** in the next year. This pattern is repeated each year that the money is in the account.

The most efficient way to calculate the total amount in the account after one or more years is to use a multiplier.

**Example 1**

Elizabeth has £400 to invest. She has a choice of two accounts.

Account A pays 6.5% simple interest if she leaves her money in the account for 3 years.

Account B pays 6% compound interest.

Which account will give her more money after 3 years?

Work out the amount in each account after 3 years.

**Account A**

Simple interest pays the same interest each year.

6.5% of £400 = £26

$\quad$ 3 × £26 = £78

After 3 years, she will have £400 + £78 = £478 with Account A.

**Account B**

With compound interest the amount in the account increases by 6% each year, so the multiplier is 1.06.

$\quad$ After 1 year she will have £400 × 1.06 = £424.

$\quad$ After 2 years she will have £424 × 1.06 = £449.44.

After 3 years she will have £449.44 × 1.06 = £476.41.

Compare the amounts.

The simple interest account gives more money after 3 years.

If you calculate the differences in the amount in Account B at the end of each year, you should see that the amount of interest increases each year (£24, £25.44 and £26.97). This shows that you could have calculated the amount in the account after 3 years as £400 × $(1.06)^3$.

You can use the formula below to calculate the total amount ($A$) at any time.

$A = P + P$ × multiplier raised to the power $n$

or $\quad A = P\left(1 + \dfrac{r}{100}\right)^n$

where $P$ is the original amount invested, $r$ is the percentage interest rate and $n$ is the number of years for which the money is invested.

> **Hint and tips** 'Raised to the power of $n$' means that $n$ lots of the expression in the brackets are multiplied together.

So, in Example 1 where $P = £400$, $\frac{r}{100} = 0.06$ and $n = 3$, $A = £400 \times (1.06)^3$.

## Using your calculator

You can do the calculation in Example 1 on your calculator without having to write down all the steps.

Instead of adding on the 6% each time, you multiply by 1.06. You can do the calculation like this:

$$\boxed{4}\;\boxed{0}\;\boxed{0}\;\boxed{\times}\;\boxed{1}\;\boxed{\cdot}\;\boxed{0}\;\boxed{6}\;\boxed{\times}\;\boxed{1}\;\boxed{\cdot}\;\boxed{0}\;\boxed{6}\;\boxed{\times}\;\boxed{1}\;\boxed{\cdot}\;\boxed{0}\;\boxed{6}\;\boxed{=}$$

or like this:

$$\boxed{4}\;\boxed{0}\;\boxed{0}\;\boxed{\times}\;\boxed{1}\;\boxed{\cdot}\;\boxed{0}\;\boxed{6}\;\boxed{x^n}\;\boxed{3}\;\boxed{=}$$

or like this:

$$\boxed{4}\;\boxed{0}\;\boxed{0}\;\boxed{\times}\;\boxed{(}\;\boxed{1}\;\boxed{0}\;\boxed{6}\;\boxed{\div}\;\boxed{1}\;\boxed{0}\;\boxed{0}\;\boxed{)}\;\boxed{x^n}\;\boxed{3}\;\boxed{=}$$

You can use any of the three methods. Choose the one you prefer.

The methods used to calculate compound interest are not only used for money. You can use the same methods to calculate, for example, growth in populations, increases in salaries or increases in body mass or height. They can also be used for a regular reduction by a fixed percentage: for example, car depreciation, population decline and even water lost from reservoirs. Some examples in the next exercise use the same method as for calculating compound interest to solve problems in non-money situations.

## Exercise 17A 🖩

**1** Work out the interest on each account.

a £2000 invested for 5 years at 4% simple interest

b £1500 invested for 3 years at 2% simple interest

c £4000 invested for 3 years at 4% compound interest

d £200 invested for 2 years at 3.2% simple interest

**2** A baby octopus increases its body weight by 5% each day for the first month of its life. A baby octopus was born weighing 10 g.

What was its weight after:

a 1 day          b 2 days          c 4 days          d 1 week?

**3** Scientists have been studying the shores of Scotland. They estimate that, due to pollution, the seal population there will decline by 15% each year. In 2010 they counted about 3000 seals on those shores.

If nothing is done about pollution, how many seals did they expect to be there in:

a 2011                    b 2012                    c 2015?

**4** The value of a new car will depreciate at the rate of 20% each year. In 2011 I bought a new car that cost £8500. What would be the value of the car in:

 **a** 2012          **b** 2013          **c** 2015?

**5** The population of a small country, Yebon, was 46 000 in 2006. Since then it has increased at a steady rate of 13% each year.

 **a** Calculate the population in:

    **i** 2007          **ii** 2011          **iii** 2015.

**6** Work out the number of years of investment for each return.

 **a** £3000 invested at 3% simple interest returns £3720

 **b** £5000 invested at 2.5% compound interest returns £6724, to the nearest pound (£)

**7** The height of my conifer hedging will increase by 17% each year for the first 20 years. When I bought my hedging, it was about 50 cm tall. How long will it take to reach 3 m?

**8** The manager of a small family business offered his staff an annual pay increase of 4% for every year they stayed with the firm.

 **a** Gareth started work at the business on a salary of £12 200. What salary will he be on after 4 years?

 **b** Julie started work at the business on a salary of £9350. How many years will it be until she is earning a salary of more than £20 000?

**9** Throughout August one year, a reservoir was losing water at the rate of 8% each day. On 1 August the reservoir held 2.1 million litres of water.

 **a** How many litres of water were in the reservoir on the following dates?

    **i** 2 August          **ii** 4 August          **iii** 8 August

 **b** On what date did the amount of water drop below 1 million litres?

**10** How long will it take for each investment to reach £1 000 000?

 **a** An investment of £100 000 at a rate of 12% compound interest

 **b** An investment of £50 000 at a rate of 16% compound interest

# 17.2 Reverse percentage (working out the original value)

This section will show you how to:

- calculate the original amount, given the final amount, after a known percentage increase or decrease.

If you know the final amount as a percentage of the original amount, you can use the method of reverse percentages to work backwards to find the original amount.

There are two methods: the unitary method and the multiplier method.

## The unitary method

The unitary method has three steps.

**Step 1**  Set the final percentage equal to the final value.

**Step 2**  Use this to calculate the value of 1%.

**Step 3**  Multiply by 100 to work out 100% (the original value).

---

**Example 2**

70 workers in a factory were given a pay rise. This was 20% of all the workers. How many workers are there altogether?

Step 1  Equate the values.　　20% represents 70 workers.

Step 2  Calculate 1%.　　70 ÷ 20 (There is no need to work out this calculation yet.)

Step 3  Calculate 100%.　　70 ÷ 20 × 100 = 350

So there are 350 workers altogether.

---

**Example 3**

The price of a car increased by 6% to £9116. Work out the price before the increase.

Step 1  Equate the values.　　106% represents £9116.

Step 2  Calculate 1%.　　£9116 ÷ 106

Step 3  Calculate 100%.　　£9116 ÷ 106 × 100 = £8600

So the price before the increase was £8600.

---

## The multiplier method

The multiplier method has two steps. By using a multiplier, step 2 combines steps 2 and 3 from the unitary method.

**Step 1**  Write down the multiplier.

**Step 2**  Divide the final value by the multiplier to give the original value.

---

**Example 4**

The price of a freezer is reduced by 12% to £220. What was the price before the reduction?

**Unitary method**

Step 1  Equate the values.　　88% = £220

Step 2  Work out 1%.　　1% = £220 ÷ 88

　　　　　　　　　　　　　 = £2.50

Step 3  Work out 100%.　　100% = 100 × £2.50

　　　　　　　　　　　　　　 = £250

**Multiplier method**

Step 1  Write down the multiplier.　　A decrease of 12% gives a multiplier of 0.88.

Step 2  Calculate the original amount.　£220 ÷ 0.88 = £250

So the price before the sale was £250.

---

The multiplier method works because 88% of £250 is 0.88 × £250 = £220.

So, working backwards, £220 ÷ 0.88 = £250.

1. Find what 100% represents in each situation.

   a 40% represents 320 g    b 14% represents 35 m    c 45% represents 27 cm

   d 4% represents £123    e 2.5% represents £5    f 8.5% represents £34

2. 28 youngsters completed an army training session. This represented 35% of the original group. How many people were in the original group?

3. VAT is a government tax added to goods and services. The cost of the items below includes 20% VAT. Work out the cost of each item before VAT is added.

   | | | | | | |
   |---|---|---|---|---|---|
   | T-shirt | £10.08 | Tights | £1.44 | Shorts | £6.24 |
   | Sweater | £12.90 | Trainers | £29.76 | Boots | £38.88 |

4. Howard spends £200 a month on food. This represents 24% of his monthly take-home pay. How much is his monthly take-home pay?

5. Tina's weekly pay is increased by 5% to £315. What was Tina's pay before the increase?

6. The number of workers in a factory fell by 5% to 228. How many workers were there originally?

7. The price of a TV is reduced to £325.50. This is a 7% reduction on the original price. What was the original price?

8. 38% of plastic bottles in a production line are blue. The remaining 7750 plastic bottles are brown. How many plastic bottles are blue?

9. I received £4.40 back from HM Revenue and Customs. This represented the 20% VAT I had paid on a piece of equipment. How much did I pay for this piece of equipment?

10. A company is in financial trouble. The workers are asked to take a 10% pay cut for each of the next two years.

    a Rob works out that his pay after the second pay cut will be £1296 per month. How much is his pay now?

    b Instead, Rob offers to take an immediate pay cut of 14% and have his pay frozen at that level for two years. Has he made the correct decision?

# 17.3 Direct proportion

This section will show you how to:

- solve problems in which two variables have a directly proportional relationship (direct variation)
- recognise the constant of proportionality
- recognise graphs that show direct variation.

**Key terms**

constant of proportionality

direct proportion

direct variation

There is **direct proportion** (sometimes called **direct variation**) between two amounts when one is always the same multiple of the other. This multiplying factor is called the **constant of proportionality**. For example:

- if you have £$P$, you must have $100P$ pennies
- the diameter, $d$, of a circle is always twice the radius, $r$ $(d = 2r)$
- 1 kilogram = 2.2 pounds so there is a multiplying factor of 2.2 between kilograms and pounds
- the circumference of a circle = $\pi d$, so there is a multiplying factor of $\pi$ ($\approx 3.142$) between the circumference of a circle and its diameter.

When you are working with direct proportion, you will be given:

- the constant of proportionality

or

- the relationship between two variables so you can work out the constant of proportionality

or

- a simple equation of proportionality.

You can then use the information you are given to solve the problem.

The symbol for proportion is $\propto$.

---

**Example 5**

A solar panel generates 5 kilowatts of power every hour when the Sun shines.

a   On a sunny day in the summer, sunrise is at 7:00 am and sunset is at 8:00 pm. How many kilowatts does the solar panel generate?

b   Ken is paid 4.4p by the electricity board for every kilowatt he puts back into the electricity grid. One month he receives £7.92. How many hours was the panel putting power back into the grid?

a   Number of hours from 7:00 am to 8:00 pm is 13.

Electricity generated = 13 × 5

= 65 kilowatts.

b   Change £7.92 to 792 pence.

Number of hours = 792 ÷ 4.4

= 180 hours.

---

**Example 6**

The cost of repairing an article is directly proportional to the time spent working on it. A repair job that takes 6 hours to complete costs £180. Work out:

a   the cost of a repair that takes 5 hours

b   the length of time it takes to complete a repair costing £240.

As a 6-hour job costs £180, the cost of a 1-hour job would be 180 ÷ 6 = £30.

This is the unit cost.

a   A 5-hour job costs 5 × 30 = £150.

b   A job costing £240 would take 240 ÷ 30 = 8 hours.

---

**Example 7**

Two variables, $D$ and $T$, are in direct proportion. The equation of proportionality is $D = 32T$.

a   Work out $D$ when $T = 20$.

b   Work out $T$ when $D = 144$.

c   Work out the increase in $D$ when $T$ increases from 12 to 30.

---

*(continued)*

**a** Substitute $T = 20$ into the proportionality equation.

$D = 32 \times 20 = 640$

**b** Substitute $D = 144$ into the proportionality equation.

$144 = 32 \times T$

$T = 144 \div 32$

$= 4.5$

**c** Work out $D$ when $T = 12$ and $T = 30$ and then subtract.

When $T = 12$, D $= 32 \times 12 = 384$

When $T = 30$, D $= 32 \times 30 = 960$

The difference is $960 - 384 = 576$.

---

**Hints and tips** Notice that you could have done part **c** of Example 7 by substituting $30 - 12 = 18$ into the proportionality equation, as the variables are in direct proportion.

---

**Example 8**

The cost, $C$ in pence, of cups of tea, $T$, are in direct proportion. The equation of proportionality is $C = 80T$.

**a** What does 80 represent?

**b** Work out $C$ when $T = 5$.

**c** Work out $T$ when $C = 640$.

**d** Baz buys 3 cups of tea and Rosie buys 7 cups of tea. How much more does Rosie spend than Baz?

**a** 80 is the cost of one cup of tea, in pence.

**b** Substitute $T = 5$ into the proportionality equation

$C = 80 \times 5 = 400$ pence or £4.00.

**c** Substitute $C = 640$ into the proportionality equation.

$640 = 80 \times T$

$T = 640 \div 80 = 8$ cups

**d** Work out $C$ when $T = 3$ and $T = 7$ and subtract the values.

When $T = 3$, $C = 80 \times 3 = 240$.

When $T = 7$, $C = 80 \times 7 = 560$.

Difference $= 560 - 240 = 320$ pence or £3.20.

---

**Hints and tips** Note that you could have done part **d** of Example 8 by substituting $7 - 3 = 4$ into the equation of proportionality as the variables are in direct proportion.

Direct proportion graphs

When two variables are in direct proportion they will produce a straight-line graph that starts at the origin.

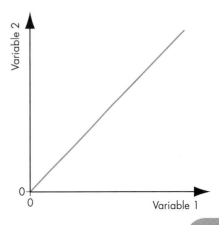

Example 9

The graph shows the relationship between $y$ and $x$.

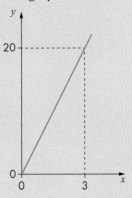

Work out:     **a** the value of $y$ when $x = 9$     **b** the value of $x$ when $y = 30$.

Use the values shown on the graph to work out the value of $y$ when $x = 1$.

When $x = 3$, $y = 20$.

When $x = 1$, $y = 20 \div 3$

$\qquad = 6\frac{2}{3}$

**a** When $x = 9$:   $y = 6\frac{2}{3} \times 9$

$\qquad\qquad = 60$

**b** When $y = 30$:   $x = 30 \div 6\frac{2}{3}$

$\qquad\qquad = 4\frac{1}{2}$

# Exercise 17C

**1** 20 chocolate bars cost £4 altogether. Work out:

**a** the cost of 17 chocolate bars

**b** how many chocolate bars you can buy for £3.40

**c** the maximum number of bars you can buy for £2.50.

**2** 12 inches is approximately 30 centimetres. Approximately how long is:

**a** 40 inches in centimetres

**b** 50 centimetres in inches?

 **c** Which is longer, 82 inches or 201 centimetres?

**3** A farmer is recommended to keep no more than 30 cows in a field with an area of 50 hectares.

**a** How many cows could the farmer keep in a field of 20 hectares?

**b** Another field has 20 cows. What is the largest area it could be?

 **c** Sheep need at least $\frac{1}{2}$ a hectare each. A farmer keeps sheep and cows in the same field. The ratio of cows to sheep must be 1 : 2. Work out the maximum number of animals that can be kept in a 40 hectares.

**4** $Q$ is directly proportional to $P$. $Q = 100$ when $P = 2$. Work out the value of:

**a** $Q$ when $P = 3$          **b** $P$ when $Q = 300$.

**(MR)** **4** This graph shows the relationship between $x$ and $y$.

Find the value of: **a** $x$ when $y = 9$  **b** $y$ when $x = 30$.

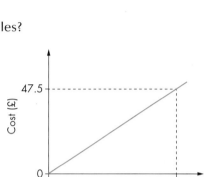

**5** The distance a train travels is directly proportional to the time taken for the journey. The train travels 105 miles in 3 hours.

**a** What distance will the train travel in 5 hours?

**b** How much time will it take the train to travel 280 miles?

**6** The cost of fuel delivery is directly proportional to its mass. The graph shows the relationship between cost and mass.

**a** What is the delivery cost for 350 kg fuel?

**b** A fuel delivery cost £33.25. How much fuel was delivered?

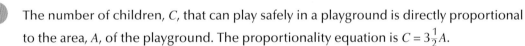

**(MR)** **7** The number of children, $C$, that can play safely in a playground is directly proportional to the area, $A$, of the playground. The proportionality equation is $C = 3\frac{1}{2}A$.

**a** How many children can safely play in a playground of area 154 m²?

**b** A playgroup has 24 children. What is the area of the smallest playground in which they could safely play?

**(EV)** **8** The number of spaces, $S$, in a car park is directly proportional to the area, $A$, of the car park. The proportionality equation is $S = \frac{1}{15}A$.

The area of the car park increases by 750 m². How many more spaces will there be?

**(PS)** **9** The number of passengers in a bus queue is directly proportional to the time that the person at the front of the queue has spent waiting.

Karen is the first to arrive at a bus stop. When she has been waiting 5 minutes the queue has 20 passengers.

When the bus arrives, there are 70 people waiting. How long was Karen waiting for the bus?

**(PS)** **10** This is a conversion graph between gallons, $G$, and litres, $L$.

**a** Write down the equation of proportionality in the form $L = kG$, where $k$ is the constant of proportionality.

**b** Will a 6 gallon tank hold 28 litres?

**c** Petrol costs £1.22 per litre. What is the cost of a gallon of petrol?

**d** Jan's car does 54 miles to the gallon. How many miles does her car do to a litre?

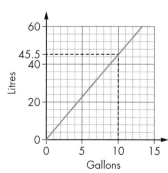

# 17.4 Inverse proportion

## This section will show you how to:

- solve problems in which two variables have an inversely proportional relationship (inverse variation)
- recognise the constant of proportionality.

There is **inverse proportion** or **inverse variation** between two amounts when as one amount increases, the other decreases.

For example, the faster you travel over a given distance, the less time it takes. As the distance is constant the given distance would be the constant of proportionality.

---

**Example 10**

Four men take three days to build a wall.

**a** How long would five men take to build the same wall, working at the same rate?

**b** What assumption have you made?

**a** If four men take three days to build a wall, then it takes 12 working days to build the wall. That means that 1 man would take 12 days to build the wall.

So if 1 man takes 12 days, 5 men take 12 ÷ 5 = 2.4 days.

**b** You have assumed that a day is not 24 hours as no one could work for three days without a break. It is more likely that a day is actually 8 hours. You have also assumed that all the men work at the same rate.

---

**Example 11**

Cars drive around a race track. Driving at an average speed of 180 km/h, a car takes $4\frac{1}{2}$ minutes to get round the track.

**a** How long would a car travelling at an average speed of 150 km/h take to get round the track?

**b** A car takes 8 minutes to get round the track. What was the average speed?

**a** Distance = speed × time

so distance = 180 × 4.5 ÷ 60 = 13.5 km.

Time = distance ÷ speed, so time = 13.5 ÷ 150 = 0.09 hours.

0.09 hours = 0.09 × 60 minutes = 5.4 minutes

> **Hints and tips** To convert minutes to hours, divide by 60. To convert hours to minutes, multiply by 60.

**b** Speed = distance ÷ time

so time = distance ÷ speed

$\qquad$ = 13.5 ÷ 8

$\qquad$ = 1.6875 km per minute.

1.6875 km per minute = 1.6875 × 60

$\qquad\qquad\qquad\qquad$ = 101.25 km/h.

---

## Recognising graphs that show inverse proportion

Graphs showing inverse proportion look different from graphs showing direct proportion.

A graph showing inverse proportion will be similar to the one shown here. The product of the $x$ and $y$ coordinates of any point on the graph will always give the constant of proportionality.

The equation is of the form $y = \dfrac{k}{x}$ or $xy = k$.

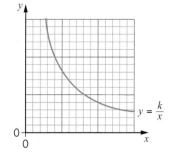

**Example 12**

The graph shows the relationship between $y$ and $x$.

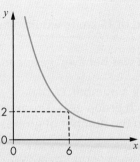

Work out:  **a** the value of $y$ when $x = 4$    **b** the value of $x$ when $y = 12$.

**a** Use the values shown on the graph to work out the constant of proportionality.

When $x = 6$ then $y = 2$.

So the value of $k = 2 \times 6 = 12$.

**a** When $x = 4$:    $y = 12 \div 4$

$= 3$

**b** When $y = 12$:    $12 = 12 \div x$

$x = 12 \div 12$

$= 1$

## Exercise 17D

**1** The distance from a light and the brightness of the light are in inverse proportion. That means that the further you are from a light, the less bright it seems. When you are at a distance, $D$, of 1 metre from a light bulb it has brightness, $B$, of 50 watts. Work out the value of:

**a** $B$ when $D = 4$    **b** $D$ when $B = 20$.

**2** Four people can paint a fence in 6 hours. How long would three people take to paint it?

**3** The area of a rectangle is fixed. As the length increases the width decreases. When the length is 7.5 cm the width is 8 cm. Work out:

**a** the width when the length is 12 cm    **b** the length when the width is 6 cm.

**4** The amount of gas in a balloon is fixed. As the outside pressure increases the volume of the balloon decreases. When the outside pressure is $\frac{1}{2}$ bar, the volume of the balloon is 2500 cm³. Work out:

**a** the volume when the outside pressure is 2 bars

**b** the outside pressure when the volume is 2000 cm³.

**5** $M$ is inversely proportional to $t$. The graph shows the relationship between $M$ and $t$.

Find the value of:

**a** $M$ when $t = 3$      **b** $t$ when $M = 1.44$.

 Remember that the product of the coordinates of any point on the graph is the constant of proportionality.

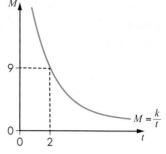

**6** At Silverstone motor-racing circuit in 2013 Mark Webber set a lap record of 1 minute 33.4 seconds at an average speed of 265.28 km/h.

**a** Work out how long a runner, running at an average speed of 15 km/h, would take to run round the circuit.

**b** A cyclist took 12 minutes to cycle round the circuit. What was her average speed?

**7** A fixed amount of money, $M$, is available to be shared among a number, $G$, of community groups. The more groups that apply, the less each group gets. This means that the amount each group gets is inversely proportional to the number of groups. When 18 groups apply, they each get £100.

**a** How much would each group have received if 12 groups applied for the money?

**b** Each group received £50. How many groups applied for the money?

**8** While doing underwater tests in one part of an ocean, a team of scientists noticed that the temperature, $T$, in Celsius degrees (°C), decreases as the depth, $D$, in kilometres (km) increases. The graph shows the relationship between $T$ and $D$.

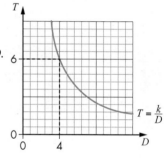

**a** What was the temperature at a depth of 8 km?

**b** At what depth was the temperature 2 °C?

**9** In the table, $y$ is inversely proportional to $x$.

Copy and complete the table, leaving your answers as fractions.

| $x$ | 8 | 24 | |
|---|---|---|---|
| $y$ | 1 | | $\frac{1}{2}$ |

**10** The fuel consumption, in miles per gallon (mpg), $M$, of a car is inversely proportional to its speed, $S$, in miles per hour (mph).

That means the faster the car goes, the lower the fuel consumption. When the speed is 60 miles per hour the fuel consumption is 30 miles a gallon.

**a** How far can the car travel at 30 mph on 2 gallons of fuel?

**b** How much further could the car travel on 1 gallon of fuel at 60 mph than at 80 mph?

# Worked exemplars

 **1** Kelly saved just enough money to buy the new TV that she wanted. When she got to the shop to buy the TV, there was a sale on and the price was reduced by 15% to £319.60. She sees a radio that costs £42.50. Will she have enough money left to buy the radio after she has paid for the TV?

| | |
|---|---|
| This is a communicating mathematically question, so you must show your working clearly and explain in words what you are working out. | |
| The price of the TV is reduced by 15%, so the multiplier is 0.85.<br>Original cost of TV is 319.60 ÷ 0.85 = 376. | Work out the original cost of the TV to find the amount of money that Kelly has saved. |
| Kelly will have 376 – 319.60 = £56.40 after buying the TV. | Work out how much she will have left after buying the TV. |
| 56.40 > 42.50, so she will have enough to buy the radio. | Write a clear conclusion. |

  **2** A plant in a greenhouse is 10 cm high. Its height increases by 13% each day. How many days does it take to double in height?

| | |
|---|---|
| This is a problem-solving question. You need to find a way to work out the number of days the plant will take to double its height. This can be solved using more than one method. The trial and improvement method is shown here. | |
| The multiplier for a 13% increase is 1.13. | Write down the multiplier. |
| Try 5 days. $10 \times 1.13^5 = 18.42$ | Try a number of days.<br>The result is too small, so try a bigger number next. |
| Try 8 days. $10 \times 1.13^8 = 26.58$ | This is too big, so try a number between your first guess and your second guess. |
| Try 6 days. $10 \times 1.13^6 = 20.82$ | This is approximately correct. |
| The plant doubles in size in 6 days. | Write a conclusion. |

  **3** Which of these calculations would give the answer to the value, after 2 years, of an investment of £6000, which increased by 5% in the first year and then 4% in the second year.

$6000 \times 1.09 \qquad 6000 \times 1.05 \times 1.04 \qquad 6000 \times 2.09 \qquad 6000 \times 1.05 + 6000 \times 1.04$

| | |
|---|---|
| This is a mathematical reasoning question with four choices. You can either eliminate those that are clearly wrong or use your knowledge of percentage change to work out the answer. All the four answers use multipliers. | |
| 6000 x 1.05 | Write down the calculation that gives the value after one year. There is no need to work it out. |
| (6000 x 1.05) x 1.04 | This value is increased by 4% (a multiplier of 1.04) in the second year. |
| 6000 x 1.05 x 1.04 | Simplify the answer to give the second option. |

# Ready to progress

I can work out simple interest problems.

I can work out compound interest.

I can solve reverse percentage problems.
I can recognise direct and inverse variation.
I know what a constant of proportionality is, and how to find it.
I can use an equation describing inverse or direct proportion.
I can solve simple problems involving direct or inverse proportion.

# Review questions

 **1** Ruby invests £2000 for two years at 3% per annum simple interest.
How much will she have in the bank after two years?

 **2** Duma invests £18 000 for three years at 3.2% per annum simple interest.
Work out the value of the investment at the end of three years.

 **3** 25 calculators cost £156.25. How much would 18 calculators cost?

 **4** Answer these questions. Read them very carefully. Some are not what you
may expect.

**a** A 10-minute mobile phone call costs 23p. How much will a 16-minute call cost?

**b** Four men take six hours to mow a meadow. How long will it take eight men,
working at the same rate, to mow the same meadow?

**c** The four men in **b** took 20 minutes to walk to the meadow. How long did it take the
eight men to walk to the meadow?

**d** Nine Collins GCSE books together have a mass of 10.8 kilograms (check this if you
like). What is the mass of 30 Collins GCSE textbooks?

**e** Four students sat a mathematics test that lasted 1 hour. How long did the same test
last when only five students took it?

**f** Eight plasterers can plaster the inside of a new house in four hours. How long
would it take two plasterers, working at the same rate, to plaster a new house?

**g** A newly plastered house takes three days for the plaster to dry. How long would it
take for the plaster in two newly plastered houses to dry?

**h** James can iron three shirts in nine minutes. How long will it take him to iron seven shirts?

**i** Three washed shirts take two hours to dry on the washing line. How long will it
take eight shirts to dry?

**j** Maxine has to take four tablets a day. She takes one every three hours. She takes
the first tablet at 9:00 am. When does she take the last tablet of the day?

 **5** Nancy wants to invest £5000 for three years. She has a choice of accounts. Account A
pays 3% simple interest. Account B pays 2.8% compound interest. Which account
should she choose?

**6** Mary invests £6000 for 12 years at 2.8% per annum compound interest.

Work out the value of the investment at the end of the 12 years.

**7** Six men take two days to build a wall.

**a** How long would it take four men to build an identical wall, assuming they all work at the same rate?

Choose your answer from this list.

3     4     6     12

**b** Six men start to build an identical wall. After one day, three of the men are called to another job. How long will it take the remaining three men to finish building the wall?

**8** After an increase of 5%, Jason's hourly rate of pay is now £9.03. What was it originally?

**9** Ahmed adds 10% as a tip to a restaurant bill. He pays £42.35. How much was the bill?

**10** $X$ and $Y$ are directly proportional. When $X = 10$, $Y = 25$.

Work out the value of:    **a** $X$ when $Y = 15$     **b** $Y$ when $X = 40$.

**11** $A$ and $B$ are inversely proportional. When $A = 8$, $B = 6$.

Work out the value of:    **a** $A$ when $B = 12$     **b** $B$ when $A = 1.5$.

**12** After a 12% decrease, the cost of a washing machine is £330. How much did it cost before the increase?

**13** The circumference, $C$, of a circle is directly proportional to the diameter, $d$. The proportionality equation is $C = \frac{22}{7}\,d$.

**a** Work out the value of:   **i**   $C$ when $d = 21$     **ii**   $d$ when $C = 5.5$.

The area of a circle is given by the formula $A = \pi r^2$.

**b** How many times bigger is the area when the circumference increases from 44 to 88?

**14** The volume of a gas, $V$, is inversely proportional to the pressure, $P$.

The proportionality equation is:

$$V = \frac{20}{P}$$

Work out the value of:    **a** $V$ when $P = 2.5$     **b** $P$ when $V = 40$.

**15** $y$ is directly proportional to $x$. Copy and complete the table.

| $x$ | 25 |    | 400 |
|-----|----|----|-----|
| $y$ | 10 | 20 |     |

**16** $y$ is inversely proportional to $x$. When $y = 8$, $x = \frac{1}{8}$.

Work out the value of:    **a** $y$ when $x = \frac{1}{125}$     **b** $x$ when $y = 2$.

**17** $P$ and $Q$ are directly proportional. When $P = 4$, $Q = 10$.

$Q$ and $R$ are inversely proportional. When $Q = 2$, $R = 8$.

Show that $R = 0.8$ when $P = 8$.

# 18 Statistics: Representation and interpretation

## This chapter is going to show you:

- how to describe the data-handling cycle
- how to collect data to obtain an unbiased sample
- how to draw and interpret pie charts
- how to identify the modal group and estimate the mean from grouped data
- how to draw scatter diagrams and lines of best fit
- how to interpret scatter diagrams and the different types of correlation.

## You should already know:

- how to draw and interpret pictograms, bar charts and line graphs
- how to extract information from tables and diagrams
- how to draw and measure angles
- how to plot coordinates
- how to work out the mode, the median, the mean and the range.

## About this chapter

This chapter extends statistical representation by introducing pie charts and scatter diagrams.

The pie chart first appeared in 1801 in a publication called *The Statistical Breviary* by William Playfair. He used graphical representations of quantitative data, such as bar charts and pie charts, because he believed that 'making an appeal to the eye when trying to show data is the best and easiest method of giving any message that might be wanted to show through such diagrams'.

The word 'scatter' comes to us from Scandinavian influences in the 12th century, but scatter diagrams first appeared in 1924 in a document from a university in what is now Pakistan. Apparently, they were one of the first establishments to use the technique of plotting points from two sources to see if any connections could be seen between the two sets of data.

Scatter diagrams were not used very much until the great energy debate in the late 1960s, when prices and sales of both gas and electricity were being studied. At that time, there was pressure for people to use more electricity, as it was thought to be an infinite power source, whereas gas would seemingly run out one day soon.

You will see pie charts and scatter diagrams every day, for example, in business presentations, social studies and polls.

# 18.1 Sampling

This section will show you how to:

- obtain a random sample from a population
- collect unbiased and reliable data for a sample.

**Key terms**

| | |
|---|---|
| bias | hypothesis |
| population | primary data |
| random sample | sample size |
| secondary data | survey |
| unbiased | |

## Data collection

There is more than one way to collect data. Data that you collect yourself is **primary data**. You control it, in terms of accuracy and amount. Data collected by someone else is **secondary data**. A lot of this type of data is available on the internet or in newspapers. It is useful because you can access a huge volume of data, but you have to rely on its sources being accurate.

Statisticians often carry out **surveys** to collect information and test hypotheses about a **population** for a wide variety of purposes. In statistics, a population may be a group of objects, events or people.

It may not be possible to survey a whole population. It might be physically impossible; for example, suppose a team of marine biologists wanted to find the average length of eels in the North Sea. It would be impossible to find and measure every eel, so they would choose a small part of the population to survey. Then they would assume that the results for this sample are representative of the whole population. Even when it is physically possible, it might take too long and cost too much money.

You should consider two questions to ensure the accuracy and reliability of a sample.

- Will the sample be representative of the whole population and avoid **bias**?
- How large should the sample be to give results that are valid for the whole population?

In statistics, you test **hypotheses**. These are statements based on a theory. Testing a hypothesis involves a cycle of planning, collecting data, evaluating the significance of the data and then interpreting the results, which may or may not show that the hypothesis is true. This cycle often leads to a refinement of the problem, which starts the cycle all over again.

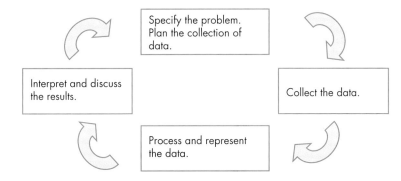

There are four steps in the data-handling cycle.

**Step 1** State the hypothesis, outlining the problem and planning what needs to be done. Plan the data collection.

**Step 2** Collect the data. Record the data collected clearly.

**Step 3** Choose the best way to process and represent the data. This will normally mean calculating averages (mean, median, mode) and measures of spread (range), then representing data in suitable diagrams.

**Step 4** Interpret the data and make conclusions.

The hypothesis can then be refined or changes made to the data collected; for example, a different type of data may be collected or the same data may be collected in a different way. So the data-handling cycle helps to improve reliability in the collection and interpretation of data.

**Example 1**

A gardener grows some tomatoes in a greenhouse and some outside. He wants to investigate this hypothesis.

*'Tomato plants grown inside the greenhouse produce more tomatoes than those grown outside.'*

Describe the data-handling cycle that may be applied to this problem.

*Plan the data collection.* Consider 30 tomato plants grown in the greenhouse and 30 plants grown outside. Count the tomatoes on each plant.

*Collect the data.* Record the numbers of tomatoes collected from the plants between June and September. Only count those that are 'fit for purpose'. This means he chooses the tomatoes that are good enough to eat. This will avoid bias.

*Choose the best way to process and represent the data.* Calculate the mean number collected per plant, as well as the range. Draw a suitable diagram to show the data. This could be a bar chart or a pictogram.

*Interpret the data and make conclusions.* Look at the statistics. What do they show? Is there a clear conclusion or does the gardener need to alter the hypothesis in any way? Discuss the results, refine the method and continue the cycle.

In describing the data-handling cycle, you must refer to each of the four parts.

## Random samples

In a **random sample**, every member of the population has an equal chance of being chosen. For example, it may be the first 100 people met in a survey, or 100 names picked from a hat, or 100 names taken at random from the electoral register or a telephone directory.

## Sample size

Before you start the sampling of a population, you must decide how much data you need to collect, to ensure that the sample is representative of the population. This is called the **sample size**.

Sample size depends on:

• how accurately the sample must represent the population

• the amount of time or money available to meet the cost of collecting the sample data.

The more representative the sample needs to be, the larger the sample size needs to be but the larger the sample size, the greater the cost and the time taken. Therefore, you will need to balance achieving high accuracy in a sample with the cost of achieving it.

The next example describes some of the problems associated with obtaining an **unbiased** sample.

**Example 2**

You are going to conduct a survey among an audience of 30 000 people at a rock concert. How would you choose the sample?

You cannot question all of them, so you might choose a sample of 200 people.

Assuming that there will be the same number of men and women at the concert, your sample should include the same number of each: 100 men and 100 women.

Assuming that about 20% of the audience will be aged under 25, your sample should include 40 people aged under 25 and 160 people aged 25 and over.

You would also need to select people from different parts of the audience, in equal proportions, to get a balanced view. So choose groups of people taken from the front, the back and the middle of the audience.

## Exercise 18A

**1** Decide whether you would use primary data or secondary data to investigate each of these.

   **a** Oliver wants to know which month of the year is the hottest.

   **b** Andrew wants to compare how well boys and girls estimate the size of an angle.

   **c** Joy thinks that more men than women go to football matches.

   **d** Sheehab wants to know if tennis is watched by more women than men.

   **e** A headteacher said that the more revision you do, the better your examination results.

   **f** A newspaper suggested that the older you are, the more likely you are to shop at a department store.

(CM) **2** Roxanne's mathematics teacher has asked her to find out if this hypothesis is true.

*'In Year 11, the girls are better than the boys at spelling.'*

Describe how she could use the data-handling cycle to test the hypothesis.

(CM) **3** Steve wants to test this hypothesis.

*'Students who play more sport will watch less TV.'*

Describe how he could use the data-handling cycle to test his hypothesis.

(MR) **4** Mr Charlton wanted to find out how often the students in his school visited a fast-food outlet. The table gives the numbers of students in each school year.

|  | Boys | Girls |
|---|---|---|
| Y9 | 122 | 129 |
| Y10 | 127 | 125 |
| Y11 | 126 | 128 |

   **a** Design a short questionnaire that Mr Charlton could use to sample the school.

   **b** Mr Charlton wanted to use a sample of 120 students. How many of each group of students should receive the questionnaire?

   **c** Explain how Mr Charlton could give out the questionnaires within each group of students.

(MR) **5** You are asked to conduct a survey at a football match where the attendance is approximately 20 000. The crowd consists of approximately 15 000 males and approximately 5000 females.

Explain how you could take a sample of the crowd.

(MR) **6** Claire makes a survey of the sixth-form students in her school. She wants to find out their opinions on the eating facilities in the school. The table shows the numbers of students in the two year groups.

| Year group | Boys | Girls | Total |
|---|---|---|---|
| 12 | 106 | 122 | 228 |
| 13 | 97 | 75 | 172 |
| Total number in the sixth form | | | 400 |

Claire decides to take a sample of 80 students.

Explain why she should not sample an equal number of boys and girls in the two years.

**7** The manager of a company carries out a survey on wages for the employees.
He decides to carry out a random 10% sample for the four groups of employees.
The table shows the number of employees in each group.

|  | Male | Female | Total |
|---|---|---|---|
| **Full time** | 132 | 68 | 200 |
| **Part time** | 43 | 57 | 100 |
| Total number of employees | | | 300 |

Draw up a similar table to show the number of employees in the sample.

# 18.2 Pie charts

This section will show you how to:

**Key term**

pie chart

- draw and interpret pie charts.

Pictograms, bar charts and line graphs are not always easy to interpret when there is a big difference between the frequencies or there are only a few categories. In these cases, it is often more convenient to illustrate the data on a **pie chart**.

In a pie chart, the whole of the data is represented by a circle (the 'pie') and each category of it is represented by a sector of the circle (a 'slice of the pie'). The angle of each sector is proportional to the frequency of the category it represents.

A pie chart can only show proportions and not individual frequencies, unlike a bar chart, for example.

Sometimes you may use a framework for a pie chart, marked off in equal sections rather than angles.

**Example 3**

20 people were surveyed about their preferred drink. The table shows their responses.

| Drink | Tea | Coffee | Milk | Pop |
|---|---|---|---|---|
| **Frequency** | 6 | 7 | 4 | 3 |

Illustrate the results on this pie chart framework.

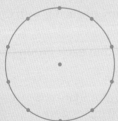

You can see that the pie chart has 10 equally-spaced divisions.

As there are 20 people, each division represents 2 people. So the sector for tea will have 3 of these divisions, the sector for coffee will have $3\frac{1}{2}$ divisions, the sector for milk will have 2 divisions and the sector for pop will have $1\frac{1}{2}$ divisions.

Preferred drinks

**Note:**

- Always give your chart a title.

- Always label the sectors of the chart. Use shading and a separate key if there is not enough space to write on the chart.

**Example 4**

In a survey, 120 people were asked the main type of transport they used to travel to their most recent holiday destination. This table shows the results.

Draw a pie chart to illustrate the data.

| Type of transport | Train | Coach | Car | Ship | Aeroplane |
|---|---|---|---|---|---|
| Frequency | 24 | 12 | 59 | 11 | 14 |

There are two methods of finding the size of the angles you need to draw a pie chart.

**Method 1: The scaling method**

Since 120 divides exactly into 360°, each person can be represented by 360° ÷ 120 = 3°. So multiply all the frequencies by 3 to give the angles of all the sectors.

| Type of transport | Frequency | Calculation | Angle |
|---|---|---|---|
| Train | 24 | 24 × 3 | 72° |
| Coach | 12 | 12 × 3 | 36° |
| Car | 59 | 59 × 3 | 177° |
| Ship | 11 | 11 × 3 | 33° |
| Aeroplane | 14 | 14 × 3 | 42° |
| Totals | 120 | | 360° |

**Method 2: The proportional method**

Work out what fraction of the pie chart represents each type of transport. To do this, divide each frequency by 120 and then multiply the result by 360°.

| Type of transport | Frequency | Calculation | Angle |
|---|---|---|---|
| Train | 24 | $\frac{24}{120} \times 360°$ | 72° |
| Coach | 12 | $\frac{12}{120} \times 360°$ | 36° |
| Car | 59 | $\frac{59}{120} \times 360°$ | 177° |
| Ship | 11 | $\frac{11}{120} \times 360°$ | 33° |
| Aeroplane | 14 | $\frac{14}{120} \times 360°$ | 42° |
| Totals | 120 | | 360° |

Use the calculated angle for each sector to draw the pie chart.

**Note:**

- Use the frequency total (120 in this case) as the denominator of each fraction.
- Check that the sum of all the angles is 360°. (You do not need to show the angles or frequencies on the pie chart.)
- Give your chart a title and label each sector.

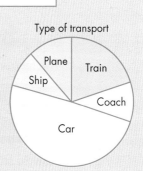

Type of transport

Example 5

The pie charts show the favourite sports for two classes in a school.

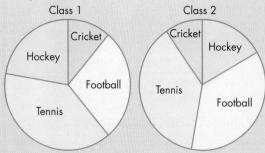

Class 1

Class 2

James says that there are more students who prefer football in class 2 than in class 1. Explain why he could be wrong.

There could be a different number of students in each class. The pie charts only show proportions so there is no way of telling the numbers represented in each pie chart.

## Exercise 18B

**1** For each set of data, copy this frame and use it to draw a pie chart.

**a** The favourite pets of 10 children

| Pet | Dog | Cat | Rabbit |
|-----|-----|-----|--------|
| Frequency | 4 | 5 | 1 |

**b** The makes of cars belonging to 20 teachers

| Make of car | Ford | Toyota | Vauxhall | Nissan | Peugeot |
|-------------|------|--------|----------|--------|---------|
| Frequency | 4 | 5 | 2 | 3 | 6 |

**c** The newspaper read by 40 office workers

| Newspaper | Sun | Mirror | Guardian | The Times |
|-----------|-----|--------|----------|-----------|
| Frequency | 14 | 8 | 6 | 12 |

**2** Draw a pie chart to represent each set of data.

> **Hints and tips** Remember to complete a table as shown in Example 4. Check that the angles add up to 360°.

**a** The numbers of children in 40 families

| Number of children | 0 | 1 | 2 | 3 | 4 |
|--------------------|---|---|---|---|---|
| Frequency | 4 | 10 | 14 | 9 | 3 |

**b** The favourite soap opera of 60 students

| Programme | Home and Away | Neighbours | Coronation Street | Eastenders | Emmerdale |
|-----------|---------------|------------|-------------------|------------|-----------|
| Frequency | 15 | 18 | 10 | 13 | 4 |

**c** How 90 students get to school

| Journey to school | Walk | Car | Bus | Cycle |
|-------------------|------|-----|-----|-------|
| Frequency | 42 | 13 | 25 | 10 |

(MR) **3** Mariam asked 24 of her friends which sport they preferred to play. Her data is shown in this frequency table.

| Sport | Rugby | Football | Tennis | Squash | Basketball |
|---|---|---|---|---|---|
| Frequency | 4 | 11 | 3 | 1 | 5 |

a Draw a pictogram to show the data.

b Draw a bar chart to show the data.

c Draw a vertical line chart to show the data.

d Draw a pie chart to show the data.

e Which diagram best illustrates the data? Give a reason to support your answer.

(MR) **4** Andy wrote down the number of lessons he had in each subject on his weekly school timetable.

Mathematics 5　　English 5　　Science 8　　Languages 6
Humanities 6　　Arts 4　　Games 2

a How many lessons did Andy have on his timetable?

b Draw a pie chart to show the data.

c Draw a bar chart to show the data.

d Which diagram better illustrates the data? Give a reason to support your answer.

(MR) **5** In a poll during the run-up to an election, 720 people were asked which political party they would vote for. The results are given in the table.

| Conservative | 248 |
|---|---|
| Labour | 264 |
| Liberal Democrat | 152 |
| Green Party | 56 |

a Draw a pie chart to illustrate the data.

b Why do you think pie charts are used to show this sort of information during elections?

(MR) **6** This table shows the numbers of candidates, at each grade, gaining music examinations in Strings and Brass.

| | Grade | | | | | Total number of candidates |
|---|---|---|---|---|---|---|
| | 3 | 4 | 5 | 6 | 7 | |
| Strings | 300 | 980 | 1050 | 600 | 70 | 3000 |
| Brass | 250 | 360 | 300 | 120 | 70 | 1100 |

a Draw a pie chart to represent each of the two instruments.

b Compare the pie charts to decide which group of candidates, Strings or Brass, are of a higher standard. Give reasons to support your answer.

**7** In a survey, a rail company asked passengers whether their service had improved. The results are shown in this pie chart.

Explain how you would work out the probability that a person picked at random from this survey answered *Don't know*.

**Has the rail service improved?**

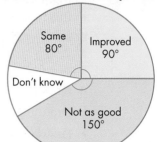

 **8** This pie chart shows the proportions of the different shoe sizes worn by 144 students in Year 11 in a London school.

**a** What is the angle of the sector representing shoe sizes 11 and 12?

**b** How many students had a shoe size of 11 or 12?

**c** What percentage of students wore the modal size?

**Shoe sizes worn by 144 students in Year 11**

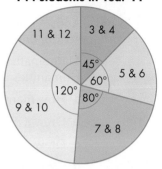

# 18.3 Scatter diagrams

This section will show you how to:

- draw, interpret and use scatter diagrams
- draw and use a line of best fit.

A **scatter diagram** (also called a scattergraph or scattergram) is one way of comparing two variables by plotting their corresponding values on a graph. The variables are usually taken from a table and are treated just like a set of (*x*, *y*) coordinates.

This scatter diagram shows the marks scored by students in an English test plotted against the marks they scored in a mathematics test. It shows that the students who had high marks in the mathematics test also tended to score high marks in the English test.

| Key terms |
| --- |
| correlation |
| extrapolation |
| interpolation |
| line of best fit |
| negative correlation |
| no correlation |
| positive correlation |
| scatter diagram |

**Comparison of English and mathematics marks**

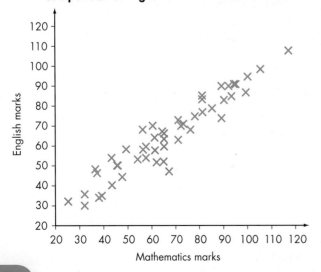

## Correlation

Here are three statements that may or may not be true.

- The taller people are, the wider their arm span is likely to be.
- The older a car is, the lower its value will be.
- The distance you live from your place of work will affect how much you earn.

You could test these relationships by collecting data and plotting each set of data on a scatter diagram.

The first statement may give a scatter diagram like this one. This diagram has good **positive correlation** because as one quantity increases so does the other. From this scatter diagram you could say that the taller someone is, the wider their arm span.

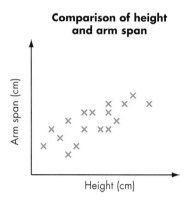

**Comparison of height and arm span**

**Good positive correlation**

Testing the second statement may give a scatter diagram like this one. This diagram has strong **negative correlation** because as one quantity increases, the other quantity decreases. From this scatter diagram you could say that as a car gets older, its value decreases.

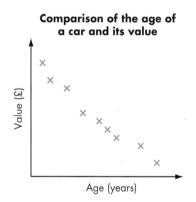

**Comparison of the age of a car and its value**

**Strong negative correlation**

Testing the third statement may give a scatter diagram like this one. This scatter diagram has **no correlation**. From this scatter diagram you could say there is no relationship between the distance a person lives from work and how much that person earns.

You can describe the **correlation** more fully by using words such as *weak*, *good* or *strong*.

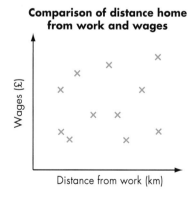

**Comparison of distance home from work and wages**

**No correlation**

**Example 6**

These two scatter diagrams show the relationships between:
- the temperature and the amount of ice cream sold
- a person's age and the amount of ice cream they eat.

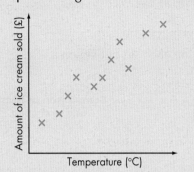 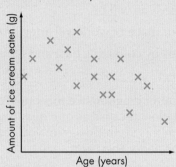

**a** Comment on the correlation of each graph. **b** What does each graph tell you?

**a** The first diagram shows strong positive correlation.

The second diagram shows weak negative correlation.

**b** The first diagram tells you that as the temperature increases, the amount of ice cream sold increases.

The second diagram tells you that as people get older, they eat less ice cream.

**Beware!** Correlation does not give you any indication of the underlying causes or reasons for a trend. For example, you cannot make the assumption that older people do not like ice cream.

## Line of best fit

A **line of best fit** is a straight line drawn between all the points on a scatter diagram. It passes as close as possible to all of them. You should try to have the same number of points on both sides of the line. When drawing a line of best fit, ignore any point that is outside the main spread of values. Such a point is called an outlier.

This scatter diagram shows the marks gained when a class took tests in mathematics and English. Notice that the teacher ignored the outliers when he drew the line of best fit.

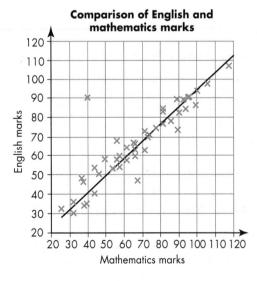 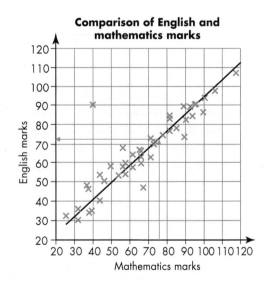

You can use the line of best fit to answer questions such as: 'A girl took the mathematics test and scored 76 marks. She was ill and missed the English test. How many marks was she likely to have scored?'

To answer this question, look at the second graph. The teacher drew a line from 76 on the mathematics axis, up to the line of best fit. Then he drew a line across to the English axis. This gives 73, so this is the mark she is likely to have scored in the English test.

**Beware!** When you are reading data from a line of best fit, the point will only be an indication of what might happen, not an exact answer. (This is called **interpolation**.) When you predict trends beyond a line of best fit, you cannot assume the trend will continue. (This is called **extrapolation**.)

## Exercise 18C

**1** For each scatter diagram:

   **i** describe the correlation           **ii** write down what it shows you.

**a**

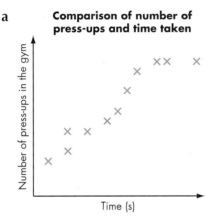

Comparison of number of press-ups and time taken

**b**

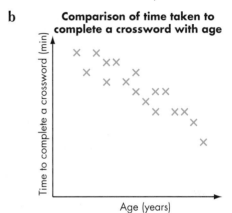

Comparison of time taken to complete a crossword with age

**c**

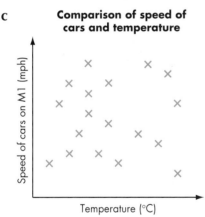

Comparison of speed of cars and temperature

**d**

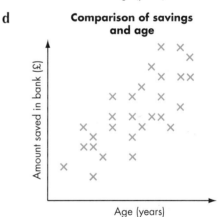

Comparison of savings and age

**2** In a science experiment, a ball is rolled along a desk top. The speed of the ball is measured at various points. The table shows the results.

| Distance from start (cm) | 10 | 20 | 30 | 40 | 50 | 60 | 70 | 80 |
|---|---|---|---|---|---|---|---|---|
| Speed (cm/s) | 18 | 16 | 13 | 10 | 7 | 5 | 3 | 0 |

   **a** Plot the data on a scatter diagram.

   **b** Draw a line of best fit.

   **c** What was the ball's speed likely to have been 5 cm from the start?

   **d** Estimate how far from the start the ball was when its speed was 12 cm/s.

**3** The table shows the marks for 10 students in their mathematics and geography examinations.

| Student | Anna | Bella | Chloe | Deia | Evie | Fatima | Grace | Hannah | Imogen | Jaya |
|---|---|---|---|---|---|---|---|---|---|---|
| Maths | 57 | 65 | 34 | 87 | 42 | 35 | 59 | 61 | 25 | 35 |
| Geog | 45 | 61 | 30 | 78 | 41 | 36 | 35 | 57 | 23 | 34 |

a Plot the data on a scatter diagram. Use the horizontal axis for the mathematics scores and mark it from 20 to 100. Use the vertical axis for the geography scores and mark it from 20 to 100.

b Draw a line of best fit.

c One of the students was ill when she took the geography examination. Which student was it most likely to be?

d Another student, Katya, was absent for the geography examination. She scored 75 in mathematics. What mark would you expect her to have scored in geography?

e Lynne was absent for the mathematics examination but scored 65 in geography. What mark would you expect her to have scored in mathematics?

**4** These are the heights, in centimetres, of 20 mothers and their 15-year-old daughters.

| Mother | 153 | 162 | 147 | 183 | 174 | 169 | 152 | 164 | 186 | 178 |
|---|---|---|---|---|---|---|---|---|---|---|
| Daughter | 145 | 155 | 142 | 167 | 167 | 151 | 145 | 152 | 163 | 168 |
| Mother | 175 | 173 | 158 | 168 | 181 | 173 | 166 | 162 | 180 | 156 |
| Daughter | 172 | 167 | 160 | 154 | 170 | 164 | 156 | 150 | 160 | 152 |

a Plot these results on a scatter diagram. Use the horizontal axis for the mothers' heights and mark it from 140 to 200. Use the vertical axis for the daughters' heights and mark it from 140 to 200.

b Is it true that the tall mothers have tall daughters?

**5** A teacher carried out a survey of his class. He asked students to say how many hours per week they spent playing sport and how many hours per week they spent watching TV. This table shows the results of the survey.

| Student | 1 | 2 | 3 | 4 | 5 | 6 | 7 | 8 | 9 | 10 |
|---|---|---|---|---|---|---|---|---|---|---|
| Hours playing sport | 12 | 3 | 5 | 15 | 11 | 0 | 9 | 7 | 6 | 12 |
| Hours watching TV | 18 | 26 | 24 | 16 | 19 | 27 | 12 | 13 | 17 | 14 |
| Student | 11 | 12 | 13 | 14 | 15 | 16 | 17 | 18 | 19 | 20 |
| Hours playing sport | 12 | 10 | 7 | 6 | 7 | 3 | 1 | 2 | 0 | 12 |
| Hours watching TV | 22 | 16 | 18 | 22 | 12 | 28 | 18 | 20 | 25 | 13 |

a Plot these results on a scatter diagram. Use the horizontal axis for the number of hours playing sport and the vertical axis for the number of hours watching TV.

b If you knew that another student from the class watched 8 hours of TV per week, would you be able to predict how long she or he spent playing sport? Explain why.

**6** The table shows the time taken and distance travelled by a taxi driver for 10 journeys one day.

| Distance (km) | 1.6 | 8.3 | 5.2 | 6.6 | 4.8 | 7.2 | 3.9 | 5.8 | 8.8 | 5.4 |
|---|---|---|---|---|---|---|---|---|---|---|
| Time (minutes) | 3 | 17 | 11 | 13 | 9 | 15 | 8 | 11 | 16 | 10 |

**a** Draw a scatter diagram with time on the horizontal axis.

**b** Draw a line of best fit on your diagram.

**c** Another taxi journey takes 5 minutes. How many kilometres would you expect the journey to have been?

**d** How long would you expect a journey of 4 km to take?

**e** Explain why you cannot give a time for a journey of 12 km.

 **7** Oliver records the time taken, in hours, and the average speed, in mph, for several different journeys.

| Time (h) | 0.5 | 0.8 | 1.1 | 1.3 | 1.6 | 1.75 | 2 | 2.4 | 2.6 |
|---|---|---|---|---|---|---|---|---|---|
| Speed (mph) | 42 | 38 | 27 | 30 | 22 | 23 | 21 | 9 | 8 |

Estimate the average speed for a journey of 90 minutes.

 **8** Describe what you would expect the scatter graph to look like if someone said that it showed negative correlation.

# 18.4 Grouped data and averages

This section will show you how to:

- identify the modal group
- calculate an estimate of the mean from a grouped table.

Data can either be discrete or continuous. **Discrete data** can only have certain values, for example, goals scored, marks in a test, number of children and shoe sizes. **Continuous data** can have any value within a range of values, for example, height, mass, time, area and capacity.

Sometimes the information you are given is grouped in some way, for example, because there are too many individual values to represent easily. This is called **grouped data**. In Example 7, the table shows the range of weekly pocket money given to Year 7 students in a particular class. In this grouped continuous data, values that are more than £1 and up to £2 are counted as one group.

When data is grouped, you can only *estimate* the mean as you do not have the actual values. Example 7 shows you how to do this.

Example 7

From the data in the table:

**a** write down the **modal group**

**b** calculate an estimate of the mean weekly pocket money.

| Pocket money, $p$ (£) | $0 < p \leqslant 1$ | $1 < p \leqslant 2$ | $2 < p \leqslant 3$ | $3 < p \leqslant 4$ | $4 < p \leqslant 5$ |
|---|---|---|---|---|---|
| Number of students | 2 | 5 | 5 | 9 | 15 |

**a** The modal group is the one with the largest frequency. Here the modal group is £4 to £5.

**b** The mean can only be estimated, since you do not have all the information. To estimate the mean, you assume that each person in each group has the **mid-class value**. Then you can build up the table as follows.

To find the mid-class value, add the two end values and divide the total by two.

| Pocket money, $p$ (£) | Frequency, $f$ | Mid-class value, $m$ | $f \times m$ |
|---|---|---|---|
| $0 < p \leqslant 1$ | 2 | 0.50 | 1.00 |
| $1 < p \leqslant 2$ | 5 | 1.50 | 7.50 |
| $2 < p \leqslant 3$ | 5 | 2.50 | 12.50 |
| $3 < p \leqslant 4$ | 9 | 3.50 | 31.50 |
| $4 < p \leqslant 5$ | 15 | 4.50 | 67.50 |
| Totals | 36 | | 120 |

The **estimated mean** is £120 ÷ 36 = £3.33 (correct to 2 decimal places).

Note the notation used for the groups.

$0 < p \leqslant 1$ means any amount above 0p up to and including £1.

$1 < p \leqslant 2$ means any amount above £1 up to and including £2.

You cannot work out the exact median from a grouped frequency table as you do not know the actual values. You can work out the group or class that includes the median. In the example, the median is the $18\frac{1}{2}$th value. This is in the $3 < p \leqslant 4$ class, so the median is between £3 and £4.

## Exercise 18D

**1** For each table of values:

**i** write down the modal class

**ii** calculate an estimate for the mean

**iii** write down the class in which the median lies.

**a**

| $x$ | $0 < x \leqslant 10$ | $10 < x \leqslant 20$ | $20 < x \leqslant 30$ | $30 < x \leqslant 40$ | $40 < x \leqslant 50$ |
|---|---|---|---|---|---|
| Frequency | 4 | 6 | 11 | 17 | 9 |

**b**

| $y$ | $0 < y \leqslant 100$ | $100 < y \leqslant 200$ | $200 < y \leqslant 300$ | $300 < y \leqslant 400$ | $400 < y \leqslant 500$ | $500 < y \leqslant 600$ |
|---|---|---|---|---|---|---|
| Frequency | 95 | 56 | 32 | 21 | 9 | 3 |

**c**

| $z$ | $0 < z \leqslant 5$ | $5 < z \leqslant 10$ | $10 < z \leqslant 15$ | $15 < z \leqslant 20$ |
|---|---|---|---|---|
| Frequency | 16 | 27 | 19 | 13 |

**d**

| Weeks | 1–3 | 4–6 | 7–9 | 10–12 | 13–15 |
|---|---|---|---|---|---|
| Frequency | 5 | 8 | 14 | 10 | 7 |

> **Hints and tips** When you copy the tables, draw them vertically, as in Example 7.

**2** Jason brought 100 pebbles back from the beach and weighed them all, to the nearest gram. His results are summarised in this table.

| Mass, $m$ (grams) | $40 < m \leqslant 60$ | $60 < m \leqslant 80$ | $80 < m \leqslant 100$ | $100 < m \leqslant 120$ | $120 < m \leqslant 140$ | $140 < m \leqslant 160$ |
|---|---|---|---|---|---|---|
| Frequency | 5 | 9 | 22 | 27 | 26 | 11 |

Work out:

**a** the modal mass of the pebbles

**b** an estimate for the total mass of the pebbles

**c** an estimate for the mean mass of the pebbles.

**3** A manufacturer tested 100 light bulbs to see whether the average life span of the bulbs was over 200 hours. The table summarises the results.

| Life span, $h$ (hours) | $150 < h \leqslant 175$ | $175 < h \leqslant 200$ | $200 < h \leqslant 225$ | $225 < h \leqslant 250$ | $250 < h \leqslant 275$ |
|---|---|---|---|---|---|
| Frequency | 24 | 45 | 18 | 10 | 3 |

**a** What is the modal length of time a bulb lasts?

**b** What percentage of bulbs last longer than 200 hours?

**c** Estimate the mean life span of the light bulbs.

**d** Do you think the test shows that the average life span is over 200 hours? Explain your answer fully.

**4** The table shows the distances run by an athlete who is training for a marathon.

| Distance, $d$ (miles) | $0 < d \leqslant 5$ | $5 < d \leqslant 10$ | $10 < d \leqslant 15$ | $15 < d \leqslant 20$ | $20 < d \leqslant 25$ |
|---|---|---|---|---|---|
| Frequency | 3 | 8 | 13 | 5 | 2 |

**a** It is recommended that an athlete's daily average mileage should be at least one-third of the distance of the race being trained for. A marathon is 26.2 miles. Is this athlete doing enough training?

**b** The athlete records the times of some runs and calculates that her average pace for all runs is $6\frac{1}{2}$ minutes for a mile. Explain why she is wrong to expect a finishing time of $26.2 \times 6\frac{1}{2}$ minutes $\approx 170$ minutes for the marathon.

**c** The athlete claims that the difference between her shortest and longest run is 21 miles. Could she be correct? Explain your answer.

**5** The owners of a boutique did a survey to find the average age of its customers. The table summarises the results.

| Age (years) | 14–18 | 19–20 | 21–26 | 27–35 | 36–50 |
|---|---|---|---|---|---|
| Frequency | 26 | 24 | 19 | 16 | 11 |

Calculate the average age of the boutique's customers.

**CM** **6** Three supermarkets each claimed to have the lowest average price increase over the year. The table summarises their average price increases.

| Price increase (pence) | 1–5 | 6–10 | 11–15 | 16–20 | 21–25 | 26–30 | 31–35 |
|---|---|---|---|---|---|---|---|
| Soundbuy | 4 | 10 | 14 | 23 | 19 | 8 | 2 |
| Springfields | 5 | 11 | 12 | 19 | 25 | 9 | 6 |
| Setco | 3 | 8 | 15 | 31 | 21 | 7 | 3 |

Compare the average price increases at each supermarket. Write a short report stating which supermarket you think has the lowest price increases over the year. Remember to use the data to support your claims.

**7** This table summarises the results of a survey about how quickly the AOne breakdown service attended calls that were not on a motorway. The times are rounded to the nearest minute.

| Time (minutes) | 1–15 | 16–30 | 31–45 | 46–60 | 61–75 | 76–90 | 91–105 |
|---|---|---|---|---|---|---|---|
| Frequency | 2 | 23 | 48 | 31 | 27 | 18 | 11 |

**a** How many calls were used in the survey?

**b** Estimate the mean time taken per call.

**c** Which average would the AOne use to advertise their average call-out time?

**d** What percentage of calls do the AOne get to within the hour?

**PS** **8** The table shows the numbers of runs scored by all the batsmen in a cricket competition.

| Runs | 0–9 | 10–19 | 20–29 | 30–39 | 40–49 |
|---|---|---|---|---|---|
| Frequency | 8 | 5 | 10 | 5 | 2 |

Helen noticed that two numbers were in the wrong part of the table and that this made a difference of 1.7 to the estimated mean.

Which two numbers were the wrong way round?

**CM** **9** The profit made each week by a charity shop is shown in the table below.

| Profit (£) | 0–500 | 501–1000 | 1001–1500 | 1501–2000 |
|---|---|---|---|---|
| Frequency | 15 | 26 | 8 | 3 |

Explain how you would estimate the mean profit made each week.

# Worked exemplars

 **1** The table shows the numbers of learners at each level for two practice driving tests, theory and practical.

| | Level | | | | | |
|---|---|---|---|---|---|---|
| | **Excellent** | **Very good** | **Good** | **Pass** | **Fail** | **Total number of learners** |
| **Theory** | 208 | 888 | 1032 | 696 | 56 | 2880 |
| **Practical** | 240 | 351 | 291 | 108 | 90 | 1080 |

**a** Represent the data for each of the two practice tests in a separate pie chart.

**b** On which test (theory or practical) do you think learners did better overall? Give a reason to justify your answer.

---

This question requires you to communicate your mathematical skills. Take care to interpret the information you are given accurately and show how you use it.

| **a** Theory | | | | Work out the angle for each level. |
|---|---|---|---|---|
| **Level** | **Frequency** | **Calculation** | **Angle** | Remember to check that all the angles add up to 360°. |
| Excellent | 208 | $\dfrac{208}{2880} \times 360°$ | 26° | |
| Very good | 888 | $\dfrac{888}{2880} \times 360°$ | 111° | Take care to interpret the information you are given accurately. |
| Good | 1032 | $\dfrac{1032}{2880} \times 360°$ | 129° | |
| Pass | 696 | $\dfrac{696}{2880} \times 360°$ | 87° | |
| Fail | 56 | $\dfrac{56}{2880} \times 360°$ | 7° | |

**Levels for driving test: theory**

Remember to label the pie chart. You do not need to show the angles.

| Practical | | | |
|---|---|---|---|
| Level | Frequency | Calculation | Angle |
| Excellent | 240 | $\frac{240}{1080} \times 360°$ | 80° |
| Very good | 351 | $\frac{351}{1080} \times 360°$ | 117° |
| Good | 291 | $\frac{291}{1080} \times 360°$ | 97° |
| Pass | 108 | $\frac{108}{1080} \times 360°$ | 36° |
| Fail | 90 | $\frac{90}{1080} \times 360°$ | 30° |

Work out the angle for each level.

Remember to check that all the angles add up to 360°.

Take care to interpret the information you are given accurately.

**Levels for driving test: practical**

Again, remember to label the pie chart. You do not need to show the angles.

**b** Overall the learners did better on the practical as 55% obtained Excellent or Very good, whereas only 38% obtained Excellent or Very good on the theory.

You must justify your answer. This is for interpreting and communicating the information accurately.

  Read these three statements.

- The older you are, the higher you score on a speed test.
- The higher the score on the speed test, the less TV you watch.
- The more TV you watch, the more hours you will sleep.

**a** Sketch a scatter diagram to illustrate the relationship described in each one.

**b** Draw a line of best fit on each diagram and describe the relationship that it shows.

This question is about communicating mathematically, so you need to show your reasoning and working and express your answer in a mathematical way.

**a**

**Comparison of test speed and age**

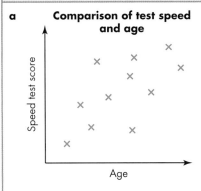

**Comparison of test speed and time watching TV**

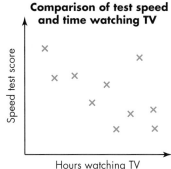

**Comparison of test speed and hours of sleep**

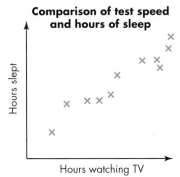

First you will need to decide on how to label the axes.

Then plot at least 10 points on the scatter diagram.

To show that you understand correlation, try to use a different type of correlation for each one where appropriate so that in part **b** you can use the terms: positive correlation, negative correlation, weak correlation, good correlation and strong correlation.

This translates the problem into a mathematical context.

**b**

**Comparison of speed test and age**

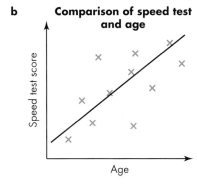

weak positive correlation

**Comparison of speed test and time watching TV**

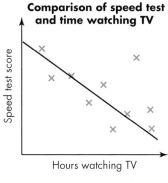

good negative correlation

**Comparison of speed test and hours of sleep**

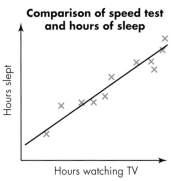

strong positive correlation

Draw a line of best fit on each diagram and identify the correct relationship for each one. Remember to make sure you leave about equal numbers of points on either side of the line.

# Ready to progress?

I know how to obtain a random sample from a population.
I know how to collect unbiased and reliable data for a sample.
I can draw and interpret pie charts.

I can write down the modal class from a grouped frequency table.
I can estimate the mean from a grouped frequency table.
I can draw and interpret a scatter diagram.
I can draw a line of best fit on a scatter diagram.

# Review questions

**1** Some male students were asked to choose their favourite sport.

The pie chart shows information about the results.

**a** 12 male students chose cricket.

Work out the number of male students who chose tennis.

**b** A second pie chart is to be drawn for 90 female students.

20 of the female students chose hockey.

Calculate the angle that will represent the female students that chose hockey in this pie chart.

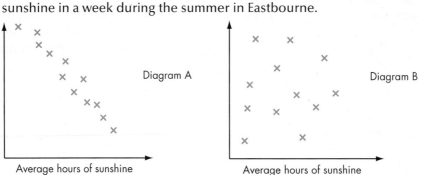

**EV** **2** The scatter diagrams show the results of a survey on the average number of hours of sunshine in a week during the summer in Eastbourne.

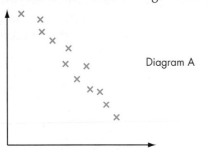

Diagram A

Average hours of sunshine

Diagram B

Average hours of sunshine

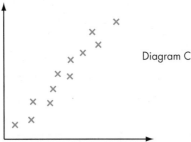

Diagram C

Average hours of sunshine

**a** Which scatter diagram do you think shows the average hours of sunshine plotted against:

 **i** the number of ice creams sold   **ii** the number of umbrellas sold

 **iii** the number of births in the town?

**b** Describe the correlation shown in each diagram.

**3** The table shows the time taken and distance travelled by a taxi driver for 10 journeys one day.

| Time (minutes) | Distance (km) |
|---|---|
| 3 | 1.7 |
| 17 | 8.3 |
| 11 | 5.1 |
| 13 | 6.7 |
| 9 | 4.7 |
| 15 | 7.3 |
| 8 | 3.8 |
| 11 | 5.7 |
| 16 | 8.7 |
| 10 | 5.3 |

**a** Plot a scatter diagram on a grid with time on the horizontal axis, from 0 to 20, and distance on the vertical axis, from 0 to 10.

**b** Draw a line of best fit on your diagram.

**c** A taxi journey takes 4 minutes. What distance is it likely to be?

**d** A taxi journey is 10 kilometres. How many minutes is it likely to take?

**4** Josh asked 30 students how many minutes they each took to get to school.

The table shows some information about his results.

| Time taken, $t$ (minutes) | Frequency |
|---|---|
| $0 < t \leqslant 10$ | 6 |
| $10 < t \leqslant 20$ | 11 |
| $20 < t \leqslant 30$ | 8 |
| $30 < t \leqslant 40$ | 5 |

**a** Write down the modal group.

**b** Which class interval contains the median time?

**c** Calculate an estimate for the mean time.

(MR) **5** The table shows the weekly pocket money of the students in one class. The values are rounded to the nearest pound.

| Pocket money (£) | 0–4 | 5–9 | 10–14 | 15–19 |
|---|---|---|---|---|
| Frequency | 4 | 6 | 12 | 8 |

**a** Sean says that he has estimated the mean amount of pocket money as £9.50.

Explain how you can tell Sean must be wrong without having to calculate the estimated mean.

**b** Calculate the correct estimate for the mean amount of pocket money.

(PS) **6** Naysha's school has 1200 students. She is in a class of 30 students. One day she noticed that that the headteacher was doing a survey over the whole school. Three boys and five girls in her class were involved in the survey.

Estimate the numbers of boys and girls in the whole school that were involved in the survey.

# 19 Geometry and measures: Constructions and loci

## This chapter is going to show you:

- how to construct a triangle from given data
- how to bisect a line and an angle
- how to construct angles of 60° and 90°
- how to define a locus
- how to solve locus problems.

## You should already know:

- how to measure lines and angles
- how to use scale drawings.

## About this chapter

When a major new train line is planned, people often object. Their reasons can be that it is too close to their village or that it ruins the countryside. This can mean that the path traced out for the new route has to meet certain conditions, such as missing a village or a forest.

In mathematics, when a point moves according to certain conditions, the path traced out is called a *locus* (plural *loci*). This is a Latin word that means 'place'. Loci have a range of practical applications, including helping to decide on suitable routes for a new train line.

This chapter is going to show you how to make accurate drawings and constructions from given conditions, including loci.

# 19.1 Constructing triangles

This section will show you how to:

- construct accurate drawings of triangles, using a pair of compasses, a protractor and a straight edge.

**Key terms**

| construct | included angle |

When drawing or constructing in mathematics, always use a sharp pencil (grade 2H is better than HB) to give you thin, clear lines.

The method you use to **construct** a triangle is decided by the information you have about the triangle. The three situations and methods are shown in the examples below.

**Example 1**

**All three sides known**

Construct a triangle with sides that are 5 cm, 4 cm and 6 cm long.

**Step 1:** Use a ruler to draw the longest side as the base.

In this case, the base is 6 cm, so draw a 6 cm line using a ruler.

**Step 2:** With the help of a pair of compasses, mark the length of the second longest side.

For the side length of 5 cm, set the compasses to 5 cm. Then place the point at one end of the 6 cm line and draw an arc.

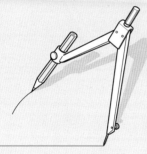

**Step 3:** With the help of a pair of compasses, mark the length of the shortest side.

Repeat for the 4 cm side at the other end of the 6 cm line.

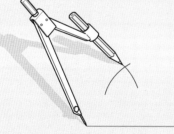

**Step 4:** Complete the triangle using the point of intersection of the arcs.

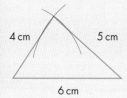

4 cm    5 cm

6 cm

Example 2

**Two sides and the included angle known**

Construct a triangle ABC, in which AB is 6 cm, BC is 5 cm and the **included angle** ABC is 55°.

**Step 1:** Use a ruler to draw AB (the longest side) as the base and label the ends of your line.

A ——————— B

**Step 2:** Place the protractor along AB with its centre on B and make a point on the diagram at the 55° mark.

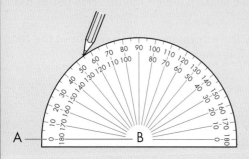

**Step 3:** Draw a line from B through the 55° point. From B, use a pair of compasses to mark 5 cm along this line.

Label the point where the arc cuts the line as C.

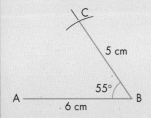

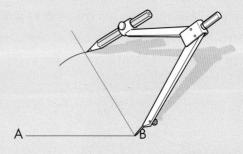

**Step 4:** Complete the triangle by joining A and C.

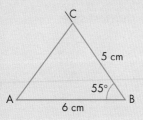

Example 3

**Two angles and a side known**

Construct a triangle ABC, in which AB is 7 cm, angle BAC is 40° and angle ABC is 65°.

> **Hints and tips**   When you know two angles of a triangle, you also know the third.

**Step 1:** Use a ruler to draw and label the known side, AB. In this case, AB is 7 cm.

A ————————————————— B

(continued)

**Step 2:** Centre the protractor on A and mark the angle of 40°. Draw a clear, clean line from A through this point.

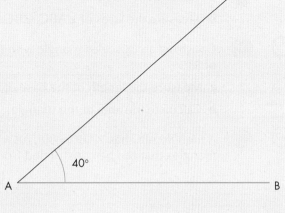

**Step 3:** Centre the protractor on B and mark the angle of 65°. Draw a clear, clean line from B through this point, to intersect the 40° line drawn from A. Label the point of intersection as C.

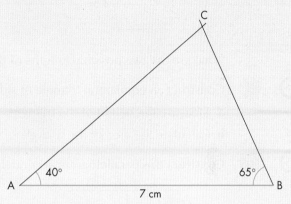

## Exercise 19A

**1** Construct each of these triangles accurately and then measure the sides and angles not marked in the diagrams.

**a**
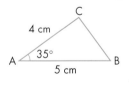

**b**
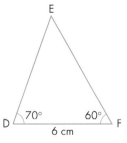

**c**
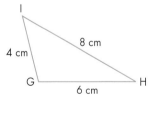

**d**
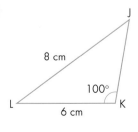

**e**
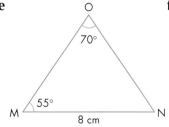

**f**
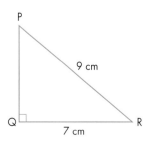

**2**   **a** Construct triangle ABC, where AB = 7 cm, BC = 6 cm and AC = 5 cm.

    **b** Measure the sizes of ∠ABC, ∠BCA and ∠CAB.

(PS)  **3**   Construct an isosceles triangle with two sides of length 7 cm and an included angle of 50°.

    **a** Measure the length of the base of the triangle.

    **b** Calculate the area of the triangle.

(MR)  **4**   A triangle ABC has ∠ABC = 30°, AB = 6 cm and AC = 4 cm. There are two different triangles that can be constructed from this information.

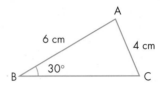

What are the two possible lengths of BC?

(PS)  **5**   Construct an equilateral triangle of side length 5 cm.

    **a** Measure the height of the triangle.

    **b** Calculate the area of the triangle.

(PS)  **6**   Construct this parallelogram accurately.

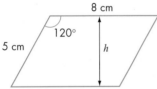

    **a** Measure the height of the parallelogram.

    **b** Calculate the area of the parallelogram.

(PS)  **7**   Construct the triangle with the largest possible area that has a perimeter of 12 cm.

(CM)  **8**   Anil says that, as long as he knows all three angles of a triangle, he can draw it. Why is Anil wrong?

# 19.2 Bisectors

This section will show you how to:

- construct the bisectors of lines and angles
- construct angles of 60° and 90°.

To bisect means to divide in half. So a bisector divides something into two equal parts.

A **perpendicular bisector** of a straight line divides the line into two equal lengths and is at right angles to it.

An **angle bisector** is the straight line that divides an angle into two equal angles.

## To construct the perpendicular bisector of a line

**Step 1:** Here is a line to bisect.

———————

**Step 2:** Open your compasses to a radius of more than half the length of the line. Using each end of the line as a centre, draw two intersecting arcs without changing the radius of your compasses.

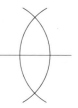

**Step 3:** Join the two points where the arcs intersect. This line is the perpendicular bisector of the original line.

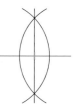

## To construct an angle bisector

**Step 1:** Here is an angle to bisect.

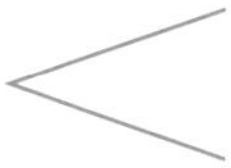

**Step 2:** Open your compasses to any radius that is less than the length of the lines forming the angle. Using the vertex of the angle as the centre, draw an arc through both lines.

**Step 3:** Using the two points where this arc intersects the arms as centres, draw two arcs that intersect (without changing the radius of your compasses).

**Step 4:** Join the point where these two arcs intersect to the vertex of the angle.

This line is the angle bisector.

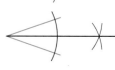

## To construct an angle of 60°

**Step 1:** Draw a line and mark a point on it.

———•————————

**Step 2:** Open the compasses and, using the point as the centre, draw an arc that crosses the line and extends almost above the point.

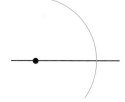

**Step 3:** Keep the compasses set to the same radius. Using the point where the first arc crosses the line as a centre, draw another arc that intersects the first one.

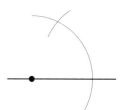

**Step 4:** Join the original point to the point where the two arcs intersect.

**Step 5:** Use a protractor to check that the angle is 60°.

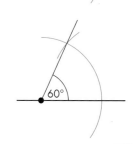

## To construct a perpendicular from a point on a line (an angle of 90°)

**Step 1:** Draw a line and mark a point on it.

**Step 2:** Open your compasses and, with point A as the centre, draw two short arcs to intersect the line at each side of the point.

**Step 3:** Open your compasses wider. From the intersections, draw equal arcs to intersect at B above the line.

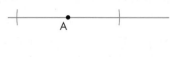

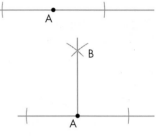

**Step 4:** Join AB.

AB is perpendicular to the line.

Note that to construct a 90° angle at the end of a line, you would first have to extend the line.

You could be even more accurate by also drawing two arcs underneath the line. This would give three points in-line.

## To construct a perpendicular from a point to a line

Note that this perpendicular distance from a point to a line is the shortest distance to the line.

**Step 1:** With point A as the centre, draw an arc that intersects the line at two points.

**Step 2:** Using these two points of intersection as centres, draw two arcs to intersect each other both above and below the line.

**Step 3:** Join the two points where the arcs intersect.

You can now draw the line that passes through point A and is perpendicular to the line.

**Note:** When a question says *construct*, you must *only* use a pair of compasses, not a protractor. When it says *draw*, you may use whatever you can to produce an accurate diagram. However, when constructing you may use your protractor to check your accuracy.

## Exercise 19B 🖩

 **1** Draw a line 7 cm long and construct the perpendicular bisector of the line. Check your accuracy by measuring each half.

> Hints and tips | Remember to show your construction lines.

 **2** Construct a circle of about 4 cm radius.

Draw a triangle inside the circle so that the corners of the triangle touch the circle.

Construct the perpendicular bisector of each side of the triangle.

The bisectors should all meet at the centre of the circle.

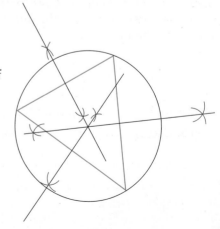

**3** **a** Draw any triangle with sides that are between 5 cm and 10 cm.

    **b** Construct the perpendicular bisector of each side.

      Your bisectors should all intersect at the same point.

    **c** Using this point as the centre, draw a circle that goes through every vertex of the triangle.

**4** Repeat question **3** with a different triangle and check that you get a similar result.

**5** **a** Draw the following quadrilateral.

    **b** Construct the perpendicular bisector of each side. These should all intersect at the same point.

    **c** Using this point as the centre, draw a circle that goes through the quadrilateral at each vertex.

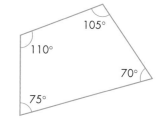

**(EV)** **6** **a** Draw an angle of 50°.

    **b** Construct the angle bisector.

    **c** Check your accuracy by measuring each half.

**7** **a** Draw a circle with a radius of 3 cm.

    **b** A tangent is a line that passes a circle to just touch it at one point.

      Draw a triangle around the circle so that each side is a tangent to the circle, as shown.

    **c** Bisect each angle of the triangle.

    **d** Where do all these bisectors meet?

**8** **a** Draw any triangle with sides that are between 5 cm and 10 cm.

    **b** At each angle construct the angle bisector. All three bisectors should intersect at the same point.

    **c** Using this point as the centre, draw a circle that just touches the sides of the triangle.

**9** Repeat question **8** with a different triangle.

**(PS)** **10** Draw a circle with radius about 4 cm.

Draw a quadrilateral, *not* a rectangle, inside the circle so that each vertex is on the circumference.

Construct the perpendicular bisector of each side of the quadrilateral.

Where is the point where these bisectors all meet?

**(CM)** **11** Briefly outline how you would *construct* a triangle with angles 90°, 60° and 30°.

**12** **a** Draw a line AB, 6 cm long, and construct an angle of 90° at A.

    **b** Bisect this angle to construct an angle of 45°.

**13** **a** Draw a line AB, 6 cm long, and construct an angle of 60° at A.

    **b** Bisect this angle to construct an angle of 30°.

**14** Draw a line AB, 6 cm long, and mark a point C, 4 cm above the middle of the line.

Construct the perpendicular from the point C to the line AB.

# 19.3 Defining a locus

This section will show you how to:

- draw a locus for a given rule.

**Key terms**

| equidistant | loci (locus) |

A **locus** (plural **loci**) is the movement of a point according to a given rule.

---

**Example 4**

What is the locus of a point that is always 5 cm away from a fixed point A?

The locus of the point (P) is such that AP = 5 cm. This will give a circle of radius 5 cm, centre A.

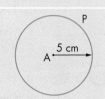

---

**Example 5**

What is the locus of a point that is always the same distance from two fixed points A and B?

The locus of the point P is such that AP = BP.

This will have a locus that is the perpendicular bisector of the line joining A and B.

Note that a point that is always the same distance from two points is **equidistant** from the two points.

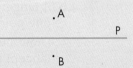

---

**Example 6**

What is the locus of a point that is always 5 cm from a line AB?

A point that moves so that it is always 5 cm from a line AB will have a locus that is a racetrack shape around the line.

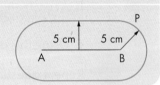

---

Questions will often ask about practical situations.

---

**Example 7**

In a grassy, flat field, a horse is tied to a stake by a rope that is 10 m long. What is the shape of the area that the horse can graze?

In reality, the horse may not be able to reach the full 10 m if the rope is tied around its neck but ignore details like that. You 'model' the situation by saying that the horse can move around in a 10 m circle and graze all the grass within that circle.

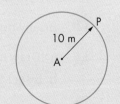

In this example, the locus is the whole of the area inside the circle.

You can express this mathematically as:

the locus of the point P is such that AP ⩽ 10 m

where A is the position of the stake and P is any point on the circumference of the circle.

**1** A is a fixed point. Construct the locus of the point P in each situation.

**a** AP = 2 cm      **b** AP = 4 cm      **c** AP = 5 cm

> Hints and tips   Sketch the situation before doing an accurate drawing.

**2** A and B are two fixed points 5 cm apart. Draw the locus of the point P for each situation.

**a** AP = BP      **b** AP = 4 cm and BP = 4 cm

**c** P is always within 2 cm of the line AB.

**(PS) 3** **a** A horse is tied in a field on a rope 4 m long. Describe or sketch the area that the horse can graze.

**b** The horse is still tied by the same rope but there is now a long, straight fence running 2 m from the stake. Draw the area that the horse can now graze.

**4** ABCD is a square of side 4 cm. In each of these loci, the point P moves only inside the square. Sketch the locus in each case.

**a** AP = BP      **b** AP < BP      **c** AP = CP

**d** CP < 4 cm      **e** CP > 2 cm      **f** CP > 5 cm

**(MR) 5** Which one of these diagrams shows the locus of a point on the rim of a bicycle wheel as it moves along a flat road?

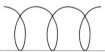 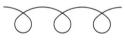  

**6** Draw the locus of the centre of the wheel for the bicycle in question **5**.

**(PS) 7** ABC is a triangle.

The region R is defined as the set of points inside the triangle such that:

- they are closer to the line AB than the line AC
- they are closer to the point A than the point C.

Using a straight edge and a pair of compasses, construct the region R.

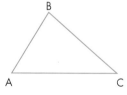

**(PS) 8** ABCD is a rectangle.

Copy the diagram and draw the locus of all points that are 2 cm from the edges of the rectangle.

**(CM) 9** How can you draw an equilateral triangle with sides of 3 cm using only a ruler?

**(EV) 10** Marcus drew a triangle and asked Gary to describe the locus of all the points that were 1 cm away from the sides of the triangle.

Gary said, "It will be a triangle inside and a triangle outside the original triangle."

Comment on Gary's statement.

# 19.4 Loci problems

This section will show you how to:

- solve practical problems using loci.

Most loci problems you come across will be practical, as in the next example.

---

**Example 8**

A radio company wants to find a site for a transmitter. The transmitter must be the same distance from Doncaster and Leeds and within 20 miles of Sheffield. Illustrate, on a scale diagram, the possible locations of the transmitter.

In mathematical terms, the possible sites for the transmitter will be along the perpendicular bisector between Leeds and Doncaster and the area within a circle of radius 20 miles from Sheffield.

Draw this on a diagram. This diagram is drawn to a scale of 1 cm = 10 miles. The transmitter can be built anywhere along the thick part of the blue line.

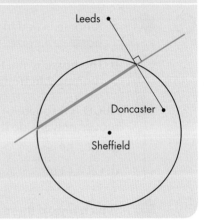

---

**Example 9**

A radar station in Birmingham has a range of 150 km (that is, it can pick up any aircraft within a radius of 150 km). Another radar station in Norwich has a range of 100 km.

Can an aircraft be picked up by both radar stations at the same time?

Draw a diagram to represent the situation. You need to use an appropriate scale to draw a circle of radius 150 km around Birmingham and another circle of radius 100 km around Norwich. An aircraft could be picked up by both radar stations in the shaded area where the two circles overlap.

---

**Example 10**

A dog is tied by a rope, 3 m long, to the corner of a shed, 4 m by 2 m.
Draw the area that the dog can guard effectively.

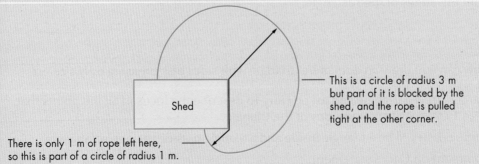

This is a circle of radius 3 m but part of it is blocked by the shed, and the rope is pulled tight at the other corner.

There is only 1 m of rope left here, so this is part of a circle of radius 1 m.

---

**1** In a field, a horse is tied to a stake by a rope 6 m long. Draw the locus of the area that the horse can graze. Use a scale of 1 cm to 2 m.

For questions **2** to **6**, you should start by sketching the picture given in each question on a 6 × 6 grid, where each square is 1 cm by 1 cm. The scale for each question is given.

**2** **a** A goat is tied by a rope, 7 m long, in the top right-hand corner of a field with a fence at each side.

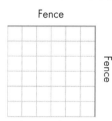

What is the locus of the area that the goat can graze? Use a scale of 1 cm to 2 m.

**b** A horse is tied to a stake near a corner of a fenced field, at a point 4 m from each fence. The rope is 6 m long. Sketch the area that the horse can graze. Use a scale of 1 cm to 2 m.

**3** A cow is tied to a rail at the top of a fence 6 m long. The rope is 3 m long.

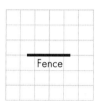

Sketch the area that the cow can graze. Use a scale of 1 cm to 2 m.

**4** A horse is tied to a corner of a shed, 2 m by 1 m. The rope is 2 m long.

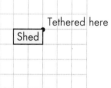

Sketch the area that the horse can graze. Use a scale of 1 cm to 1 m.

**5** A goat is tied by a 4 m rope to a stake at one corner of a pen, 4 m by 3 m.

Sketch the area of the pen where the goat cannot graze. Use a scale of 1 cm to 1 m.

**(PS)** **6** A puppy is tied to a stake by a rope, 1.5 m long, on a flat lawn. There are two raised brick flower beds on the lawn. The stake is at one corner of a bed, as shown.

Sketch the area that the puppy is free to move in. Use a scale of 1 cm to 1 m.

For questions **7** to **15** and **17**, you should use a copy of the map opposite. For each question, trace the map and mark on the points that are relevant to that question.

**7** A radio station broadcasts from London on a frequency of 1000 kHz with a range of 300 km. Another radio station broadcasts from Glasgow on the same frequency with a range of 200 km.

   **a** Sketch the area to which each station can broadcast.

   **b** Will they interfere with each other?

   **c** If the Glasgow station increases its range to 400 km, will they then interfere with each other?

**8** The radar at Leeds Bradford International Airport has a range of 200 km. The radar at Exeter International Airport has a range of 200 km.

   **a** Will a plane flying over Birmingham be detected by the Leeds radar?

   **b** Sketch the area where a plane can be picked up by both radars at the same time.

**9** A radio transmitter is to be built according to these rules.
   • It must be the same distance from York and Birmingham.
   • It must be within 350 km of Glasgow.
   • It must be within 250 km of London.

   **a** Sketch the line that is the same distance from York and Birmingham.

   **b** Sketch the area that is within 350 km of Glasgow and 250 km of London.

   **c** Show clearly the possible places where the transmitter could be built.

**10** A radio transmitter centred at Birmingham is designed to give good reception in an area greater than 150 km and less than 250 km from the transmitter. Sketch the area of good reception.

**11** Three radio stations pick up a distress call from a boat in the Irish Sea. The station at Glasgow can tell from the strength of the signal that the boat is within 300 km of the station. The station at York can tell that the boat is between 200 km and 300 km from York. The station at London can tell that it is less than 400 km from London. Sketch the area where the boat could be.

**12** Sketch the area that is between 200 km and 300 km from Newcastle upon Tyne, and between 150 km and 250 km from Bristol.

**13** An oil rig is positioned in the North Sea so that it is the same distance from Newcastle upon Tyne and Manchester. It is also the same distance from Sheffield and Norwich. Draw the line that shows all the points that are the same distance from Newcastle upon Tyne and Manchester. Repeat for the points that are the same distance from Sheffield and Norwich and work out where the oil rig is located.

(PS) **14** While looking at a map, Fred notices that his house is the same distance from Glasgow, Norwich and Exeter. Where is it?

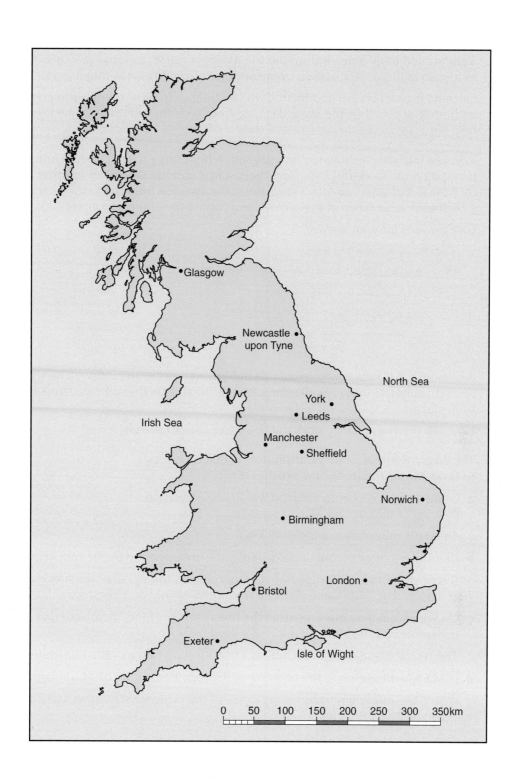

**15** Tariq wanted to fly himself from the Isle of Wight north, towards Scotland. He wanted to remain at the same distance from London as from Bristol as much as possible.

Once he is past London and Bristol, which city should he aim toward to keep him, as accurately as possible, the same distance from London and Bristol? Use the map to help you.

 **16** Wathsea Harbour is shown in the diagram. A boat sets off from point A and steers so that it stays the same distance from the sea wall and the West Pier. Another boat sets off from B and steers so that it stays the same distance from the East Pier and the sea wall. If each boat sailed at the same speed, would they hit each other?

Give reasons for your answer.

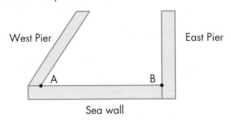

 **17** A distress call is heard by coastguards in both Newcastle and Bristol. The signal strength suggests that the call comes from a ship that is the same distance from both places.

How can the coastguards work out the area of sea to search?

 **18** The diagram shows a radio transmitter (T) that broadcasts to two towns, Arnold (A) and Beeston (B).

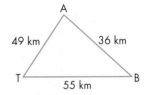

**a** Using a scale of 1 cm to 10 km, draw an accurate scale drawing of the triangle ATB.

The transmitter has a range of 40 km.

**b** Draw accurately, on your scale drawing, the area covered by the transmitter.

It is planned to build a repeater station to repeat the signal of the first transmitter, at a point that is the same distance from both Arnold and Beeston.

**c** On your scale drawing, construct the line that represents the places that the new transmitter can be built.

The repeater station is to be built at the maximum range of the transmitter.

**d i** Mark, with a letter R, the position of the repeater station on your diagram.

**ii** Find the minimum transmitting range of the repeater station so that both towns can receive the signal.

 **19** The diagram below is drawn to scale. It shows a wheel, centre A, of radius 25 cm, which rolls along the ground and then mounts a step of height 15 cm.

Draw the locus of A as the wheel approaches the step, mounts it and moves on.

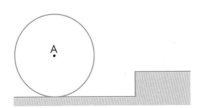

# Worked exemplars

 **1** A tangent to a circle is a line that just touches the circle at a point and is perpendicular to the radius of the circle.

Construct a tangent to a circle and clearly show the method you used.

> This is a mathematical reasoning question so you need to show the construction of the tangent and explain how you constructed it.

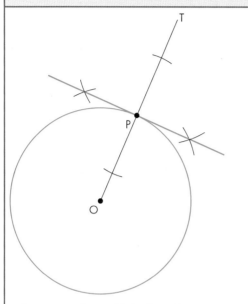

Draw the diagram first and then explain each stage clearly.

Make sure all your construction lines are clearly visible.

I drew a circle and labelled the centre O. I chose and labelled a point, P, on the circumference.
Next I drew a line, OT, passing through P.
I then used a pair of compasses to make equidistant arcs on OT from point P. I extended the width of the compasses and using the arcs on OT as the centres, drew intersecting arcs. I then drew a straight line between the intersecting arcs. This passed through P and was perpendicular to the radius OP and so is a tangent to the circle.

2 The map shows three boats, A, B and C, on a lake. There is a straight path along one edge of the lake and treasure hidden at the bottom of the lake.

The treasure is:

- between 150 m and 250 m from B
- closer to A than C
- more than 100 m from the path.

Shade the region where the treasure lies.

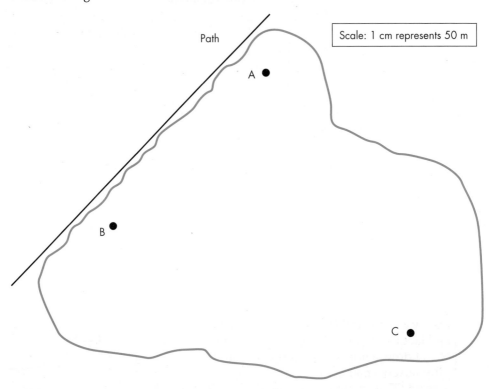

| This is a problem-solving question so you need to work out how to use the information given to find the treasure. | |
|---|---|
| You need to first construct the locus of each condition accurately using the given scale and then identify and shade the correct region. | For the first condition, draw two circles with their centres at B. The scale is 1 cm to 50 m so the first circle should have radius 150 ÷ 50 = 3 cm and the second should have 250 ÷ 50 = 5 cm. |
| | For the second condition, construct the perpendicular bisector of AC. Your construction arcs should be clearly visible. |
| | For the third condition, draw a parallel line 100 ÷ 50 = 2 cm from the path. |

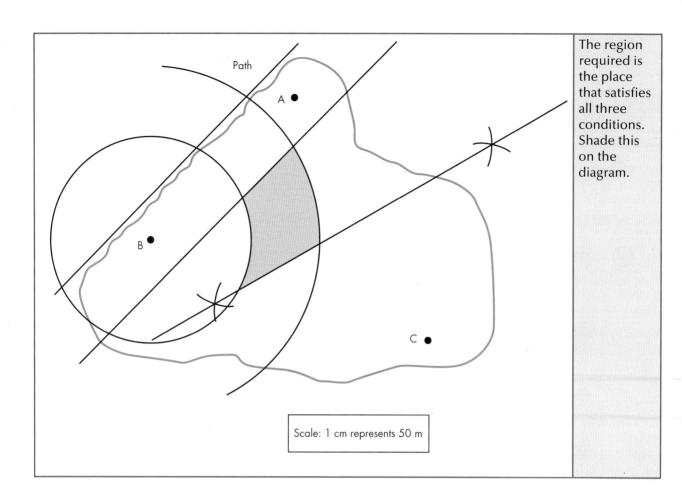

The region required is the place that satisfies all three conditions. Shade this on the diagram.

Path

A ●

B ●

C ●

Scale: 1 cm represents 50 m

# Ready to progress?

I can make accurate drawings of or construct triangles.
I can construct perpendiculars as well as perpendicular bisectors of lines and angle bisectors.
I can construct certain angles without using a protractor.
I understand what is meant by a locus and can use loci to solve problems.

# Review questions

**1**   **a**   Use a straight edge and a pair of compasses to construct an angle of 90°.

     **b**   Construct a bisector of this right angle to create an angle of 45°.

     **c**   Comment on how accurate you have been.

**2**   Use a straight edge and a pair of compasses to construct an equilateral triangle with side length 5 cm.

     Show all your construction lines.

**3**   **a**   Draw a line AB that is 6 cm long.

     **b**   Draw the locus of all points that are exactly 2 cm from the line AB.

**4**   **a**   *Construct* a rectangle ABCD with side length 4 cm.

     **b**   Shade the set of points inside the rectangle that are less than 3 cm from either point A or D or both points A and D and less than 1.5 cm from the line BC.

**5**   **a**   Using a pencil, ruler and compasses only, draw a line AB 8 cm long and construct its perpendicular bisector.

     **b**   Copy and complete this sentence.

     The perpendicular bisector of the line AB is the locus of points that are…

**6**   **a**   With straight edge and compasses only, construct a hexagon ABCDEF, with side length 3 cm.

     **b**   The region R is defined as the set of points inside the octagon that are:

       • closer to the side AB than the side BC    and

       • closer to the point C than the point F.

     Accurately construct the region R.

**7**   **a**   Construct the triangle with sides of length 8 cm, 7 cm and 6 cm.

     **b**   Calculate the area of this triangle.

**8**   **a**   Construct the parallelogram with sides of length 6 cm and 4 cm. The included angle is 130°.

     **b**   Calculate the area of the parallelogram.

**9** The diagram shows an L shape, with one line 7 cm and the other 3 cm.

Oliver said, "If I draw the locus of all points 2 cm from the L shape, it will give a solid L shape with all straight lines."

Show that Oliver is incorrect.

**10** Two lifeboat stations P and Q receive a distress call from a boat.

The boat is within 5 km of station P.

The boat is within 7 km of station Q.

On a copy of the diagram, shade the possible locations of the boat.

Scale: 1 cm represents 2 km

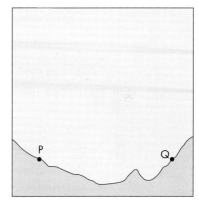

**11** A new call-centre is to be built, in line with these rules.

- It must be the same distance from Glasgow and London.
- It must be within 200 miles of Birmingham.
- It must be within 100 miles of Sheffield.

**a** Sketch the line that is the same distance from Glasgow and London.

**b** Sketch the region that is within 200 miles of Birmingham.

**c** Sketch the region that is within 100 miles of Sheffield.

**d** Show the possible places where the call-centre can be built.

# 20 Geometry and measures: Curved shapes and pyramids

## This chapter is going to show you:

- how to calculate the length of an arc
- how to calculate the area and angle of a sector
- how to calculate the volume and surface area of a pyramid
- how to calculate the volume and surface area of a cone and a sphere.

## You should already know:

- the formula for the area of a rectangle (area = length × width *or A = lw*)
- the formula for the area of a triangle (area = $\frac{1}{2}$ × base × height *or A = $\frac{1}{2}$bh*)
- the formula for the area of a circle (area = π × radius² *or A = πr²*)
- the formula for the circumference of a circle (circumference = π × diameter *or C = πd*)
- the formula for the volume of a prism (volume = area of cross section × length).

## About this chapter

About 70% of the Earth's surface is covered with water. What area of water is that? How big is the Earth's surface? What is the volume of air in the atmosphere above the Earth's surface?

The Earth is roughly the shape of a sphere with a circumference of approximately 40 000 km, but how can you use that to work out the surface area? On a smaller scale, consider the amount of material needed to make a football. The circumference of a football is approximately 70 cm, but can you use that to work out the area of its surface?

Volume calculations apply to all sorts of 3D shapes. You might be able to imagine how to calculate the surface area of a volcano, but a geologist would need to know the amount of rock that could be thrown out of the volcano by an eruption. This means they would need to know the volume of the volcano.

In this chapter, you will learn more about calculating the surface areas and volumes of prisms by learning how to calculate the volumes and surface areas of some more complex shapes.

# 20.1 Sectors

This section will show you how to:

- calculate the length of an arc
- calculate the area and angle of a sector.

An **arc** is any part of the edge of a circle. A **sector** is formed by two radii and an arc.

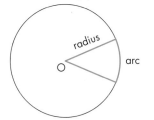

The angle **subtended** at the centre of the circle by the arc of a sector is called the angle of the sector.

When you divide a circle into only two sectors, the larger one is called the major sector and the smaller one is called the minor sector. Their equivalent arcs are called the major arc and the minor arc.

### Length of an arc and area of a sector

A sector is a fraction of the whole circle. The size of the fraction depends on the size of angle of the sector. The angle is often written as $\theta$, a Greek letter pronounced *theta*.

For example, the sector shown in the diagram represents the fraction $\frac{\theta}{360}$.

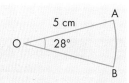

You can use this to work out both the length of an arc and the area of a sector.

For the length of an arc, multiply the fraction by the formula for the circumference.

length of an arc $= \frac{\theta}{360} \times 2\pi r$   or   $\frac{\theta}{360} \times \pi d$

For the area of a sector, multiply the fraction by the formula for the area of a circle.

area of a sector $= \frac{\theta}{360} \times \pi r^2$

---

**Example 1**

Work out the length of the arc AB and the area of the sector AOB in the diagram.

The angle of the sector is 28° and the radius is 5 cm. So:

length of arc $\quad = \frac{\theta}{360} \times 2\pi r$

$\qquad\qquad\quad = \frac{28}{360} \times \pi \times 2 \times 5$

$\qquad\qquad\quad = 2.4$ cm (1 dp)

area of sector $\quad = \frac{\theta}{360} \times \pi r^2$

$\qquad\qquad\quad = \frac{28}{360} \times \pi \times 5^2$

$\qquad\qquad\quad = 6.1$ cm² (1 dp)

**Example 2**

A sector has a radius of 8 cm and the length of the arc is 5 cm.

What is the angle of the sector?

Use the formula for the length of the arc.

Length of arc $= \dfrac{\theta}{360} \times 2\pi r$

$5 = \dfrac{\theta}{360} \times 2 \times \pi \times 8$     Substitute $r = 8$ and length $= 5$.

$5 \times 360 = \theta \times 2 \times \pi \times 8$

$\theta = \dfrac{5 \times 360}{2 \times \pi \times 8}$

$= 35.8°$ (1 dp)

## Exercise 20A 🖩

**1** For each sector, calculate:

   **i** the length of the arc   **ii** the area of the sector.

**a**
40°
8 cm

**b**
60°
5 cm

**c**
180°
7 cm

**d**
78°
12 cm

**2** For each sector, calculate:

   **i** the length of the arc   **ii** the perimeter of the sector.

**a**
65°
6 cm

**b**
50°
9 cm

**c**
120°
4 cm

**d**
100°
10 cm

**e**
22°
8.5 cm

**3** An arc subtends an angle of 60° at the centre of a circle with a diameter of 12 cm.
Calculate the length of the arc and the area of a sector. Leave your answer in terms of π.

**4** **a** Write down the angle of the sector.

   **b** Calculate the length of the arc.

   **c** Calculate the total perimeter of the sector.

11 cm

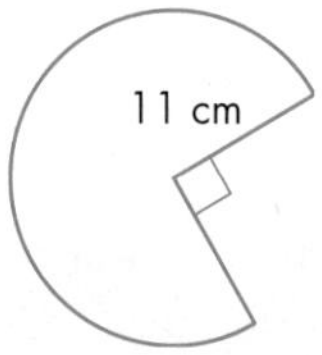

**(MR)** **5** Show that the area of the sector shown has the same area as a circle
with diameter 8 cm.

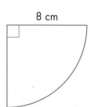
8 cm

**6** Calculate the size of θ in each diagram. Give your answers to 1 decimal place.

**a**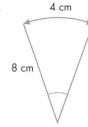

4 cm

8 cm

**b**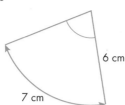

6 cm

7 cm

**c**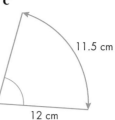

11.5 cm

12 cm

EV **7** Gabby said: 'The area of sector A is half the size of sector B.'

Is Gabby's statement true?

**A**

110°

7 cm

**B**

50°

8 cm

**8** O is the centre of a circle of radius 12.5 cm.

Calculate the length of the arc ACB.

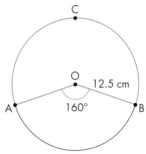

C

O  12.5 cm

A  160°  B

PS **9** **a** A sector has a radius of 6 cm and the length of the arc is 7 cm. What is the angle of the sector?

**b** A sector has a radius of 5 cm and the length of the arc is 4 cm. What is the area of the sector?

# 20.2 Pyramids

## This section will show you how to:

- calculate the volume and surface area of a pyramid.

| Key terms | |
|---|---|
| apex | pyramid |
| slant height | vertical height |

A **pyramid** is a 3D shape with a base from which triangular faces rise to a common vertex. This is called the **apex**.

## Volume

The volume of a pyramid is:

volume = $\frac{1}{3}$ × area of base × **vertical height**

$V = \frac{1}{3}Ah$

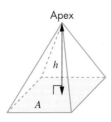

Apex

h

A

Example 3

Calculate the volume of the pyramid.

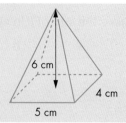

First work out the area of the base.

$$\text{area of the base} = lw$$
$$= 5 \times 4$$
$$= 20 \text{ cm}^2$$

Then use the formula to calculate the volume.

$$\text{Volume} = \frac{1}{3}Ah$$
$$= \frac{1}{3} \times 20 \times 6$$
$$= 40 \text{ cm}^3$$

Example 4

Calculate the volume of this pyramid.

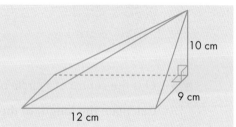

First work out the area of the base.

$$\text{area of the base} = lw$$
$$= 12 \times 9$$
$$= 108 \text{ cm}^2$$

Then use the formula to calculate the volume.

$$\text{Volume} = \frac{1}{3}Ah$$
$$= \frac{1}{3} \times 108 \times 10$$
$$= 360 \text{ cm}^3$$

Notice that the apex is directly above one vertex of the rectangular base.

## Exercise 20B 🖩

**1** Calculate the volume of each of these rectangular-based pyramids.

**a**
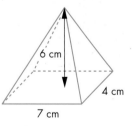

**b**
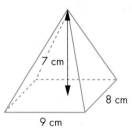

**c**
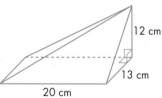

**d**
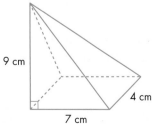

**e**
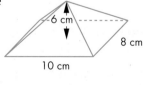

**MR** **2** A pyramid has a square base of side 9 cm and a vertical height of 10 cm.

Show that the volume of the pyramid is 270 cm³.

**3** Calculate the volume of each shape.

a

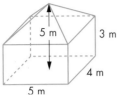
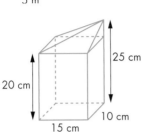

b

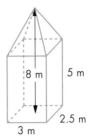

c

**PS** **4** A solid, plastic pyramid has a square base of side 4 cm and a height of 3 cm.

The plastic has a density of 13 g/cm³.

Calculate the mass of the pyramid.

> **Hints and tips** 1 cm³ of the plastic has a mass of 13 g.

**MR** **5** A crystal is formed from two square-based pyramids, joined at their bases (see diagram).

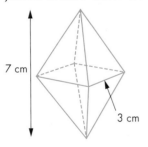

The crystal has a mass of 31.5 g.

What is the density of the crystal?

> **Hints and tips** The density is how heavy each cubic centimetre of the crystal is, measured in g/cm³.

**MR** **6** A pyramid has a square base of side 6.4 cm.

Its volume is 81.3 cm³.

Show that the height of the pyramid is 6.0 cm.

**PS** **7** A square-based pyramid has the same volume as a cube of side 10 cm.

The height of the pyramid is the same as the side of the square base.

Calculate the height of the pyramid.

## Surface area

You can calculate the total surface area of a pyramid by adding together the areas of all its faces. This is the sum of the area of the base and each of the triangular faces.

The pyramid shown has a square base of side 5 cm and a **slant height** of 6 cm.

Work out the total surface area of the pyramid.

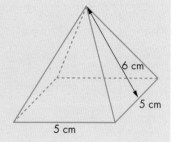

Calculate the area of each face.

Area of base = $5^2$

$= 25$ cm$^2$

The base is a square so all the triangular faces are the same.

Area of each triangular face $= \frac{1}{2} \times b \times h$

$= \frac{1}{2} \times 5 \times 6$

$= 15$ cm$^2$

Add them together.

Total area $= (4 \times 15) + 25$

$= 85$ cm$^2$

## Exercise 20C

**1**  Calculate the total surface area of each pyramid.

**a**

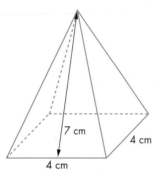

**b**

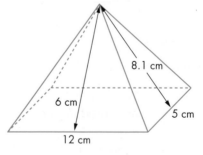

  **2**  A pyramid has a square base of side 9 cm and a slant height of 10 cm. Show that the total surface area of the pyramid is 261 cm$^2$.

**3** Pete wants to paint the roof and the sides of the small hut shown in the diagram. He knows that the hut will need three coats of paint and that one tin of paint is enough to cover 12 m². How many tins of paint will Pete need?

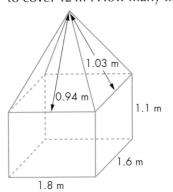

**4** Andrew said: 'The larger a pyramid's volume, the larger its total surface area.'

Calculate the surface areas of the pyramids below and use your answers to check Andrew's statement.

**a**

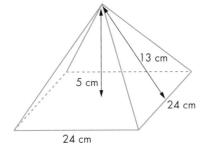

**b**

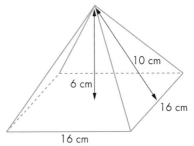

**c**

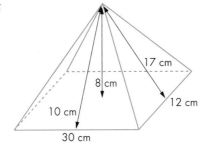

**d**

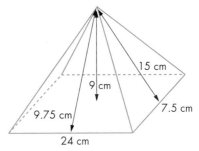

**5** The diagram shows the roof of Jessica's house.

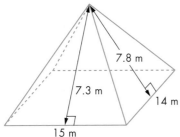

A roofer has given her a quote to retile the whole roof for £11.50 per m².

She cannot afford to spend more than £2600 on retiling the roof.

Could Jessica afford to pay this quote?

Explain your answer.

# 20.3 Cones

This section will show you how to:

• calculate the volume and surface area of a cone.

You will discover some interesting information about cones as you work through Exercise 20D.

## Exercise 20D 🖩

**1** Work out the area of each sector. Give your answer to 3 significant figures.

**a**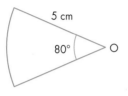
5 cm, 80°

**b**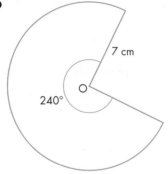
7 cm, 240°

**c**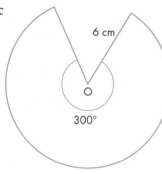
6 cm, 300°

**2** **a** Make an accurate drawing of each sector from question **1** on card and carefully cut it out.

**b** Make each sector into a cone by taping the edges together. The cones should stand on their own.

**c** Measure accurately the diameter of the base of each cone. Divide it by 2 to find the radius of the base of the cone.

**3** Copy and complete the table, using the information from your answers to questions **1** and **2**.

| Sector | Area of sector | Base radius of cone, $r$ | Slant height, $l$ | $\pi \times r \times l$ |
|--------|---------------|--------------------------|-------------------|-------------------------|
| a | | | 5 cm | |
| b | | | 7 cm | |
| c | | | 6 cm | |

> **Hints and tips** As shown in the diagram, the slant height of the cone is labelled $l$. This is the same length as the radius of the original sector from question **1**.

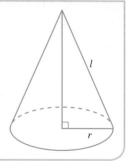

**CM** **4** **a** What do you notice about the area of the sector and the value of $\pi \times r \times l$?

**b** Write down a formula for calculating the curved surface area of a cone, with radius $r$ and slant height $l$.

## Volume and total surface area

You can treat a cone as a pyramid with a circular base. So the formula for the volume of a cone is the same as the formula for the volume of a pyramid:

volume = $\frac{1}{3}$ × area of base × vertical height

$V = \frac{1}{3}\pi r^2 h$

where $r$ is the radius of the base and $h$ is the vertical height of the cone.

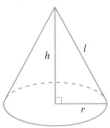

You should have found that the formula for the curved surface area of a cone is:

curved surface area = $\pi$ × radius × slant height

$S = \pi r l$

where $l$ is the slant height of the cone.

The total surface area of a cone ($A$) is the curved surface area plus the area of its circular base.

$A = \pi r l + \pi r^2$

---

**Example 6**

For the cone in the diagram, calculate:

  **i**  its volume
  **ii**  its total surface area.

Give your answers in terms of $\pi$.

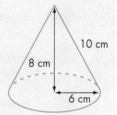

  **i**  The volume is given by:

$$V = \frac{1}{3}\pi r^2 h$$
$$= \frac{1}{3} \times \pi \times 36 \times 8$$
$$= 96\pi \text{ cm}^3$$

  **ii**  The total surface area is given by:

$$A = \pi r l + \pi r^2$$
$$= \pi \times 6 \times 10 + \pi \times 36$$
$$= 96\pi \text{ cm}^2$$

**1** For each cone, calculate:

   i   its volume          ii   its total surface area.

Give your answers to the nearest whole number.

**a**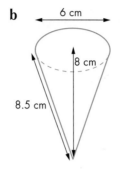
10 cm
9 cm
8 cm

**b**
6 cm
8 cm
8.5 cm

**c**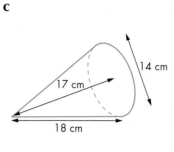
14 cm
17 cm
18 cm

**d**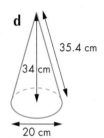
35.4 cm
34 cm
20 cm

**e**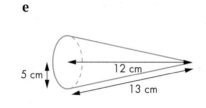
5 cm
12 cm
13 cm

**f**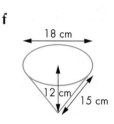
18 cm
12 cm
15 cm

**(PS) 2** A solid, metal cone has base radius 6 cm and vertical height 8 cm. The density of the metal is 3.1 g/cm³. Work out the mass of the cone.

**(MR) 3** The total surface area of a cone with base radius 3 cm is 75.4 cm². Show that its slant height is approximately 5 cm.

**4** Calculate the volume of each shape. Leave your answers in terms of π.

**a**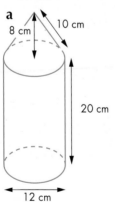
8 cm   10 cm
20 cm
12 cm

**b**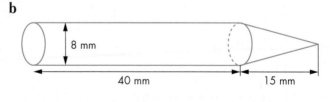
8 mm
40 mm
15 mm

**(EV) 5** A cone has the dimensions shown in the diagram.

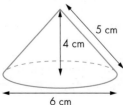
5 cm
4 cm
6 cm

Mandy said the total surface area was 80 cm². Neil said it was 70 cm².

Who is correct? What has the other person done wrong?

# 20.4 Spheres

This section will show you how to:

- calculate the volume and surface area of a sphere.

**Key term**

sphere

The volume of a **sphere**, radius $r$, is:

$V = \frac{4}{3}\pi r^3$

Its surface area is:

$A = 4\pi r^2$

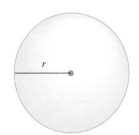

---

**Example 7**

For a sphere of radius of 8 cm, calculate:   **i** its volume    **ii** its surface area.

**i**   The volume is given by:

$$V = \frac{4}{3}\pi r^3$$
$$= \frac{4}{3} \times \pi \times 8^3$$
$$= \frac{2048}{3} \times \pi$$
$$= 2140 \text{ cm}^3 \text{ (3 sf)}$$

**ii**   The surface area is given by:

$$A = 4\pi r^2$$
$$= 4 \times \pi \times 8^2$$
$$= 256 \times \pi$$
$$= 804 \text{ cm}^2 \text{ (3 sf)}$$

---

## Exercise 20F 🖩

**1**   Calculate the volume of spheres with these dimensions.

   **a** Radius 3 cm           **b** Radius 6 cm           **c** Diameter 20 cm

**2**   Calculate the surface area of spheres with these dimensions.

   **a** Radius 3 cm           **b** Radius 5 cm           **c** Diameter 14 cm

**3**   Calculate the volume and surface area of a sphere with a diameter of 30 cm.
Give your answers in terms of $\pi$.

**(PS)** **4**   A solid sphere fits exactly into an open box. The box is a cuboid with side 25 cm.

   **a** What is the surface area of the sphere?

   **b** The sphere is in the box. How much water can be poured into the box before it spills over?

**(MR)** **5**   A metal sphere of radius 15 cm is melted down and made into a solid cylinder of radius 6 cm. Calculate the height of the cylinder.

**(PS)** **6**   Lead pellets are spherical and have a radius of 1.5 mm. The density of lead is 11.35 g/cm³. Calculate the maximum number of pellets that can be made from 1 kg of lead.

# Worked exemplars

**1** The key hole shape shown is made up of a circle of radius 1 cm and a sector of angle 30°.

Show that the area of this shape is 7.1 cm² (2 sf).

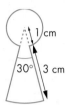

| | |
|---|---|
| This is a communicating mathematics question where you need to construct and communicate chains of reasoning to achieve a given result. | |
| Area of sector $= \dfrac{\theta}{360} \times \pi \times r^2$ <br><br> $= \dfrac{30}{360} \times \pi \times 4^2$ <br><br> $= \dfrac{480}{360} \times \pi$ <br><br> $= 4.188\ 790\ 205$ | Plan your solution by considering how to work out the total area of the shape. Remember that in this type of question you have the final result to work towards, so if your answer is not correct, look again to see what you have done wrong. |
| Area of major sector in circle $= \dfrac{\theta}{360} \times \pi \times r^2$ <br><br> $= \dfrac{330}{360} \times \pi \times 1^2$ <br><br> $= 2.879\ 793\ 266$ | Write down all the digits in your calculator display. You need to show that when you add the two areas together, the final answer rounds to the given answer to 2 sf. |
| Total area $= 4.188\ 790\ 205 + 2.879\ 793\ 266$ <br><br> $= 7.068\ 583\ 471$ <br><br> $= 7.1$ (2 sf) | |

**2** Three balls of diameter 8.2 cm just fit inside a cylindrical container.

Andrew said: 'The volume of that container is 1300 cm³.'

Sophia said: 'No it's not.'

They could both be correct. Show how this is possible.

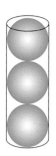

| This is an evaluating question where you need to see how each claim could be correct. | |
|---|---|
| The diameter of the cylinder is 8.2 cm and the height is 8.2 × 3 = 24.6 cm.<br><br>The radius of the cylinder is 4.1 cm.<br><br>The internal volume = $\pi \times 4.1^2 \times 24.6$<br><br>$\qquad\qquad = 1299.13\ldots$<br><br>$\qquad\qquad = 1300 \text{ cm}^3 \text{ (2 sf)}$ | Before you can evaluate the comments, you first need to calculate the actual volume. |
| Andrew has correctly stated the volume to 2 (or 3) significant figures, which is 1300 cm².<br><br>Sophia could also be correct as the area is only 1300 cm² as a rounded answer. | Once you have found the volume, you should see what makes their statement correct.<br><br>Here, you need to make sure you state Andrew's accuracy is to 2 or 3 sf and also state a reason why Sophia is correct. You could also argue that since the original data was 2 sf, the accuracy of the answer could be given to 1 sf which would be 1000 cm³. |

# Ready to progress?

I can calculate the length of an arc and the angle and area of a sector.
I can calculate the volume of pyramids, cones and spheres.
I can calculate the surface area of pyramids, cones and spheres.

# Review questions

**1** The diagram shows a sector of a circle, centre O. The radius of the circle is 8 cm. The angle of the sector is 200°.

Calculate:

**a** the total perimeter of the sector

**b** the area of the sector.

Give your answers to 3 significant figures

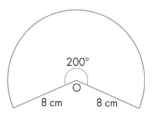

**(PS) 2** A tunnel is being created in the shape of a half cylinder. It is 1.25 km long and its cross-section is a semi-circle with diameter 10 m. What volume of rock must be removed to create the tunnel?

**(CM) 3** The diagram shows a container in the shape of a 30-cm cube on top of a pyramid.

The total height of the container is 40 cm.
Show that the volume of the container is 30 000 cm³.

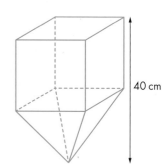

**(EV) 4** A cylindrical bottle with radius 4.5 cm is filled with fruit punch to a height of 25 cm, as shown.

At a party, 45 glasses are laid out to be filled from the bottle. The glasses are the shape of a cone with radius 2.5 cm and height 6 cm.

David said: 'You will run out of fruit punch before 40 of the glasses are filled.'

Is David's comment correct? Give reasons for your answer.

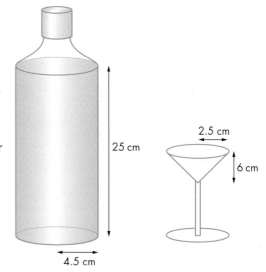

**MR** **5** A solid metal ball bearing has a radius of 0.75 cm. The metal has a density of 7.3 g/cm³.
Show that a pack of 20 of these ball bearings has a mass of 258 g.

**PS** **6** A drink is in a cylindrical glass, with radius 4 cm. Six spherical balls of ice, each 2 cm
in diameter, are put into the drink and sink to the bottom. By how many centimetres
will the level of the liquid in the glass rise?

**PS** **7** A tent is in the shape of a cylinder with a cone on top.
The radius of the base of the tent is 2 m. The heights
are shown in the diagram.

What is the surface area of the material on the outside of the tent?

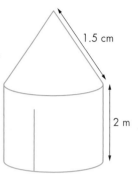

**8** **a** Which of these shapes has the largest total surface area?

    **A** a sphere of radius 9 cm

    **B** a square-based pyramid with side length 17 cm and slant height 21 cm

    **C** a cone of radius 11 cm and slant height 18.4 cm

  **b** Which shape has the smallest total surface area?

**9** A one-person tent is in the shape of a triangular prism, 2 m long.

The zip opening at the front of the tent is 75 cm long.

The width of the tent is 1.1 m.

Calculate the volume of air inside the tent.

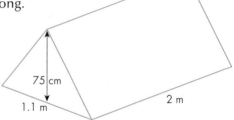

**PS** **10** A football is made from 33 hexagons all having an area of 20 cm².

Calculate the radius of the football.

**MR** **11** Two brothers, Tom and James were looking at these shapes: a sphere, a cube and a cone.

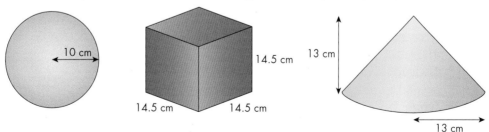

Tom said: 'The cube has the largest surface area,' but James said, 'No, it's the sphere.'

Comment on the brothers' answers.

**PS** **12** The volume of a cone is given as $4.88 \times 10^3$ cm³.

You are told the base of the cone has an area of $1.6 \times 10^2$ cm².

Calculate the height of the cone.

# 21 Algebra: Number and sequences

## This chapter is going to show you:

- how to recognise rules for sequences
- how to express a rule for a sequence, in words and algebraically
- how to generate the terms of a linear sequence, given a formula for the $n$th term
- how to find the $n$th term of a linear sequence
- some common sequences of numbers.

## You should already know:

- how to substitute numbers into an algebraic expression
- how to state a rule for a simple linear sequence in words
- how to factorise simple linear expressions.

## About this chapter

Mathematicians enjoy finding patterns; you have already seen patterns in sequences such as square numbers and multiples. As well as being important in mathematics, number patterns can also help in the study of nature and geometric patterns.

Many mathematical patterns are found in nature. The most famous of these is probably the Fibonacci series, in which each term after the second is formed by adding the two previous terms.

1  1  2  3  5  8  13  21  …

The sequence was discovered by an Italian, Leonardo Fibonacci, in 1202, when he was investigating the breeding patterns of rabbits!
Since then, the pattern has been found in many other places in nature. The spirals found in a nautilus shell and in the seed heads of a sunflower plant also follow the Fibonacci series.

# 21.1 Patterns in number

This section will show you how to:

- recognise patterns in number sequences.

Look at these number **patterns**.

$0 \times 9 + 1 = 1$

$1 \times 9 + 2 = 11$

$12 \times 9 + 3 = 111$

$123 \times 9 + 4 = 1111$

$1234 \times 9 + 5 = 11\,111$

$1 \times 8 + 1 = 9$

$12 \times 8 + 2 = 98$

$123 \times 8 + 3 = 987$

$1234 \times 8 + 4 = 9876$

$12\,345 \times 8 + 5 = 98\,765$

$1 \times 3 \times 37 = 111$

$2 \times 3 \times 37 = 222$

$3 \times 3 \times 37 = 333$

$4 \times 3 \times 37 = 444$

$7 \times 7 = 49$

$67 \times 67 = 4489$

$667 \times 667 = 444\,889$

$6667 \times 6667 = 44\,448\,889$

The numbers form a **sequence**. Check that the patterns you can see are correct, then try to continue each pattern without using a calculator. Check them with a calculator afterwards.

Spotting patterns is an important part of mathematics. It helps you to see rules for making calculations.

## Exercise 21A

In questions **1** to **5**, look for the pattern and then write the next two lines. Check your answers with a calculator afterwards.

You may find that some of the answers are too big to fit in a calculator display. This is one of the reasons why spotting patterns is important.

**Hints and tips** Look for symmetries in the number patterns.

**1**  **a**

$1 \times 1 = 1$

$11 \times 11 = 121$

$111 \times 111 = 12\,321$

$1111 \times 1111 = 1\,234\,321$

**b**

$9 \times 9 = 81$

$99 \times 99 = 9801$

$999 \times 999 = 998\,001$

$9999 \times 9999 = 99\,980\,001$

**2**  **a** $3 \times 4 = 3^2 + 3$

$4 \times 5 = 4^2 + 4$

$5 \times 6 = 5^2 + 5$

$6 \times 7 = 6^2 + 6$

**b** $10 \times 11 = 110$

$20 \times 21 = 420$

$30 \times 31 = 930$

$40 \times 41 = 1640$

**Hints and tips** Think of the answers as 1 10, 4 20, 9 30, 16 40 …

**3**  **a**

$1 = 1 = 1^2$

$1 + 2 + 1 = 4 = 2^2$

$1 + 2 + 3 + 2 + 1 = 9 = 3^2$

$1 + 2 + 3 + 4 + 3 + 2 + 1 = 16 = 4^2$

**b**

$1 = 1 = 1^3$

$3 + 5 = 8 = 2^3$

$7 + 9 + 11 = 27 = 3^3$

$13 + 15 + 17 + 19 = 64 = 4^3$

**4**  **a**

$$1 = 1$$
$$1 + 1 = 2$$
$$1 + 2 + 1 = 4$$
$$1 + 3 + 3 + 1 = 8$$
$$1 + 4 + 6 + 4 + 1 = 16$$
$$1 + 5 + 10 + 10 + 5 + 1 = 32$$

**b**

$$12\ 345\ 679 \times 9 = 111\ 111\ 111$$
$$12\ 345\ 679 \times 18 = 222\ 222\ 222$$
$$12\ 345\ 679 \times 27 = 333\ 333\ 333$$
$$12\ 345\ 679 \times 36 = 444\ 444\ 444$$

 **5**  **a**

$$1^3 = 1^2 = 1$$
$$1^3 + 2^3 = (1 + 2)^2 = 9$$
$$1^3 + 2^3 + 3^3 = (1 + 2 + 3)^2 = 36$$

**b**

$$3^2 + 4^2 = 5^2$$
$$10^2 + 11^2 + 12^2 = 13^2 + 14^2$$
$$21^2 + 22^2 + 23^2 + 24^2 = 25^2 + 26^2 + 27^2$$

> **Hints and tips**
> $$4 + 5 = 9 = 3^2$$
> $$12 + 13 = 25 = 5^2$$
> $$24 + 25 = 49 = 7^2$$

Use your observations on the number patterns in questions **1** to **5** to answer question **6** without using a calculator.

> **Hints and tips**  Look for clues in the patterns from questions **1** to **5**, for example, $1111 \times 1111 = 1\ 234\ 321$. This is four 1s times four 1s, so what will it be for nine 1s times nine 1s?

 **6**  **a**  $111\ 111\ 111 \times 111\ 111\ 111 =$

**b**  $999\ 999\ 999 \times 999\ 999\ 999 =$

**c**  $12 \times 13 =$

**d**  $90 \times 91 =$

**e**  $1 + 2 + 3 + 4 + 5 + 6 + 7 + 8 + 9 + 8 + 7 + 6 + 5 + 4 + 3 + 2 + 1 =$

**f**  $57 + 59 + 61 + 63 + 65 + 67 + 69 + 71 =$

**g**  $1 + 9 + 36 + 84 + 126 + 126 + 84 + 36 + 9 + 1 =$

**h**  $12\ 345\ 679 \times 81 =$

**i**  $1^3 + 2^3 + 3^3 + 4^3 + 5^3 + 6^3 + 7^3 + 8^3 + 9^3 =$

**7**  This is Gauss's method for working out the sum of all the numbers from 1 to 100.

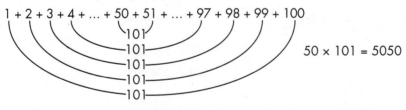

Use Gauss's method to work out the sum of all the whole numbers from 1 to 500.

# 21.2 Number sequences

## This section will show you how to:

- recognise how number sequences are built up
- generate sequences, given the $n$th term.

**Key terms**

| | |
|---|---|
| consecutive | $n$th term |
| difference | term-to-term |
| position-to-term | |

## Terms and rules

You know that a number sequence is an ordered set of numbers based on a rule. The rule that takes you from one number to the next could be a simple addition or multiplication, but it may be a more complex rule. You always need to look very carefully at the pattern of a sequence.

Each number in a sequence is called a term. Each term has a specific position in the sequence. Terms that follow on, one from another, are called **consecutive** terms. The rule that shows how to work out the terms is the **term-to-term** rule.

Look at these sequences and their rules.

| First four terms | Term-to-term rule | Next two terms |
|---|---|---|
| 3, 6, 12, 24, … | doubling the previous term each time | … 48, 96, … |
| 2, 5, 8, 11, … | adding 3 to the previous term each time | … 14, 17, … |
| 1, 10, 100, 1000, … | multiplying the previous term by 10 each time | … 10 000, 100 000, … |
| 22, 15, 8, 1, … | subtracting 7 from the previous term each time | … −6, −13, … |

## Term-to-term sequences

The first thing to do is to identify the link from one term to the next. A pattern in which you work out each term (apart from the first) from the term before it is a term-to-term sequence.

## Differences

For some sequences you need to look at the **differences** between consecutive terms to determine the pattern.

---

**Example 1**

Find the next two terms of the sequence 1, 3, 6, 10, 15, …

Look at the differences between each pair of consecutive terms.

```
1   3   6   10  15
  ↑   ↑   ↑   ↑
  2   3   4   5
```

You can continue the sequence like this.

```
1   3   6   10  15   21  28
  ↑   ↑   ↑   ↑
  2   3   4   5  +6  +7
```

The differences in this sequence form a number sequence of their own, so you need to work out the sequence of the differences before you can continue the original sequence.

---

## Position-to-term sequences

Sometimes you may need to know, for example, the 50th term in a number sequence. To do this, you need to find the rule that produces the sequence. This is called the **position-to-term** rule. To see how this works, look at the problem backwards: make up a rule and see how it produces a sequence.

Example 2

A sequence is formed by the rule $3n + 1$, where $n = 1, 2, 3, 4, 5, 6, \ldots$ Write down the first five terms of the sequence.

Substitute $n = 1, 2, 3, 4, 5$ in turn.

| $(3 \times 1 + 1)$ | $(3 \times 2 + 1)$ | $(3 \times 3 + 1)$ | $(3 \times 4 + 1)$ | $(3 \times 5 + 1)$ | $\ldots$ |
|---|---|---|---|---|---|
| 4 | 7 | 10 | 13 | 16 | $\ldots$ |

So the sequence is 4, 7, 10, 13, 16, …

In the last example, the rule that generated the sequence was the expression $3n + 1$. This expression is called the **$n$th term** of the sequence. As you have seen, you can use it to find any term, by substituting the term number for $n$.

In the sequence above, notice that the difference between one term and the next is always 3, which is the coefficient of $n$ (the number attached to $n$) in the rule. The constant term is the difference between the first term and the coefficient (in this case, $4 - 3 = 1$).

Example 3

The $n$th term of a sequence is $4n - 3$. Write down the first five terms of the sequence.

Substitute $n = 1, 2, 3, 4, 5$ in turn.

| $(4 \times 1 - 3)$ | $(4 \times 2 - 3)$ | $(4 \times 3 - 3)$ | $(4 \times 4 - 3)$ | $(4 \times 5 - 3)$ | $\ldots$ |
|---|---|---|---|---|---|
| 1 | 5 | 9 | 13 | 17 | $\ldots$ |

So the sequence is 1, 5, 9, 13, 17, …

Again, the difference between each term and the next is always 4, which is the coefficient of $n$ in the formula for the $n$th term. The constant term is the difference between the first term and the coefficient ($1 - 4 = -3$).

## Exercise 21B

**1** Look carefully at each number sequence below. Find the next two numbers in the sequence and try to explain the pattern.

**a** 1, 1, 2, 3, 5, 8, 13, …     **b** 1, 4, 9, 16, 25, 36, …     **c** 3, 4, 7, 11, 18, 29, …

> **Hints and tips** These patterns do not go up by the same value each time so you will need to find another connection between the terms.

**2** The pattern shows how triangular numbers are formed.

| 1 | 3 | 6 | 10 |

Work out the next four triangular numbers.

**3** The pattern shows how hexagonal numbers are formed.

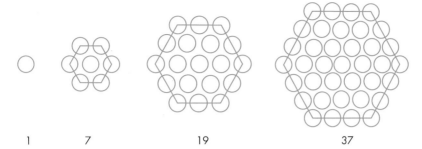

1       7              19                  37

Work out the next three hexagonal numbers.

**4** Work out the first five terms of the sequence formed by the rule $5n + 1$.

**5** The first two terms of the sequence of fractions $\frac{n-1}{n+1}$ are:

$n = 1: \frac{1-1}{1+1} = \frac{0}{2} = 0$          $n = 2: \frac{2-1}{2+1} = \frac{1}{3}$

Work out the next five terms of the sequence.

**6** A sequence is formed by the rule $\frac{1}{2} \times n \times (n + 1)$ for $n = 1, 2, 3, 4, \ldots$

The first term is given by substituting $n = 1$.      $\frac{1}{2} \times 1 \times (1 + 1) = 1$

The second term is given by substituting $n = 2$.      $\frac{1}{2} \times 2 \times (2 + 1) = 3$

**a** Work out the next five terms of this sequence.

**b** This is a well-known sequence you have met before. What is it?

**7** In mathematics, 5! means '5 factorial' or 'factorial 5', which is $5 \times 4 \times 3 \times 2 \times 1 = 120$.

In the same way 7! means $7 \times 6 \times 5 \times 4 \times 3 \times 2 \times 1 = 5040$.

**a** Calculate the values of 2!, 3!, 4! and 6!

**b** If your calculator has a factorial button, check that it gives the same answers as you get for part **a**. What is the largest factorial you can work out with your calculator before you get an error?

(PS) **8** The letters of the alphabet are written as the pattern:

ABBCCCDDDDEEEEEFFFFFFGGGGGGG …

so that the number of times each letter is written matches its place in the alphabet.

So, for example, as J is the 10th letter in the alphabet, there will be 10 Js in the list.

The pattern repeats after the 26 Zs.

What letter will be the 1000th in the list?

> **Hints and tips**   Work out how many letters there are in the sequence from ABB …
> to … ZZZ, then work out how many of these sequences you need
> to get past 1000 letters.

**9** On the first day of Christmas my true love sent to me:

a partridge in a pear tree.

On the second day of Christmas my true love sent to me:

two turtle doves

and a partridge in a pear tree.

and so on until…

On the twelfth day of Christmas my true love sent to me:

twelve drummers drumming

eleven pipers piping

ten lords a-leaping

nine ladies dancing

eight maids a-milking

seven swans a-swimming

six geese a-laying

five golden rings

four calling birds

three French hens

two turtle doves

and a partridge in a pear tree.

How many presents were given, in total, on the 12 days of Christmas?

> **Hints and tips** Work out the pattern for the number of presents each day. For example, on day 1 there was 1 present, on day 2 there were 2 + 1 = 3 presents, and so on. Total the presents after each day, so after 1 day there was a total of 1 present, after 2 days a total of 4 presents, and so on. Try to spot any patterns.

**10** Look at these two sequences.

8, 11, 14, 17, 20, …

1, 5, 9, 13, 17, …

The first term that they have in common is 17. What are the next two terms that the two sequences have in common?

**11** Look at these two sequences.

2, 5, 8, 11, 14, …

3, 6, 9, 12, 15, …

Will the two sequences ever have a term in common?

Give a reason for your answer.

**12** The $n$th term of a sequence is $3n + 7$.

The $n$th term of another sequence is $4n - 2$.

These two sequences have several terms in common but only one term that is common and has the same position in the sequence.

Without writing out the sequences, show how you can tell, using the expressions for the $n$th term, that it is the 9th term.

# 21.3 Finding the $n$th term of a linear sequence

This section will show you how to:

- find the $n$th term of a linear sequence.

**Key terms**

arithmetic sequence

linear sequence

In a **linear** or **arithmetic sequence**, the *difference* between each term and the next is always the same. For example, look at this sequence.

> 2, 5, 8, 11, 14, …          difference of 3

You could call this a term-to-term sequence, with the rule 'add three'.

You can also call it a position-to-term sequence, where the $n$th term is given by $3n - 1$.

**Remember:** The $n$th term of a sequence is a rule, written algebraically, that gives any term based on its position ($n$) in the sequence. Here is another linear sequence.

> 5, 7, 9, 11, 13, …          difference of 2

The $n$th term of this sequence is $2n + 3$.

The $n$th term of a linear sequence is *always* of the form $An \pm b$, where:

- $A$, the coefficient of $n$, is the difference between each term and the next term (consecutive terms).

- $b$ is the difference between the first term and $A$.

---

**Example 4**

Find the $n$th term of each sequence.

**a**   5, 7, 9, 11, 13, …          **b**   95, 90, 85, 80, 75, …

**a**   The difference between consecutive terms is 2. So the coefficient of $n$ in the $n$th term is 2 and the $n$th term starts with $2n$.

Subtract $A$, 2, from the first term, 5, which gives $5 - 2 = 3$.

So the $n$th term is given by $2n + 3$.

(You can test it by substituting $n = 1, 2, 3, 4, …$)

**b**   The difference between consecutive terms is 5 and the terms in the sequence are getting smaller. So the coefficient of $n$ in the $n$th term is $-5$ and the $n$th term starts with $-5n$.

Subtract $A$, $-5$, from the first term, 95, which gives $95 - -5 = 100$.

So the $n$th term is given by $-5n + 100$.

(You can test it by substituting $n = 1, 2, 3, 4, …$)

---

**Example 5**

Find the $n$th term of the sequence 3, 7, 11, 15, 19, …

The difference between consecutive terms is 4. So the coefficient of $n$ in the $n$th term is 4 and the $n$th term starts with $4n$.

Subtract $A$, 4, from the first term, 3, which gives $3 - 4 = -1$.

So the $n$th term is given by $4n - 1$.

---

**Hints and tips**   When you work out an $n$th term, always check by substituting $n = 1, 2$ and 3 to make sure the first three terms agree.

Example 6

From the sequence 5, 12, 19, 26, 33, ... work out:

**a** the $n$th term      **b** the 50th term      **c** the first term that is greater than 1000.

**a** The difference between consecutive terms is 7. So the coefficient of $n$ in the $n$th term is 7 and the $n$th term starts with $7n$.

Subtract $A$, 7, from the first term, 5, which gives $5 - 7 = -2$.

So the $n$th term is given by $7n - 2$.

**b** Find the 50th term by substituting $n = 50$ into the rule, $7n - 2$.

$7 \times 50 - 2$

$= 350 - 2$

$= 348$

**c** The first term that is greater than 1000 is given by:

$7n - 2 > 1000$

$\Rightarrow 7n > 1000 + 2$

$\Rightarrow n > \dfrac{1002}{7}$

$\quad n > 143.14$

So the first term (which has to be a whole number) that is greater than 1000 is the 144th term, 1006.

**Note:** The symbol > means 'is greater than', and follows similar rules as for the = sign most of the time. You will learn more about this and other signs in a later chapter, when you study inequalities.

## Exercise 21C

**1** Find the next two terms and the $n$th term in each linear sequence.

     **a** 3, 5, 7, 9, 11, ...          **b** 5, 9, 13, 17, 21, ...          **c** 8, 13, 18, 23, 28, ...

     **d** 2, 8, 14, 20, 26, ...        **e** 5, 8, 11, 14, 17, ...         **f** 2, 9, 16, 23, 30, ...

     **g** 1, 5, 9, 13, 17, ...          **h** 3, 7, 11, 15, 19, ...         **i** 2, 5, 8, 11, 14, ...

     **j** 32, 22, 12, 2, ...            **k** 20, 16, 12, 8, ...           **l** 24, 19, 14, 9, 4, ...

> **Hints and tips**   Remember to look at the differences and the first term.

**2** Find the $n$th term and the 50th term in each linear sequence.

     **a** 4, 7, 10, 13, 16, ...        **b** 7, 9, 11, 13, 15, ...         **c** 3, 8, 13, 18, 23, ...

     **d** 1, 5, 9, 13, 17, ...          **e** 2, 10, 18, 26, ...           **f** 5, 6, 7, 8, 9, ...

     **g** 6, 11, 16, 21, 26, ...      **h** 3, 11, 19, 27, 35, ...       **i** 1, 4, 7, 10, 13, ...

     **j** 21, 24, 27, 30, 33, ...     **k** 40, 33, 26, 19, 12, ...     **l** 33, 25, 17, 9, 1, ...

**3**   **a** Which term of the sequence 5, 8, 11, 14, 17, ... is the first that will be greater than 100?

     **b** Which term of the sequence 1, 8, 15, 22, 29, ... is the first that will be greater than 200?

     **c** Which term of the sequence 4, 9, 14, 19, 24, ... is the closest to 500?

**4** For each sequence **a** to **j**, find:

    **i** the $n$th term       **ii** the 100th term       **iii** the term closest to 100.

    **a** 5, 9, 13, 17, 21, …       **b** 3, 5, 7, 9, 11, 13, …       **c** 4, 7, 10, 13, 16, …

    **d** 8, 10, 12, 14, 16, …       **e** 9, 13, 17, 21, …       **f** 6, 11, 16, 21, …

    **g** 0, 3, 6, 9, 12, …       **h** 2, 8, 14, 20, 26, …       **i** 197, 189, 181, 173, …

    **j** 225, 223, 221, 219, …

**5** A sequence of fractions is $\frac{3}{4}, \frac{5}{7}, \frac{7}{10}, \frac{9}{13}, \frac{11}{16}, \ldots$

    **a** Find the $n$th term in the sequence.

    **b** Change each fraction to a decimal. What happens to the decimal values?

    **c** What, as a decimal, will be the value of:

      **i** the 100th term       **ii** the 1000th term?

    **d** Use your answers to part **c** to predict what the 10 000th term and the millionth term are. (Check these on your calculator.)

**6** Repeat question **5** for the sequence $\frac{3}{6}, \frac{7}{11}, \frac{11}{16}, \frac{15}{21}, \frac{19}{26}, \ldots$

**7** A2B Haulage uses this formula to calculate the cost of transporting $n$ pallets.

    For $n \leqslant 5$, the cost will be £$(40n + 50)$.

    For $6 \leqslant n \leqslant 10$, the cost will be £$(40n + 25)$.

    For $n \geqslant 11$, the cost will be £$40n$.

    **a** How much will the company charge to transport 7 pallets?

    **b** How much will the company charge to transport 15 pallets?

    **c** A2B charged £170 for transporting pallets. How many pallets did they transport?

    **d** Speedy Haulage uses the formula £$50n$ to calculate the costs for transporting $n$ pallets. At what value of $n$ do the two companies charge the same amount?

**(EV)**

**8** The formula for working out a series of fractions is $\frac{2n + 1}{3n + 1}$.

    **a** Work out the first three fractions in the series.

    **b i** Work out the value of the fraction as a decimal when $n = 1\,000\,000$.

      **ii** What fraction is equivalent to this decimal?

    **c** How can you tell this from the original formula?

**(CM)**

**9** This chart is used by an online retailer for the charges for buying $n$ T-shirts, including any postage and packing charges.

| $n$ | 1 | 2 | 3 | 4 | 5 | 6 | 7 | 8 | 9 | 10 | 11 | 12 | 13 | 14 | 15 |
|---|---|---|---|---|---|---|---|---|---|---|---|---|---|---|---|
| Charge (£) | 10 | 18 | 26 | 34 | 42 | 49 | 57 | 65 | 73 | 81 | 88 | 96 | 104 | 112 | 120 |

    **a** Using the charges for 1 to 5 T-shirts, work out an expression for the $n$th term.

    **b** Using the charges for 6 to 10 T-shirts, work out an expression for the $n$th term.

    **c** Using the charges for 11 to 15 T-shirts, work out an expression for the $n$th term.

    **d** What is the basic charge for a T-shirt?

**10** Look at this series of fractions.

$$\frac{31}{109}, \frac{33}{110}, \frac{35}{111}, \frac{37}{112}, \frac{39}{113}, \dots$$

    **a** Show that the $n$th term of the sequence of the numerators is $2n + 29$.

    **b** Write down the $n$th term of the sequence of the denominators.

    **c** Show that the terms of the series will eventually get very close to 2.

     **d** Which term of the series has a value equal to 1?

> **Hints and tips**   Use algebra to set up an equation.

 **11** The $n$th term of the sequence of even numbers is $2n$. The $n$th term of the sequence of odd numbers is $2n - 1$.

    Show why these expressions tell you that there can never be a number in both sequences.

 **12** The $n$th term of a linear sequence is given by $An + b$, where $A$ and $b$ are integers.

    The 5th term is 10 and the 8th term is 19. Work out the values of $A$ and $b$.

# 21.4 Special sequences

## This section will show you how to:

- recognise and continue some special number sequences.

Most of the sequences you have been working with so far have been arithmetic or linear. In these sequences the difference between consecutive terms has a constant value. There are many other number sequences that you need to know about. Many of them do not have a constant difference.

> **Key terms**
>
> geometric sequence
>
> powers of 2
>
> powers of 10
>
> quadratic sequence

- The even numbers are 2, 4, 6, 8, 10, 12, …      The $n$th term of this sequence is $2n$.
- The odd numbers are 1, 3, 5, 7, 9, 11, …      The $n$th term of this sequence is $2n - 1$.
- The square numbers are 1, 4, 9, 16, 25, 36, …      The $n$th term of this sequence is $n^2$.
- The cube numbers are 1, 8, 27, 64, 125, 216, …      The $n$th term of this sequence is $n^3$.
- The triangular numbers are 1, 3, 6, 10, 15, 21, …      The $n$th term of this sequence is $\frac{1}{2}n(n + 1)$.
- The **powers of 2** are 2, 4, 8, 16, 32, 64, …      The $n$th term of this sequence is $2^n$.
- The **powers of 10** are 10, 100, 1000, 10 000, 100 000, 1 000 000, …      The $n$th term of this sequence is $10^n$.

## Geometric sequences

A sequence in which you find each term by multiplying the previous term by a fixed value is a **geometric sequence**. The $n$th term of a geometric sequence is given by $a \times r^{n-1}$, where $a$ is the first term and $r$ is the multiplier. **Note:** Any number raised to a power of 0 is 1.

For example:      2, 6, 18, 54, 162, …      The $n$th term for this sequence is $2 \times 3^{n-1}$.

                 12, 48, 192, 768, 3072, …      The $n$th term for this sequence is $12 \times 4^{n-1}$.

Note that, as 12 has a factor of 4, the $n$th term of the last sequence could have been written as $3 \times 4^n$.

## Fibonacci sequences

The sequence 1, 1, 2, 3, 5, 8, 13, 21, ... was discovered by Leonardo Fibonacci in about 1202.
Starting with the third term, each term is the sum of the previous two terms:

$$1 + 1 = 2, 1 + 2 = 3, 2 + 3 = 5 \text{ and so on.}$$

There is no algebraic expression for the $n$th term.

## Prime numbers

A prime number is a number that only has two factors, 1 and itself.

The first 20 prime numbers are 2, 3, 5, 7, 11, 13, 17, 19, 23, 29, 31, 37, 41, 43, 47, 53, 59, 61, 67 and 71.

There is no pattern to the sequence of prime numbers.

There is no formula for the $n$th prime number.

**Remember:** 2 is the only even prime number.

---

**Example 7**

$p$ is a prime number, $q$ is an odd number and $r$ is an even number.
Say if each expression is always odd (O), always even (E) or could be either odd or even (X).

**a** $pq$      **b** $p + q + r$      **c** $pqr$      **d** $q^2 + r^2$

**a** One way to answer this question is to substitute numbers and see whether the outcome is odd or even.

For example, let $p = 2$ and $q = 3$. Then $pq = 6$ and is even; but $p$ could also be 3 or 5, which are odd, so $pq = 3 \times 5 = 15$, which is odd.

So $pq$ could be either (X).

**b** Let $p = 2$ or 3, $q = 5$ and $r = 4$; so $p + q + r = 2 + 5 + 4 = 11$, or $3 + 5 + 4 = 12$

So $p + q + r$ could be either (X).

**c** Let $p = 2$ or 3, $q = 5$ and $r = 4$; so $pqr = 2 \times 5 \times 4 = 40$ or $3 \times 5 \times 4 = 60$

Both are even, so $pqr$ is always even (E).

**d** Let $q = 5$ and $r = 4$; $q^2 + r^2 = 5^2 + 4^2 = 25 + 16 = 41$

This is odd, so $q^2 + r^2$ is always odd (O).

---

**Example 8**

**a** Work out the first five terms of the sequences with these $n$th terms.

     **i** $3^n$          **ii** $6^n$

**b** Work out the 8th term of the sequence with an $n$th term of $4^n$.

**a** **i** Substitute $n = 1$ into the $n$th term expression, then multiply each term by the multiplier 3 to find the other terms.

$3^1 = 3 \times 3 = 9, 9 \times 3 = 27 \ldots$

The sequence is 3, 9, 27, 81, 243, ...

     **ii** Substitute $n = 1$ into the $n$th term expression, then multiply each term by the multiplier 6 to find the other terms.

$6^1 = 6, 6 \times 6 = 36, 36 \times 6 = 216\ldots$

The sequence is 6, 36, 216, 1296, 7776, ...

**b** Substitute $n = 8$ into the $n$th term. You can use the power button [xʸ] on your calculator.

[ 48 ] [ = ] [ 4 ] [ xʸ ] [ 8 ] [ = ] [ 65536 ]

The 8th term is 65 536.

Example 9

The $n$th term of a sequence is $n^2 + 1$. Write down the first five terms of the sequence.

Substitute $n = 1, 2, 3, 4, 5$ in turn.

| $(1^2 + 1)$ | $(2^2 + 1)$ | $(3^2 + 1)$ | $(4^2 + 1)$ | $(5^2 + 1)$ | ... |
|---|---|---|---|---|---|
| 2 | 5 | 10 | 17 | 26 | ... |

So the sequence is 2, 5, 10, 17, 26, ...

This sequence is different to those in examples 1, 2 and 3 because it is a **quadratic sequence**. These are sequences based on the square numbers. That means that the $n$th term contains $n^2$. You could still use a term-to-term rule, as the terms increase by 3, 5, 7, 9, ..., so the differences between consecutive terms are the odd numbers. This is not always easy to write so usually it is better to use a position-to-term rule for quadratic sequences.

## Exercise 21D

**1**

**a** Pick any odd number. Pick another odd number. Add the two numbers together. Is the answer odd or even?

Copy and complete this table.

| + | Odd | Even |
|---|---|---|
| **Odd** | Even | |
| **Even** | | |

**b** Pick any odd number. Pick another odd number. Multiply the two numbers together. Is the answer odd or even?

Copy and complete this table.

| × | Odd | Even |
|---|---|---|
| **Odd** | Odd | |
| **Even** | | |

 **2**

**a** Write down the next two lines of this number pattern.

$1 = 1 = 1^2$

$1 + 3 = 4 = 2^2$

$1 + 3 + 5 = 9 = 3^2$

**b** Use the pattern in part **a** to write down the totals of these numbers.

**i** $1 + 3 + 5 + 7 + 9 + 11 + 13 + 15 + 17 + 19$

**ii** $2 + 4 + 6 + 8 + 10 + 12 + 14$

 **3**

**a** Work out the first 12 terms of the Fibonacci sequence 1, 1, 2, 3, 5, ...

**b** Explain why the Fibonacci sequence always has a repeated pattern of two odd terms followed by one even term.

**c** The first three terms of a Fibonacci sequence are $a$, $b$, $a + b$, ...

**i** Write out the next five terms.     **ii** Describe the pattern of the coefficients of $a$ and $b$.

 **4**

$p$ is an odd number; $q$ is an even number. State whether each expression is odd or even.

**a** $p + 1$      **b** $q + 1$      **c** $p + q$

**d** $p^2$       **e** $qp + 1$     **f** $(p + q)(p - q)$

**g** $q^2 + 4$     **h** $p^2 + q^2$    **i** $p^3$

**5** $p$ is a prime number; $q$ is an even number.

State whether each expression is odd, even or could be either odd or even.

**a** $p + 1$    **b** $p + q$    **c** $p^2$

**d** $qp + 1$    **e** $(p + q)(p - q)$    **f** $2p + 3q$

**6** **a** $p$ is an odd number, $q$ is an even number and $r$ is an odd number. Is each expression odd or even?

**i** $pq + r$    **ii** $pqr$    **iii** $(p + q)^2 + r$

**b** $x$ is a prime number and both $y$ and $z$ are odd.

Write an expression using all of $x$, $y$ and $z$, and no other numbers or letters, so that the answer is always even.

**7** The powers of 2 are $2^1$, $2^2$, $2^3$, $2^4$, $2^5$, …

This gives the sequence 2, 4, 8, 16, 32, …

The $n$th term is given by $2^n$.

**a** Continue the sequence for another five terms.

**b** Give the $n$th term of these sequences.

**i** 1, 3, 7, 15, 31, …    **ii** 3, 5, 9, 17, 33, …    **iii** 6, 12, 24, 48, 96, …

**8** The powers of 10 are $10^1$, $10^2$, $10^3$, $10^4$, $10^5$, …

This gives the sequence 10, 100, 1000, 10 000, 100 000, …

The $n$th term is given by $10^n$.

**a** Describe the connection between the number of zeros in each term and the power of the term.

**b** If $10^n = 1\,000\,000$, what is the value of $n$?

**c** Give the $n$th terms of these sequences.

**i** 9, 99, 999, 9 999, 99 999, …    **ii** 20, 200, 2000, 20 000, 200 000, …

**9** The first four cube numbers are 1, 8, 27 and 64.

**a** Write down the next two cube numbers.

**b** Add consecutive cube numbers together. For example:

1 + 8, 1 + 8 + 27, 1 + 8 + 27 + 64

What do you notice about the answers?

**10** The triangular numbers are 1, 3, 6, 10, 15, 21, …

**a** Continue the sequence for another five terms.

**b** The $n$th term of this sequence is given by $\frac{1}{2}n(n + 1)$.

Use the formula to find:

**i** the 20th triangular number    **ii** the 100th triangular number.

**c** Add consecutive terms of the triangular number sequence.

For example, 1 + 3 = 4, 3 + 6 = 9, …

What do you notice?

(CM) 11 **a** $n$ is a positive integer.

    **i** Give a reason why $n(n + 1)$ must be an even number.

    **ii** Give a reason why $2n + 1$ must be an odd number.

**b** $p$ is an odd number and $q$ is an even number.

Copy this table, then tick the correct box to show whether each expression is odd or even.

| Expression | Odd | Even |
|---|---|---|
| $p^2$ | | |
| $p(q + 1)$ | | |
| $2p + 1 + 2q + 1$ | | |
| $3(q + 1) + 1$ | | |
| $(q - 1)^2$ | | |

**c** Show algebraically why, when you square an odd number, the answer is always odd and when you square an even number, the answer is always even.

(CM) 📠 12 A palindromic number is one that reads the same forwards as backwards, such as 242 and 1001.

In the triangular number series 1, 3, 6, 10, 15, …, the first palindromic number is the 10th term, 55.

Find the next two palindromic triangular numbers.

(MR) 13 The square numbers are 1, 4, 9, 16, 25, …

The $n$th term of this sequence is $n^2$.

**a** Continue the sequence for another five terms.

**b** Give the $n$th term of each sequence.

    **i** 2, 5, 10, 17, 26, …     **ii** 2, 8, 18, 32, 50, …     **iii** 0, 3, 8, 15, 24, …

📠 14 **a** Work out the first five terms of the sequences with the $n$th term as shown.

    **i** $4^n$     **ii** $7^n$     **iii** $5^n$

    **iv** $9^n$     **v** $8^n$

**b** Work out the stated term for the series with the $n$th term as shown.

    **i** the 5th term of $12n$     **ii** the 20th term of $2n$

(PS) 15 Mia adds two prime numbers. The result is another prime number. Write down the value of one of Mia's two prime numbers. Explain your answer.

(PS) 16 **a** Find two prime numbers that add together to give a square number greater than 30.

**b** Find two square numbers with a difference that is a prime number.

17 The $n$th term of a quadratic sequence is $n^2 + 2$.

**a** Work out the first five terms of the sequence.

**b** Work out the 10th term of the sequence.

**c** A term in the sequence is 402. Which number term is this?

(CM) **d** Can there ever be two consecutive even terms in the sequence? If not, why not?

# 21.5 General rules from given patterns

This section will show you how to:

- find the $n$th term from practical problems involving sequences.

Many problem-solving situations that you are likely to meet involve number sequences. So you need to be able to formulate general rules from given number patterns.

The diagram shows how a pattern of squares builds up.

a   How many squares will there be in the $n$th pattern?

b   Which pattern has 99 squares in it?

a   First, build up a table for the patterns.

| Pattern number | 1 | 2 | 3 | 4 | 5 |
|---|---|---|---|---|---|
| Number of squares | 1 | 3 | 5 | 7 | 9 |

Look at the differences between consecutive patterns. It is always 2 squares, so use $2n$.

Subtract the difference, 2, from the first number. This gives $1 - 2 = -1$.

So the number of squares in the $n$th pattern is $2n - 1$.

b   Now find $n$ when $2n - 1 = 99$.

$$2n - 1 = 99$$
$$2n = 99 + 1$$
$$= 100$$
$$n = 100 \div 2$$
$$= 50$$

The pattern with 99 squares is the 50th.

When you are trying to find a general rule from a sequence of diagrams, always set up a table to connect the pattern number with the number of items (such as squares, matches, seats) for which you are trying to find the rule. Once you have set up the table, it is easy to find the $n$th term.

## Exercise 21E

**1**   This pattern of triangles is built up from matchsticks.

      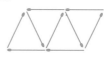

    1            2            3            4

a   Draw the fifth set of triangles in this pattern.

b   How many matchsticks would you need for the $n$th set of triangles?

c   How many matchsticks would you need to make the 60th set of triangles?

d   If you only have 100 matchsticks, what is the largest set of triangles you could make?

**2** This pattern of squares is built up from matchsticks.

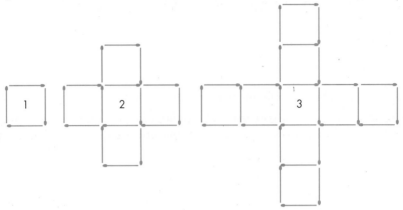

Hints and tips Write out the number sequences to help you see the patterns.

**a** Draw the fourth diagram.

**b** How many squares are there in the $n$th diagram?

**c** How many squares are there in the 25th diagram?

**d** What is the biggest diagram that you could make with 200 squares?

**3** The tables at a conference centre can each take six people. When the tables are put together, people can sit as shown.

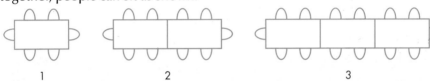

**a** How many people could be seated at four tables put together in this way?

**b** How many people could be seated at $n$ tables put together in this way?

**c** When 50 people attend a conference, they decide to use the tables in this way. How many tables do they need?

**4** This pattern is made from regular pentagons of side length 1 cm.

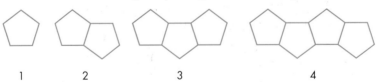

**a** Write down the perimeter of each shape.

**b** What is the perimeter of patterns like this made from:

  **i** six pentagons         **ii** $n$ pentagons         **iii** 50 pentagons?

**c** What is the largest number of pentagons that can be put together like this to have a perimeter less than 1000 cm?

**5** Prepacked fencing units come in the shape shown here.

1

Each is made up from four pieces of wood. When you put them together in stages to make a fence, you also need an extra joining piece, so the fence will start to build up like this.

2

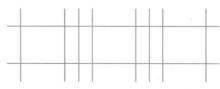

3

**a** How many pieces of wood would you have in a fence made up in:

   **i** 5 stages            **ii** $n$ stages            **iii** 45 stages?

**b** I made a fence out of 124 pieces of wood. How many stages did I use?

**6** Lamp-posts are put at the end of every 100-m stretch of a motorway.

1

2

3

**a** How many lamp-posts are needed for:

   **i** 900 m of this motorway            **ii** 8 km of this motorway?

**b** The contractor building the M99 motorway has ordered 1598 lamp-posts. How long is this motorway?

**7** A school dining hall had trapezium-shaped tables.

Each table could seat five people, as shown. When the tables are joined together, as shown below, fewer people can sit at each table.

1

2

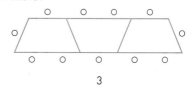

3

**a** In this arrangement, how many could be seated if there were:

   **i** 4 tables            **ii** $n$ tables            **iii** 13 tables?

**b** For an outside charity event, up to 200 people had to be seated. How many tables arranged like this did they need?

**8** When setting out tins to make a display of a certain height, you need to know how many tins to start with at the bottom.

**a** How many tins are needed on the bottom if you wish the display to be:

   **i** 5 tins high               **ii** $n$ tins high             **iii** 18 tins high?

**b** Albi started to build a display with 20 tins on the bottom. How high was the display when it was finished?

**9** This pattern is formed from small squares.

Pattern 1         Pattern 2          Pattern 3             Pattern 4

Work out the number of small squares in pattern $n$.

**10** Consecutive patterns of triangular numbers are put together.

Pattern 1 and 2      Pattern 2 and 3      Pattern 3 and 4

How many counters will there be in the combined pattern formed by patterns $n$ and $n + 1$?

**11** **a** The values of 2 raised to a positive whole-number power are 2, 4, 8, 16, 32, …

What is the $n$th term of this sequence?

**b** A supermarket sells four different-sized bottles of water: pocket size, 100 ml; standard size, 200 ml; family size, 400 ml; giant size, 800 ml.

   **i** Describe the number pattern that the contents form.

   **ii** The supermarket wants to sell a super-giant-sized bottle, which is the next-sized bottle in the pattern. How much does this bottle hold?

**c** A litre of water weighs 1 kg. The supermarket estimates that the heaviest they could possibly make a bottle of water is 10 kg. Assuming that the plastic bottle has a negligible weight and that the pattern of bottles continues, what is the largest size of bottle the supermarket could have?

**12** Draw an equilateral triangle.

Mark the midpoints of each side and draw and shade in the equilateral triangle formed by these points.

Repeat this with the three triangles that remain unshaded.

Keep on doing this with the unshaded triangles that are left.

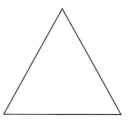 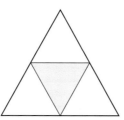 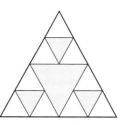 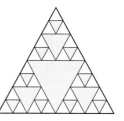

This is called a Sierpinski triangle. It is one of the earliest examples of a fractal pattern.

The shaded areas in each triangle are $\frac{1}{4}, \frac{7}{16}, \frac{37}{64}, \frac{175}{256}, \ldots$

It is very difficult to work out an $n$th term for this series of fractions.

Use your calculator to work out the area left unshaded, for example, $\frac{3}{4}, \frac{9}{16}, \ldots$

You should be able to write down a formula for the $n$th term of this pattern.

Pick a large value for $n$.

Will the shaded area ever cover all of the original triangle?

**13** Thom is using matchsticks to build three different patterns.

He builds the patterns in steps.

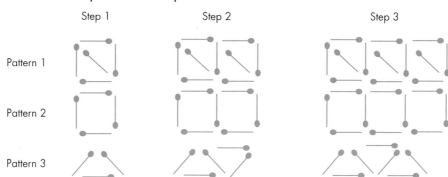

Thom has five boxes of matches, each labelled 'Average contents 42 matches'.

Can Thom build the pattern for the 20th step?

Show your working.

**14** A supermarket manager wants to display grapefruit stacked in layers, each of which is a triangle. These are the first four layers.

**a** If the display is four layers deep, how many grapefruit will there be in the display?

**b** The manager tells her staff that there should not be any more than eight layers, as otherwise the fruit will get squashed.

What is the most grapefruit that could be stacked?

**EV** **15** Some students are making hollow patterns from small squares.

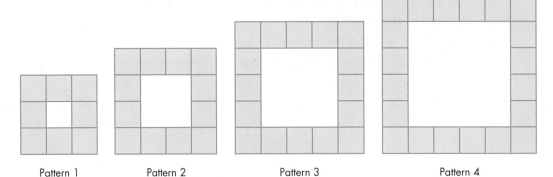

| Pattern 1 | Pattern 2 | Pattern 3 | Pattern 4 |

Four students write down their methods for finding the number of squares in the $n$th pattern.

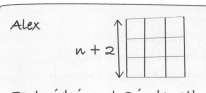

**Alex**

$n + 2$

Each side is $n + 2$ in length. There are 4 sides so just times by 4.

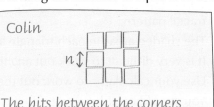

**Colin**

$n$

The bits between the corners are $n$ in length. There are 4 of these. Add the four corners.

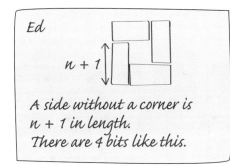

**Ed**

$n + 1$

A side without a corner is $n + 1$ in length. There are 4 bits like this.

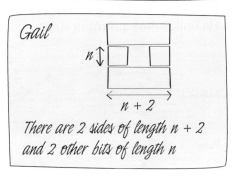

**Gail**

$n$

$n + 2$

There are 2 sides of length $n + 2$ and 2 other bits of length $n$

Evaluate each student's method. Do they give the correct answer?

Are any of their methods wrong? What mistake have they made? What should they do to correct it?

**16** In a charity event, buckets are lined up like this.

£1    £2    £4    £8

People are asked to give £1 each. The aim is that the first bucket will contain £1, the second will contain £2, the third will contain £4 and so on, doubling the amount in the bucket each time.

**a** How many pounds (£) will there be in the 8th bucket?

**b** How much will there be altogether in the first five buckets?

**c** By the end of the day the organisers have managed to complete the sequence up to and including the 12th bucket. How much have they raised altogether?

# Worked exemplars

 **1** These are expressions for the $n$th terms of three sequences.

Sequence 1:     $4n + 1$

Sequence 2:     $5n - 2$

Sequence 3:     $5n + 10$

Write down whether the sequences generated by the $n$th terms always give multiples of 5 (A), never give multiples of 5 (N) or sometimes give multiples of 5 (S).

| This is a 'communicating mathematically' question so you must show how you arrive at your decisions. | |
|---|---|
| Sequence 1 $\rightarrow$ 5, 9, 13, 17, 21, 25, 29, … <br><br> Sequence 2 $\rightarrow$ 3, 8, 13, 18, 23, 28, … <br><br> Sequence 3 $\rightarrow$ 15, 20, 25, 30, 35, 40, … | Substitute $n = 1, 2, 3, …$ until you can be sure of the sequences. <br><br> The sign $\rightarrow$ means 'gives'. |
| Sequence 1: Sometimes (S) <br><br> Sequence 2: Never (N) <br><br> Sequence 3: Always (A) | The series generated will show whether their terms are never, sometimes or always multiples of 5. |

**2** Look at this number pattern.

Line 1                    $2 \times 1 = 1 \times 2$

Line 2                $2 \times (1 + 2) = 2 \times 3$

Line 3            $2 \times (1 + 2 + 3) = 3 \times 4$

Line 4        $2 \times (1 + 2 + 3 + 4) = 4 \times 5$

**a** Write down line 5 of the pattern.

 **b** Write down line 10 of the pattern.

**c** Use the pattern to find the sum of the whole numbers from 1 to 100.

You *must* show your working.

| This question involves mathematical reasoning and problem solving. Make sure you show clearly how you reach your answers. | |
|---|---|
| **a** $2 \times (1 + 2 + 3 + 4 + 5) = 5 \times 6$ | You are extending the given patterns, so use line 4 and add '+5' into the brackets on the left, and increase the numbers in the product on the right by 1. |
| **b** $2 \times (1 + 2 + 3 + 4 + 5 + 6 + 7 + 8 + 9 + 10)$ <br> $= 10 \times 11$ | This is where you use mathematical reasoning to spot that each line number is the last value in the brackets on the left and the first value in the product on the right. |
| **c** $2 \times (1 + 2 + … + 99 + 100) = 100 \times 101$ <br><br> $\therefore 1 + 2 + … + 99 + 100 = \dfrac{100 \times 101}{2}$ <br><br> $\therefore 1 + 2 + … + 99 + 100 = 50 \times 101$ <br><br> $\qquad\qquad\qquad\qquad\quad = 5050$ | This is the problem-solving part. Notice that the expression in the brackets on the left is the sum of all the numbers from 1 to $n$. Use the same idea as in **b** to write down the 100th line. There is no need to write out all the terms. <br><br> Divide both sides by 2 and work out the answer. |

# Ready to progress?

I can substitute numbers into an $n$th-term rule for a linear sequence.
I can understand how prime, odd and even numbers interact in addition, subtraction and multiplication problems.

I can give the $n$th term of a linear sequence.
I can recognise special patterns, such as the Fibonacci sequence and geometric sequences.

I can give the $n$th term of a sequence of powers of 2 or 10.

# Review questions

1. The $n$th term of a sequence is $3n + 4$.

   Beth says: 'Every odd term of this sequence is a prime number.'

   Is Beth correct?

   Justify your answer.

2. **a** Write down the next two terms in this sequence.

   33, 29, 25, 21, 17, …

   **b** What is the rule for continuing the sequence?

   **c** Jaime says that −4 will be a term in this sequence.

   Give a reason why Jaime is wrong.

3. **a** Write down the next two terms in each sequence.

   **i** 70    64    58    52    46    …    …

   **ii** 1    2    4    8    16    …    …

   **b** The two sequences in part **a** have three terms in common. What are they?

4. The first ten prime numbers are 2, 3, 5, 7, 11, 13, 17, 19, 23 and 29.

   $P$ is a prime number.

   $Q$ is an odd number.

   State whether each expression is always odd, always even or could be either odd or even.

   **a** $P(Q + 3)$              **b** $Q - P + 1$

5. This expression is the $n$th term of a sequence.

   $n^2 - 1$

   Write down the first three terms of the sequence.

**6**

**a** This is the start of a sequence of numbers.

13    11    9    7    …    …

Write down the next two numbers in this sequence.

**b** Another sequence uses this rule.

'Add the last two numbers then halve the result.'

The sequence of numbers starts:    20    12    16    14    …    …

Write down the next two numbers in this sequence.

(EV) **c** Ed says: 'As the sequence in **b** continues, the terms will never be less than 14.'

Is Ed right? Explain your answer.

(CM) **7**    **a** The $n$th term of a sequence is $5n + 1$.

**i** Write down the first three terms of the sequence.

**ii** Tim says that 2015 is a term in this sequence. Explain why he is wrong.

**b** Another sequence has $(n + 2)^2 - 9$ as its $n$th term.

Will any of the terms of this sequence be negative numbers? Explain your answer.

**8**    These are the first five terms of an arithmetic sequence.

5    9    13    17    21

**a** Find, in terms of $n$, an expression for the $n$th term of the sequence.

(CM) **b** Give a reason why 112 is a not a term in this arithmetic sequence.

(MR) **c** Work out the term in the sequence that is closest to 112.

**9**    These are the first five terms of an arithmetic sequence.

6    11    16    21    26

Work out the 150th term in the sequence.

(MR) **10**    These are the first five terms of an arithmetic sequence.

9    15    21    27    33

**a** Find, in terms of $n$, an expression for the $n$th term of the sequence.

**b** Another arithmetic sequence has an $n$th term of $3n + 2$.

Will the two sequences ever have a term in common?

Show working to support your answer.

**11**    $p$ and $q$ are odd numbers. State whether each of these expressions is always odd, always even or could be either.

**a** $p + 2q$                    **b** $p^2q$

(MR) **12**    Martin says that the square of any number is always bigger than the number. Give an example to show that Martin is wrong.

(CM) **13**    It is known that $n$ is an integer.

**a** Explain why $2n + 1$ is always an odd number for all values of $n$.

**b** Explain why $n^2$ could be either odd or even.

# 22 Geometry and measures: Right-angled triangles

## This chapter is going to show you:

- how to use Pythagoras' theorem in right-angled triangles
- how to solve problems using Pythagoras' theorem
- how to use trigonometric ratios in right-angled triangles
- how to use trigonometry to solve problems.

## You should already know:

- how to calculate the square and square root of a number
- how to solve equations.

## About this chapter

When a builder needs to build two walls at right angles to one another, how does he make sure the angle between them is 90°? He will probably use a wooden right-angled triangle. What the builder may not know is that thousands of years ago, builders used a length of rope, tied at various places, to pull tight and form a right angle. This was used as far back as the construction of the pyramids of Egypt and maybe even before that.

The rope 'trick' works because of Pythagoras' theorem.

Pythagoras was a Greek who lived about 2600 years ago but, although the rule has been named after him, he was certainly not the first person to discover it. There is written evidence that the theorem was known in ancient Mesopotamia, China and India and it was probably discovered independently at different times in different parts of the world.

This chapter will show you how you can use Pythagoras' rule and trigonometric functions to solve a variety of problems using right-angled triangles.

# 22.1 Pythagoras' theorem

This section will show you how to:

- discover Pythagoras' theorem
- calculate the length of the hypotenuse in a right-angled triangle.

**Key terms**

hypotenuse

Pythagoras' theorem

Pythagoras was a philosopher and a mathematician. He was born in 580 BC, on the island of Samos in Greece and later moved to Crotona (Italy). It was here that he established the Pythagorean Brotherhood, a secret society to discuss politics, mathematics and astronomy. It is said that when he discovered his famous theorem, he was so full of joy that he showed his gratitude to the gods by sacrificing one hundred oxen.

## Rediscovering Pythagoras' theorem

## Exercise 22A

  **1**   **a**   Construct a triangle with sides of 3 cm, 4 cm and 5 cm.

      **b**   Measure the size of the largest angle.

      **c**   Square the length of each side ($3^2$, $4^2$ and $5^2$).

      **d**   Can you see a connection between these squares?

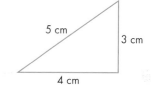

 **2**   **a**   Construct a triangle with sides of 5 cm, 12 cm and 13 cm.

      **b**   Measure the size of the largest angle.

      **c**   Square the length of each side.

      **d**   Can you see a connection between these squares?

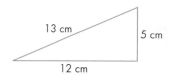

  **3**   **a**   Construct a triangle with sides of 6 cm, 8 cm and 10 cm.

      **b**   Measure the size of the largest angle.

      **c**   Square the length of each side.

      **d**   Can you see a connection between these squares?

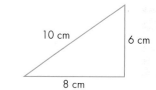

  **4**   **a**   Use your answers to questions **1** to **3** to work out the connection between the sides $a$, $b$ and $c$ for this triangle. Write down your answer.

> You have just rediscovered **Pythagoras' theorem**!
>
> In any right-angled triangle, the square of the length of the longest side (the **hypotenuse**) is equal to the sum of the squares of the lengths of the other two sides.
> The hypotenuse is always opposite the right angle.
>
> Pythagoras' theorem is usually written as: $c^2 = a^2 + b^2$.
>
>

**5**  **a** Construct this right angled triangle.

   **b** Measure the hypotenuse, $c$.

   **c** Use Pythagoras' theorem, $c^2 = a^2 + b^2$, to check that your calculation and your measurement of the hypotenuse are the same.

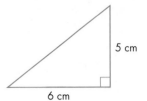

Pythagoras' theorem can also be stated as follows:

*For any right-angled triangle, the area of the square drawn on the hypotenuse is equal to the sum of the areas of the squares drawn on the other two sides.*

The form of the rule that most people remember is:

*In any right-angled triangle, the square of the hypotenuse is equal to the sum of the squares of the other two sides.*

Pythagoras' theorem is only true in right-angled triangles.

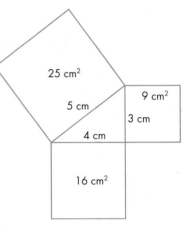

## Calculating the length of the hypotenuse

If you know the two smaller sides of a triangle, you can use Pythagoras' theorem to calculate the length of the hypotenuse. You can square each of the smaller sides, add them together, then calculate the square root.

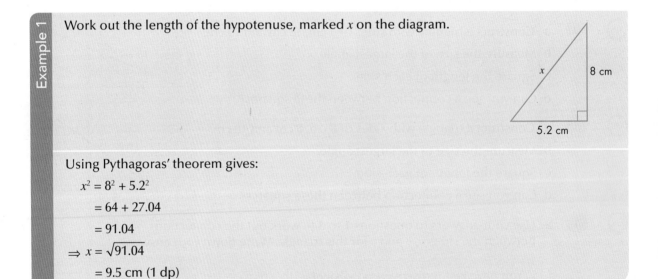

**Example 1**

Work out the length of the hypotenuse, marked $x$ on the diagram.

Using Pythagoras' theorem gives:

$x^2 = 8^2 + 5.2^2$

$\quad = 64 + 27.04$

$\quad = 91.04$

$\Rightarrow x = \sqrt{91.04}$

$\quad = 9.5$ cm (1 dp)

You can also use it to test if a triangle is right-angled.

To test if the triangle with sides 7 cm, 12 cm and 13 cm is right-angled, square each side. If the sum of the squares of the two shorter sides is the same as the square of the longest side, it is right-angled.

Sum the squares of the two shorter sides.　　　$7^2 + 12^2 = 49 + 144$

$\qquad\qquad\qquad\qquad\qquad\qquad\qquad\qquad = 193$

Square the longest side.　　　　　　　　　　$13^2 = 169$

These are not the same, so this triangle is not right-angled.

Example 2

Find the distance between the points (1, 7) and (7, 2).

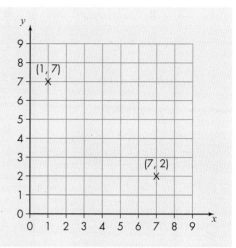

Start by drawing a right-angled triangle. (1, 7) and (7, 2) are at the ends of the hypotenuse.

Work out the lengths of the shorter sides of the right-angled triangle.

For the base: $7 - 1 = 6$

For the height: $7 - 2 = 5$

Using Pythagoras' theorem gives:

$a^2 = 6^2 + 5^2$

$\quad = 36 + 25$

$\quad = 61$

$a = \sqrt{61}$

$\quad = 7.8$ units

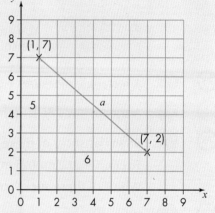

## Exercise 22B ▦

 **1** Calculate the length of the hypotenuse, $x$, for each triangle. Give your answers correct to 1 decimal place.

**a**

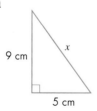

9 cm
$x$
5 cm

**b**

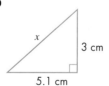

$x$
3 cm
5.1 cm

**c**

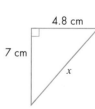

4.8 cm
7 cm
$x$

**d**

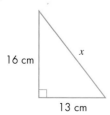

16 cm
$x$
13 cm

**e**

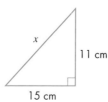

$x$
11 cm
15 cm

**f**

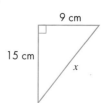

9 cm
15 cm
$x$

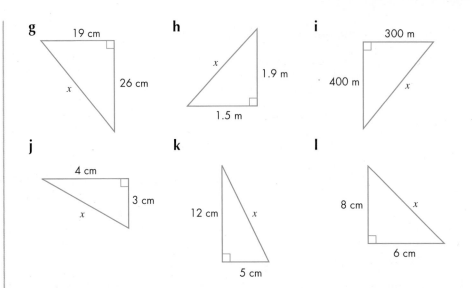

**g** 19 cm, 26 cm, $x$

**h** $x$, 1.9 m, 1.5 m

**i** 300 m, 400 m, $x$

**j** 4 cm, 3 cm, $x$

**k** 12 cm, $x$, 5 cm

**l** 8 cm, $x$, 6 cm

> **Hints and tips** You are calculating the hypotenuse in all these questions so add the squares of the two shorter sides each time.

**(PS) 2** A rectangle measures 8 cm by 12 cm. How long is the diagonal?

**(MR) 3** A square has side length 6 cm. Show that its diagonal is 8.5 cm.

**(PS) 4** Brian has sticks of these lengths.

50 cm, 1 m, 1.2 m, 1.25 m, 1.3 m, 1.5 m, 2 m, 2.5 m

Which of these sticks could he use to make a right-angled triangle?

**(CM) 5** Why isn't a triangle with sides of 6 cm, 7 cm and 10 cm a right-angled triangle?

**6** Draw a set of coordinate axes with the $x$-axis numbered from –2 to 19 and the $y$-axis numbered from –3 to 12.

Use your grid to find the distance…

Find the distance between each pair of points:

**a** (6, 2) and (18, 11)

**b** (14, 10) and (2, 5)

**c** (2, 9) and (4, 2)

**d** (10, –2) and (–1, 7)

**(MR) 7** The triangles in question **1** parts **j**, **k** and **l** give whole-number answers. Sets of whole numbers that obey Pythagoras' theorem are called **Pythagorean triples**. Examples of these are 3, 4 and 5 and 5, 12 and 13.

If you multiply the Pythagorean triple 3, 4 and 5 by two, you get the Pythagorean triple 6, 8 and 10.

**a** What Pythagorean triple do you get if you multiply 3, 4 and 5 by three?

**b** What Pythagorean triple do you get if you multiply 5, 12 and 13 by five?

**c** Write down six more Pythagorean triples.

# 22.2 Calculating the length of a shorter side

This section will show you how to:

- calculate the length of a shorter side in a right-angled triangle.

You can calculate the length of one of the shorter sides $b$, by rearranging the formula for Pythagoras' theorem.

$$c^2 = a^2 + b^2$$

$$\Rightarrow a^2 = c^2 - b^2 \quad \text{and} \quad b^2 = c^2 - a^2$$

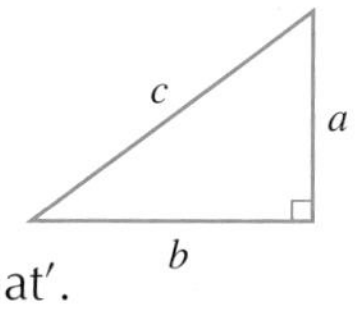

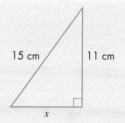

**Note:** The symbol $\Rightarrow$ means 'implies' or means 'that'.

---

**Example 3**

Calculate the length of $x$.

$x$ is one of the shorter sides.

Use Pythagoras' theorem starting from $a^2 = c^2 - b^2$

$$x^2 = 15^2 - 11^2$$
$$= 225 - 121$$
$$= 104$$
$$\Rightarrow x = \sqrt{104}$$
$$= 10.2 \text{ cm (1 dp)}$$

---

You can also substitute the numbers in before you rearrange.

---

**Example 4**

Calculate the length of $x$.

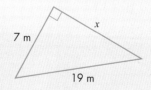

$x$ is one of the shorter sides.

Use Pythagoras' theorem, starting from $c^2 = a^2 + b^2$.

$$19^2 = x^2 + 7^2$$
$$361 = x^2 + 49$$
$$x^2 = 361 - 49$$
$$x^2 = 312$$
$$\Rightarrow x = \sqrt{312}$$
$$= 17.7 \text{ cm (1 dp)}$$

---

# Exercise 22C

**1** Calculate the length of $x$ for each triangle. Give your answers to 1 decimal place.

**a**
17 cm
$x$
8 cm

**b**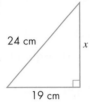
24 cm
$x$
19 cm

**c**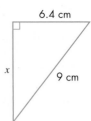
6.4 cm
$x$
9 cm

**d**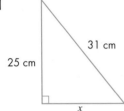
31 cm
25 cm
$x$

**e**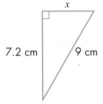
$x$
7.2 cm
9 cm

**f**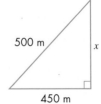
500 m
$x$
450 m

**g**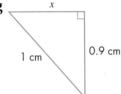
$x$
1 cm
0.9 cm

**h**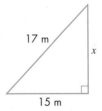
17 m
$x$
15 m

**2** Calculate the length of $x$ in each triangle. Give your answers to 1 decimal place.

**a**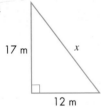
17 m
$x$
12 m

**b**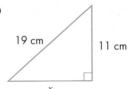
19 cm
11 cm
$x$

**c**
17 m
$x$
23 m

**d**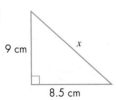
9 cm
$x$
8.5 cm

**e**
34 m
$x$
41 m

**f**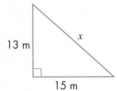
$x$
13 m
15 m

**g**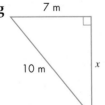
7 m
$x$
10 m

**h**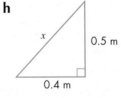
$x$
0.5 m
0.4 m

> **Hints and tips** In this question, sometimes $x$ is a shorter side and sometimes it is the hypotenuse. Make sure you combine the squares of the sides correctly.

**3** Calculate the length of $x$ in each triangle. Give your answers to 1 decimal place.

**a**  $x$, 13 m, 12 m

**b** 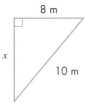 8 m, $x$, 10 m

**c** 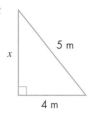 $x$, 5 m, 4 m

**d** 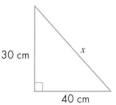 30 cm, $x$, 40 cm

**PS** **4** Bain wants to buy a TV that has a screen with a diagonal length of 32 cm.

He looks at TVs with these screen sizes.

Which one has a diagonal length closer to 32 cm?

Show your working.

 12 cm, 30 cm, 11 cm, 30 cm

**PS** **5** Give three possible pairs of lengths for the sides marked $a$ and $b$ in this triangle.

14 cm, $a$, $b$

**CM** **6** Show that a triangle with sides of 80 m, 60 m and 100 m is a right-angled triangle.

**CM** **7** This is a visual proof of Pythagoras' theorem.

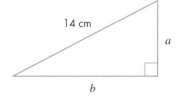

Here is an explanation of how the diagram illustrates Pythagoras' theorem.

The large square is 7 by 7 giving 49 square units.

The red and yellow triangles all have shorter lengths of 3 and 4 and so each has an area of 6 square units.

The area of the inner square (green and yellow) must be $49 - 4 \times 6 = 25$, so the hypotenuse of the red and yellow triangles must be $\sqrt{25} = 5$.

This shows that $3^2 + 4^2 = 5^2$.

**a** Now follow these steps for a triangle with sides 5, 12 and 13.

* Start with a grid measuring 17 squares by 17 squares.

* Now draw the four red triangles with shorter sides of 5 squares and 12 squares. There should be one triangle in each corner of the large square.

* Next draw the four yellow triangles.

* Finally there should be a smaller square in the middle. Colour this green.

**b** Work out:

  **i** the area of the large square

  **ii** the total area of the four red triangles

  **iii** the total area of the yellow triangles and the green square

  **iv** the hypotenuse of the red and yellow triangles.

**c** Now show that $5^2 + 12^2 = 13^2$.

# 22.3 Applying Pythagoras' theorem in real-life situations

This section will show you how to:

- solve problems using Pythagoras' theorem.

Follow these steps to solve practical problems using Pythagoras' theorem.

- Draw a diagram for the problem that includes a right-angled triangle.
- Label the unknown side $x$.
- Look at the diagram and decide whether $x$ is the hypotenuse or one of the shorter sides.
- If it's the hypotenuse, square both numbers, then add the squares and take the square root of the sum. If it's one of the shorter sides, square both numbers, then subtract the smaller square from the larger square and take the square root of the difference.

A plane leaves Manchester airport and heads due east. It flies 160 km before turning due north. It then flies a further 280 km and lands. On its return journey, the plane flies straight back to Manchester airport. What is the distance of the return flight?

First, sketch the situation.

Then use Pythagoras' theorem.

$$x^2 = 160^2 + 280^2$$
$$= 25\,600 + 78\,400$$
$$= 104\,000$$
$$\Rightarrow x = \sqrt{104\,000}$$
$$= 322 \text{ km (3 sf)}$$

The distance of the return flight is 322 km.

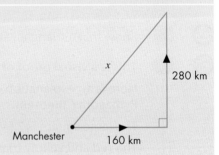

## Exercise 22D

> **Hints and tips** Show all your working, as in Example 5.

 **1** A 12-m ladder is safe if the foot of the ladder is about 2.5 m away from the wall. A ladder, 12 m long, leans against a wall. It reaches 10 m up the wall. Is this ladder safe?

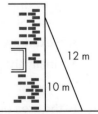

**2** A model football pitch is 2 m long and 0.5 m wide. How long is the diagonal?

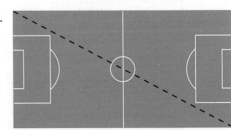

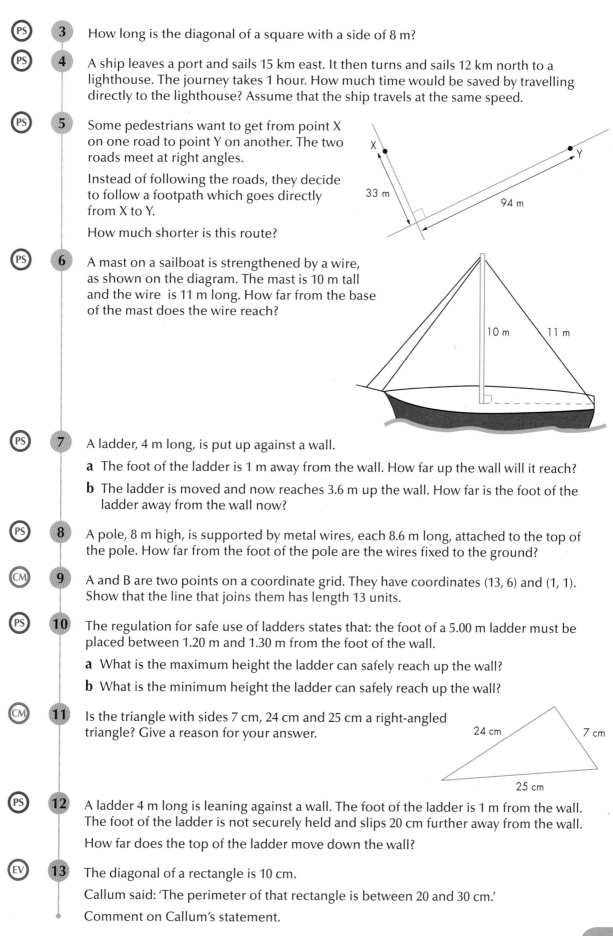

(PS) **3** How long is the diagonal of a square with a side of 8 m?

(PS) **4** A ship leaves a port and sails 15 km east. It then turns and sails 12 km north to a lighthouse. The journey takes 1 hour. How much time would be saved by travelling directly to the lighthouse? Assume that the ship travels at the same speed.

(PS) **5** Some pedestrians want to get from point X on one road to point Y on another. The two roads meet at right angles.

X

Y

33 m

94 m

Instead of following the roads, they decide to follow a footpath which goes directly from X to Y.

How much shorter is this route?

(PS) **6** A mast on a sailboat is strengthened by a wire, as shown on the diagram. The mast is 10 m tall and the wire is 11 m long. How far from the base of the mast does the wire reach?

10 m

11 m

(PS) **7** A ladder, 4 m long, is put up against a wall.

   **a** The foot of the ladder is 1 m away from the wall. How far up the wall will it reach?

   **b** The ladder is moved and now reaches 3.6 m up the wall. How far is the foot of the ladder away from the wall now?

(PS) **8** A pole, 8 m high, is supported by metal wires, each 8.6 m long, attached to the top of the pole. How far from the foot of the pole are the wires fixed to the ground?

(CM) **9** A and B are two points on a coordinate grid. They have coordinates (13, 6) and (1, 1). Show that the line that joins them has length 13 units.

(PS) **10** The regulation for safe use of ladders states that: the foot of a 5.00 m ladder must be placed between 1.20 m and 1.30 m from the foot of the wall.

   **a** What is the maximum height the ladder can safely reach up the wall?

   **b** What is the minimum height the ladder can safely reach up the wall?

(CM) **11** Is the triangle with sides 7 cm, 24 cm and 25 cm a right-angled triangle? Give a reason for your answer.

24 cm

7 cm

25 cm

(PS) **12** A ladder 4 m long is leaning against a wall. The foot of the ladder is 1 m from the wall. The foot of the ladder is not securely held and slips 20 cm further away from the wall.

How far does the top of the ladder move down the wall?

(EV) **13** The diagonal of a rectangle is 10 cm.

Callum said: 'The perimeter of that rectangle is between 20 and 30 cm.'

Comment on Callum's statement.

# 22.4 Pythagoras' theorem and isosceles triangles

This section will show you how to:

- use Pythagoras' theorem in isosceles triangles.

Every isosceles triangle has a line of symmetry that divides the triangle into two matching right-angled triangles. So when you have a problem involving an isosceles triangle, you can split that triangle down the middle to create a right-angled triangle.

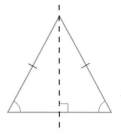

**Example 6**

Calculate the area of this triangle.

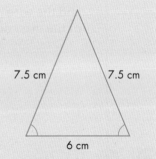

It is an isosceles triangle and you need to calculate its height to work out its area.

First split the triangle into two right-angled triangles to work out its height.

Let the height be $x$.

Then use Pythagoras' theorem.

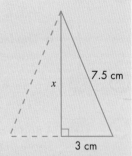

$$x^2 = 7.5^2 - 3^2$$
$$= 56.25 - 9$$
$$= 47.25$$
$$\Rightarrow x = \sqrt{47.25} \text{ cm}$$
$$= 6.873863542 \text{ cm}$$

Keep the accurate figure in the calculator memory and use it to calculate the area of the triangle.

$\frac{1}{2} \times 6 \times 6.873863542$ cm² (from the calculator memory)

$= 20.6$ cm² (1 dp)

Keeping a figure in a calculator memory is very helpful when you have a long decimal in questions on Pythagoras' theorem or circles. Most calculators have an [Ans] button, which you can press instead of rekeying the previous answer. In the above example, you would key in

$\frac{1}{2} \times 6 \times$ [Ans] [=] [0] [•] [5] [×] [6] [×] [Ans] [=]

# Exercise 22E

**1** Calculate the areas of these isosceles triangles.

   **a** 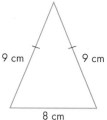    **b** 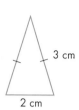    **c**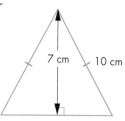

**2** Calculate the area of an isosceles triangle with sides of 8 cm, 8 cm and 6 cm.

**3** Calculate the area of an equilateral triangle of side 6 cm.

**4** An isosceles triangle has sides of 5 cm and 6 cm.

  **a** Sketch the two different isosceles triangles that fit this data.

  **b** Which of the two triangles has the greater area?

**5** **a** Sketch a regular hexagon, showing all its lines of symmetry.

  **b** Calculate the area of the hexagon if its side is 8 cm.

**6** Calculate the area of a hexagon of side 10 cm.

**7** These isosceles triangles have the same perimeter.

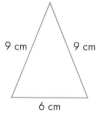

  **a** Do the three triangles have the same area?

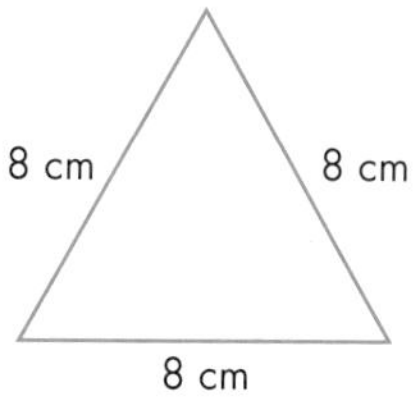

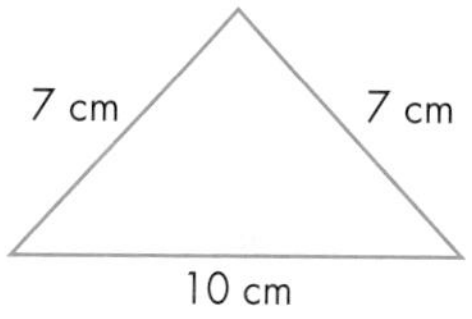

  **b** Can you describe an isosceles triangle with the same perimeter but a larger area?

  **c** Can you describe the pattern in your findings?

**8** A piece of land is in the shape of an isosceles triangle with sides 6.5 m, 6.5 m and 7.4 m.

Show that the area of the land is 19.8 m².

**9** The diagram shows an isosceles triangle ABC.

Calculate the area of triangle ABC.

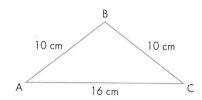

(PS) **10** Calculate the lengths marked $x$ in these isosceles triangles.

**a**

**b**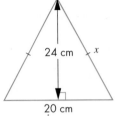

**c**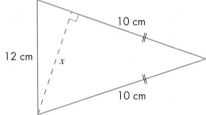

> **Hints and tips** For part **c**, work out the area first.

(PS) **11** An isosceles triangle with two sides of 5 cm has an area of 12 cm². Work out the length of the base of the triangle.

(EV) **12** A kite had two lengths of 8 cm and two lengths of 10 cm. One of the diagonals was 12 cm long.

Andrew and Olly both calculated the area of the kite. Andrew calculated the area to be 79.7 cm² and Olly calculated it to be 79.8 cm².

Check and comment on the accuracy of each answer.

# 22.5 Trigonometric ratios

## This section will show you how to:

- define, understand and use the three trigonometric ratios.

**Trigonometry** is concerned with the calculation of sides and angles in triangles.

Work through Exercise 22F to discover some interesting ratios.

| Key terms |
| --- |
| adjacent |
| cosine |
| opposite |
| sine |
| tangent |
| trigonometry |
| trigonometric functions |
| trignometric ratios |

## Exercise 22F 🖩

**1** **a** Choose an angle between 25° and 80°.

Draw six different right-angled triangles that also contain your chosen angle. Draw the right angle and your chosen angle in the positions shown by the marked angles in the diagram. Label your triangles A to F.

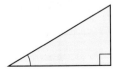

> **Hints and tips** Draw the base of your triangle as a whole number of centimetres.

**b i** Measure and then label the lengths of all three sides of each triangle.

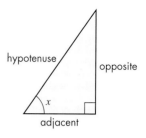

**ii** You already know that the hypotenuse is the longest side and is opposite the right angle.

The **opposite** side is the side that is opposite the angle you are using. The **adjacent** side is the side between the right angle *and* the angle you are using.

Label the opposite (opp), the adjacent (adj) and the hypotenuse (hyp) on each of your triangles.

**c i** Copy and complete the table below. Give your answers in the last three columns to 3 decimal places.

| Triangle | Opposite | Adjacent | Hypotenuse | $\dfrac{\text{Opposite}}{\text{Hypotenuse}}$ | $\dfrac{\text{Adjacent}}{\text{Hypotenuse}}$ | $\dfrac{\text{Opposite}}{\text{Adjacent}}$ |
|---|---|---|---|---|---|---|
| A | | | | | | |
| B | | | | | | |
| C | | | | | | |
| D | | | | | | |
| E | | | | | | |
| F | | | | | | |

**ii** What do you notice about the values in each of the last three columns?

**2 a** Repeat question **1** with a different chosen angle.

**b** Is the pattern you saw in **1cii** the same for both angles?

**3** Follow these instructions on your calculator. You may find that you need to enter sin after the angle, depending on your type of calculator.

**a** Enter the sin key followed by the first angle you chose. What do you notice?

**b** Enter the cos key followed by the first angle you chose. What do you notice?

**c** Enter the tan key followed by the first angle you chose. What do you notice?

**d** Repeat steps **a** to **c** for the second angle you chose. Do you get the same results?

**e** Match each button ( sin , cos and tan ) to one of the final three columns in your table from question 3 ( $\dfrac{\text{opposite}}{\text{hypotenuse}}$ , $\dfrac{\text{adjacent}}{\text{hypotenuse}}$ or $\dfrac{\text{opposite}}{\text{adjacent}}$ ).

In any right-angled triangle, you can identify the hypotenuse, the opposite side and the adjacent side in relation to an angle $x$ or $\theta$.

In the triangles you drew, the adjacent side was always at the base of the triangle. Look at these examples.

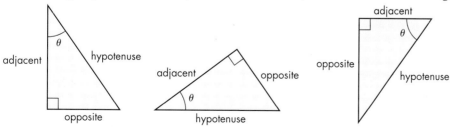

The position of the adjacent and the opposite depend on which angle is $\theta$. There is only one possible adjacent side and one possible opposite side for each angle.

In a right-angled triangle:

- the side opposite the right angle is called the hypotenuse and is the longest side
- the side opposite the angle $\theta$ is called the opposite side
- the other side next to both the right angle and the angle $\theta$ is called the adjacent side.

**Sine, cosine** and **tangent are trigonometrical functions or ratios.** As you have already discovered, they are defined in terms of the sides of a right-angled triangle and an angle. The sine, cosine and tangent ratios for an angle $\theta$ are defined as:

$$\text{sine } \theta = \frac{\text{Opposite}}{\text{Hypotenuse}} \qquad \text{cosine } \theta = \frac{\text{Adjacent}}{\text{Hypotenuse}} \qquad \text{tangent } \theta = \frac{\text{Opposite}}{\text{Adjacent}}$$

These ratios are usually abbreviated as:

$$\sin \theta = \frac{O}{H} \qquad \cos \theta = \frac{A}{H} \qquad \tan \theta = \frac{O}{A}$$

These abbreviated forms are also used on calculator keys.

To memorise these formulae, you can use a mnemonic such as,

**S**ome **O**ld **H**ens **C**ackle **A**ll **H**ours **T**il **O**ld **A**ge

in which the first letter of each word is taken in order to give:

$$S = \frac{O}{H} \qquad C = \frac{A}{H} \qquad T = \frac{O}{A}$$

or **T**ommy **O**n **A** **S**hip **O**f **H**is **C**aught **A** **H**erring

$$T = \frac{O}{A} \qquad S = \frac{O}{H} \qquad C = \frac{A}{H}$$

Make up some of your own mnemonics. Use family names or words you can    easily remember.

---

Example 7

For each triangle, write down a trigonometrical ratio of angle $\theta$ that uses the marked lengths.

**a**

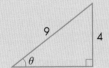

**b**

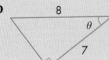

**c**

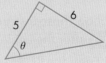

**a** 4 is opposite $\theta$ and 9 is the hypotenuse, so you can write down the sine ratio.

$$\sin \theta = \frac{\text{opp}}{\text{hyp}}$$

$$= \frac{4}{9}$$

**b** 7 is adjacent to $\theta$ and 8 is the hypotenuse, so you can write down the cosine ratio.

$$\cos \theta = \frac{\text{adj}}{\text{hyp}}$$

$$= \frac{7}{8}$$

**c** 6 is opposite $\theta$ and 5 is adjacent to $\theta$, so you can write down the tangent ratio.

$$\tan \theta = \frac{\text{opp}}{\text{adj}}$$

$$= \frac{6}{5}$$

Sketch a triangle for each trigonometric ratio.

**a** $\sin \theta = \dfrac{4}{7}$  **b** $\cos \theta = \dfrac{5}{8}$  **c** $\tan \theta = \dfrac{3}{4}$

**a** $\sin \theta = \dfrac{\text{opp}}{\text{hyp}}$   So 4 is opposite $\theta$ and 7 is the hypotenuse.

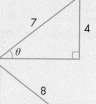

**b** $\cos \theta = \dfrac{\text{adj}}{\text{hyp}}$   So 5 is adjacent to $\theta$ and 8 is the hypotenuse

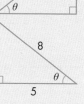

**c** $\tan \theta = \dfrac{\text{opp}}{\text{adj}}$   So 3 is opposite $\theta$ and 4 is adjacent.

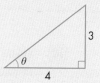

## Exercise 22G

**1** For each triangle, identify the hypotenuse and the sides opposite and adjacent to $\theta$.

**a** 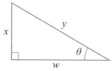  **b**   **c**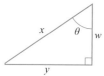

**d** 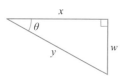  **e** 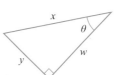  **f**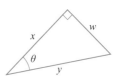

**2** Copy and complete the trigonometric ratios for each triangle. Part **a** has been done for you.

**a**   **b** 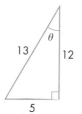  **c**

**a**

$\sin \theta = \dfrac{\text{opp}}{\text{hyp}} = \dfrac{3}{5}$

$\cos \theta = \dfrac{\text{adj}}{\text{hyp}} = \dfrac{4}{5}$

$\tan \theta = \dfrac{\text{opp}}{\text{adj}} = \dfrac{3}{4}$

**b**

$\sin \theta = \dfrac{\square}{\square} = \dfrac{\square}{\square}$

$\cos \theta = \dfrac{\square}{\square} = \dfrac{\square}{\square}$

$\tan \theta = \dfrac{\square}{\square} = \dfrac{\square}{\square}$

**c**

$\sin \theta = \dfrac{\square}{\square} = \dfrac{\square}{\square}$

$\cos \theta = \dfrac{\square}{\square} = \dfrac{\square}{\square}$

$\tan \theta = \dfrac{\square}{\square} = \dfrac{\square}{\square}$

**3** For each triangle:

   **i** identify the hypotenuse and the sides that are opposite and adjacent to θ

   **ii** define the three trignometric ratios (sin, cos and tan) for angle θ.

**a**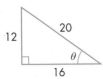

**b**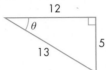

**c**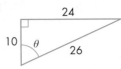

**d**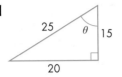

**e**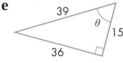

**f**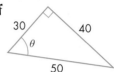

**4** Write down the fraction for the trigonometric ratio and fraction that can be identified in each triangle.

The first one has been done for you.

**a**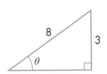

**b**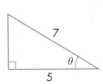

**c**

**d**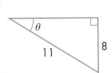

**e**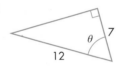

**f**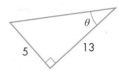

**a** $\sin\theta = \dfrac{\text{opp}}{\text{hyp}} = \dfrac{3}{8}$

**5** Jezz looked at a triangle with an angle marked as θ.

He said: 'Sine θ is three-fifths and cosine θ is four-fifths.'

Comment on his statement.

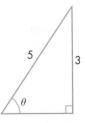

**6** For each trigonometric ratio, sketch a right-angled triangle and label the known sides.

   **a** $\tan\theta = \dfrac{5}{8}$        **b** $\cos\theta = \dfrac{3}{7}$        **c** $\sin\theta = \dfrac{5}{6}$

   **d** $\tan\theta = \dfrac{7}{3}$        **e** $\cos\theta = \dfrac{3}{7}$        **f** $\sin\theta = \dfrac{5}{9}$

   **g** $\cos\theta = \dfrac{8}{9}$        **h** $\tan\theta = \dfrac{3}{8}$

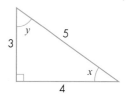

**7** Look at this triangle.

   **a** Write each of these as a fraction.

     **i** $\tan x$         **ii** $\sin x$    **iii** $\cos x$

   **b** Write down the fraction for each of these.

     **i** $\tan y$         **ii** $\sin y$    **iii** $\cos y$

   **c** Write down what you notice about:

     **i** $\sin x$ and $\cos y$    **ii** $\sin y$ and $\cos x$.

   **d** What do you notice about $\tan x$ and $\tan y$?

## Trigonometry using your calculator

- You will need to use a scientific calculator to work out trigonometric ratios. Different calculators work in different ways, so make sure you know how to use your model.

- Angles can be measured in degrees, radians or grads. Calculators can be set to operate in any of these three units, so make sure your calculator is set up to work in degrees.

- To work out the sine of 60 degrees you will probably press the keys `sin` `6` `0` `=` in that order, but it might be different on your calculator. The answer should be 0.8660… or $\frac{\sqrt{3}}{2}$. If your calculator gives answers in the form $\frac{\sqrt{3}}{2}$, make sure you can convert that to the decimal form.

- $3\cos 57°$ is a shorthand way of writing $3 \times \cos 57°$. On some calculators you do not need to use the × button and you can just press the keys in the way it is written: `3` `cos` `5` `7` `=`

- Check to see whether your calculator works this way. The answer should be 1.63.

---

**Example 9**

Calculate the value of $5.6 \sin 30°$.

This means $5.6 \times$ sine of 30 degrees.

$5.6 \sin 30° = 2.8$

---

# Exercise 22H 🖩

**1** Use a calculator to work out these values. Give your answers to 3 significant figures.

   **a** $\sin 43°$     **b** $\sin 56°$     **c** $\sin 67.2°$     **d** $\sin 90°$

   **e** $\sin 45°$     **f** $\sin 20°$     **g** $\sin 22°$     **h** $\sin 0°$

**2** Calculate these values. Give your answers to 3 significant figures.

   **a** $\cos 43°$     **b** $\cos 56°$     **c** $\cos 67.2°$     **d** $\cos 90°$

   **e** $\cos 45°$     **f** $\cos 20°$     **g** $\cos 22°$     **h** $\cos 0°$

 **3** From your answers to questions **1** and **2**, what angle has the same value for sine and cosine?

 **4**  **a**  **i** What is $\sin 35°$?    **ii** What is $\cos 55°$?

   **b**  **i** What is $\sin 12°$?    **ii** What is $\cos 78°$?

   **c**  **i** What is $\cos 67°$?    **ii** What is $\sin 23°$?

   **d** What connects the values in parts **a**, **b** and **c**?

   **e** Copy and complete these sentences.

     **i** $\sin 15°$ is the same as $\cos$ …    **ii** $\cos 82°$ is the same as $\sin$ …

     **iii** $\sin x$ is the same as $\cos$ …

**5** Use your calculator to work out these values.

    **a** tan 43°    **b** tan 56°    **c** tan 67.2°    **d** tan 90°

    **e** tan 45°    **f** tan 20°    **g** tan 22°    **h** tan 0°

**6** Use your calculator to work out these values.

    **a** sin 73°    **b** cos 26°    **c** tan 65.2°    **d** sin 88°

    **e** cos 35°    **f** tan 30°    **g** sin 28°    **h** cos 5°

(EV) **7** What is different about tan compared with both sin and cos?

**8** Use your calculator to work out these values.

    **a** 5 tan 65°    **b** 6 tan 42°    **c** 6 tan 90°    **d** 5 tan 0°

**9** Use your calculator to work out these values.

    **a** 4 sin 63°    **b** 7 tan 52°    **c** 5 tan 80°    **d** 9 cos 8°

(MR) **10** Show that (5 sin 65° + 6 cos 42°) − (6 sin 90° + 5 sin 0°) is 3.

(PS) **11** What is the sine of the smallest angle in a triangle with sides of 3, 4 and 5?

# 22.6 Calculating lengths using trigonometry

This section will show you how to:

- use trigonometric ratios to calculate a length in a right-angled triangle.

Look at this right-angled triangle.

You are given the angle 50° and the hypotenuse (8 cm).

How could you work out the length of the opposite side, BC?

You know that $\sin 50 = \dfrac{\text{opp}}{\text{hyp}}$

So   $\sin 50 = \dfrac{BC}{8}$

Multiply both sides by 8.   $8 \sin 50 = BC$

$$BC = 6.1 \text{ cm (1 dp)}$$

How could you work out the length of the adjacent side, AC?

You know that $\cos 50 = \dfrac{\text{adj}}{\text{hyp}}$

So   $\cos 50 = \dfrac{AC}{8}$

Multiply both sides by 8.   $8 \cos 50 = AC$

$$AC = 5.1 \text{ cm (1 dp)}$$

The first step to solving this type of question is to identify the information you have and the information you need to find. Then you can decide which ratio you need to use.

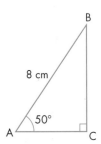

Example 10

Calculate the length of $x$ in this triangle.

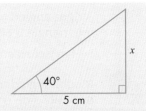

Identify the information you have been given and the information you need to find.

In this case, $x$ is opposite the angle you are given and 5 cm is the adjacent.

Decide which ratio to use.

In this case, use the tangent ratio as it uses the opposite and the adjacent.

Remember, $\tan \theta = \dfrac{\text{opp}}{\text{adj}}$

So,        $\tan 40 = \dfrac{x}{5}$

Rearrange the equation and work out the answer.

$x = 5 \tan 40$            (Multiply both sides by 5)

    $= 4.2$ cm (1 dp)

## Exercise 22I

1  Calculate the length marked $x$ in each triangle. Give your answers to 1 decimal place.

**a**
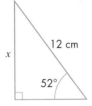

**b**
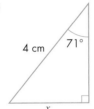

**c**

**d**

**e**
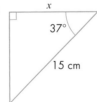

**f**
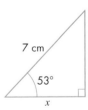

**g**
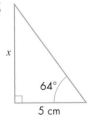

**h**
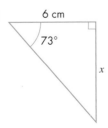

**i**
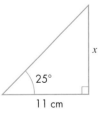

**j**

**k**

**l**
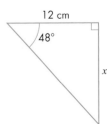

**2** A right-angled triangle has a hypotenuse of 10 cm. Harry says: 'If one of the angles in the triangle is 60°, then the other two sides are 5 cm and 8.7 cm.'

Is Harry correct? Why?

 **3** A right-angled triangle has an angle of 38°. The adjacent side is 8 cm. Show that the opposite side is 6.25 cm.

**4** Andrew said: 'In a right-angled triangle, the sine of one acute angle is the same as the cosine of the other acute angle.' Evaluate this statement.

# 22.7 Calculating angles using trigonometry

This section will show you how to:

• use the trigonometric ratios to calculate an angle.

What angle has a cosine of 0.6? You can use a calculator to work it out.

You write 'the angle with a cosine of 0·6' as $\cos^{-1} 0.6$. This is called the '**inverse** cosine of 0.6'.

Find out where $\cos^{-1}$ is on your calculator.

You will probably find it on the same key as cos, but you will need to press `shift` or `INV` or `2ndF` first.

Check that $\cos^{-1} 0.6 = 53.1301… = 53.1°$ (1 dp).

Check that $\cos 53.1° = 0.600$ (3 dp).

Check that you can calculate the inverse sine and the inverse tangent in the same way.

**Example 11**

What angle has a sine of $\frac{3}{8}$ ?

You need to calculate the value of $\sin^{-1} \frac{3}{8}$ .

Use the fraction button on your calculator.

If you use the fraction key you may not need a bracket, or your calculator may put one in automatically.

The answer is 22.0°.

## Exercise 22J

Use your calculator to work out the answers to the following. Give your answers to questions **1** to **6** to 1 decimal place.

 **1** Work out the size of angle $x$.

**a** $\sin x = 0.5$      **b** $\sin x = 0.785$      **c** $\sin x = 0.64$

**d** $\sin x = 0.877$      **e** $\sin x = 0.999$      **f** $\sin x = 0.707$

 **2** Work out the size of angle $x$.

**a** $\cos x = 0.5$      **b** $\cos x = 0.64$      **c** $\cos x = 0.999$

**d** $\cos x = 0.707$      **e** $\cos x = 0.2$      **f** $\cos x = 0.7$

**3** Work out the size of angle $x$.

   **a** $\tan x = 0.6$            **b** $\tan x = 0.38$           **c** $\tan x = 0.895$

   **d** $\tan x = 1.05$           **e** $\tan x = 2.67$            **f** $\tan x = 4.38$

**4** Work out the size of angle $x$.

   **a** $\sin x = \dfrac{4}{5}$           **b** $\sin x = \dfrac{2}{3}$           **c** $\sin x = \dfrac{7}{10}$

   **d** $\sin x = \dfrac{5}{6}$           **e** $\sin x = \dfrac{1}{24}$          **f** $\sin x = \dfrac{5}{13}$

**5** Work out the size of angle $x$.

   **a** $\cos x = \dfrac{4}{5}$           **b** $\cos x = \dfrac{2}{3}$           **c** $\cos x = \dfrac{7}{10}$

   **d** $\cos x = \dfrac{5}{6}$           **e** $\cos x = \dfrac{1}{24}$          **f** $\cos x = \dfrac{5}{13}$

**6** Work out the size of angle $x$.

   **a** $\tan x = \dfrac{3}{5}$           **b** $\tan x = \dfrac{7}{9}$           **c** $\tan x = \dfrac{2}{7}$

   **d** $\tan x = \dfrac{9}{5}$           **e** $\tan x = \dfrac{11}{7}$         **f** $\tan x = \dfrac{6}{5}$

**(EV)** **7** **a** Jack says: 'The opposite side is always shorter than the hypotenuse.'

     This means that the sine of any acute angle is always less than 1.

     Comment on his statement.

   **b** Is it also true for cosine and tangent of any acute angle?

     Give reasons for your answers.

   **c** What is the largest value of sine you can put into your calculator without getting an error when you press the inverse sine key?

**(EV)** **8** **a** **i** What angle has a sine of 0.3? (Keep the answer in your calculator memory.)

     **ii** What angle has a cosine of 0.3?

     **iii** Add the two accurate answers of parts **i** and **ii** together.

   **b** Will you always get the same answer to **aiii**, no matter what number you start with?

**9** Work out the size of the angle marked $\theta$ in each triangle. Give your answers to 1 decimal place.

**a**

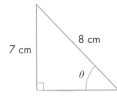

**b**

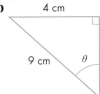

**c**

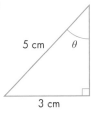

**d**

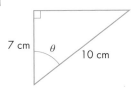

**e**

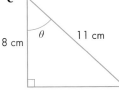

**f**

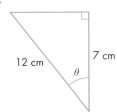

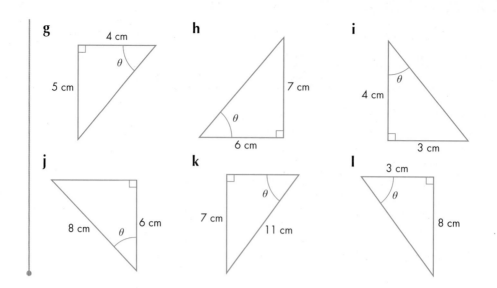

# 22.8 Trigonometry without a calculator

This section will show you how to:

- work out and remember trigonometric values for angles of 30°, 45°, 60° and 90°.

This triangle has angles of 90°, 45° and 45°. As well as being a right-angled triangle, it is also an isosceles triangle, so the adjacent side and the opposite side are the same length.

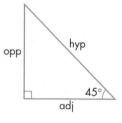

Suppose that the opposite and adjacent sides in this triangle are 1.

Then $\tan 45° = \dfrac{\text{opp}}{\text{adj}} = \dfrac{1}{1}$

$\qquad\qquad = 1$

You can then use Pythagoras' theorem to calculate the length of the hypotenuse.

$\qquad 1^2 + 1^2 = \text{hypotenuse}^2$

$\qquad\qquad = 2$

$\Rightarrow \text{hypotenuse} = \sqrt{2}$

Then $\sin\theta = \dfrac{\text{opp}}{\text{hyp}}$ and $\cos\theta = \dfrac{\text{adj}}{\text{hyp}}$

$\sin 45° = \dfrac{1}{\sqrt{2}}$ and $\cos 45° = \dfrac{1}{\sqrt{2}}$

Note that $\dfrac{1}{\sqrt{2}}$ is the exact value of $\sin 45°$ and $\cos 45°$. This means that it is correct to an infinite number of decimal places.

You can use an equilateral triangle divided into two matching right-angled triangles to work out trigonometric values for angles of 30° and 60°.

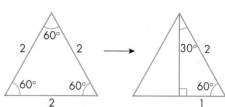

Using the 60° angle, you can then see that $\cos 60° = \dfrac{1}{2}$.

Using the 30° angle, you can then see that $\sin 30° = \dfrac{1}{2}$

It is useful to remember that $\cos 60°$ and $\sin 30°$ are both $\dfrac{1}{2}$.

# Exercise 22K

**CM** **1** You are told that $\cos 60° = \frac{1}{2}$.

Use this information to show that the exact value of:

**a** $\tan 60°$ is $\sqrt{3}$      **b** $\sin 60°$ is $\frac{\sqrt{3}}{2}$.

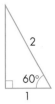

**CM** **2** You are told that $\sin 30° = \frac{1}{2}$.

Use this information to show that the exact value of:

**a** $\tan 30°$ is $\frac{1}{\sqrt{3}}$      **b** $\cos 30°$ is $\frac{\sqrt{3}}{2}$.

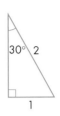

**3** Calculate the length marked $x$ in each diagram.

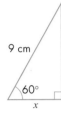 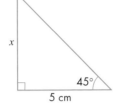 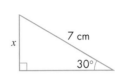

**EV** **4** In the triangle ABC shown, Iain said that AC was 7 cm.

Comment on Iain's statement.

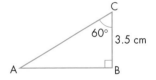

**MR** **5** Use the series of diagrams to help you to show that:

**i** $\sin 90° = 1$      **ii** $\cos 0° = 1$      **iii** $\sin 0° = 0$      **iv** $\cos 90° = 0$.

**CM** **6** **a** Draw a diagram to show that $\tan 45° = 1$.

**b** Use your diagram to show that $\sin 45° = \cos 45° = \frac{1}{\sqrt{2}}$.

# 22.9 Solving problems using trigonometry

This section will show you how to:

- solve practical problems using trigonometry
- solve problems using an angle of elevation or an angle of depression.

**Key terms**

angle of depression

angle of elevation

Sometimes you need to use trigonometry in a triangle as part of solving a practical problem. You should follow these steps.

- Draw the triangle required.
- Write on the information given (angles and sides).
- Label the unknown angle or side $x$.
- Mark on two of O, A or H as appropriate.
- Decide which ratio you need to use.
- Write out the equation with the numbers in.
- Rearrange the equation, then work out the answer.

---

**Example 12**

A window cleaner has a ladder that is 7 m long. She leans it against a wall so that the foot of the ladder is 3 m from the wall. What angle does the ladder make with the wall?

Draw the situation as a right-angled triangle.

Then mark the sides and angle.

 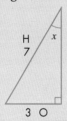

Recognise it is a sine problem because you have O and H.

So, $\sin x = \dfrac{3}{7}$

$\Rightarrow \quad x = \sin^{-1} \dfrac{3}{7}$

$\qquad = 25°$ (to the nearest degree)

---

## Exercise 22L

In these questions, give answers involving angles to the nearest degree.

**1** A ladder, 6 m long, rests against a wall. The foot of the ladder is 2.5 m from the base of the wall. What angle does the ladder make with the ground?

 **2** The ladder in question **1** has a 'safe angle' with the ground of between 70° and 80°. What are the safe limits for the distance of the foot of this ladder from the wall? What is the highest point the ladder reaches on the wall?

**3** A 10-m ladder is placed so that it reaches 7 m up the wall. What angle does it make with the ground?

**4** A ladder is placed so that it makes an angle of 76° with the ground. The foot of the ladder is 1.7 m from the foot of the wall. How high up the wall does the ladder reach?

(PS) **5** Calculate the angle that the diagonal makes with the long side of a rectangle which measures 10 cm by 6 cm.

(EV) **6** This diagram shows a frame for a bookcase.

a What angle does the diagonal strut make with the long side?

b Calculate the length of the strut.

c Why might your answers be inaccurate in this case?

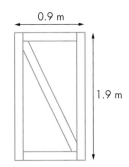

**7** This diagram shows a roof truss.

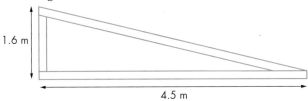

a What angle will the roof make with the horizontal?

b Calculate the length of the sloping strut.

(CM) **8** Alicia paces out 100 m from the base of a church. She then measures the angle to the top of the spire as 23°. How would Alicia work out the height of the church spire?

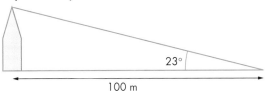

(PS) **9** A girl is flying a kite on a string 32 m long. The string, which is being held 1 m above the ground, makes an angle of 39° with the horizontal. How high is the kite above the ground?

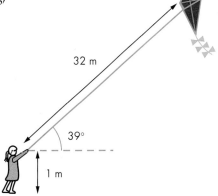

(MR) **10** An aircraft takes off at an angle of 5.1° to the ground. From the point of take off it travels 3000 m in a straight line. Show that at this point it is 267 m off the ground.

(PS) **11** A plank 9 m long is leaning against a wall. It makes an angle of 25° with the wall. How far up the wall does the plank reach?

## Angles of elevation and depression

When you look *up* at an aircraft in the sky, the angle that your line of sight turns through, from looking straight ahead (the horizontal), is called the **angle of elevation**.

When you are standing on a high point and look *down* at a boat, the angle that your line of sight turns through, from looking straight ahead (the horizontal) is called the **angle of depression**.

Example 13

From the top of a vertical cliff, 100 m high, Andrew sees a boat out at sea. The angle of depression from Andrew to the boat is 42°. How far from the base of the cliff is the boat?

The diagram of the situation is shown in figure **i**.

From this, you get the triangle shown in figure **ii**.

From figure **ii**, you see that this is a tangent problem.

So, $\tan 42° = \dfrac{100}{x}$

$\Rightarrow \qquad x = \dfrac{100}{\tan 42°}$

$\qquad\qquad = 111$ m (3 sf)

## Exercise 22M 🖩

In these questions, give any answers involving angles to the nearest degree.

**1**   Eric sees an aircraft in the sky. The aircraft is at a horizontal distance of 25 km from Eric. The angle of elevation is 22°. How high is the aircraft?

**2**   An aircraft is flying at an altitude height of 4000 m and is 10 km from the airport. If a passenger can see the airport, what is the angle of depression?

**3**   A man is standing 200 m from the base of a television transmitter. He notices that the angle of elevation to the top of the transmitter is 65°. How high is the transmitter?

**CM**   **4**   **a**   A boat has an angle of depression of 52° from the top of a 200-m high cliff. How far from the base of the cliff is the boat?

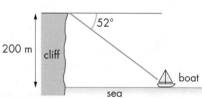

         **b**   The boat now sails away from the cliff so that the distance is doubled. Does that mean that the angle of depression is halved? Give a reason for your answer.

**5** From a boat, the angle of elevation of the foot of a lighthouse on the edge of a cliff is 34°.

   **a** If the cliff is 150 m high, how far from the base of the cliff is the boat?

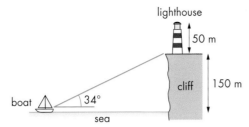

   **b** If the lighthouse is 50 m high, what is the angle of elevation of the top of the lighthouse from the boat?

**6** A bird flies from the top of a 12-m tall tree, at an angle of depression of 34°, to catch a worm on the ground.

   **a** How far does the bird actually fly?

   **b** How far was the worm from the base of the tree?

**7** Sunil wants to calculate the height of a building. He stands about 50 m away from a building. The angle of elevation from Sunil to the top of the building is about 15°.

   How tall is the building?

(CM) **8** The top of a ski run is 100 m above the finishing line. The run is 300 m long. Show that the angle of depression of the ski run is 19.5°.

(EV) **9** Nessie and Cara are standing on opposite sides of a tree.

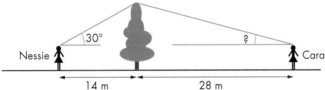

   Nessie is 14 m away and the angle of elevation of the top of the tree is 30°.

   Cara is 28 m away. She says the angle of elevation for her must be 15° because she is twice as far away.

   Is Cara correct? If not, give the actual angle of elevation.

# 22.10 Trigonometry and bearings

This section will show you how to:

- solve bearing problems using trigonometry.

A bearing is the direction to one place from another. You give a bearing as an angle measured from north in a clockwise direction. This is how a navigational compass and a surveyor's compass measure bearings.

A bearing is always written as a three-digit number, known as a three-figure bearing.

The diagram shows how this works, using the main compass points as examples.

When working with bearings, remember:

- always work clockwise from north
- always give a bearing in degrees as a three-figure bearing.

Always look for a right-angled triangle that you can use to solve trigonometry problems involving bearings.

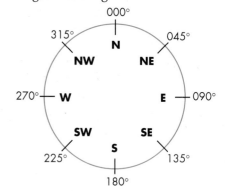

**Example 14**

A ship sails on a bearing of 120° for 50 km. How far east has it travelled?

The diagram of the situation is shown in figure **i**. From this, you can get the acute-angled triangle shown in figure **ii**.

From figure **ii**, you see that this is a cosine problem.

So, $\cos 30° = \frac{x}{50}$

$\Rightarrow x = 50 \cos 30°$

$= 43.301$

The ship has sailed 43.3 km east (to 3 significant figures).

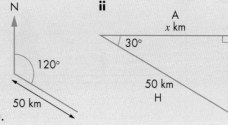

## Exercise 22N 🖩

1 A ship sails for 75 km on a bearing of 078°.

   **a** How far east has it travelled?

   **b** How far north has it travelled?

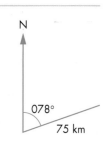

**2** Lopham is 17 miles from Wath on a bearing of 210°.

    **a** How far south of Wath is Lopham?

    **b** How far east of Lopham is Wath?

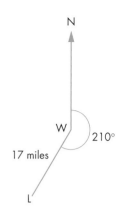

(CM) **3** A plane sets off from an airport and flies due east for 120 km. It turns and flies due south for 70 km before landing at Seddeth. Another pilot decides to fly the direct route from the airport to Seddeth.

    Show that he should fly on an approximate bearing of 120°.

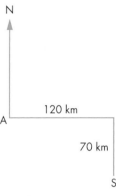

**4** A helicopter leaves an army base and flies 60 km on a bearing of 278°.

    **a** How far west has the helicopter flown?

    **b** How far north has the helicopter flown?

**5** A ship sails from a port for 35 km on a bearing of 117°. It then heads due north for 40 km and docks at Angle Bay.

    **a** How far south had the ship sailed before turning?

    **b** How far north had the ship sailed from the port to Angle Bay?

    **c** How far east of the port is Angle Bay?

    **d** What is the bearing of Angle Bay from the port?

(PS) **6** Mountain A is due west of a walker. Mountain B is due north of the walker. The guidebook says that mountain B is 4.3 km from mountain A, on a bearing of 058°. How far is the walker from mountain B?

**7** The shopping centre is 5.5 km east of my house and the supermarket is 3.8 km south. What is the bearing of the supermarket from the shopping centre?

(EV) **8** Joe sailed for 120 km on a bearing of 035°.

    He calculated he had now sailed 100 km north.

    Comment on Joe's calculation.

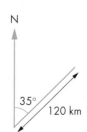

# 22.11 Trigonometry and isosceles triangles

This section will show you how to:

- use trigonometry to solve problems involving isosceles triangles.

Isosceles triangles often feature in trigonometry problems because they can be split into two matching right-angled triangles.

**a** Work out the length marked $x$ in this isosceles triangle.

**b** Calculate the area of the triangle.

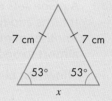

Draw a perpendicular from the apex of the triangle to its base, splitting the triangle into two matching, right-angled triangles.

**a** To calculate the length $y$, which is half of $x$, use cosine.

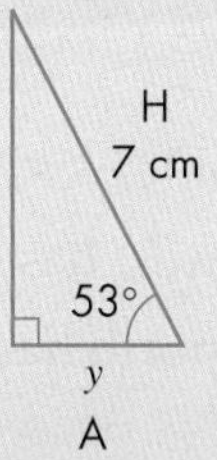

So, $\cos 53° = \dfrac{y}{7}$

$\Rightarrow y = 7 \cos 53°$

$\quad = 4.2127051$ cm

$x$ is $2y$, so $= x = 8.43$ cm (3 sf)

**b** To calculate the area of the original triangle, you first need to work out its vertical height, $h$.

You have two choices, both involving the right-angled triangle of part **a**. You can use either Pythagoras' theorem ($h^2 + y^2 = 7^2$) or trigonometry. It is safer to use trigonometry again, since you are then still using known information.

This is a sine problem.          $\sin 53° = \dfrac{h}{7}$

$\Rightarrow h = 7 \sin 53°$

$\quad = 5.5904486$ cm

(Keep the accurate figure in the calculator.)

The area of the triangle is $\frac{1}{2} \times$ base $\times$ height. You should use the most accurate figures you have for this calculation.

$A = \dfrac{1}{2} \times 8.4254103 \times 5.5904486$

$\quad = 23.6$ cm$^2$ (3 sf)

You do not need to write down these eight-figure numbers, just to use them.

# Exercise 22O

1. Work out the value of $x$ in each triangle.

a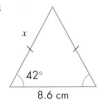

b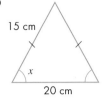

c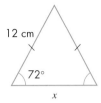

d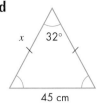

2. The diagram below shows a roof truss. How wide is the roof?

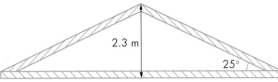

(PS) 3. Calculate the area of each triangle.

a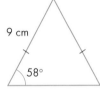

b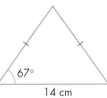

c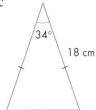

d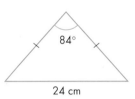

(PS) 4. An equilateral triangle has sides of length 10 cm.

A square is drawn on each side.

The corners of the squares are joined as shown.

What is the area of the hexagon this creates?

# Worked exemplars

  **a** Calculate the area of a regular hexagon of side 6 cm.

**b** Comment on the accuracy of your answer.

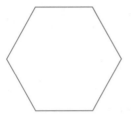

| This is an evaluating question where you are required to comment on a result. | |
|---|---|
| **a** 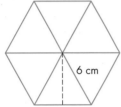 <br> Base of right-angled triangle is $\frac{1}{2}$ of 6 cm = 3 cm <br><br> Using Pythagoras' theorem: <br> height of triangle$^2$ + $3^2$ = $6^2$ <br> height of triangle$^2$ = $6^2 - 3^2$ <br> height of triangle = $\sqrt{(6^2 - 3^2)}$ <br> Area of one triangle = $\frac{1}{2} \times 6 \times$ height <br> Area of hexagon = 6 × area of triangle <br> $\qquad = 6 \times \frac{1}{2} \times 6 \times \sqrt{(6^2 - 3^2)}$ <br> $\qquad = 93.530743\ldots$ <br> $\qquad = 94$ cm$^2$ (2 sf) | You need to show how you have divided the hexagon into six equilateral triangles and then divided one of these triangles in half to find a right-angled triangle. <br><br><br> You need to show how you are accurately calculating the area of the shape without rounding too early. You could simplify to $\sqrt{27}$, but this isn't necessary. <br><br> Then give a final answer with suitable rounding. |
| **b** The accuracy was kept by not rounding until the last stage. The initial data was assumed to be accurate, and so 2 sf gives an appropriate degree of accuracy. | You should make a suitable comment reflecting the accuracy, giving a clear reason why you selected the accuracy you did. |

**2** A clock is designed to have circular face on a triangular surround.
The triangle is equilateral.

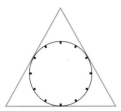

The face extends to the edge of the triangle.

The diameter of the clock face is 18 cm.

Show that the perimeter of the triangle is 94 cm.

> This is a communicating mathematics question where you have to construct a chain of reasoning to achieve a given result.
>
> You have to find the strategy of getting to the given result of 94 cm, clearly showing your method at each stage.

| | |
|---|---|
| 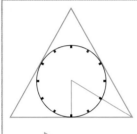<br><br>The radius is 9 cm.<br><br>The bottom angle is 30° because it is half the angle of an equilateral triangle. The other angle is 60°.<br><br>$\tan 60° = \dfrac{x}{9}$<br>$\Rightarrow x = 9 \tan 60°$ | You need to show a correct trigonometry ratio that can be used to calculate $x$.<br><br><br><br><br>If you use the 60° angle in your ratio, $x$ will be on the top. |
| Perimeter of triangle = $6 \times 9 \tan 60°$<br>$\qquad\qquad\qquad\quad = 93.530…$ cm<br>$\qquad\qquad\qquad\quad = 94$ cm (2 sf) | Show the correct value of 93·53 and how you rounded to 2 sf in order to get the given solution of 94 cm. |

# Ready to progress?

I can use Pythagoras' theorem in 2D.
I can solve problems using Pythagoras' theorem.

I can use the trigonometric ratios for sine, cosine and tangent in right-angled triangles.
I can solve problems using trigonometry.
I can solve problems using angles of elevation, angles of depression and bearings.

# Review questions

**1** Calculate the length of $x$. Give your answer to 1 decimal place.

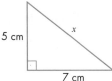

5 cm

$x$

7 cm

**2** Calculate the length of $x$. Give your answer to 1 decimal place.

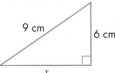

9 cm

6 cm

$x$

**(PS) 3** Calculate the diagonal length of a square of area 16 cm².

**(EV) 4** Michael always puts his 5-metre ladder against a wall so that it makes an angle of 75° with the floor. He says this means it will reach just over 4.8 m up the wall.

Show that Michael is correct.

**(PS) 5** ABC is a right-angled triangle.

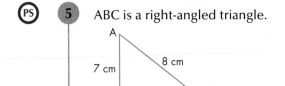

A

8 cm

7 cm

B

C

Calculate the area of the triangle.

**(PS) 6** A lighthouse, L, is 15 km due west of a port, P. A ship, S, is 8 km due north of the lighthouse, L.

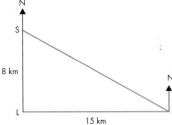

N

S

8 km

N

L

P

15 km

The ship leaves its position at 12 noon. It sails directly to the port at a speed of 8.5 km/h.

What time will it reach the port?

**7** A lighthouse, L, is 4.3 km due east of a port, P. A ship, S, is 2.8 km due north of the lighthouse, L. Work out the bearing of the port, P, from the ship, S.

**8** The diagram is made up of two right-angled triangles ABC and BCD.

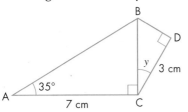

Calculate the value of $y$.

**9** A ladder that is 6 m long leans against a wall. Its foot is 80 cm away from the base of the wall.

Select the correct answer for each part.

**a** The height the ladder reaches up the wall is:

   **i** 5.8 m        **ii** 5.9 m        **iii** 6.0 m        **iv** 6.1 m.

**b** The angle between the ladder and the floor is:

   **i** 76°        **ii** 77°        **iii** 82°        **iv** 83°.

**10** An isosceles triangle has two sides of length 14 cm and one side of length 7 cm.

**a** Calculate the size of the two identical angles.

**b** What is the area of the triangle?

**11** These shapes are possible motif designs.

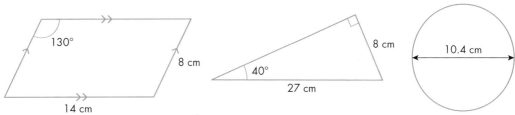

Joe said that the parallelogram motif had the greatest area.

Hannah said: 'No, the triangle has the greatest area.'

Comment on their statements.

**12** The diagram shows some of the beams supporting a roof.

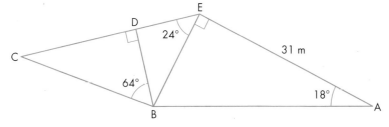

Calculate, to two decimal places, the length of:

**a** AB        **b** BE        **c** BD        **d** BC.

# 23 Geometry and measures: Congruency and similarity

## This chapter is going to show you:

- how to show that two triangles are congruent
- how to work out the scale factor for two similar shapes
- how to work out lengths of sides in similar shapes.

## You should already know:

- how to enlarge a shape by a given scale factor
- how to solve equations.

## About this chapter

Thales of Miletus (624–547 BC) was a Greek philosopher and one of the Seven Sages of Greece. Historians think he was the first person to use similar triangles to work out the heights of tall objects.

Thales discovered that, at a particular time of day, the height of an object and the length of its shadow were the same. He used this observation to calculate the height of the Egyptian pyramids. Later, he took this knowledge back to Greece. His observations are thought to be the origins of using similar triangles to solve problems like this.

Astronomers use the geometry of triangles to measure the distance to nearby stars. They take advantage of the Earth's orbit around the Sun to calculate the maximum distance between two measurements. They observe the star twice, from the same point on Earth and at the same time of day, but six months apart.

This chapter will show you what similar triangles and shapes are and how we can use the scale factor of enlargement to solve different sorts of problems.

# 23.1 Congruent triangles

## This section will show you how to:

- demonstrate that two triangles are congruent.

Two shapes are **congruent** if they are exactly the same size and shape.

For example, these triangles are all congruent.

Notice that the triangles can be positioned differently (reflected, translated or rotated).

## Conditions for congruent triangles

Two triangles are congruent if they meet any one of the four conditions below.

### Condition 1

All three sides of one triangle are equal to the corresponding sides of the other triangle.

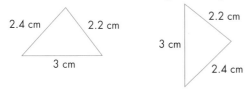

This condition is known as SSS (side, side, side).

### Condition 2

Two sides, and the angle between them, of one triangle are equal to the corresponding sides and angle of the other triangle.

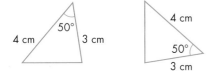

This condition is known as SAS (side, angle, side).

### Condition 3

Two angles and a side of one triangle are equal to the angles and corresponding side of the other triangle.

This condition is known as ASA (angle, side, angle) or AAS (angle, angle, side).

## Condition 4

Both triangles have a right angle, an equal hypotenuse and another equal side.

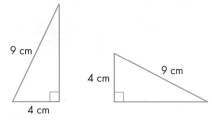

This condition is known as RHS (right angle, hypotenuse, side).

### Notation

Once you have shown that triangle ABC is congruent to triangle PQR by one of the above conditions, it means the points ABC correspond exactly to the points PQR in that order.

 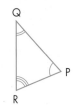

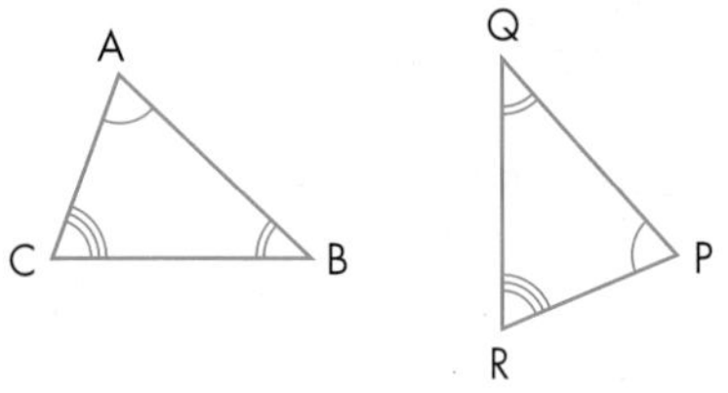

$$\angle A = \angle P \quad AB = PQ$$
$$\angle B = \angle Q \quad BC = QR$$
$$\angle C = \angle R \quad AC = PR$$

Notice how to indicate the corresponding angles of each triangle.
You can write triangle ABC is congruent to triangle PQR as $\triangle ABC \equiv \triangle PQR$.

---

**Example 1**

Show that triangle ABC is congruent to triangle EFD.

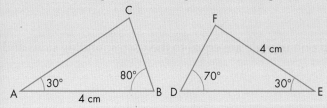

If you know two angles in a triangle, you can work out the third angle.

Angle C = 180 – (30 + 80)

      = 70°

Angle F = 180 – (70 + 30)

      = 80°

Both triangles have angles of 30°, 70° and 80°. (This is not enough to show that they are congruent.)

Side AB, 4 cm, is between the 30° and the 80° angles and side EF, 4 cm, is between the 30° and the 80° angles.

So AB and EF correspond.

So $\triangle ABC \equiv \triangle EFD$ (ASA).

---

Example 2

ABCD is a kite. Show that triangle ABC is congruent to triangle ADC.

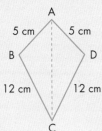

AB = AD

BC = DC

AC is common.

So △ABC ≡ △ADC (SSS).

## Exercise 23A

**1** State the condition that shows that the triangles in each pair are congruent.

**a**

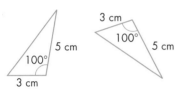

**b**

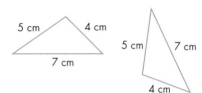

**c**

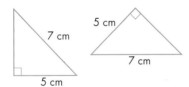

**d**

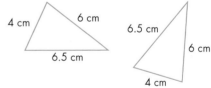

**e**

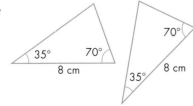

**f**

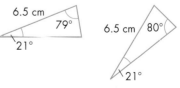

**MR** **2** The triangles in each pair are congruent. Write down:

    **i** the condition that shows that the triangles are congruent

    **ii** the corresponding points.

    **a** ABC where AB = 8 cm, BC = 9 cm, AC = 7.4 cm

       PQR where PQ = 9 cm, QR = 7.4 cm, PR = 8 cm

    **b** ABC where AB = 5 cm, BC = 6 cm, angle B = 35°

       PQR where PQ = 6 cm, QR = 50 mm, angle Q = 35°

**3** Triangle ABC is congruent to triangle PQR. $\angle A = 60°$, $\angle B = 80°$ and AB = 5 cm. Calculate these.

    **a** $\angle P$           **b** $\angle Q$           **c** $\angle R$           **d** length PQ

**EV** **4** This diagram is made up of triangles and squares.

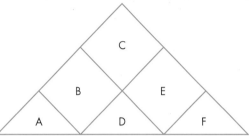

    Angie said: 'Triangles A, D and F are congruent.'

    Is Angie correct? Give reasons for your answer.

**CM** **5** Draw a rectangle EFGH. Draw in the diagonal EG. Show that triangle EFG is congruent to triangle EHG.

**CM** **6** Draw an isosceles triangle ABC where AB = AC. Draw the line from A to X, the midpoint of BC. Show that triangle ABX is congruent to triangle ACX.

**EV** **7** Write down whether each statement is always true, never true or sometimes true.

    **a** When a triangle is reflected in a line, the object and image are congruent.

    **b** When a triangle is rotated about a point, the object and image are congruent.

    **c** When a triangle is translated with a column vector, the object and image are congruent.

    **d** When a triangle is enlarged from a point, the object and image are congruent.

**MR** **8** ABCD and DEFG are squares.

    **a** Show that $\angle CDG$ and $\angle ADE$ have the same value, $x$.

    **b** Show that that:        **i** ED = DG           **ii** CD = AD.

    **c** Show that triangle ADE is congruent to triangle CDG.

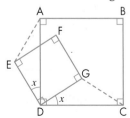

(EV) (9) Jez says that these two triangles are congruent because two angles and a side are the same.

Show that he is wrong.

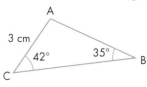
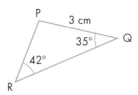

# 23.2 Similarity

This section will show you how to:

- recognise similarity in any two shapes
- show that two shapes are similar
- work out the scale factor between similar shapes.

If two people are similar, you expect them to be alike in lots of ways, but not exactly the same.

The mathematical meaning of **similar** is more precise. Work through the next exercise to discover more about the meaning of this term.

## Exercise 23B

(MR) (1)  a  Construct the triangles shown below.

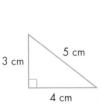
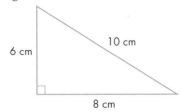

   b  Measure and label the angles in each triangle.

   c  What do you notice about the angles of each triangle?

   d  What do you notice about the corresponding sides of the larger triangle and the smaller triangle?

   e  What is the scale factor of enlargement from the smaller triangle to the larger triangle?

(MR) (2)  a  Construct the triangles shown below.

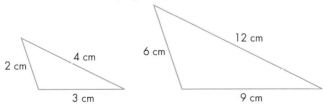

   b  Measure and label the angles in each triangle.

   c  What do you notice about the angles of each triangle?

   d  What do you notice about the corresponding sides of the larger triangle and the smaller triangle?

   e  What is the scale factor of enlargement from the smaller triangle to the larger triangle?

**3**  **a** Construct the triangles shown below.

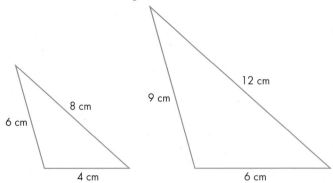

**b** Measure and label the angles in each triangle.

**c** What do you notice about the angles of each triangle?

**d** The base of the larger triangle can be found by multiplying the base of the smaller triangle by 1.5. Does this multiplier work for the other two sides?

**e** What is the scale factor of enlargement from the smaller triangle to the larger triangle?

**4**  **a** Construct a triangle with sides 4 cm, 5 cm and 6 cm.

**b** Construct another triangle with sides double the length of those in part **a**.

**c** Measure and label the angles in each triangle.

**d** What do you notice about the angles of each triangle?

**e** What is the scale factor of enlargement from the smaller triangle to the larger triangle?

**5**  **a** Construct a triangle with sides 3 cm, 2 cm and 2.5 cm.

**b** Construct another triangle with sides three times the length of those in part **a**.

**c** Measure and label the angles in each triangle.

**d** What do you notice about the angles of each triangle?

**e** What is the scale factor of enlargement from the smaller triangle to the larger triangle?

In Exercise 23B, you should have found that when one triangle is an enlargement of another, the angles in both triangles are the same. The two triangles are mathematically similar, with a scale factor of enlargement between them.

This is true for any shape, not just triangles. If one shape is an enlargement of another, then the two shapes are similar.

Example 3

These triangles are similar.

**a** What is the scale factor of the enlargement?

**b** What is the length of PQ?

**c** What is the length of AC?

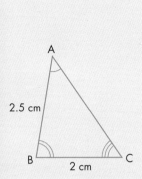
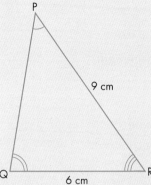

**a** To calculate the scale factor of an enlargement, look at two known sides that correspond. In the triangles above, the base lengths BC and QR correspond.

So the scale factor is $6 \div 2 = 3$.

**b** Use the corresponding side from triangle ABC and multiply it by the scale factor.

$PQ = AB \times 3$

$\quad = 2.5 \times 3$

$\quad = 7.5$ cm

**c** Use the corresponding side from triangle PQR and divide it by the scale factor.

$AC = PR \div 3$

$\quad = 9 \div 3$

$\quad = 3$ cm

Example 4

Triangles ABC and PQR are similar.

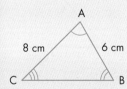
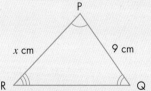

**a** What is the scale factor of the enlargement?

**b** Calculate the length of side PR.

**c** QR = 6 cm. What is the length of BC?

**a** Use two known corresponding sides.

So the scale factor is $9 \div 6 = 1.5$.

**b** $PR = AC \times 1.5$

$\quad = 8 \times 1.5$

$\quad = 12$ cm

**c** $BC = QR \div 1.5$

$\quad = 6 \div 1.5$

$\quad = 4$ cm

Example 5

Rectangles ABCD and PQRS are similar.

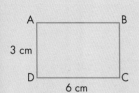
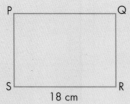

A      B

3 cm

D      C

6 cm

P      Q

S      R

18 cm

**a** What is the scale factor of the enlargement?

**b** Calculate the length of side PS.

**c** The length of PR is 20.1 cm.

How long would you expect AC to be?

**a** Use two known corresponding sides.

So the scale factor is 18 ÷ 6 = 3.

**b** PS = AD × 3

= 3 × 3

= 9 cm

**c** AC = PR ÷ 3

= 20.1 ÷ 3

= 6.7 cm

## Exercise 23C

**1** Are these pairs of triangles similar? If so, give the scale factor. If not, give a reason.

**a**

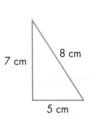

7 cm   8 cm

5 cm

21 cm   24 cm

15 cm

**b**

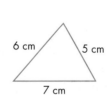

6 cm   5 cm

7 cm

12 cm   10 cm

14 cm

**c**

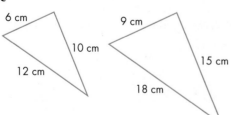

6 cm

10 cm

12 cm

9 cm

15 cm

18 cm

**d**

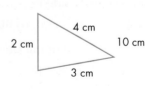

2 cm   4 cm

3 cm

10 cm   20 cm

16 cm

**2** These triangles are similar.

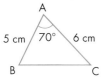 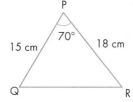

**a** What is the scale factor of the enlargement from ABC to PQR?

**b** Which angle corresponds to angle C?

**c** Which side corresponds to side QP?

 **3** **a** Show that triangle ABC is similar to triangle PQR.

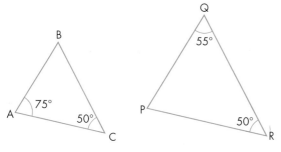

**b** Which side corresponds to side AC?

**c** Which side corresponds to side QR?

**d** The scale factor of the enlargement from ABC to PQR is 4.

 **i** If AB is 7 cm, what is the length of PQ?

 **ii** If PR is 24 cm, what is the length of AC?

 **4** Triangle ABC is similar to triangle PQR.

**a** What is the size of angle B?

**b** What is the scale factor of the enlargement from ABC to PQR?

**c** The length of side AB is 7.8 cm.
How can you use this to work out the
length of side PQ?

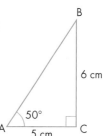 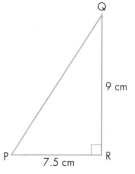

**5** In diagrams **a** and **b**, the pairs of shapes are similar but not drawn to scale.

 **i** Work out *x*.   **ii** Work out PQ.

**a**

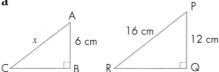

**b**

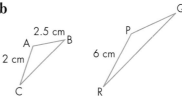

**6** In diagrams **a** to **d**, each pair of shapes is similar. For each pair:

  **i** write down the pairs of corresponding points

  **ii** work out the scale factor of the enlargement

  **iii** calculate the marked lengths $x$ and $y$.

**a**

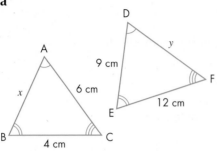

**b**

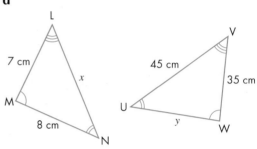

**c**

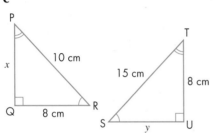

**d**

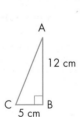
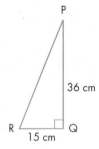

**7** **a** Show that these two triangles are similar.

  **b** What is the scale factor of the enlargement from ABC to PQR?

  **c** Use Pythagoras' theorem to calculate the length of side AC.

  **d** Use your answer to part **c** to calculate the length of the side PR.

**8** The diagram shows two regular hexagons.

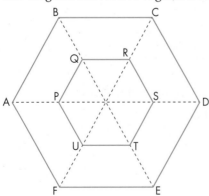

  **a** Show that hexagon PQRSTU is similar to hexagon ABCDEF.

    AB is 6 cm, PQ is 4 cm and PS is 8 cm.

  **b** What is the length of AD?

  **c** Hussain calculated BF as 10.392 cm. If Hussain is correct, what is the length of QU?

## Solving problems with similar triangles

Work through the next two examples. They show you how you can use similar triangles to solve different problems.

Example 6

Sean is standing near a tree.

He is 1.6 m tall and he casts a shadow that measures 0.5 m.

The tree casts a shadow that measures 4 m.

Work out the height of the tree.

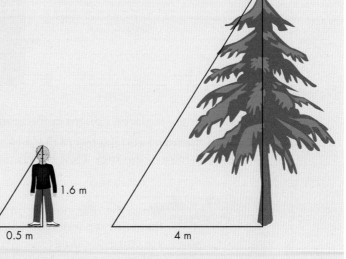

In problems like this, you need to identify similar triangles. The sun will cast a shadow at the same angle to both the tree and Sean, so the two triangles are similar.

The scale factor between the two triangles is $4 \div 0.5 = 8$.

The height of the tree is therefore $1.6 \times 8 = 12.8$ m.

Example 7

Ahmed is lying in a park at A. As he looks at a tall building, the top of a fence is exactly in line with the top of the building, as shown in the diagram.

He walks 10 paces to the fence and measures the height of the fence as 2 m.

From the other side of the fence, he walks 100 paces to the building.

How can you use this information to work out the height of the building?

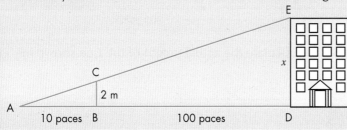

Identify a pair of similar triangles.

Triangles ABC and ADE are similar triangles as all their angles correspond.

Draw the triangles separately so the corresponding sides and lengths are clear.

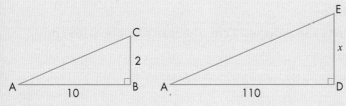

Work out the scale factor.                           $110 \div 10 = 11$

Use this to work out the height of the building, $x$.      $2 \times 11 = 22$ m

**1** This diagram shows a method of working out the height of a tower.

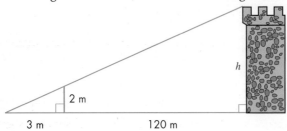

A stick, 2 m high, is placed upright in the ground 120 m from the base of a tower.

The top of the tower and the top of the stick are in line with a point on the ground 3 m from the base of the stick. How high is the tower?

> **Hints and tips** When the triangles are on the same diagram, always redraw them out separately so you can clearly see which sides are corresponding.

**2** A factory chimney is 330 feet high.

Patrick paces out the distances shown in the diagram: from the chimney, past the flagpole, to the point where the top of the chimney and the top of the flag pole are in line with each other.

Show that the flag pole is 220 feet tall.

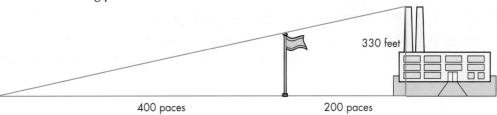

**3** The height of a golf flag is 1.5 m. Calculate the actual height of the tree shown in the diagram.

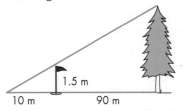

**4** A pole casts a shadow of 1.5 m.

At the same time, a man casts a shadow of 75 cm. The man's height is 165 cm.

Work out the height of the pole.

**5** Bob, a builder, is making this wooden frame for a roof.

In the diagram, triangle ABC is similar to triangle AXY.

AB = 1.5 m, BX = 3.5 m and XY = 6 m.

**a** Draw out the two similar triangles that Bob is using.

**b** What length of wood does Bob need to make BC?

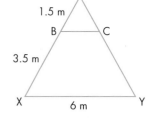

**6** For each part:

**i** state a pair of similar triangles

**ii** calculate the length marked $x$.

You can separate the similar triangles if it helps.

**a**

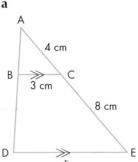

**b**

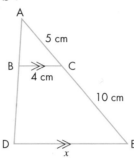

**7** Triangle ABC is similar to triangle DAC.

AC = 9 cm and CD = 6 cm.

**a** Redraw the triangles separately. Make sure that the corresponding angles are in the same order. Write down the corresponding sides.

**b** Show that BC is 13.5 cm.

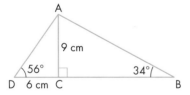

**8** Triangle ABC is similar to triangle AXY.

Which of the following is the correct length of BX?

How did you decide?

**a** 2 cm **b** 3 cm **c** 4 cm **d** 5 cm

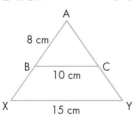

# Worked exemplars

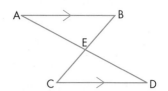

**1** AB and CD are parallel.

E is the midpoint of AD.

Show that triangle ABE is congruent to triangle CDE.

| This is a communicating mathematics question so you must show clear reasons at each stage. ||
|---|---|
| 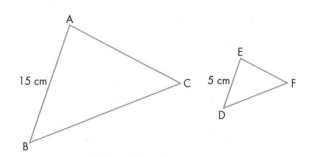 | Identify the elements that are identical in both triangles. It can help to draw out the two triangles separately and match up the sides and angles this way. |
| AE = DE (E is midpoint of AD) <br><br> ∠BAE = ∠CDE (alternate angles) <br><br> ∠AEB = ∠CED (opposite angles) | Clearly state the reason why each pair of sides or angles is identical. |
| So △ABE ≡ △CDE (ASA). | Finish with the clear statement that you have used ASA to show congruency. |

**2** Triangles ABC and EDF are similar.

AB = 15 cm and DE = 5 cm

When you add the lengths of BC and DF, the total is 24 cm.

Maggie said that BC is 12 cm longer than DF.

Is Maggie correct?

| This is an evaluation question so you need to evaluate Maggie's statement to see if it is correct. Show each step of your working clearly. ||
|---|---|
| AB ÷ DE = 15 ÷ 5 <br><br> $= 3$ <br><br> So the scale factor is 3. | First show how you calculate the scale factor between the two similar shapes. |
| BC = 3 × DF <br><br> Let DF = $x$ <br><br> then BC = $3x$ <br><br> BC + DF = $x + 3x$ <br><br> $= 4x$ <br><br> $\Rightarrow 4x = 24$ <br><br> $x = 6$ <br><br> So DF = 6 cm and BC = 3 × 6 = 18 cm | Use the scale factor and the given information to create an equation and solve it to work out the lengths DF and BC. |
| 18 − 6 = 12 <br><br> $\Rightarrow$ BC is 12 cm longer than DF <br><br> So Maggie is correct. | Finish by showing that the difference is 12 cm and so Maggie is correct. |

# Ready to progress?

I can work out the scale factor between two similar shapes.

I can show that two triangles are congruent.
I can solve problems using similar shapes.

# Review questions

**1**  **a**  Which triangles are congruent to triangle A?

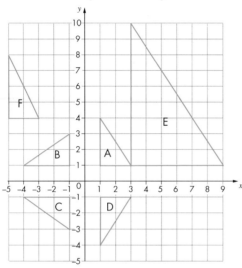

  **b**  Describe the transformation that takes A to:

   **i** B      **ii** C      **iii** D      **iv** E      **v** F.

  **c**  Lexie said: 'All the transformations in part **b** produce triangles that are congruent to triangle A.'

   Comment on Lexie's statement.

**2**  Ellie said that both of the pairs of shapes shown below are similar.

  **a**

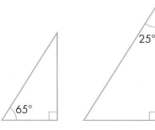

  **b**

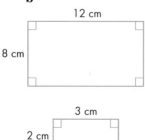

  Comment on Ellie's statement.

**(EV)** **3** Write down whether each statement is always true, never true or sometimes true.

    **a** Two circles are similar.

    **b** Two regular pentagons are similar.

    **c** Two parallelograms are similar.

    **d** Two isosceles triangles are similar.

    **e** Two equilateral triangles are similar to each other.

**4** Which of the following pairs of triangles are congruent? For the pairs that are congruent, state the condition that shows this.

    **a** 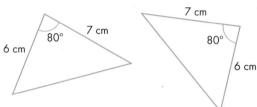     **b**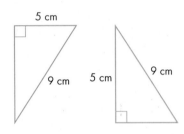

    **c** 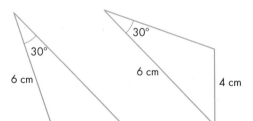     **d**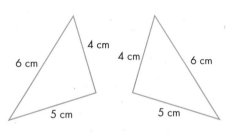

**(CM)** **5** A tree casts a shadow of 10 m. At the same time a girl casts a shadow of 40 cm.
Show that, if the girl's height is 140 cm, the height of the tree is 35 m.

**(PS)** **6** Construct a triangle that is congruent to the one shown below.

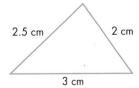

**7** Draw a rectangle (not a square) and label the vertices ABCD. Draw in the diagonals AC and BD. Label the point where the diagonals intersect X. Which triangles are congruent?

(EV) **8** Andrew calculated BE as 5.25 cm.

Eve said that ED is 8 cm.

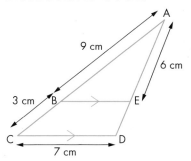

Are they both correct? Give reasons for your answers.

> **Hints and tips** Redraw the triangles separately so you can clearly see which sides are corresponding.

(PS) **9** Triangles ABC and PQR are similar.

Triangle PQR has an area of 32 cm².

What is the area of triangle ABC?

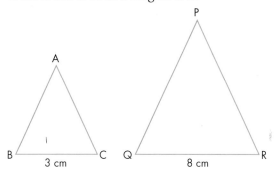

# 24 Probability: Combined events

## This chapter is going to show you:

- how to work out the probabilities for two or more events
- how to use two-way tables to solve probability problems
- how to use Venn diagrams to solve probability problems
- how to draw and use frequency tree diagrams
- how to draw and use probability tree diagrams to solve probability problems.

## You should already know:

- how to use theoretical or experimental models to work out the probabilities of outcomes of events.

## About this chapter

Chance is a part of our everyday lives. Judgements are frequently made based on probability. A good example is the weather forecast. You are likely to hear, for example: 'There is a 40 per cent chance of rain today.'

How do they know that?

- Records of data that predict possibility of rainfall go back as far as 1854, when meteorologists regarded the presence of nimbus clouds as an indication that there was a good chance of rain.
- Barometers were used to predict the chance of rainfall. A sign of falling pressure on the barometer was taken as an indication of a good chance of rain.
- Finally, the direction of wind was used to determine the chances of rainfall. If the wind blew from a rainy part of the country, the chance of rain would be high.

The occurrence of all these three indicators would almost certainly mean that rain would come.

# 24.1 Combined events

This section will show you how to:

- work out the probabilities when two or more events occur at the same time.

**Key terms**

probability space diagram

sample space diagram

There are many situations in which two events occur at the same time. Here are three examples.

## Throwing two dice

Suppose you throw two dice, one red and one blue. Each dice can give a score from 1 to 6. Both diagrams show that there are 6 × 6 = 36 equally likely outcomes. In the diagram on the left-hand side, the first number in each pair is the score on the blue dice and the second number is the score on the red dice.

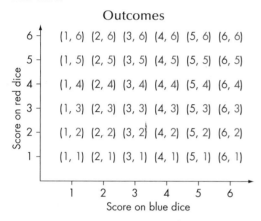

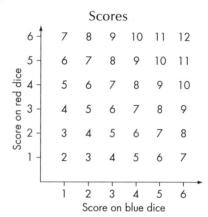

The outcome (2, 3) gives a total score of 5. The total scores for all the outcomes are shown in the diagram on the right-hand side. Diagrams that show all the outcomes of combined events are called **sample space diagrams** or sometimes **probability space diagrams**.

---

**Example 1**

Rachel throws two dice. Work out:

**a** P(score of 3)      **b** P(double).

Refer to the diagrams above.

**a** From the right-hand diagram, you can see that there are two ways to get a score of 3.

So P(score of 3) is $\frac{2}{36} = \frac{1}{18}$.

**b** From the left-hand diagram, you can see that there are six ways to get a 'double'.

So P(double) is $\frac{6}{36} = \frac{1}{6}$.

---

## Throwing two coins

There are four equally likely outcomes: (H, H), (H, T), (T, H), (T, T).

**Example 2**

Tanya throws two coins. Work out:

**a** P(two heads)      **b** P(a head and a tail).

Look back at the diagram showing all the possible outcomes.

**a** P(two heads) = $\frac{1}{4}$

**b** There are 2 ways out of 4 for P(a head and a tail), so the probability is $\frac{2}{4} = \frac{1}{2}$.

## Throwing a dice and a coin

The sample space diagram for these events looks like this.

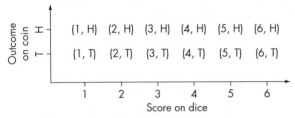

There are 12 equally likely outcomes.

**Example 3**

Omar throws a dice and a coin. Work out:

**a** P(a head and an even number)      **b** P(a tail and a number greater than 2).

**a** There are 3 ways out of 12 for P(a head and an even number), so the probability is $\frac{3}{12} = \frac{1}{4}$.

**b** There are 4 ways out of 12 for P(a tail and a number greater than 2) so the probability is $\frac{4}{12} = \frac{1}{3}$.

# Exercise 24A

**1**   Sasha throws two fair dice, each numbered from 1 to 6.

   **a** What is the most likely score?

   **b** Which two scores are least likely?

> **Hints and tips**   Use the right-hand sample space diagram for throwing two dice, at the start of this section.

   **c** Copy and complete the table to show the probabilities of all scores from 2 to 12.

| Score | 2 | 3 | 4 | 5 | 6 | 7 | 8 | 9 | 10 | 11 | 12 |
|---|---|---|---|---|---|---|---|---|---|---|---|
| Probability | | | | | | | | | | | |

   **d** What is the probability of a score that is:

     **i** bigger than 10      **ii** from 4 to 6 inclusive      **iii** even

     **iv** a square number      **v** a prime number      **vi** a triangular number?

**2** When two fair dice are thrown together, what is the probability that:

   **a** the score is an even 'double'

   **b** at least one of the dice shows a 2

   **c** the score on one dice is twice the score on the other dice

   **d** at least one of the dice shows a multiple of 3?

> **Hints and tips** Use the left-hand sample space diagram at the start of this section.

**3** When two fair dice are thrown together, what is the probability that:

   **a** both dice show a 6

   **b** at least one of the dice will show a 6

   **c** exactly one dice shows a 6?

**4** When two fair coins are thrown together, what is the probability of scoring:

   **a** two heads     **b** a head and a tail     **c** at least one tail     **d** no tails?

**5** A dice and a coin are thrown together. What is the probability of scoring:

   **a** a head on the coin and a 6 on the dice

   **b** a tail on the coin and an even number on the dice

   **c** a head on the coin and a square number on the dice?

**6** This sample space diagram shows some possible outcomes of the event 'the difference between the scores when two fair dice are thrown'.

   **a** Copy and complete the diagram.

   **b** What is the probability that the difference is:

     **i** 1     **ii** 0     **iii** 4     **iv** 6     **v** an odd number?

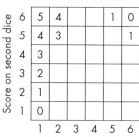

**Difference in scores**

| | | | | | | |
|---|---|---|---|---|---|---|
| 6 | 5 | 4 | | | 1 | 0 |
| 5 | 4 | 3 | | | | 1 |
| 4 | 3 | | | | | |
| 3 | 2 | | | | | |
| 2 | 1 | | | | | |
| 1 | 0 | | | | | |
| | 1 | 2 | 3 | 4 | 5 | 6 |

Score on second dice / Score on first dice

**7** Luka spins two fair five-sided spinners together. He records the total scores of the sides that they land on in a sample space diagram, like this.

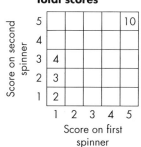

**Total scores**

| | | | | | |
|---|---|---|---|---|---|
| 5 | | | | | 10 |
| 4 | | | | | |
| 3 | 4 | | | | |
| 2 | 3 | | | | |
| 1 | 2 | | | | |
| | 1 | 2 | 3 | 4 | 5 |

Score on second spinner / Score on first spinner

   **a** Copy and complete the sample space diagram.

   **b** What is the most likely score?

   **c** When Luka spins two fair five-sided spinners together, what is the probability that his total score is:

     **i** 5     **ii** an even number     **iii** a 'double'     **iv** less than 7?

**PS** **8** Tanya has two fair eight-sided dice showing the numbers 1 to 8. She throws them at the same time.

What is the probability that the product of the scores is an even square number?

**MR** **9** Isaac throws two dice and multiplies the scores together to give their product. He wants to know the probability of scoring two numbers that will give a product between 19 and 35 inclusive.

Show how a probability space diagram will help him.

**10** **a** List all the possible outcomes when three fair coins are thrown together.

**b** Write down the probability of each outcome.

   **i** P(throwing three heads)         **ii** P(throwing two heads and one tail)

   **iii** P(throwing no heads)          **iv** P(throwing at least one tail).

**MR** **11** When one coin is thrown, there are two possible outcomes. When two coins are thrown, there are four possible outcomes. When three coins are thrown, there are eight possible outcomes.

**a** How many possible outcomes will there be when four coins are thrown?

**b** How many possible outcomes will there be when five coins are thrown?

**c** How many possible outcomes will there be when 10 coins are thrown?

**d** How many possible outcomes will there be when $n$ coins are thrown?

**PS** **12** When Mel walked into her local shopping mall, she saw a competition taking place. Mel decided to have a go.

Roll 2 dice!

Score a total of 11 and win a prize!

**a** Draw the sample space diagram for this event.

**b** What is the probability of winning a prize?

**c** How many goes should she have in order to expect to win at least once?

**d** If she had 40 goes, how many times could she expect to have won?

# 24.2 Two-way tables

This section will show you how to:

- read two-way tables and use them to work out probabilities.

A **two-way table** is a table that links two variables. This two-way table shows the colours and makes of cars in the school car park.

|          | Red | Blue | White |
|----------|-----|------|-------|
| Ford     | 2   | 4    | 1     |
| Vauxhall | 0   | 1    | 2     |
| Toyota   | 3   | 3    | 4     |
| Peugeot  | 2   | 0    | 3     |

One variable (the make of car) is written in the rows of the table and the other variable (the colour) is written in the columns of the table.

This two-way table shows the numbers of boys and girls in a class and whether they are left-handed or right-handed.

|              | Boys | Girls |
|--------------|------|-------|
| Left-handed  | 2    | 4     |
| Right-handed | 10   | 13    |

**a** What is the probability that a student selected at random will be a left-handed boy?

**b** A student selected at random is a girl. What is the probability that she is right-handed?

**a** The total number of students is 29.

So, P(left-handed boy) = $\frac{2}{29}$.

**b** The total number of girls is 17.

So, P(right-handed girl) = $\frac{13}{17}$.

## Exercise 24B

**1** Look at this set of cards.

**a** Copy and complete the two-way table.

|       |           | Shaded | Unshaded |
|-------|-----------|--------|----------|
| Shape | Circles   |        |          |
|       | Triangles |        |          |

**b** One card is taken at random.

What is the probability that it shows:

**i** an unshaded triangle    **ii** either a shaded triangle or an unshaded circle?

**2** This two-way table shows the numbers of doors and windows in each room of a school.

| | | Number of doors | | |
|---|---|---|---|---|
| | | **1** | **2** | **3** |
| **Number of windows** | **1** | 5 | 4 | 2 |
| | **2** | 4 | 7 | 4 |
| | **3** | 1 | 4 | 3 |
| | **4** | 1 | 3 | 2 |

a How many rooms are there in the school altogether?

b What percentage of the rooms have two doors?

c The headteacher walks into one of the rooms at random. What is the probability that he walks into a room with three windows and three doors?

**3** In a sample of 80 people, it was found that they were all born in England, Scotland or Wales.

This two-way table shows some information about these people.

| | England | Scotland | Wales | Total |
|---|---|---|---|---|
| **Female** | | | 4 | 42 |
| **Male** | | 8 | | |
| **Total** | 50 | 20 | | 80 |

a Copy and complete the table.

b A person is chosen at random. What is the probability that they were born in Scotland?

c A male is chosen at random. What is the probability that he was born in Wales?

**4** 100 students went on a school trip.

They could choose to go to either Blackpool or Chester Zoo.

32 boys and 28 girls went to Blackpool.

14 boys went to Chester Zoo.

a Copy and complete the two-way table to show this information.

| | Blackpool | Chester Zoo | Total |
|---|---|---|---|
| **Boys** | | | |
| **Girls** | | | |
| **Total** | | | |

b A student is chosen at random. What is the probability that they went to Blackpool?

c A girl is chosen at random. What is the probability that she went to Chester Zoo?

**5** This two-way table shows the ages and genders of a sample of 50 students in a school.

| | | Age (years) | | | | | |
|---|---|---|---|---|---|---|---|
| | | 11 | 12 | 13 | 14 | 15 | 16 |
| Gender | Boy | 4 | 3 | 5 | 3 | 5 | 4 |
| | Girl | 2 | 5 | 4 | 5 | 4 | 6 |

  **a** How many students in the sample are aged 13 or less?

  **b** What percentage of the students in the sample are aged 16?

  **c** A student in the sample is selected at random. What is the probability that the student is a boy aged 14?

  **d** There are 1000 students in the school. Use the table to estimate the number of girls in the school.

**6** This two-way table shows the numbers of adults and the numbers of cars in a random sample of 40 households in a village.

| | | Number of adults | | | |
|---|---|---|---|---|---|
| | | 1 | 2 | 3 | 4 |
| Number of cars | 0 | 2 | 1 | 0 | 0 |
| | 1 | 3 | 9 | 3 | 1 |
| | 2 | 0 | 7 | 5 | 3 |
| | 3 | 0 | 0 | 4 | 2 |

  **a** What percentage of the households have three cars?.

  **b** One of the households in the sample is chosen at random to complete a questionnaire.

  **c** There are 200 households in total in the village. Use the table to estimate the number of households that have two cars.

**(PS)** **7** Nicki went to a garden centre to buy some roses.

She found they came in six different colours – white, red, orange, yellow, pink and copper.

She also found they came in five different sizes – dwarf, small, medium, large climbing and rambling.

  **a** Draw a two-way table to show all the options.

  **b** She buys a rose, at random, for her aunt. Her aunt only likes red and pink roses and she does not like climbing or rambling roses.

  What is the probability that Nicki has bought a rose:

  **i** that her aunt likes

  **ii** that her aunt does not like?

**MR** **8** The table shows the wages for the men and women in a factory.

| Wage, per week (£$w$) | Men | Women |
|---|---|---|
| £100 < $w$ ≤ £150 | 3 | 4 |
| £150 < $w$ ≤ £200 | 7 | 5 |
| £200 < $w$ ≤ £250 | 23 | 12 |
| £250 < $w$ ≤ £300 | 48 | 27 |
| £300 < $w$ ≤ £350 | 32 | 11 |
| More than £350 | 7 | 1 |

**a** Work out the probability that a person chosen at random will earn more than £350 per week.

**b** What percentage of the men earn between £250 and £300 per week?

**c** What percentage of the women earn between £250 and £300 per week?

**d** Is it possible to work out the mean wage of the men and women? Give a reason to justify your answer.

# 24.3 Probability and Venn diagrams

This section will show you how to:

- use Venn diagrams to solve probability questions.

A **set** is a collection of objects or **elements**. Capital letters are often used to represent a set. For example, the set of odd numbers less than 10 could be represented by $A$.

$A$ = {1, 3, 5, 7, 9}

Notice that in set notation the elements of a set are written inside curly brackets.

You already know that the probability of outcome $A$ occurring is written P($A$).

Suppose outcome $A$ does not happen. You write this as $A'$. This is the **complement** of $A$ and you read it as '$A$ dash'.

You write the probability of $A$ not happening as P($A'$).

The **universal set** is a set that contains all elements used and is represented as ξ. This is the Greek letter *xi* (pronounced *ksi*).

| Key terms |
|---|
| complement |
| element |
| intersection |
| set |
| union |
| universal set |
| Venn diagram |

**Example 5**

Given that P($A$) = 0.6, write down P($A'$).

P($A'$) = 1 – P($A$)

       = 1 – 0.6

       = 0.4

Diagrams that represent connections between different sets are called **Venn diagrams**. They are named after John Venn who introduced them in about 1880. In Venn diagrams the universal set, ξ, is usually represented by a rectangle.

All the elements of $A$ are in the shaded area.

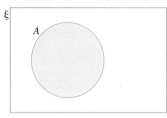

$A$ is shaded.

All the elements that are in the complement of $A$, so they are not in $A$, are in the shaded area.

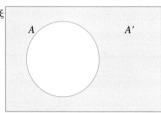

$A'$ is shaded.

This Venn diagram shows two sets $A$ and $B$.

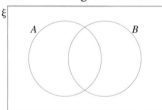

The region where the sets overlap represents the elements that are in both sets. It is called the **intersection** and is written as $A \cap B$.

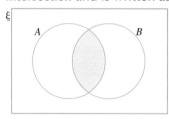

$A \cap B$ is shaded.

The combined set that contains all of $A$ and all of $B$ is called the **union** and is written $A \cup B$.

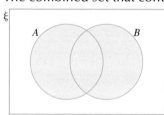

$A \cup B$ is shaded.

This Venn diagram shows the numbers of students who study French (*F*) and Spanish (*S*).

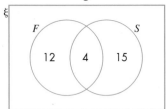

12 students study French only.                16 students study French.

15 students study Spanish only.               19 students study Spanish.

4 students study both French and Spanish.     All students study at least one language.

**Example 6**

The Venn diagram shows the number of gym club members with fair hair (*A*) and the number with blue eyes (*B*).

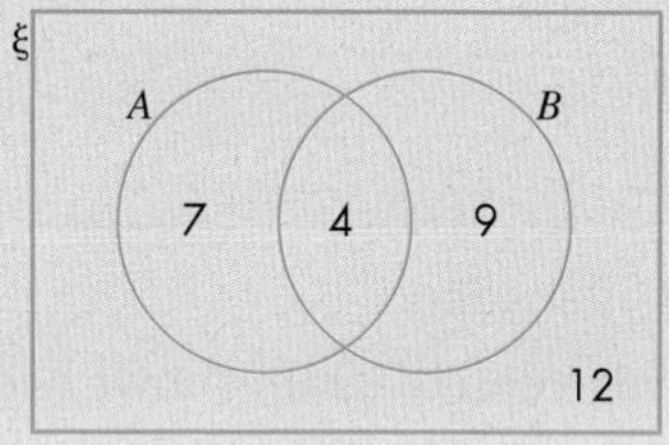

**a** How many gym club members are there altogether?

**b** What is the probability that a gym club member chosen at random has blue eyes?

**c** Work out P(*A'*).

**d** Work out P(*A* ∩ *B*).

**e** Work out P(*A* ∪ *B*).

**f** Work out the probability that a gym club member chosen at random has fair hair but does not have blue eyes.

**a** Add up the numbers in all the parts of the diagram.

7 + 4 + 9 + 12 = 32

There are 32 gym club members altogether.

**b** There are 4 + 9 = 13 people with blue eyes, so P(*B*) = $\frac{13}{32}$ .

**c** *A'* means 'not in *A*' or 'does not have fair hair'.

9 + 12 = 21 do not have fair hair, so P(*A'*) = $\frac{21}{32}$ .

**d** *A* ∩ *B* means 'has fair hair *and* has blue eyes'. There are four of these, so P(*A* ∩ *B*) = $\frac{4}{32}$ = $\frac{1}{8}$.

**e** *A* ∪ *B* means 'has fair hair or blue eyes or both'. There are 20 of these, so P(*A* ∪ *B*) = $\frac{20}{32}$ which cancels to $\frac{5}{8}$ .

**f** P(fair hair but *not* blue eyes) = $\frac{7}{32}$

## Exercise 24C

**1** P(*A*) = 0.1 and P(*B*) = 0.3. Write down:

**a** P(*A'*)                    **b** P(*B'*).

**2** P(*A*) = 0.25 and P(*B*) = 0.55. Write down:

**a** P(*A'*)                    **b** P(*B'*).

**3**  $\xi = \{1, 2, 3, 4, 5, 6, 7, 8, 9, 10\}$ $\qquad A = \{1, 2, 4, 8\}$ $\quad B = \{1, 3, 4, 9, 10\}$

**a** Show this information in a Venn diagram.

**b** Use your Venn diagram to work out:

   **i** P(A)                        **ii** (P(A′)                **iii** P(B)

   **iv** P(B′)                    **v** P(A ∪ B)             **vi** P(A ∩ B).

**4**  In a survey, Polly asked 100 people if they liked cats (C) and dogs (D).

The results are shown in the Venn diagram.

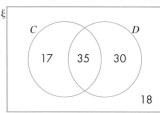

A person is chosen at random.

**a** Work out:

   **i** P(C)            **ii** P(C′)           **iii** P(D)

   **iv** P(D′)         **v** P(C ∪ D)     **vi** P(C ∩ D).

**b** Work out the probability that a person likes dogs but does not like cats.

**5**  The Venn diagram shows some probabilities.

**a** Copy and complete the Venn diagram.

**b** Work out:    **i** P(B)    **ii** P(A ∪ B)    **iii** P(A ∩ B).

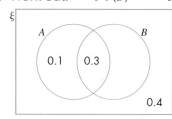

**6**  The Venn diagram shows some probabilities.

**a** Copy and complete the Venn diagram.

**b** Work out:    **i** P(A)  **ii** P(B)  **iii** P(A ∪ B)  **iv** P(A ∩ B).

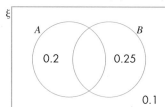

(MR) **7** The Venn diagram shows the number of students who walk to school ($S$) and the number of students who walk home from school ($H$).

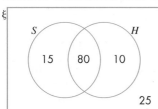

a How many students are there altogether?

b i Work out P($S \cap H$).

   ii Describe in words what P($S \cap H$) represents.

c What is the probability that a student chosen at random only walks one way, either to or from school?

(PS) **8** $\xi = \{1, 2, 3, 4, 5, 6, 7, 8, 9, 10\}$

$A = \{$even numbers$\}$    $B = \{$numbers greater than 6$\}$

Work out:

a P($A$)        b P($B$)        c P($A \cup B$)    d P($A \cap B$).

(PS) **9** P($A$) = 0.7    P($B$) = 0.6    P($A \cup B$) = 0.9

Work out P($A \cap B$).

(PS) **10** P($A$) = 0.12    P($B$) = 0.45    P($A \cap B$) = 0.07

Work out P($A \cup B$).

(CM) **11** Use set notation to describe the shaded area in each Venn diagram.

a
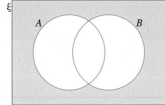

b
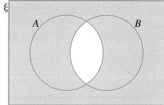

(CM) **12** For each part of this question, copy the Venn diagram and shade the appropriate area.

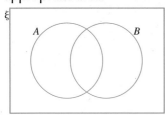

a $A' \cup B$        b $(A' \cap B)'$        c $(A' \cup B)'$

(PS) **13** There are 80 cars in a car park. 43 have traction control. 31 are black. 24 are black and have traction control.

Work out the probability that the first car to leave the car park does not have traction control.

# 24.4 Tree diagrams

## This section will show you how to:

- understand frequency tree diagrams and probability tree diagrams
- use probability tree diagrams to work out the probabilities involved in combined events.

**Key terms**

frequency tree diagram

probability tree diagram

In mathematics, you have seen that difficult ideas can be expressed in diagrams. These can be very helpful in the study of probability.

### Frequency tree diagrams

Suppose a bag contains coloured discs. You are going to take discs at random from the bag. The probability of taking a red disc is $\frac{1}{8}$, the probability of taking a blue disc is $\frac{1}{4}$ and the probability of taking a green disc is $\frac{5}{8}$.

You take a disc and then replace it. You do this 80 times.

How many discs of each colour would you expect to take?

P(red) = $\frac{1}{8}$, so you would expect to choose $\frac{1}{8} \times 80 = 10$ red discs.

P(blue) = $\frac{1}{4}$, so you would expect to choose $\frac{1}{4} \times 80 = 20$ blue discs.

P(green) = $\frac{5}{8}$, so you would expect to choose $\frac{5}{8} \times 80 = 50$ green discs.

You can show this information on a **frequency tree diagram**.

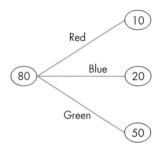

---

**Example 7**

400 people take a two-part test. In the first part of the test, the probability that someone passes is 0.8. In the second part of the test, the probability that someone passes is 0.6. Show the outcomes on a frequency tree.

First part:    P(pass) = 0.8, so 400 × 0.8 = 320 pass. So 80 fail.

Second part:    P(pass) = 0.6, so if someone passes the first part, then 0.6 × 320 = 192 pass the second part and 128 fail the second part.

Second part:    P(pass) = 0.6, so if someone fails the first part, then 0.6 × 80 = 48 pass the second part and 32 fail the second part.

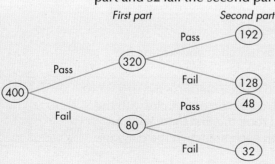

## Probability tree diagrams

Suppose you take two cards from this pack of six cards, but you replace the first card before you take the second card.

One way to show all the possible outcomes of this experiment is in a probability space diagram.

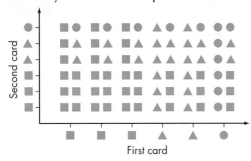

From the diagram, you can see immediately that the probability of taking, for example, two squares, is 9 out of 36 pairs of cards. So:

P(two squares) is $\frac{9}{36} = \frac{1}{4}$

**Example 8**

Look again at the probability space diagram above. What is the probability of taking:

**a** a square and a triangle (in any order)    **b** two circles    **c** two shapes that are the same?

**a** There are 6 combinations that give a square then a triangle and 6 that give a triangle then a square. So there are 12 combinations that give a square and a triangle altogether. So:

P(square and triangle, in any order) is $\frac{12}{36} = \frac{1}{3}$

**b** There is only 1 combination that gives two circles. So:

P(two circles) = $\frac{1}{36}$

**c** There are 9 combinations of two squares together, 4 combinations of two triangles together and 1 combination of two circles together. These give a total of 14 combinations with two shapes the same. So:

P(two shapes the same) is $\frac{14}{36} = \frac{7}{18}$

You can also use **probability tree diagrams** to solve problems involving combined events. Returning to the 'six-card' problem above, when you take the first card, there are three possible outcomes: a square, a triangle or a circle. For a single event:

P(square) = $\frac{3}{6}$     P(triangle) = $\frac{2}{6}$     P(circle) = $\frac{1}{6}$

You can show this by representing each outcome as a branch and writing its probability on that branch.

Then you can extend the diagram to show a second choice. Because the first card has been replaced, you can still take a square, a triangle or a circle, with the same probabilities. This is true no matter what you took the first time. You can demonstrate this by adding three more branches to each branch in the diagram.

Here is the complete probability tree diagram.

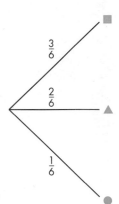

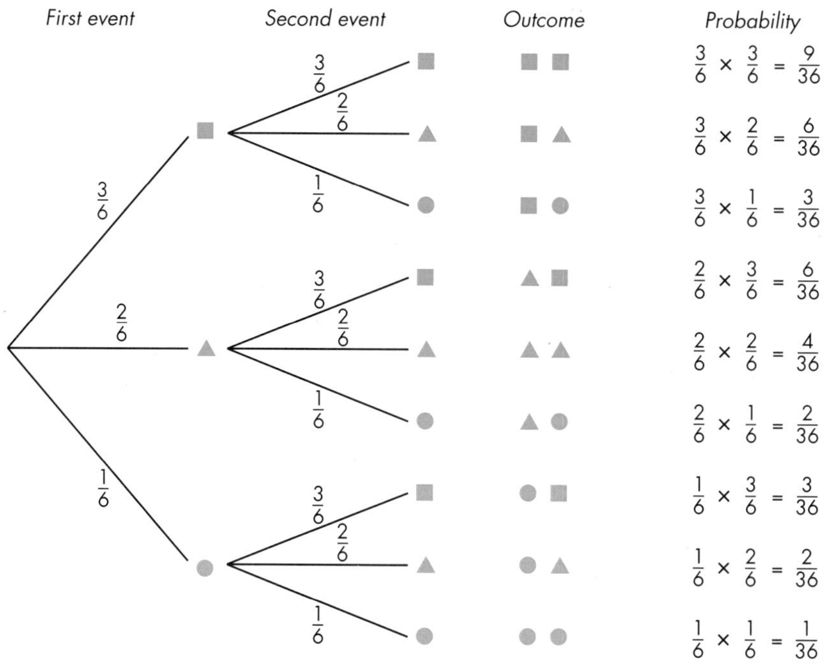

First event    Second event    Outcome    Probability

$\frac{3}{6} \times \frac{3}{6} = \frac{9}{36}$

$\frac{3}{6} \times \frac{2}{6} = \frac{6}{36}$

$\frac{3}{6} \times \frac{1}{6} = \frac{3}{36}$

$\frac{2}{6} \times \frac{3}{6} = \frac{6}{36}$

$\frac{2}{6} \times \frac{2}{6} = \frac{4}{36}$

$\frac{2}{6} \times \frac{1}{6} = \frac{2}{36}$

$\frac{1}{6} \times \frac{3}{6} = \frac{3}{36}$

$\frac{1}{6} \times \frac{2}{6} = \frac{2}{36}$

$\frac{1}{6} \times \frac{1}{6} = \frac{1}{36}$

Notice that the sum of all the probabilities is 1, as the outcomes are exhaustive.

You can calculate the probability of any final outcome by multiplying all the probabilities on its branches. For instance:

P(two squares) is $\frac{3}{6} \times \frac{3}{6} = \frac{9}{36}$ which cancels to $\frac{1}{4}$

P(triangle followed by circle) is $\frac{2}{6} \times \frac{1}{6} = \frac{2}{36}$ which cancels to $\frac{1}{18}$.

**Example 9**

Look again at the probability tree diagram above. What is the probability of taking:

**a** two triangles    **b** a circle followed by a triangle    **c** a square and a triangle, in any order

**d** two circles    **e** two shapes that are the same?

**a** P(two triangles) is $\frac{4}{36} = \frac{1}{9}$

**b** P(circle followed by triangle) is $\frac{2}{36} = \frac{1}{18}$

**c** There are 2 results in the outcome column that show a square and a triangle. These are in the second and fourth rows. The probability of each is $\frac{6}{36}$. Their combined probability is given by the addition rule.

$$P(\text{square and triangle, in any order}) \text{ is } \frac{6}{36} + \frac{6}{36} = \frac{12}{36}$$
$$= \frac{1}{3}$$

**d** P(two circles) = $\frac{1}{36}$

**e** There are 3 final outcomes that have two shapes the same. These are on the first, fifth and last rows. The probabilities are respectively $\frac{9}{36}$, $\frac{4}{36}$ and $\frac{1}{36}$. Their combined probability is given by the addition rule.

$$P(\text{two shapes the same}) \text{ is } \frac{9}{36} + \frac{4}{36} + \frac{1}{36} = \frac{14}{36}$$
$$= \frac{7}{18}$$

Note that the answers to parts **c**, **d** and **e** are the same as the answers found in Example 8.

# Exercise 24D

**1** 80 students took a driving test.

Before the test each student predicted whether they would pass or fail.

50 students predicted they would pass.

After the test, 42 students who predicted they would pass did actually pass.

56 students passed the test.

Copy and complete the frequency tree diagram.

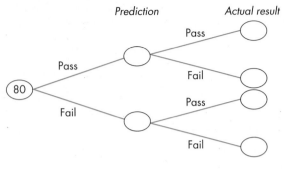

**2** Mia throws a fair coin twice.

**a** Copy and complete this probability tree diagram to show all the possible outcomes.

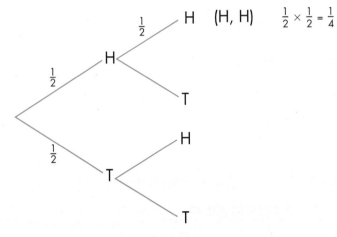

First event    Second event    Outcome    Probability

H    (H, H)    $\frac{1}{2} \times \frac{1}{2} = \frac{1}{4}$

**b** Use the tree diagram to work out the probability that Mia throws:

   **i** two tails    **ii** a head and a tail in any order    **iii** at least one tail.

**3** Zak takes a card at random from a standard pack of playing cards.

**a** What is the probability that the card was an ace?

**b** What is the probability that the card was not an ace?

   He replaces the card, shuffles the pack and takes another card.

**c** Draw a probability tree diagram to show the outcomes of two cards being taken as described (ace, not an ace).

**d** Use your diagram to work out the probability that:

   **i** both cards will be aces

   **ii** only one card will be an ace

   **iii** at least one of the cards will be an ace.

**4** On her way to work, Cheryl drives through two sets of road works with traffic lights that only show green or red. She knows that the probability of the first set being green is $\frac{1}{3}$ and the probability of the second set being green is $\frac{1}{2}$.

**a** What is the probability that the first set of lights will be red?

**b** What is the probability that the second set of lights will be red?

**c** Copy and complete the probability tree diagram, showing the possible outcomes when passing through both sets of lights.

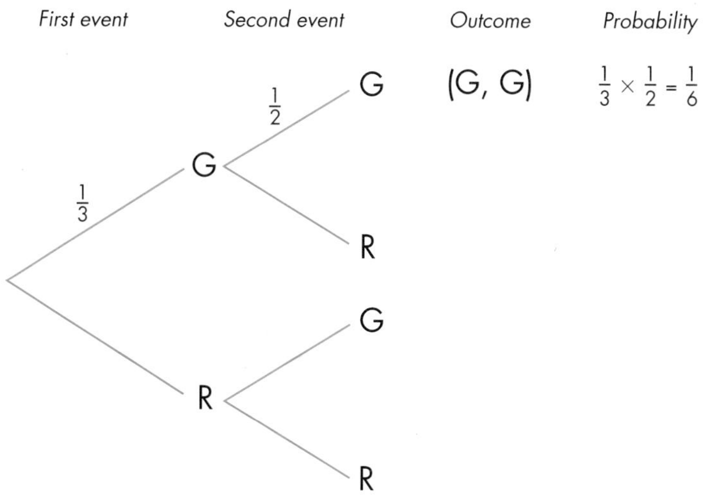

| First event | Second event | Outcome | Probability |
|---|---|---|---|

$$\frac{1}{3} \times \frac{1}{2} = \frac{1}{6}$$

(G, G)

**d** Use your tree diagram to work out the probability that Cheryl:

**i** does not get held up at either set of lights

**ii** gets held up at exactly one set of lights

**iii** gets held up at least once.

**e** Over a school term Cheryl makes 90 journeys to work. On how many days can she expect to get two green lights?

**5** Six out of every 10 cars in Britain are made abroad.

**a** What is the probability that any car in Britain will be British made?

**b** Two cars can be seen approaching in the distance. Draw a probability tree diagram and use it to work out the probability that:

**i** both cars are British made

**ii** one car is British and the other was made abroad.

**6**  **a** Jack throws three fair coins.

Copy and complete the probability tree diagram to show the possible outcomes.

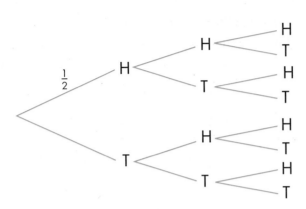

First event    Second event    Third event    Outcome    Probability

(H, H, H)  $\frac{1}{2} \times \frac{1}{2} \times \frac{1}{2} = \frac{1}{8}$

**b** Use your diagram to work out the probability that Jack throws:

**i** three tails    **ii** two heads and a tail    **iii** at least one tail.

**(PS)** **7**  Aziz takes a three-part language examination paper. He has a 0.4 chance of passing the first part, on 'speaking'. He has a 0.5 chance of passing the second part, on 'listening'. He has a 0.7 chance of passing the third part, on 'writing'.

**a** Draw a probability tree diagram where the first event is passing or failing the 'speaking' part, the second event is passing or failing the 'listening' part and the third event is passing or failing the 'writing' part.

**b** If he passes all three parts, his father will give him £50. What is the probability that he gets the money?

**c** If he passes two parts only, he can resit the other part. What is the probability that he will have to resit the examination?

**d** If he fails all three parts, he will be not be able to continue the course. What is the probability that he will not continue the course?

**(PS)** **8**  Look at all the probability tree diagrams that you have seen so far.

**a** What do the probabilities across any set of branches (such as those outlined in the blue rectangle on the left of this diagram) always add up to?

First event    Second event    Outcome    Probability    **b**

$\frac{3}{4}$  C    (A, C)    $\frac{2}{5} \times \frac{3}{4}$    $= \frac{3}{10}$

$\frac{2}{5}$  A

?  D    (A, D)    $\frac{2}{5} \times$ ?    $=$ ?

?  E    (B, E)    ? $\times$ ?    $= \frac{1}{5}$

?  B

?  F    (B, F)    ? $\times$ ?    $=$ ?

**b** What do the final probabilities (outlined in the blue rectangle on the right of the diagram) always add up to?

**c** Now copy the diagram and fill in all of the missing values.

**9** I have a bag containing red, yellow and green jelly babies.

I take three jelly babies at random from the bag.

Without drawing a probability tree diagram, write down:

**a** how many different outcomes there are

**b** how many of these outcomes will have three jelly babies of different colours.

**10 a** Suppose you throw a fair coin four times. Show how you could work out the probability of getting four heads without drawing a probability tree diagram.

**b** A fair coin is thrown $n$ times. Write down the probability of scoring $n$ heads.

# Worked exemplars

 **1** The Venn diagram shows the number of students who study geography ($G$) and the number who study history ($H$).

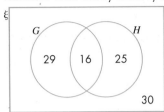

**a** Explain how you would work out P(a student chosen at random studies history).

**b** Describe in words what $P(G \cap H)$ represents.

**c** Describe in words what $P(G \cup H)$ represents.

| | |
|---|---|
| This question assesses how you communicate and interpret information accurately. | |
| **a** There are 29 + 16 + 25 + 30 = 100 students altogether.<br><br>16 + 25 = 41 study history.<br><br>So P($H$) is $\frac{41}{100}$. | This part of the question assesses your written communication. Do not just write down a probability. You must explain in words how you arrived at your answer. |
| **b** $G \cap H$ represents those students who study both geography and history.<br><br>$P(G \cap H)$ is the probability that a student chosen at random studies both geography and history. | Make sure you interpret the questions carefully. You need to explain what each probability represents.<br><br>It is important to remember that the students are chosen at random. |
| **c** $G \cup H$ represents those students who study geography, history or both.<br><br>$P(G \cup H)$ is the probability that a student chosen at random studies geography, history or both. | |

Susie is taking a driving test. The test is made up of two parts, a practical and a theory. She is told that the probability of passing only one of the two parts is 0.44 and the probability of passing the practical part of the test is 0.8.

**a** If P(passing the theory part) = $x$, write down P(not passing the theory part).

**b** Draw a probability tree diagram to show this information.

**c** Set up an equation, in terms of $x$, to calculate the probability of passing the theory part.

| This is a problem-solving question. You need to process the problem into a series of algebraic steps. | |
|---|---|
| **a** $1 - x$ | Remember, P(event not happening)<br><br>= 1 – P(event happening). |
| **b**<br><br>*Practical*      *Theory*      *Outcome*   *Probability*<br><br>0.8 — Pass<br>    $x$ — Pass   PP    $0.8x$<br>    $1 - x$ — Fail   PF    $0.8(1 - x)$<br>0.2 — Fail<br>    $x$ — Pass   FP    $0.2x$<br>    $1 - x$ — Fail   FF    $0.2(1 - x)$ | Remember to multiply the probabilities on the branches for each outcome. |
| **c**   P(only pass one part of the test)<br>    = P(PF) + P(FP)<br>    = $0.8(1 - x) + 0.2x$<br>  So $0.8 - 0.8x + 0.2x = 0.44$<br>      $0.8 - 0.6x = 0.44$<br>        $0.6x = 0.8 - 0.44$<br>          $x = \dfrac{0.8 - 0.44}{0.6}$<br>          $x = 0.6$<br>  So the probability of passing the theory part is 0.6. | Use the probability tree diagram to work out P(only pass one part of the test).<br><br>There are two outcomes on the diagram – PF and FP. Remember to add the two probabilities.<br><br>This is the equation to solve.<br><br>Rearrange the equation to calculate $x$. |

# Ready to progress?

I can calculate probabilities when two events happen at the same time.
I can read two-way tables and use them to work out probabilities.

I can understand set notation.
I can use Venn diagrams to work out probabilities.
I can use frequency tree diagrams to solve problems.

I can draw and use probability tree diagrams to work out probabilities.

# Review questions

**1** Ivy spins a fair five-sided spinner and throws a fair coin.

   **a** Draw a two-way table to show the possible outcomes.

   **b** Ivy spins the spinner once and throws the coin once. Write down the probability that she will get an odd number and a tail.

**2** The two-way table shows the ages and genders of a random sample of 50 students in a school.

| | | Age (years) | | | | | |
|---|---|---|---|---|---|---|---|
| | | 11 | 12 | 13 | 14 | 15 | 16 |
| Gender | Boys | 4 | 3 | 6 | 2 | 5 | 4 |
| | Girls | 2 | 5 | 3 | 6 | 4 | 6 |

   **a** How many students are aged 13 years or less?

   **b** What percentage of the students are 16?

   **c** A student is selected at random from the sample. What is the probability that the student will be 14 years of age? Give your answer as a fraction in its lowest form.

   **d** There are 1000 students in the school. Use the results for the sample to estimate how many boys are in the school altogether.

**3** This Venn diagram shows the number of students who like football (F) and the number who like hockey (H).

   **a** How many students are there altogether?

   **b** Work out P(F).

   **c** Work out P(F ∩ H).

   **d** What is the probability that a student chosen at random does not like football?

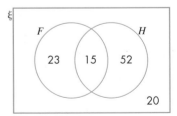

**4** There are 200 students in the first year of a college mathematics course. The probability that a male is on the course in the first year is 0.6.

   The probability that a male stays on the course into the second year is 0.8. The probability that a female stays on the course into the second year is 0.7.

   Show the outcomes on a frequency tree diagram.

**5** This Venn diagram shows some probabilities for two events, $A$ and $B$.

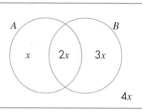

**a** Write down the value of $x$.

**b** Work out:  **i** P($A$)  **ii** P($B'$)  **iii** P($A \cup B$)  **iv** P($A' \cap B$).

**6** In a sixth form of 260 students, 93 study Spanish, 95 study chemistry, 165 study mathematics, 18 study Spanish and chemistry, 75 study chemistry and mathematics, 20 study mathematics and Spanish and 15 study all three subjects.

**a** Draw a Venn diagram to illustrate the data.

**b** Use your diagram to calculate the probability that a student selected at random studies:

**i** only Spanish  **ii** mathematics and chemistry but not Spanish

**iii** none of these subjects  **iv** Spanish, given that they study mathematics.

Give your answers to 3 significant figures.

**7** Adel puts eight red counters and four blue counters into a bag.

She takes a counter, at random, from the bag, writes down its colour and puts the counter back in the bag.

She then takes, at random, a second counter from the bag.

**a** Copy and complete the probability tree diagram.

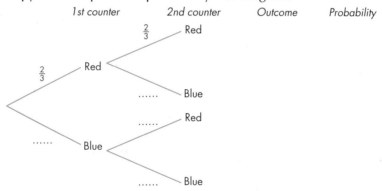

**b** Work out the probability that Adel takes two counters that are the same colour.

**c** Work out the probability that Adel takes two counters that are different colours.

**8** At the end of a course, army cadets have to pass an examination to gain a certificate. The probability of passing at the first attempt is 0.6. Those who fail are allowed to resit.

The probability of passing the resit is 0.7. No further attempts are allowed.

**a** Copy and complete the probability tree diagram.

| 1st attempt | 2nd attempt | Outcome | Probability |
|---|---|---|---|
| 0.6  Pass | | Pass | 0.6 |
| | 0.7  Pass | Pass | |
| .........  Fail | | | |
| | .........  Fail | Fail | |

**b i** What is the probability that a cadet gains a certificate?

**ii** What is the probability that a cadet fails to gain a certificate after two attempts?

**c** Two cadets take the exam. Explain how to calculate the probability that both of them gain a certificate.

# 25 Number: Powers and standard form

## This chapter is going to show you:

- how to calculate with powers (indices)
- how to write numbers in standard form
- how to calculate with standard form.

## You should already know:

- how to multiply and divide by 10, 100, 1000, …

## About this chapter

Scientists use standard form as a short way of writing very large and very small numbers. For example, the planets and the Sun are huge distances away from Earth. The furthest known galaxy is about 110 000 000 000 000 000 000 000 000 km from Earth. If you always have to write this in full, you could put down the wrong number of zeros. This could be a problem if you were doing a calculation about space travel.

Writing numbers in standard form can help you and people who you work with to follow your calculations. In standard form, the number above is written as $1.1 \times 10^{26}$, which is much neater.

Electrons, unlike planets, are very small. The mass of an electron is about 0.000 000 000 000 000 000 000 000 000 000 91 kg. You can write it as $9.1 \times 10^{-31}$ kg.

This chapter will show you how to do calculations with numbers in standard form.

# 25.1 Powers (indices)

This section will show you how to:

- write a number as a power of another number
- use powers (also known as indices)
- multiply and divide by powers of 10.

**Key terms**

index (indices)

power

**Powers** are a short way of writing repeated multiplications. A power is also called an **index** (plural **indices**).

The power tells you how many 'lots' of a number to multiply together. For example:

$4^6 = 4 \times 4 \times 4 \times 4 \times 4 \times 4$       six lots of 4 multiplied together

$6^4 = 6 \times 6 \times 6 \times 6$       four lots of 6 multiplied together

$7^3 = 7 \times 7 \times 7$

$12^2 = 12 \times 12$

You can write a million as $1\,000\,000 = 10^6$.

A number multiplied by itself is the number 'squared'. You need to know the square numbers (power 2) up to $15^2 = 225$. How many of them can you write down from memory? Learn any that you did not remember.

You will also find it helpful to know the cubes of some numbers (power 3).

$1^3 = 1$, $2^3 = 8$, $3^3 = 27$, $4^3 = 64$, $5^3 = 125$ and $10^3 = 1000$

---

**Example 1**

**a**   What is the value of:   **i** 7 squared    **ii** 5 cubed?

**b**   Write each number out in full.
Do not evaluate them.

   **i** $2^5$    **ii** $8^4$    **iii** $7^3$    **iv** $12^2$

**c**   Write these multiplications as powers.

   **i** $3 \times 3 \times 3 \times 3 \times 3 \times 3 \times 3 \times 3$      **ii** $13 \times 13 \times 13 \times 13 \times 13$

   **iii** $7 \times 7 \times 7 \times 7$              **iv** $5 \times 5 \times 5 \times 5 \times 5 \times 5 \times 5$

**a**   **i**   7 squared $(7^2)$ is $7 \times 7 = 49$     **ii** 5 cubed $(5^3)$ is $5 \times 5 \times 5 = 125$

**b**   **i**   $2^5 = 2 \times 2 \times 2 \times 2 \times 2$       **ii** $8^4 = 8 \times 8 \times 8 \times 8$

   **iii** $7^3 = 7 \times 7 \times 7$            **iv** $12^2 = 12 \times 12$

**c**   **i**   $3 \times 3 \times 3 \times 3 \times 3 \times 3 \times 3 \times 3 = 3^8$    **ii** $13 \times 13 \times 13 \times 13 \times 13 = 13^5$

   **iii** $7 \times 7 \times 7 \times 7 = 7^4$         **iv** $5 \times 5 \times 5 \times 5 \times 5 \times 5 \times 5 = 5^7$

---

## Working out powers on your calculator

The power button on your calculator will probably look like this $\boxed{x^\blacksquare}$ . You can use the power button to work out $5^7$ on your calculator.

$5^7 =$ $\boxed{5}$ $\boxed{x^\blacksquare}$ $\boxed{7}$ $= 78\,125$

## Two special powers

| Power 1 | Power 0 (zero) |
|---|---|
| Any number to the power 1 is the same as the number itself. This is always true, so normally you do not write the power 1.<br><br>For example: $5^1 = 5$    $32^1 = 32$    $(-8)^1 = -8$ | Any number to the power 0 is equal to 1.<br><br>For example: $5^0 = 1$    $32^0 = 1$    $(-8)^0 = 1$ |

Use your calculator to check these results.

## Exercise 25A

**1** Write these expressions in index notation. Do not work them out yet.

   **a** $2 \times 2 \times 2 \times 2$                  **b** $3 \times 3 \times 3 \times 3 \times 3$

   **c** $7 \times 7$                           **d** $5 \times 5 \times 5$

   **e** $10 \times 10 \times 10 \times 10 \times 10 \times 10 \times 10$    **f** $6 \times 6 \times 6 \times 6$

   **g** $4$                             **h** $1 \times 1 \times 1 \times 1 \times 1 \times 1 \times 1$

   **i** $0.5 \times 0.5 \times 0.5 \times 0.5$           **j** $100 \times 100 \times 100$

**2** Write each of these power terms out in full. Do not work them out yet.

   **a** $3^4$          **b** $9^3$          **c** $6^2$          **d** $10^5$          **e** $2^{10}$

   **f** $8^1$          **g** $0.1^3$        **h** $2.5^2$        **i** $0.7^3$        **j** $1000^2$

**3** Use the power key on your calculator (or any method you prefer) to work out the value of each power term in question **1**.

**4** Use the power key on your calculator (or any method you prefer) to work out the value of each power term in question **2**.

**5** A storage container is in the shape of a cube. The length of the container is 5 m.

Work out the total storage space in the container. Use the formula for the volume of a cube.

     volume = (length of edge)$^3$

**(PS) 6** Write each number as a power of a different number. The first one has been done for you.

   **a** $32 = 2^5$     **b** $100$           **c** $8$            **d** $25$

**7** Work out the value of each power term. Do not use a calculator.

   **a** $2^0$          **b** $4^1$          **c** $5^0$          **d** $1^9$          **e** $1^{235}$

**(EV) 8** What do the answers to question **7 d** and **e** tell you about 1 raised to any power?

**9** Write your answer to question **1 j** as a power of 10.

**10** Write your answer to question **2 j** as a power of 10.

**11** Using your calculator, or any other method, work out the value of each power term.

   **a** $(-1)^0$       **b** $(-1)^1$       **c** $(-1)^2$       **d** $(-1)^4$       **e** $(-1)^5$

**(MR) 12** Using your answers to question **11**, write down the value of each power term.

   **a** $(-1)^8$       **b** $(-1)^{11}$       **c** $(-1)^{99}$       **d** $(-1)^{80}$       **e** $(-1)^{126}$

**13** The number 16 777 216 is a power of 2. It is also a power of 4, a power of 8 and a power of 16.

Write the number 16 777 216 in terms of each of the powers.

**14** Solve this equation. $2^x = 128$

# 25.2 Rules for multiplying and dividing powers

## This section will show you how to:

- use rules for multiplying and dividing powers
- multiply and divide numbers by powers of 10.

### Positive indices

What happens when you multiply numbers that are written as powers of the same number or variable (letter)?

$$3^3 \times 3^5 = (3 \times 3 \times 3) \times (3 \times 3 \times 3 \times 3 \times 3)$$
$$= 3^8$$

$$a^2 \times a^3 = (a \times a) \times (a \times a \times a)$$
$$= a^5$$

Can you see the rule? You can find these products by adding the powers.

$$2^3 \times 2^4 \times 2^5 = 2^{3+4+5}$$
$$= 2^{12}$$

$$a^3 \times a^4 = a^{3+4}$$
$$= a^7$$

What happens when you divide numbers that are written as powers of the same number or letter (variable)?

$$7^6 \div 7 = (7 \times 7 \times 7 \times 7 \times 7 \times 7) \div (7)$$
$$= 7 \times 7 \times 7 \times 7 \times 7$$
$$= 7^5$$

$$a^5 \div a^2 = (a \times a \times a \times a \times a) \div (a \times a)$$
$$= a \times a \times a$$
$$= a^3$$

Can you see the rule? You can complete these divisions by subtracting the powers.

$$a^4 \div a^3 = a^{4-3}$$
$$= a^1$$
$$= a$$

$$b^7 \div b^4 = b^{7-4}$$
$$= b^3$$

### Negative indices

Now you can divide numbers that are written as powers.

But what happens when you are dividing numbers that are written as powers, and the power of the second number is higher than the power of the first?

$$c^5 \div c^7 = c^{5-7}$$ You can think of this as $\dfrac{c \times c \times c \times c \times c}{c \times c \times c \times c \times c \times c \times c}$
$$= c^{-2}$$
$$= \dfrac{1}{c \times c}$$
$$= \dfrac{1}{c^2}$$

You can write $\dfrac{1}{c^2}$ as $c^{-2}$. The negative power is a short way of writing the reciprocal of the positive power.

> **Hints and tips** The reciprocal of a number is the result of dividing 1 by that number.

When you *multiply* powers of the same number or variable, you *add* the indices, even if you are working with negative indices.

$3^4 \times 3^5 = 3^{(4+5)} = 3^9 \qquad 2^3 \times 2^4 \times 2^5 = 2^{12} \qquad 10^4 \times 10^{-2} = 10^2 \qquad 10^{-3} \times 10^{-1} = 10^{-4} \qquad a^x \times a^y = a^{(x+y)}$

When you *divide* powers of the same number or variable, you *subtract* the indices.

$a^4 \div a^3 = a^{(4-3)} \qquad b^4 \div b^7 = b^{-3} \qquad 10^4 \div 10^{-2} = 10^6 \qquad 10^{-2} \div 10^{-4} = 10^2 \qquad a^x \div a^y = a^{(x-y)}$
$\qquad\;\; = a^1$
$\qquad\;\; = a$

## Powers of powers

Look at this equation.

$(a^2)^3 = (a \times a)^3$

The term $a^2$ has been raised to the power 3.

$(a^2)^3 = (a \times a)^3$
$\qquad\;\; = (a \times a) \times (a \times a) \times (a \times a)$
$\qquad\;\; = a \times a \times a \times a \times a \times a$
$\qquad\;\; = a^6$

So $(a^2)^3 = a^6$

Similarly: $\qquad (a^2)^6 = a^{12} \qquad (a^{-2})^4 = a^{-8} \qquad (a^x)^y = a^{xy}$

When you raise a power to a further power, it is a 'power of a power'. To simplify a power of a power, you *multiply* the indices.

Here are some examples of different kinds of expression that include numbers and powers. Separate the numbers and powers to reduce the chance of making mistakes.

$2a^2 \times 3a^4 = (2 \times 3) \times (a^2 \times a^4) \qquad 4a^2b^3 \times 2ab^2 = (4 \times 2) \times (a^2 \times a) \times (b^3 \times b^2)$
$\qquad\quad = 6 \times a^6 \qquad\qquad\qquad\qquad\qquad = 8a^3b^5$
$\qquad\quad = 6a^6$

$12a^5 \div 3a^2 = (12 \div 3) \times (a^5 \div a^2) \qquad (2a^2)^3 = (2)^3 \times (a^2)^3$
$\qquad\quad = 4a^3 \qquad\qquad\qquad\qquad\qquad = 8 \times a^6$
$\qquad\qquad\qquad\qquad\qquad\qquad\qquad\qquad = 8a^6$

# Exercise 25B

**1** Write each expression as a single power of 5.

   **a** $5^2 \times 5^2$     **b** $5^4 \times 5^6$     **c** $5^2 \times 5^3$     **d** $5 \times 5^2$     **e** $5^6 \times 5^9$

   **f** $5 \times 5^8$     **g** $5^2 \times 5^4$     **h** $5^6 \times 5^3$     **i** $5^2 \times 5^6$

**2** Write each expression as a single power of $x$.

   **a** $x^2 \times x^6$     **b** $x^5 \times x^4$     **c** $x^6 \times x^2$     **d** $x^3 \times x^2$     **e** $x^6 \times x^6$

   **f** $x^5 \times x^8$     **g** $x^7 \times x^4$     **h** $x^2 \times x^8$     **i** $x^{12} \times x^4$

**3** Write each expression as a single power of 6.

    **a** $6^5 \div 6^2$         **b** $6^7 \div 6^2$         **c** $6^3 \div 6^2$         **d** $6^4 \div 6^4$         **e** $6^5 \div 6^4$

    **f** $6^5 \div 6^2$         **g** $6^4 \div 6^2$         **h** $6^4 \div 6^3$         **i** $6^5 \div 6^3$

**4** Write each expression as a single power of $x$.

    **a** $x^7 \div x^3$         **b** $x^8 \div x^3$         **c** $x^4 \div x$         **d** $x^6 \div x^3$         **e** $x^{10} \div x^4$

    **f** $x^6 \div x$         **g** $x^8 \div x^6$         **h** $x^8 \div x^2$         **i** $x^{12} \div x^3$

(MR) **5**   **a** Write down the value of $216 \div 216$.

    **b** Write $6^3 \div 6^3$ as a single power of 6.

       $6^3 = 216$

    **c** Use your answers to parts **a** and **b** to write down the value of $6^0$.

(MR) **6**   **a** Write down the value of $625 \div 625$.

    **b** Write $5^4 \div 5^4$ as a single power of 5.

       $5^4 = 625$

    **c** Use your answers to parts **a** and **b** to write down the value of $5^0$.

(CM) **7** What do you notice about your answers to questions **5c** and **6c**?

(PS) **8** $4^a \times 4^b = 4^7$

    $a$ and $b$ are both less than 5. Write down a pair of possible values for $a$ and $b$.

(PS) **9**   **a** A common error is to write $x^a + x^a = x^{2a}$.

      Find two whole numbers for $a$ and $x$ for which this is true.

    **b** Another common error is to write $x^a \times x^b = x^{ab}$.

      Find a whole number for $x$ for which this is true.

**10** Simplify each expression.

    **a** $2a^2 \times 3a^3$         **b** $3a^4 \times 3a^{-2}$         **c** $(2a^2)^3$

    **d** $-2a^2 \times 3a^2$       **e** $-4a^3 \times -2a^5$      **f** $-2a^4 \times 5a^{-7}$

**11** Simplify each expression.

    **a** $6a^3 \div 2a^2$         **b** $12a^5 \div 3a^2$        **c** $15a^5 \div 5a$

    **d** $18a^2 \div 3a$        **e** $24a^5 \div 6a^2$       **f** $30a \div 6a^5$

> **Hints and tips** Deal with numbers and indices separately and do not confuse the rules. For example: $12a^5 \div 4a^2 = (12 \div 4) \times (a^5 \div a^2)$.

(MR) **12** Write down two different:

    **a** multiplication questions with an answer of $12x^2y^5$

    **b** division questions with an answer of $12x^2y^5$.

**13** Simplify each expression.

    **a** $\dfrac{6a^4b^3}{2ab}$         **b** $\dfrac{2a^2bc^2 \times 6abc^3}{4ab^2c}$         **c** $\dfrac{3abc \times 4a^3b^2c \times 6c^2}{9a^2bc}$

**14** Simplify each expression.

    **a** $2a^2b^3 \times 4a^3b$          **b** $5a^2b^4 \times 2ab^{-3}$        **c** $6a^2b^3 \times 5a^{-4}b^{-5}$

    **d** $12a^2b^4 \div 6ab$             **e** $24a^{-3}b^4 \div 3a^2b^{-3}$      **f** $16a^7b^{-2} \div 4a^2b^3$

**15** Write each of these as a single power of 4.

    **a** $(4^2)^3$               **b** $(4^3)^5$               **c** $(4^1)^6$

    **d** $(4^3)^{-2}$            **e** $(4^{-2})^{-3}$         **f** $(4^7)^0$

 **16** $a$, $b$ and $c$ are three different positive integers.

    What is the smallest possible value of $a^2b^3c$?

## Multiplying and dividing by powers of 10

When you write a million in figures, how many zeros does it have? What is a million as a power of 10? This table shows some of the pattern of the powers of 10.

| Number | 0.001 | 0.01 | 0.1 | 1 | 10 | 100 | 1000 | 10000 | 100 000 | 1 000 000 |
|---|---|---|---|---|---|---|---|---|---|---|
| Power of 10 | $10^{-3}$ | $10^{-2}$ | $10^{-1}$ | $10^0$ | $10^1$ | $10^2$ | $10^3$ | $10^4$ | $10^5$ | $10^6$ |

What is the pattern in the top row? What is the pattern in the powers in the bottom row?

Note that the negative indices give decimal values. A negative index means 'divide that power of 10 into 1'.

$10^{-1} = \frac{1}{10^1} = \frac{1}{10} = 0.1$ and $10^{-2} = \frac{1}{10^2} = \frac{1}{100} = 0.01$

## Multiplication

The result of a multiplication calculation is called a *product*. For example, the product of 5 and 7 is $5 \times 7 = 35$.

**Remember:**

* multiplying any number by 0 gives 0
* multiplying any number by 1 gives the original number.

What happens when you multiply by 10? Try these on your calculator.

$7.34 \times 10$        $0.678 \times 10$        $0.007 \times 10$

Can you see the rule for multiplying by 10? You may have learned that when you multiply a number by 10, you 'add a zero' to the end of the number. This is only true when you start with a whole number. It is not true for a decimal.

When you multiply a number by 10, the place value of each digit is increased. For example, 0.07 becomes 0.7, 0.3 becomes 3, and so on. All the digits move one place to the left.

What happens when you multiply by 100? Try these on your calculator.

$7.34 \times 100$        $0.678 \times 100$        $0.007 \times 100$

This time you should find that the digits move two places to the left. When you multiply by 100, the place value of each digit increases by 2 places, so 0.07 becomes 7, and 0.3 becomes 30, and so on.

To multiply by any power of 10, you must move the digits according to these two rules.

* When the index is positive, move the digits to the left by the same number of places as the value of the index.
* When the index is negative, move the digits to the right by the same number of places as the value of the index.

Check that this happened with the examples you tried.

Work out the value of each multiplication.

**a** $12.356 \times 10^2$      **b** $3.45 \times 10^1$      **c** $753.4 \times 10^{-2}$      **d** $6789 \times 10^{-1}$

**a** $12.356 \times 10^2 = 1235.6$      **b** $3.45 \times 10^1 = 34.5$

**c** $753.4 \times 10^{-2} = 7.534$      **d** $6789 \times 10^{-1} = 678.9$

Sometimes, you have to insert zeros to make up the required number of digits.

Write these as ordinary numbers.

**a** $75 \times 10^4$      **b** $2.04 \times 10^5$      **c** $6.78 \times 10^{-3}$      **d** $8.97 \times 10^{-4}$

**a** $75 \times 10^4 = 750\ 000$      **b** $2.04 \times 10^5 = 204\ 000$

**c** $6.78 \times 10^{-3} = 0.006\ 78$      **d** $8.97 \times 10^{-4} = 0.000\ 897$

## Division

Is there a similar connection for division by multiples of 10? Try these on your calculator. Look for the connection between the calculation and the answer.

$12.3 \div 10$      $3.45 \div 1000$      $3.45 \div 10^3$

$0.075 \div 100$      $2.045 \div 10^2$      $6.78 \div 1000$

To divide by any power of 10, you must move the digits according to these two rules.

- When the index is positive, move the digits to the right by the same number of places as the value of the index.
- When the index is negative, move the digits to the left by the same number of places as the value of the index.

## Working with multiples of powers of 10

You can use this principle to multiply multiples of 10, 100, … You also use this method in estimation. You can do this in your head to check that your answers to calculations are about right.

Use a calculator to work out these multiplications.

$200 \times 300$      $100 \times 40$      $2000 \times 3000$

Can you see a way of working them out without using a calculator or pencil and paper?

What about division? Use a calculator to do these divisions.

$400 \div 20$      $250 \div 50$      $30\ 000 \div 600$

Can you see a way of doing these 'in your head'? Look at these examples.

$300 \times 4000 = 1\ 200\ 000$      $5000 \div 200 = 25$      $200 \times 50 = 10\ 000$

$60 \times 5000 = 300\ 000$      $400 \div 20 = 20$      $30\ 000 \div 600 = 50$

To multiply $200 \times 3000$, for example, you multiply the non-zero digits ($2 \times 3 = 6$) and then write the total number of zeros in both numbers at the end, to give 600 000.

$200 \times 3000 = 2 \times 100 \times 3 \times 1000 = 6 \times 100\ 000 = 600\ 000$

For division, you divide the non-zero digits and then cancel the zeros. For example:

$$400\ 000 \div 80 = \frac{400\ 000}{80}$$

$$= \frac{{}^5 \cancel{400\ 000}}{\cancel{80}_{\ 1}}$$

$$= 5000$$

Example 4

Work out the value of each division.

**a** $712.35 \div 10^2$    **b** $38.45 \div 10^1$    **c** $3.463 \div 10^{-2}$    **d** $6.789 \div 10^{-1}$

**a** $712.35 \div 10^2 = 7.1235$    **b** $38.45 \div 10^1 = 3.845$

**c** $3.463 \div 10^{-2} = 346.3$    **d** $6.789 \div 10^{-1} = 67.89$

Sometimes, you have to insert zeros to make up the required number of digits.

Example 5

Work out the value of each division.

**a** $75 \div 10^4$    **b** $2.04 \div 10^5$    **c** $6.78 \div 10^{-3}$    **d** $0.08 \div 10^{-4}$

**a** $75 \div 10^4 = 0.0075$    **b** $2.04 \div 10^5 = 0.000\ 020\ 4$

**c** $6.78 \div 10^{-3} = 6780$    **d** $0.08 \div 10^{-4} = 800$

When you work through the next exercise, remember:

$$10\ 000 = 10 \times 10 \times 10 \times 10 = 10^4 \qquad 1 = 10^0$$
$$1000 = 10 \times 10 \times 10 \qquad = 10^3 \qquad 0.1 = 1 \div 10 \qquad = 10^{-1}$$
$$100 = 10 \times 10 \qquad = 10^2 \qquad 0.01 = 1 \div 100 \qquad = 10^{-2}$$
$$10 = 10 \qquad = 10^1 \qquad 0.001 = 1 \div 1000 = 10^{-3}$$

# Exercise 25C

**1** Write down the answers.

**a** $200 \times 300$    **b** $30 \times 4000$    **c** $3 \times 50$    **d** $60 \times 700$

**e** $200 \times 7$    **f** $10 \times 30$    **g** $(20)^2$    **h** $(20)^3$

**i** $(400)^2$    **j** $30 \times 150$    **k** $40 \times 200$    **l** $50 \times 5000$

**2** Write down the answers.

**a** $2000 \div 400$    **b** $3000 \div 60$    **c** $5000 \div 200$    **d** $6000 \div 200$

**e** $2100 \div 300$    **f** $9000 \div 30$    **g** $300 \div 50$    **h** $2100 \div 70$

**i** $5000 \div 5000$    **j** $30\ 000 \div 2000$    **k** $2000 \times 40 \div 2000$    **l** $200 \times 20 \div 800$

**m** $200 \times 6000 \div 30\ 000$    **n** $20 \times 80 \times 600 \div 3000$

**3** You are given that $16 \times 34 = 544$.

**a** Write down the value of $160 \times 340$.    **b** What is $544\ 000 \div 34$?

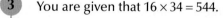

**4** Write these calculations in order, starting with the one that gives the smallest answer.

$5000 \times 4000$    $600 \times 8000$    $200\ 000 \times 700$    $30 \times 90\ 000$

**5** One year there were £20 notes to the value of £28 000 million in circulation. How many £20 notes is this?

**6** Write down the value of each expression.

**a** $3.1 \times 10$    **b** $3.1 \times 100$    **c** $3.1 \times 1000$    **d** $3.1 \times 10\ 000$

**7** Write down the value of each expression.

**a** $6.5 \times 10$    **b** $6.5 \times 10^2$    **c** $6.5 \times 10^3$    **d** $6.5 \times 10^4$

**8** Write down the value of each expression.

    **a** $3.1 \div 10$      **b** $3.1 \div 100$      **c** $3.1 \div 1000$      **d** $3.1 \div 10\,000$

**9** Write down the value of each expression.

    **a** $6.5 \div 10$      **b** $6.5 \div 10^2$      **c** $6.5 \div 10^3$      **d** $6.5 \div 10^4$

**10** Evaluate each expression.

    **a** $2.5 \times 100$      **b** $3.45 \times 10$      **c** $4.67 \times 1000$      **d** $34.6 \times 10$

    **e** $20.789 \times 10$      **f** $56.78 \times 1000$      **g** $2.46 \times 10^2$      **h** $0.076 \times 10$

    **i** $0.999 \times 10^6$      **j** $234.56 \times 10^2$      **k** $98.7654 \times 10^3$      **l** $43.23 \times 10^6$

    **m** $0.003\,4578 \times 10^5$      **n** $0.0006 \times 10^7$      **o** $0.005\,67 \times 10^4$      **p** $56.0045 \times 10^4$

**11** Evaluate each expression.

    **a** $2.5 \div 100$      **b** $3.45 \div 10$      **c** $4.67 \div 1000$      **d** $34.6 \div 10$

    **e** $20.789 \div 100$      **f** $56.78 \div 1000$      **g** $2.46 \div 10^2$      **h** $0.076 \div 10$

    **i** $0.999 \div 10^6$      **j** $234.56 \div 10^2$      **k** $98.7654 \div 10^3$      **l** $43.23 \div 10^6$

    **m** $0.003\,4578 \div 10^5$      **n** $0.0006 \div 10^7$      **o** $0.005\,67 \div 10^4$      **p** $56.0045 \div 10^4$

> **Hints and tips** Even though you are really moving digits left or right, you may think of it as the decimal point moving right or left.

**12** Work these out without using a calculator.

    **a** $2.3 \times 10^2$      **b** $5.789 \times 10^5$      **c** $4.79 \times 10^3$      **d** $5.7 \times 10^7$

    **e** $2.16 \times 10^2$      **f** $1.05 \times 10^4$      **g** $3.2 \times 10^{-4}$      **h** $9.87 \times 10^3$

(MR) **13** Which of these statements is true about the numbers in question **12**?

    **a** The first part is always a number between 1 and 10.

    **b** There is always a multiplication sign in the middle of the expression.

    **c** There is always a power of 10 at the end.

    **d** Calculator displays sometimes show numbers in this form.

# 25.3 Standard form

## This section will show you how to:

- write a number in standard form
- calculate with numbers in standard form.

**Standard form** is also known as **standard index form**. Any number can be written as a value in the range from 1 to 10 multiplied by a power of 10. The general form of a number in standard form is:

$A \times 10^n$ where $1 \leqslant A < 10$, and $n$ is a whole number.

Standard form is often used for very large or very small numbers.

Read through these examples to see how to write numbers in standard form.

$52 \quad = 5.2 \times 10 \qquad \rightarrow \quad 5.2 \times 10^1$

$73 \quad = 7.3 \times 10 \qquad \rightarrow \quad 7.3 \times 10^1$

$625 \ = 6.25 \times 100 \qquad \rightarrow \quad 6.25 \times 10^2$ These numbers are in standard form.

$389 \ = 3.89 \times 100 \qquad \rightarrow \quad 3.89 \times 10^2$

$3147 = 3.147 \times 1000 \quad \rightarrow \quad 3.147 \times 10^3$

When you are writing a number in this way, you must always follow two rules.

- The first part must be a number between 1 and 10 (1 is allowed but 10 isn't).

- The second part must be a whole-number (negative or positive) power of 10. Note that you would *not normally* write the power 1.

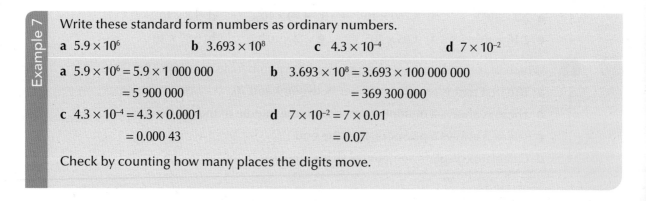

**Example 6**

Write these numbers in standard form.

**a** 34 700 **b** 560 000 **c** 0.005 **d** 0.000 672

**a** $34\,700 = 3.47 \times 10\,000$ **b** $560\,000 = 5.6 \times 100\,000$

$\qquad = 3.47 \times 10^4$ $\qquad = 5.6 \times 10^5$

**c** $0.005 = 5 \times 0.001$ **d** $0.000\,672 = 6.72 \times 0.0001$

$\qquad = 5 \times 10^{-3}$ $\qquad = 6.72 \times 10^{-4}$

Check by counting how many places the digits move.

**Example 7**

Write these standard form numbers as ordinary numbers.

**a** $5.9 \times 10^6$ **b** $3.693 \times 10^8$ **c** $4.3 \times 10^{-4}$ **d** $7 \times 10^{-2}$

**a** $5.9 \times 10^6 = 5.9 \times 1\,000\,000$ **b** $3.693 \times 10^8 = 3.693 \times 100\,000\,000$

$\qquad = 5\,900\,000$ $\qquad = 369\,300\,000$

**c** $4.3 \times 10^{-4} = 4.3 \times 0.0001$ **d** $7 \times 10^{-2} = 7 \times 0.01$

$\qquad = 0.000\,43$ $\qquad = 0.07$

Check by counting how many places the digits move.

## Standard form on a calculator

It is difficult to key a number such as 123 000 000 000 into a calculator. Instead, you can enter it in standard form (assuming you are using a scientific calculator).

$123\,000\,000\,000 = 1.23 \times 10^{11}$

These are the keystrokes to enter this number into a typical calculator.

`1` `.` `2` `3` `×10ˣ` `1` `1`

Your calculator display will display the number either as an ordinary number, if there is enough space in the display, or in standard form if there is not.

## Standard form for numbers less than 1

The numbers on the right-hand side of each equals sign are written in standard form. Make sure that you understand how they are formed.

$$0.4 = 4 \times 10^{-1} \qquad 0.05 = 5 \times 10^{-2} \qquad 0.007 = 7 \times 10^{-3}$$

$$0.123 = 1.23 \times 10^{-1} \qquad 0.007\,65 = 7.65 \times 10^{-3} \qquad 0.9804 = 9.804 \times 10^{-1}$$

$$0.0098 = 9.8 \times 10^{-3} \qquad 0.000\,007\,8 = 7.8 \times 10^{-6}$$

On a typical calculator you would enter $1.23 \times 10^{-6}$, for example, as:

[ 1 ] [ • ] [ 2 ] [ 3 ] [ ×10ˣ ] [ (−) ] [ 6 ]

Practise entering some of the numbers above into your calculator.

## Calculating with standard form

Calculations involving very large or very small numbers can be done more efficiently if you use standard form. These examples show you how to work out the area of a pixel on a computer screen, and how long it takes light to reach Earth from a distant star.

You should not need to use a calculator to do the calculation in Example 8.

**Example 8**

A pixel on a computer screen is $2 \times 10^{-2}$ cm long by $7 \times 10^{-3}$ cm wide.

What is the area of the pixel? Give your answer in standard form.

Area = length × width

$= (2 \times 10^{-2}) \times (7 \times 10^{-3})$ cm²

$= (2 \times 7) \times (10^{-2} \times 10^{-3})$ cm²       (Multiply the numbers and add the powers of 10.)

$= 14 \times 10^{-5}$ cm²

Now change the answer to standard form.

$14 = 1.4 \times 10^{1}$, so:     $14 \times 10^{-5}$ cm² $= 1.4 \times 10^{1} \times 10^{-5}$ cm²

$= 1.4 \times 10^{-4}$ cm²

**Example 9**

The star *Betelgeuse* is $3.8 \times 10^{15}$ miles from Earth. Light travels at $1.86 \times 10^{5}$ miles per second.

**a** How many seconds does it take light to travel from *Betelgeuse* to Earth? Give your answer in standard form to one decimal place.

**b** How many years does it take light to travel from *Betelgeuse* to Earth?

**a** Time = distance ÷ speed

$= (3.8 \times 10^{15}$ miles$) \div (1.86 \times 10^{5}$ miles per second$)$

$= (3.8 \div 1.86) \times (10^{15} \div 10^{5})$ seconds       (Divide the numbers and subtract the powers of 10.)

$= 2.043\,010\,753 \times 10^{10}$ seconds

$= 2.04 \times 10^{10}$       (Round the answer.)

$= 2.0 \times 10^{10}$ seconds (1 sf)       (Check it is in standard form.)

**b** To convert from seconds to years, first divide by 3600 to change seconds to hours, then divide by 24 to change hours to days, and finally by 365 to change days to years.

$2.0 \times 10^{10} \div (3600 \times 24 \times 365) = 634.2$ years

You will have noticed that the bigger the power of 10, the bigger the number. So if you are asked to order some standard form numbers, first look at the power of 10. If this is the same for some numbers, then look at the number part.

Example 10

Put these numbers in order of size, starting with the smallest.

$3.45 \times 10^4$      $34 \times 10^3$      $349\ 000$      $15 \times 2.3 \times 10^4$

First write all the numbers in the same format. You could write them out in full or write them all in standard form. Standard form is neater.

$3.45 \times 10^4$

$34 \times 10^3 = 3.4 \times 10^4$

$349\ 000 = 3.49 \times 10^5$

$15 \times 2.3 \times 10^4 = 34.5 \times 10^4 = 3.45 \times 10^5$

Then compare the powers of 10. There are two with power 4 and two with power 5, so compare the number parts in each pair.

$3.4 \times 10^4$      $3.45 \times 10^4$      $3.45 \times 10^5$      $3.49 \times 10^5$

## Exercise 25D

**1**   Write down the value of each expression.

    **a** $3.1 \times 0.1$      **b** $3.1 \times 0.01$      **c** $3.1 \times 0.001$      **d** $3.1 \times 0.0001$

**2**   Write down the value of each expression.

    **a** $6.5 \times 10^{-1}$      **b** $6.5 \times 10^{-2}$      **c** $6.5 \times 10^{-3}$      **d** $6.5 \times 10^{-4}$

 **3**   Find out what your calculator can do.

    **a** What is the largest number you can enter into your calculator?

    **b** What is the smallest number you can enter into your calculator?

**4**   Work out the value of each expression.

    **a** $3.1 \div 0.1$      **b** $3.1 \div 0.01$      **c** $3.1 \div 0.001$      **d** $3.1 \div 0.0001$

**5**   Work out the value of each expression.

    **a** $6.5 \div 10^{-1}$      **b** $6.5 \div 10^{-2}$      **c** $6.5 \div 10^{-3}$      **d** $6.5 \div 10^{-4}$

**6**   Write these numbers out in full.

    **a** $2.5 \times 10^2$      **b** $3.45 \times 10$      **c** $4.67 \times 10^{-3}$      **d** $3.46 \times 10$

    **e** $2.0789 \times 10^{-2}$      **f** $5.678 \times 10^3$      **g** $2.46 \times 10^2$      **h** $7.6 \times 10^3$

    **i** $8.97 \times 10^5$      **j** $8.65 \times 10^{-3}$      **k** $6 \times 10^7$      **l** $5.67 \times 10^{-4}$

**7**   Write each number in standard form.

    **a** 250                **b** 0.345             **c** 46 700

    **d** 3 400 000 000     **e** 20 780 000 000     **f** 0.000 567 8

    **g** 2460             **h** 0.076            **i** 0.000 76

    **j** 0.999            **k** 234.56          **l** 98.7654

    **m** 0.0006          **n** 0.005 67       **o** 56.0045

For questions **8** to **10**, write each of the given numbers in standard form.

**8** One year, 27 797 runners completed the New York marathon.

**9** The largest number of dominoes ever toppled by one person is 281 581, although 30 people set up and toppled 1 382 101.

**10** The asteroid *Phaethon* comes within 12 980 000 miles of the Sun, whilst the asteroid *Pholus*, at its furthest point, is a distance of 2997 million miles from Earth. In 2011 a small asteroid came within 74 000 miles of Earth.

**11** These numbers are not in standard form. Write them in standard form.

**a** $56.7 \times 10^2$      **b** $0.06 \times 10^4$      **c** $34.6 \times 10^{-2}$

**d** $0.07 \times 10^{-2}$      **e** $56 \times 10$      **f** $2 \times 3 \times 10^5$

**g** $2 \times 10^2 \times 35$      **h** $160 \times 10^{-2}$      **i** 23 million

**j** $0.0003 \times 10^{-2}$      **k** $25.6 \times 10^5$      **l** $16 \times 10^2 \times 3 \times 10^{-1}$

**m** $2 \times 10^4 \times 56 \times 10^{-4}$      **n** $(18 \times 10^2) \div (3 \times 10^3)$      **o** $(56 \times 10^3) \div (2 \times 10^{-2})$

**12** Work these out. Give your answers in standard form.

**a** $2 \times 10^4 \times 5.4 \times 10^3$      **b** $1.6 \times 10^2 \times 3 \times 10^4$      **c** $2 \times 10^4 \times 6 \times 10^4$

**d** $2 \times 10^{-4} \times 5.4 \times 10^3$      **e** $1.6 \times 10^{-2} \times 4 \times 10^4$      **f** $2 \times 10^4 \times 6 \times 10^{-4}$

**g** $7.2 \times 10^{-3} \times 4 \times 10^2$      **h** $(5 \times 10^3)^2$      **i** $(2 \times 10^{-2})^3$

**13** A typical adult has about 20 000 000 000 000 red corpuscles. Each red corpuscle has a mass of about 0.000 000 000 1 g.

**a** Write both of these numbers in standard form.

**b** Multiply the two numbers to calculate the total mass of red corpuscles in a typical adult.

**14** Work out the value of $\dfrac{E}{M}$ when $E = 1.5 \times 10^3$ and $M = 3 \times 10^{-2}$. Give your answer in standard form.

**15** Put these numbers in order of size, starting with the smallest.

$79 \times 10^3$      $3.7 \times 2.1 \times 10^4$      7800      $7.85 \times 10^4$

**16** Put these numbers in order of size, starting with the smallest.

0.0005      $6 \times 10^{-2}$      $4 \times 10^{-3}$      $0.52 \times 10^{-2}$

**17** Use your calculator to work these out.

Give your answer in standard form, correct to 2 significant figures.

**a** $5.7 \times 10^8 \div 3.2 \times 10^4$      **b** $6 \times 10^9 \div 3.6 \times 10^2$

**c** $7 \times 10^{11} \div 8 \times 10^3$      **d** $5.6 \times 10^8 \div 3 \times 10^3$

**e** $2.4 \times 10^{12} \div 3.5 \times 10^5$      **f** $1.64 \times 10^{-2} \div 2.3 \times 10^6$

**g** $3 \times 10^{-2} \div 2.5 \times 10^5$      **h** $1.6 \times 10^2 \div 3 \times 10^{-7}$

**18** **a** Write 1 in standard form.

**b** Write 10 in standard form.

**c** Give a reason why you would not normally use standard form to do this calculation.

$2 \times 10^1 \times 3 \times 10^2$

# Worked exemplars

  **a** Which of these is *not* equivalent to $3 \times 10^2 \times 6 \times 10^3$?

$18 \times 10^5$     $1.8 \times 10^6$     $180\,000$     $1\,800\,000$

**b** Arrange these expressions in order of size, starting with the smallest.

$6.3 \times 10^3$     $64 \times 10^2$     $6250$     $130\,000 \div 20$

| This is a problem-solving question, so you need to show a clear strategy in your solution, and show all the steps in your answer. | |
|---|---|
| **a** $3 \times 10^2 \times 6 \times 10^3$ <br><br> $= (3 \times 6) \times (10^2 \times 10^3)$ <br><br> $= 18 \times 10^5$ <br><br> $= 1\,800\,000$ <br><br> $= 1.8 \times 10^6$ <br><br> So, $180\,000$ is not equivalent to $3 \times 10^2 \times 6 \times 10^3$. | Work out the value of $3 \times 10^2 \times 6 \times 10^3$ and then compare it to the values given. |
| **b** $6.3 \times 10^3 = 6300$ <br><br> $64 \times 10^2 = 6400$ <br><br> $6250$ <br><br> $130\,000 \div 20 = 13\,000 \div 2$ <br><br> $\qquad\qquad\qquad = 6500$ <br><br> So, in order: <br><br> $6250, 6.3 \times 10^3, 64 \times 10^2, 130\,000 \div 20$ | Work out the value of each expression in the same format (either an ordinary number or standard form). The ordinary numbers do not have that many digits, so this is the more straightforward format in this case. |

  This is a table of powers of 3.

| $3^1$ | $3^2$ | $3^3$ | $3^4$ | $3^5$ | $3^6$ | $3^7$ |
|---|---|---|---|---|---|---|
| 3 | 9 | 27 | 81 | 243 | 729 | 2187 |

**a** Use your calculator to work out $27 \div 243$. Give the answer as a fraction.

**b** Use the rules of indices to write $3^3 \div 3^5$ as a single power of 3.

**c** Deduce the value, as a fraction, of $3^{-3}$.

| This is a mathematical reasoning question. The first two parts set up the information you will need. Make sure that you show your steps and use connecting words or symbols such as $\Rightarrow$ (implies) and $\therefore$ (therefore). | |
|---|---|
| **a** $27 \div 243 = \frac{1}{9}$ | Enter $27 \div 243$ as a fraction. This should cancel to the simplest form. <br><br> Make sure you know how to change an answer into a fraction if the display shows a decimal, in this case $0.111\ldots$ |
| **b** $3^3 \div 3^5 = 3^{3\,-\,5}$ <br><br> $= 3^{-2}$ | Apply the rules of indices. When dividing powers with the same base, subtract them. |
| **c** Since $3^3 = 27$ and $3^5 = 243$, parts **a** and **b** $\Rightarrow \frac{1}{9} = 3^{-2}$ <br><br> $\therefore 3^{-3} = \frac{1}{27}$ | This is the mathematical reasoning section. Parts **a** and **b** are linked in that they are the same calculation in different forms, so the answers must be the same. So if $\frac{1}{9} = 3^{-2}$, then $3^{-3}$ must be $\frac{1}{27}$. |

# Ready to progress?

I can multiply and divide by powers of 10.
I can write and calculate with numbers written in index form.
I can multiply and divide numbers written in index form.

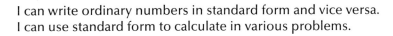

I can write ordinary numbers in standard form and vice versa.
I can use standard form to calculate in various problems.

# Review questions

**1** Work out the value of each power term.

   **a** $2^6$      **b** $4^3$      **c** $10^7$

**2** Work out the value of each expression.

   **a** $6.2 \times 10$    **b** $4.9 \times 1000$    **c** $34.7 \div 10$    **d** $123 \div 1000$

**3** Write down the answer to each of these.

   **a** $200 \times 30$    **b** $50 \times 6000$    **c** $3200 \div 40$    **d** $8000 \div 200$

**4** **a** Write down the value of $14^2$.

   **b** Explain how you know that $35^2$ is not equal to 1220.

**5** Simplify each expression.

   **a** $7^4 \times 7^5$      **b** $x^8 \div x^4$

   **c** Sammi writes: $3x^2 \times 5x^7 = 15x^{14}$

   Explain the mistakes he has made.

   Write down the correct answer to $3x^2 \times 5x^7$.

**6** Simplify each expression.

   **a** $t^5 \times t^3$      **b** $\dfrac{m^8}{m^3}$      **c** $(3x^3)^2$      **d** $2a^2h \times 5a^5h^4$

**7** Simplify each expression.

   **a** $x^5 \times x^6$      **b** $\dfrac{m}{m^6}$      **c** $(2k^3m^2) \times (4k^2m)$

**8** **a** Write the number 75 000 in standard form.

   **b** Write $9 \times 10^{-3}$ as an ordinary number.

**9** Simplify each expression. **a** $\dfrac{4x^3y^2 \times 3xy^2}{6x^4y^3}$    **b** $(2m^3p^4)^3$

**10** **a** Work out the value of: $(4 \times 10)^2 - 4 \times 10^2$.

   **b** Show that $a \times 10^b = (a \times 10)^b$ only when $a = 1$ or $b = 1$.

**11** $n$ is a whole number. When $n \times 10^n$ is written as an ordinary number, it has four digits. What is the value of $n$?

# 26 Algebra: Simultaneous equations and linear inequalities

## This chapter is going to show you:

- how to solve linear simultaneous equations algebraically
- how to solve a linear inequality and represent the solution on a number line.

## You should already know:

- the basic language of algebra
- how to collect together like terms
- how to solve basic linear equations
- how to substitute into formulae.

## About this chapter

In many real-world situations, equations have more than one variable. Some examples are mobile phone tariffs and systems to avoid vehicle collisions. These equations are called simultaneous equations because there are two or more variables 'at the same time'.

If I buy three cakes and two buns for £8, it is not possible to work out how much one cake or one bun costs. There is more than one answer. But if you also buy one cake and four buns for £6, then there is enough information to find the price of each one.

If my three cakes and two buns cost £8, then six cakes and four buns would cost twice as much, £16. I would have bought the same number of buns as you but five more cakes and I would have spent £10 more.

So, five cakes cost £10 and, therefore, one cake costs £2. I can then work out that one bun costs £1.

# 26.1 Elimination method for simultaneous equations

This section will show you how to:

**Key term**

eliminate

- solve simultaneous linear equations in two variables using the elimination method.

A pair of simultaneous equations is two equations for which you want the same solution and so you must solve them together.

For example, $x + y = 10$ has many solutions: $\quad\quad x = 2, y = 8 \quad\quad x = 4, y = 6 \quad\quad x = 5, y = 5 \dots$

and $2x + y = 14$ has many solutions: $\quad\quad x = 2, y = 10 \quad\quad x = 3, y = 8 \quad\quad x = 4, y = 6 \dots$

but only one solution satisfies both equations at the same time: $x = 4$ and $y = 6$.

The elimination method is one way to solve simultaneous equations. It has six steps.

**Step 1:** <u>Balance</u> the coefficients of one of the variables.

**Step 2:** **<u>Eliminate</u>** this variable by adding or subtracting the equations.

**Step 3:** <u>Solve</u> the linear equation you get using the other variable.

**Step 4:** <u>Substitute</u> the value you found back into one of the previous equations.

**Step 5:** <u>Solve</u> the equation you get.

**Step 6:** <u>Check</u> that the two values you found satisfy the original equations.

---

**Example 1**

Solve this pair of simultaneous equations. $\quad\quad 6x + y = 15$ and $4x + y = 11$

First label the equations so that you can clearly explain the method.

$$6x + y = 15 \quad\quad (1)$$
$$4x + y = 11 \quad\quad (2)$$

**Step 1:** Since the $y$-term in both equations has the same coefficient there is no need to balance them. (You will learn how to do this later in the chapter.)

**Step 2:** The coefficients of $y$ have the same sign, so subtract one equation from the other. In this case, subtracting equation (2) from equation (1) gives positive values.

$$6x + y = 15 \quad\quad (1) \quad\quad -$$
$$4x + y = 11 \quad\quad (2)$$
$$2x = 4 \quad\quad (1) - (2)$$

**Step 3:** Solve this equation. $\quad\quad x = 2$

**Step 4:** Substitute $x = 2$ into equation (2), as this has smaller numbers than equation (1).

$$4x + y = 11$$
$$4 \times 2 + y = 11$$
$$8 + y = 11$$

**Step 5:** Solve this equation. $\quad\quad y = 3$

**Step 6:** Check the solution by substituting in the original equations.

Substituting $x = 2$ and $y = 3$ into $6x + y$ gives $12 + 3 = 15$.

Substituting $x = 2$ and $y = 3$ into $4x + y$ gives $8 + 3 = 11$.

These are correct, so you can confidently say the solution is $x = 2$ and $y = 3$.

Example 2

Solve this pair of simultaneous equations.     $3p + 2q = 29$ and $5p - 2q = 27$

**Step 1:** Since the $q$-term in both equations has the same coefficient, there is no need to balance them.

**Step 2:** The coefficients of $q$ have different signs, so add the equations.

$$3p + 2q = 29 \qquad (1) \qquad +$$
$$5p - 2q = 27 \qquad (2)$$
$$8p \quad\;\; = 56 \qquad (1) + (2)$$

**Step 3:** Solve this equation.     $p \quad = 7$

**Step 4:** Substitute $p = 7$ into equation (1).     $3p + 2q = 29$
$$3 \times 7 + 2q = 29$$
$$21 + 2q = 29$$

**Step 5:** Solve this equation.     $2q = 8$
$$q = 4$$

**Step 6:** Check the solution by substituting in the original equations.

$$3 \times 7 + 2 \times 4 = 29$$
$$5 \times 7 - 2 \times 4 = 27$$

## Exercise 26A

> **Hints and tips**  Remember the key words:
> balance, eliminate, solve, substitute, solve, check.

 Solve each pair of simultaneous equations by the elimination method.

**a** $x + 3y = 9$
$x + y = 6$

**b** $2x + 5y = 16$
$2x + 3y = 8$

**c** $3x - y = 9$
$5x + y = 11$

**d** $4x + y = 22$
$4x + 3y = 42$

**e** $3x + 2y = 23$
$7x - 2y = 67$

**f** $10x + 3y = 17$
$6x + 3y = 15$

 Solve each pair of simultaneous equations by the elimination method.

**a** $3a + b = 31$
$3a + 5b = 71$

**b** $7c + d = 39$
$7c - d = 17$

**c** $5e - 2f = 19$
$e + 2f = 11$

**d** $8g + 5h = 22$
$6g + 5h = 24$

**e** $j + 4k = 21$
$j - 3k = 7$

**f** $4m - n = 13$
$12m - n = 25$

 Solve this pair of simultaneous equations.

$19x - 15y = 198$
$8x - 15y = 66$

# 26.2 Substitution method for simultaneous equations

This section will show you how to:

- solve simultaneous linear equations in two variables using the substitution method.

The substitution method is an alternative way of solving simultaneous equations. The method you use depends on the coefficients in the equation and the way the equations are written. There are five steps in this method.

**Step 1:** Rearrange one of the equations into the form $y = \ldots$ or $x = \ldots$

**Step 2:** Substitute the right-hand side of the rearranged equation into the other equation, in place of the variable on the left-hand side.

**Step 3:** Expand and solve this equation.

**Step 4:** Substitute the value into the $y = \ldots$ or $x = \ldots$ equation.

**Step 5:** Check that the values work in the original equations.

---

**Example 3**

Solve this pair of simultaneous equations.  $y = 2x + 3$ and $3x + 4y = 1$

The first equation is in the form $y = \ldots$, so this suggests that the substitution method should be used.

Label the equations to help explain the method.

$$y = 2x + 3 \qquad (1)$$
$$3x + 4y = 1 \qquad (2)$$

**Step 1:** As equation (1) is in the form $y = \ldots$ there is no need to rearrange an equation.

**Step 2:** Substitute the right-hand side of equation (1) into equation (2) for the variable $y$.

$$3x + 4(2x + 3) = 1$$

**Step 3:** Expand and solve the equation.

$$3x + 8x + 12 = 1$$
$$11x = -11$$
$$x = -1$$

**Step 4:** Substitute $x = -1$ into equation (1).

$$y = 2x + 3$$
$$y = 2 \times -1 + 3$$
$$y = -2 + 3$$
$$= 1$$

**Step 5:** Check the solutions by substituting $x = -1$ and $y = -1$ into the original equations.

Substituting into $y = 2x + 3$ gives $1 = -2 + 3$.

Substituting into $3x + 4y = 1$ gives $-3 + 4 = 1$.

These are correct, so the solution is $x = -1$ and $y = 1$.

**1** Solve each pair of simultaneous equations by the substitution method.

  **a** $3x + 7y = 13$
  $y = x - 11$

  **b** $2x + y = 6$
  $y = 4x + 3$

  **c** $4x - 3y = 18$
  $y = x - 7$

  **d** $2x - 5y = 6$
  $y = x + 6$

  **e** $7x - 2y = 11$
  $y = 2x - 1$

  **f** $3x + y = 8$
  $y = 3x + 5$

**2** Solve each pair of simultaneous equations by the substitution method.

  **a** $2x + 5y = 37$
  $y = 11 - 2x$

  **b** $4x - 3y = 7$
  $x = 13 - 3y$

  **c** $4x - y = 17$
  $x = 2 + y$

  **d** $x + 2y = 14$
  $y = 17 - 3x$

  **e** $9x - 2y = 105$
  $x = y + 14$

  **f** $6y - x = 19$
  $x = 1 - 4y$

**3** Solve each pair of simultaneous equations by either elimination or substitution.

  **a** $3x + 11y = 36$
  $y = 5 - 2x$

  **b** $8x + y = 49$
  $8x - 3y = 13$

# 26.3 Balancing coefficients to solve simultaneous equations

This section will show you how to:

- solve simultaneous linear equations by balancing coefficients.

You were able to solve the pairs of equations in Examples **1**, **2** and **3** by adding or subtracting the equations in each pair, or by substituting without rearranging. This does not always happen. The next examples show what to do when there are no identical terms to begin with, or when you need to rearrange.

In this example, you start by balancing the coefficients of one of the variables, and then use the elimination method as before.

Solve this pair of simultaneous equations.

  $5x + 2y = 24$ (1)
  $2x - y = 6$ (2)

**Step 1:** Neither $x$ nor $y$ has the same coefficient in both equations. You need to balance one of the variables. If you choose to balance the $y$-coefficients, you only need to multiply one equation. Multiply the second equation by 2 and label it with a new equation number.

  $5x + 2y = 24$ (1)
  $4x - 2y = 12$ (3)

**Step 2:** The signs are different, so add the two equations to eliminate the $y$-terms.

  (1) + (3)      $9x = 36$

**Step 3:** Solve this equation.      $x = 4$

**Step 4:** Substitute $x = 4$ into one of the original equations.

$$5x + 2y = 24$$
$$5 \times 4 + 2y = 24$$
$$20 + 2y = 24$$

**Step 5:** Solve this equation. $\qquad y = 2$

**Step 6:** Check the solution by substituting $x = 4$ and $y = 2$ into the original equations.

Substituting into (1) gives $20 + 4 = 24$ and substituting into (2) gives $8 - 2 = 6$. These are correct, so the solution is $x = 4$ and $y = 2$.

## Balancing coefficients in both equations

When you have a pair of simultaneous equations in which neither coefficient is a factor of the other, you need to change both equations to get identical terms. You do this by finding the lowest common multiple and then multiplying both equations, as shown in the next example.

**Note:** The substitution method is not suitable for these types of equations, as you end up with fractional terms.

---

**Example 4**

Solve this pair of simultaneous equations.

$$4x + 3y = 27 \quad (1)$$
$$5x + 2y = 25 \quad (2)$$

Both equations have to be changed to obtain identical terms in either $x$ or $y$.

You can make either the $x$- or $y$-coefficients the same. Since the $y$-coefficients are smaller, solving the equation will be more straightforward if you make these the same.

**Step 1:** Multiply equation (1) by 2 (the $y$-coefficient of equation (2)).

$$2 \times (1) \text{ or } 2 \times (4x + 3y = 27) \quad \rightarrow \quad 8x + 6y = 54 \quad (3)$$

Multiply equation (2) by 3 (the $y$-coefficient of the equation (1)).

$$3 \times (2) \text{ or } 3 \times (5x + 2y = 25) \quad \rightarrow \quad 15x + 6y = 75 \quad (4)$$

Label the new equations (3) and (4).

It is helpful to write the equations so that the one with the larger coefficient comes first.

**Step 2:** Eliminate one of the variables. $(4) - (3) \qquad 7x = 21$

**Step 3:** Solve the equation. $\qquad x = 3$

**Step 4:** Substitute $x = 3$ into equation (1).
$$4 \times 3 + 3y = 27$$
$$12 + 3y = 27$$

**Step 5:** Solve the equation. $\qquad y = 5$

**Step 6:** Check by substituting. (1) $\quad 4 \times 3 + 3 \times 5 = 12 + 15 = 27$

$\qquad\qquad\qquad\qquad\qquad$ (2) $\quad 5 \times 3 + 2 \times 5 = 15 + 10 = 25$

These are correct, so the solution is $x = 3$ and $y = 5$.

---

## Exercise 26C

**1** Solve each pair of simultaneous equations.

**a** $2x + 3y = 19$
$\quad 6x + 2y = 22$

**b** $5x - 2y = 26$
$\quad 3x - y = 15$

**c** $10x - y = 3$
$\quad 3x + 2y = 17$

**d** $5x - 2y = 4$
$\quad 3x - 6y = 6$

**e** $2x + 3y = 13$
$\quad 4x + 7y = 31$

**f** $3x - 2y = 3$
$\quad 5x + 6y = 12$

**2** Solve each pair of simultaneous equations.

**a** $2x + 5y = 15$
$3x - 2y = 13$

**b** $2x + 3y = 30$
$5x + 7y = 71$

**c** $2x - 3y = 15$
$5x + 7y = 52$

**d** $3x - 2y = 15$
$2x - 3y = 5$

**e** $5x - 3y = 14$
$4x - 5y = 6$

**f** $3x + 2y = 28$
$2x + 7y = 47$

**3** Solve each pair of simultaneous equations.

**a** $5a + 6b = 53$
$7a - 4b = 68$

**b** $2c + 3d = 9$
$c + 6d = 27$

**c** $8e + 3f = 50$
$e + 2f = 29$

**d** $g - 3h = 13$
$9g - 5h = 7$

**e** $2m - 7n = 16$
$3m + n = 24$

**f** $8p + 3q = 22$
$4p - 5q = 24$

**g** $8r - 2s = 49$
$4r + 3s = 49$

**h** $3t - 7u = 9$
$9t - 4u = 10$

**i** $3v + 2w = 232$
$2v - w = 80$

# 26.4 Using simultaneous equations to solve problems

This section will show you how to:

- solve problems using simultaneous linear equations.

You need to write some types of problem as a pair of simultaneous equations in order to solve them. The next example shows you how to do this.

---

**Example 5**

At a café, three teas and four coffees cost £9. Five teas and two coffees cost £8. How much would one tea and one coffee cost?

Write a pair of simultaneous equations from the information given.

Let $t$ be the cost of a tea and $c$ be the cost of a coffee. Then:

$$3t + 4c = 9 \qquad (1)$$
$$\text{and} \qquad 5t + 2c = 8 \qquad (2)$$

Now solve these equations just as you have done in the previous examples.

**Step 1:** Multiply equation (2) by 2 to get the same number of $c$s in each equation.

$$10t + 4c = 16 \qquad (3) \qquad \text{(Remember to label the new equation.)}$$

**Step 2:** Subtract equation (1) from equation (3). $7t = 7$

**Step 3:** Solve the equation. $\qquad\qquad\qquad\qquad\qquad t = 1$

**Step 4:** Substitute $t = 1$ into equation (1). $\qquad 3 \times 1 + 4c = 9$

**Step 5:** Solve the equation. $\qquad\qquad\qquad\qquad 3 + 4c = 9$

$$4c = 6$$
$$c = 1.5$$

**Step 6:** Check answers in both equations.

(1) $\qquad 3 \times 1 + 4 \times 1.5 = 3 + 6 = 9$

(2) $\qquad 5 \times 1 + 2 \times 1.5 = 5 + 3 = 8$

These are correct, so the solution is $t = 1$ and $c = 1.5$.

So one tea and one coffee would cost £1 + £1.50 = £2.50.

---

Example 6

Two families went to the theatre but couldn't remember how much they paid for each adult or each child ticket. They could remember what they had paid altogether.

Mr and Mrs Advani and their daughter, Rupa, paid £42.

Mrs Shaw and her two children, Len and Sue, paid £39.

How much would I have to pay for my wife, my four children and myself?

Write a pair of simultaneous equations from the information given.

Let $x$ be the cost of an adult ticket, and $y$ be the cost of a child ticket. Then:

$$\text{For the Advani family: } 2x + y = 42 \qquad (1)$$
$$\text{For the Shaw family: } \quad x + 2y = 39 \qquad (2)$$

Now solve these equations just as you have done in the previous examples.

**Step 1:** Multiply equation (1) by 2. $\qquad\qquad 4x + 2y = 84 \quad (3)$

**Step 2:** Subtract equation (2) from equation (3). $\qquad 3x = 45$

**Step 3:** Solve the equation. $\qquad\qquad\qquad\qquad x = 15$

**Step 4:** Substitute $x = 15$ into equation (1). $\qquad 2 \times 15 + y = 42$

**Step 5:** Solve the equation. $\qquad\qquad\qquad\qquad 30 + y = 42$

$$y = 12$$

So, $x = £15$ and $y = £12$.

**Step 6:** Check answers in both equations.

$\qquad$ (1) $\quad 2 \times 15 + 12 = 30 + 12 = 42$

$\qquad$ (2) $\quad 15 + 2 \times 12 = 15 + 24 = 39$

These are correct, so the solution is $x = 15$ and $y = 12$.

You can now work out the cost for my family, which will be $(2 \times £15) + (4 \times £12) = £78$.

## Exercise 26D

  **1** In this sequence, the next term is found by multiplying the previous term by $a$ and then adding $b$, where $a$ and $b$ are positive whole numbers.

$\qquad$ 3 $\qquad$ 14 $\qquad$ 47 $\qquad$ … $\qquad$ …

An equation that links the first and second terms is $3a + b = 14$.

**a** Set up another equation in $a$ and $b$.

**b** Solve the equations to work out $a$ and $b$.

**c** Work out the next two terms in the sequence.

 **2** In a tea shop it costs £8.10 for three teas and five buns.

In the same tea shop it costs £6.30 for three teas and three buns.

**a** Write a pair of simultaneous equations to represent the above information. Use $t$ to represent the cost of a tea and $b$ to represent the cost of a bun.

**b** What is the price of: **i** one tea $\qquad$ **ii** one bun?

**c** How much will I pay for four teas and six buns?

**(PS) 3** Two people bought stamps at the Post Office. One person bought 10 second-class and 5 first-class stamps at a total cost of £8.40. The other bought 8 second-class and 10 first-class stamps at a total cost of £10.44.

**a** Let $x$ be the cost of a second-class stamp and $y$ be the cost of a first-class stamp. Set up two simultaneous equations to represent the information given.

**b** How much did I pay for three second-class and four first-class stamps?

**(PS) 4** Here are four equations.

A: $5x + 2y = 1$    B: $4x + y = 9$    C: $3x - y = 5$    D: $3x + 2y = 3$

Here are four sets of $(x, y)$ values.

$(1, -2)$, $(-1, 3)$, $(2, 1)$, $(3, -3)$

Match each pair of $(x, y)$ values to a pair of equations.

> **Hints and tips** You could solve each possible set of pairs but there are six to work out. Alternatively, you can substitute values into the equations to see which work.

**(PS) 5** Three chews and four bubblies cost 72p. Five chews and two bubblies cost 64p. What would three chews and five bubblies cost?

**(PS) 6** On a production line, all the nuts produced have the same mass and all the bolts produced have the same mass. An order of 50 nuts and 60 bolts has a mass of 10.6 kg. An order of 40 nuts and 30 bolts has a mass of 6.5 kg. What should the mass of an order of 60 nuts and 50 bolts be?

**(PS) 7** Four sacks of potatoes and two sacks of carrots weigh 188 lb.

Five sacks of potatoes and one sack of carrots weigh 202 lb.

Baz buys seven sacks of potatoes and eight sacks of carrots.

Will he be able to carry them in his trailer, which can safely carry 450 lb?

> **Hints and tips** Set up two simultaneous equations using $p$ and $c$ for the mass of a sack of potatoes and carrots, respectively.

**(PS) 8** My local taxi company charges a fixed amount plus a certain amount for each mile. When I took a six-mile journey the cost was £3.70. When I took a ten-mile journey the cost was £5.10. My next journey is going to be eight miles. How much will this cost?

**(PS) 9** Five bags of bark and four trays of pansies cost £24.50.

One bag of bark and five trays of pansies cost £12.25.

Camilla wants six bags of bark and eight trays of pansies.

She has £30. Will she have enough money?

**(PS) 10** The sum of my son's age and my age this year is 72.

Six years ago my age was double that of my son.

Let my age now be $x$ and my son's age now be $y$.

**a** Show that $x - 6 = 2(y - 6)$.

**b** Work out the values of $x$ and $y$.

**11** Amul and Kim have £10.70 between them. Amul has £3.70 more than Kim. Let $x$ be the amount Amul has and $y$ be the amount Kim has. Set up a pair of simultaneous equations. How much does each have?

  **12** **a** Mary is solving the simultaneous equations $4x - 2y = 8$ and $2x - y = 4$.

She works out a solution of $x = 5$, $y = 6$ that works for both equations.

Why isn't this a unique solution?

**b** Max is solving the simultaneous equations $6x + 2y = 9$ and $3x + y = 7$.

Why is it impossible to have a solution that works for both equations?

# 26.5 Linear inequalities

## This section will show you how to:

- solve a simple linear inequality and represent it on a number line.

**Inequalities** behave similarly to equations: you use the same rules to solve linear inequalities as you use to solve linear equations. There are four inequality signs:

< which means 'less than'

> which means 'greater than'

≤ which means 'less than or equal to'

≥ which means 'greater than or equal to'.

**Key terms**

inclusive inequality

inequality

strict inequality

**Be careful:** Never replace the inequality sign with an equals sign.

Solve the inequality $2x < 11$.

Divide both sides by 2.
$$\frac{2x}{2} < \frac{11}{2}$$
$$\Rightarrow x < 5.5$$

The symbol ⇒ is a short way of writing 'which means that'.

This means that $x$ can take any value below 5.5 but not the value 5.5.

< and > are called **strict inequalities**.

**Note:** Use the same inequality sign in the answer as in the question.

 **Example 7**

Janet said: 'If I were four years older than half my age, I'd still be at least 13 years old.' How old must Janet be? Use $x$ for Janet's age.

Write this information as an inequality, using $x$ for Janet's current age. $\frac{x}{2} + 4 \geq 13$

Solve this inequality in the same way as an equation, but leave the inequality sign in place of the equals sign.

Subtract 4 from both sides. $\frac{x}{2} \geq 9$

Multiply both sides by 2. $x \geq 18$

This means that $x$ can take any value above and including 18.

Janet is at least 18 years old but could be older.

≤ and ≥ are called **inclusive inequalities**.

Example 8

Show that if $14 > \dfrac{3x + 7}{2}$, then $x$ must be below 7.

Rewrite this as $\qquad\qquad\qquad \dfrac{3x + 7}{2} < 14$

Multiply both sides by 2. $\qquad\quad 3x + 7 < 28$

Subtract 7 from both sides. $\qquad 3x < 21$

Divide both sides by 3. $\qquad\qquad x < 7$

Example 9

**a** Solve this inequality. $-5 < 3x + 4 \leqslant 13$

**b** State all the integers that solve the inequality.

**a** Divide the inequality into two parts, and treat each part separately.

$-5 < 3x + 4 \qquad\qquad\qquad\qquad 3x + 4 \leqslant 13$

$\Rightarrow \qquad -9 < 3x \qquad\qquad\qquad \Rightarrow \qquad\quad 3x \leqslant 9$

$\Rightarrow \qquad -3 < x \qquad\qquad\qquad\; \Rightarrow \qquad\quad\; x \leqslant 3$

So $-3 < x \leqslant 3$

**b** $x$ can be $-2, -1, 0, 1, 2$ or $3$.

## Exercise 26E

**1** Solve each linear inequality.

**a** $x + 4 < 7$ $\qquad\qquad$ **b** $t - 3 > 5$ $\qquad\qquad$ **c** $p + 2 > 12$

**d** $2x - 3 < 7$ $\qquad\qquad$ **e** $4y + 5 < 17$ $\qquad\quad$ **f** $3t - 4 > 11$

**g** $\dfrac{x}{2} + 4 < 7$ $\qquad\qquad$ **h** $\dfrac{y}{5} + 3 < 6$ $\qquad\quad$ **i** $\dfrac{t}{3} - 2 > 4$

**j** $3(x - 2) < 15$ $\qquad\quad$ **k** $5(2x + 1) < 35$ $\qquad$ **l** $2(4t - 3) > 34$

**2** Write down the largest integer value of $x$ that satisfies each of the following.

**a** $x - 3 < 5$, where $x$ is a positive integer

**b** $x + 2 < 9$, where $x$ is a positive even integer

**c** $3x - 11 < 40$, where $x$ is a square number

**d** $5x - 8 < 15$, where $x$ is a positive odd number

**e** $2x + 1 < 19$, where $x$ is a positive prime number

**3** Write down the smallest integer value of $x$ that satisfies each of the following.

**a** $x - 2 \geqslant 9$, where $x$ is a positive integer

**b** $x - 2 > 13$, where $x$ is a positive even integer

**c** $2x - 11 > 19$, where $x$ is a square number

**d** $3x + 7 > 15$, where $x$ is a positive odd number

**e** $4x - 1 > 23$, where $x$ is a positive prime number

**4** Ahmed went to town with £20 to buy two CDs. His bus fare was £3. The CDs were both the same price. When he reached home he still had some money in his pocket. What was the most Ahmed paid for each CD?

> **Hints and tips** Set up an inequality and solve it. Use $x$ for the price of a CD.

**5** **a** Why can't you make a triangle with three sides of length 3 cm, 4 cm and 8 cm?

**b** Three sides of a triangle are $x$, $x + 2$ and 10 cm.

$x$ is a whole number.

What is the smallest value that $x$ can take?

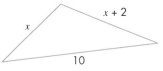

**6** Five cards have inequalities or equations marked on them.

| $x > 0$ | $x < 3$ | $x \geqslant 4$ | $x = 2$ | $x = 6$ |

The cards are shuffled and then turned over, one at a time.

If two consecutive cards have any numbers in common, then a point is scored.

If they do not have any numbers in common, then a point is lost.

**a** The first two cards below score –1 because $x = 6$ and $x < 3$ have no numbers in common.

> **Hints and tips** The next two cards to compare are $x < 3$ and $x > 0$.

Show that the total for this combination of scores is 0.

| $x = 6$ | $x < 3$ | $x > 0$ | $x = 2$ | $x \geqslant 4$ |

**b** What does this combination score?

| $x > 0$ | $x = 6$ | $x \geqslant 4$ | $x = 2$ | $x < 3$ |

**c** Arrange the cards to give a maximum score of 4.

**7** Solve each linear inequality.

**a** $3y - 12 \leqslant y - 4$     **b** $2x + 3 \geqslant x + 1$     **c** $2(4x - 1) \leqslant 3(x + 4)$

**d** $\dfrac{x - 3}{5} > 7$     **e** $\dfrac{2x + 5}{3} < 6$     **f** $\dfrac{5y + 3}{5} \leqslant 2$

**8** Solve these linear inequalities.

**a** $7 < 2x + 1 < 13$     **b** $5 < 3x - 1 < 14$     **c** $-1 < 5x + 4 \leqslant 19$

**d** $1 \leqslant 4x - 3 < 13$     **e** $11 \leqslant 3x + 5 < 17$     **f** $-3 \leqslant 2x - 3 \leqslant 7$

**9** Meg bought seven crates of pineapple juice and Arthur bought four crates of pineapple juice.

Each crate contained the same number of bottles of pineapple juice.

When Meg gave ten bottles of juice to Arthur, Arthur then had more bottles of juice than Meg.

Calculate the maximum number of bottles of pineapple juice in a crate.

## The number line

You can show the solution to a linear inequality on a number line in the following ways.

A strict inequality does not include the boundary point but an inclusive inequality does include the boundary point.

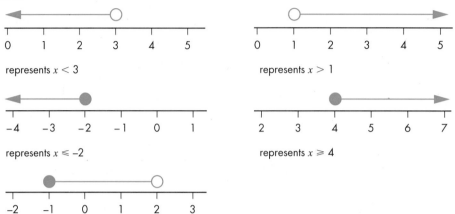

represents $x < 3$

represents $x > 1$

represents $x \leqslant -2$

represents $x \geqslant 4$

represents $-1 \leqslant x < 2$

The last example is a 'between' inequality. It can be written as $x \geqslant -1$ and $x < 2$, but the notation $-1 \leqslant x < 2$ is neater. Notice the direction of the inequality sign when you swap over the terms from $x$ [the sign] $-1$ to $-1$ [the other sign] $x$.

**Example 10**

**a** Write down the inequality shown by this diagram.

**b i** Solve the inequality $2x + 3 < 11$.

   **ii** Mark the solution on a number line.

**c** Write down the integers that satisfy both the inequalities in **a** and **b**.

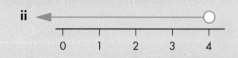

**a** The inequality shown is $x \geqslant 1$.

**b i** $2x + 3 < 11$

    $\Rightarrow 2x < 8$

    $\Rightarrow x < 4$

**ii**

**c** The integers that satisfy both inequalities are 1, 2 and 3.

**Example 11**

A rectangle has sides of $x$ cm and $(x - 2)$ cm.

If its perimeter is no longer than 16 cm, show how $x$ can be represented like this:

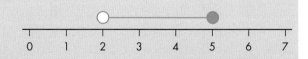

Set up an inequality.     $x + x + x - 2 + x - 2 \leqslant 16$

Simplify.                     $4x - 4 \leqslant 16$

Add 4 to both sides.             $4x \leqslant 20$

Divide by 4.                     $x \leqslant 5$

Also, since $(x - 2)$ cm is the length of one side of the rectangle, $x - 2 > 0$.

So $x > 2$.

Putting these inequalities together gives: $2 < x \leqslant 5$.

This is the inequality represented on the number line.

# Exercise 26F

**1** Write down the inequality represented by each diagram.

a

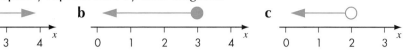
b     c

d     e     f

**2** Draw diagrams to illustrate these inequalities.

**a** $x \leqslant 3$     **b** $x > -2$     **c** $x \geqslant 0$     **d** $x < 5$

**e** $x \geqslant -1$     **f** $2 < x \leqslant 5$     **g** $-1 \leqslant x \leqslant 3$     **h** $-3 < x < 4$

**3** Solve these inequalities and illustrate their solutions on number lines.

**a** $x + 4 \geqslant 8$     **b** $x + 5 < 3$     **c** $x - 1 \leqslant 2$     **d** $x - 4 > -1$

**e** $2x > 8$     **f** $3x \leqslant 15$     **g** $4x + 3 < 9$     **h** $\dfrac{x}{2} + 3 \leqslant 2$

**i** $\dfrac{x}{5} - 2 > 8$     **j** $2x - 1 \leqslant 4$     **k** $\dfrac{x + 2}{3} > 4$     **l** $\dfrac{x - 1}{4} < 3$

**(PS) 4** Max went to the supermarket with £1.20. He bought three apples costing $x$ pence each and a chocolate bar costing 54p. When he got to the till, he found he didn't have enough money.

An inequality showing this information is $3x + 54 > 120$.

**a** Solve the inequality.

Max took one of the apples back and paid for two apples and the chocolate bar. He counted his change and found he had enough money to buy a 16p sweet.

**b** Write an inequality showing this information.

**c** Show the solution to both of these inequalities on a number line.

**d** Give all the possible prices of an apple.

**(PS) 5** What numbers are being described in the bubbles?

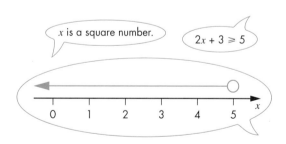

$x$ is a square number.

$2x + 3 \geqslant 5$

**6** Solve these inequalities and illustrate their solutions on number lines.

**a** $\dfrac{2x + 5}{3} > 3$     **b** $\dfrac{3x + 4}{2} \geqslant 11$     **c** $\dfrac{2x + 8}{3} \leqslant 2$     **d** $\dfrac{2x - 1}{3} \geqslant -3$

# Worked exemplars

  Georgia has a card with the expression $2x + 16$.

Amy has a card with the expression $5x - 2$.

$$\boxed{2x + 16} \quad \boxed{5x - 2}$$

**a** Show that their cards are equal when $x = 6$.

**b** Georgia says that her card has a higher value than Amy's card when $x$ is more than 6.
Use inequalities to show that Georgia is incorrect.

| This question assesses 'communicating mathematically', so whichever method you choose to answer the question, show and explain all of the steps. | |
| --- | --- |
| **a Method 1**<br><br>When $x = 6$,<br>$\quad 2x + 16 = 2 \times 6 + 16$<br>$\qquad\qquad = 12 + 16$<br>$\qquad\qquad = 28$<br><br>When $x = 6$,<br>$\quad 5x - 2 = 5 \times 6 - 2$<br>$\qquad\qquad = 30 - 2$<br>$\qquad\qquad = 28$<br>Their cards both equal 28 when $x = 6$. | There are two acceptable methods to answer part **a**. The first involves substituting $x = 6$ into both expressions and showing that they have the same value. |
| **Method 2**<br>$2x + 16 = 5x - 2$<br>$2x + 18 = 5x$    (Add 2 to both sides.)<br>$\quad\; 18 = 3x$    (Subtract $2x$ from both sides.)<br>$\qquad 6 = x$    (Divide both sides by 3.) | The other acceptable method is to make the expressions equal and show that $x = 6$.<br><br>Include an explanation of what you are doing. |
| **b** If Georgia is correct, then $x > 6$ when $2x + 16 > 5x - 2$.<br><br>$2x + 16 > 5x - 2$<br>$2x + 18 > 5x$  (Add 2 to both sides.)<br>$\quad\; 18 > 3x$   (Subtract $2x$ from both sides.)<br>$\qquad 6 > x$    (Divide both sides by 3.)<br>This means that $x < 6$, so Georgia is incorrect. | Write the inequality from her statement and solve it to show that it is not true. |

  Mark set off for town with a £20 note and £1.50 in change.

His bus fare was 50p.

He bought three CDs that each cost the same amount and a drink that cost 98p.

He then came home on the bus, paying 50p for his fare.

Work out the most that Mark could have paid for each CD. Give your answer in pounds, correct to the nearest penny.

This is a problem-solving question, so you will need to use a series of mathematical processes to solve the problem. Remember to show your strategy clearly.

| | |
|---|---|
| £20 + £1.50 = £21.50<br><br>         = 2150p<br><br>One CD costs $x$ p.<br><br>Three CDs cost $3x$ p.<br><br>Total bus fare plus drink<br>= 50p + 98p + 50p<br>= 198p | Start by writing down all the information you know. Remember to convert all the amounts so that they are written in the same units. Note that Mark's money is given in pounds but what he buys is given in pence. If you use pence throughout, then there will not be any decimals until the end. Alternatively, you could convert what he buys to pounds and work with the decimal values. |
| $3x + 198 < 2150$ | Next set up an inequality to show that three CDs plus the amount Mark spent on the other items must be less than what he set off with. |
| $3x < 1952$   (Subtract 198 from both sides.)<br><br>  $x < 650.667$    (Divide both sides by 3.) | Solve the inequality using the balancing method. |
| $x$ must be less than 650.667p.<br><br>The largest integer less than 650.667 is 650.<br><br><br>The most that Mark could have paid for each CD was 650p or £6.50. | Since the question asks for the most each CD cost (and the price must be an integer), you need to work out the largest integer below the answer.<br><br>Remember to give your answer in pounds. |
| Check by substituting £6.50 and £6.51 into your inequality.<br><br>$3 \times 6.50 + 1.98 = £21.48$<br><br>$3 \times 6.51 + 1.98 = £21.51$ | It is good practice to check your answer by substituting.<br><br>Using £6.50, Mark spends £21.48, which is less than £21.50, but using £6.51, Mark spends £21.51, which is more than £21.50. Therefore, the CDs cannot cost more than £6.50 each. |

  **3**     Nicky did a 22 km hill race. She ran $x$ km to the top of the hill at an average speed of 8 km/h. She then ran $y$ km down the hill at an average speed of 15 km/h. She finished the race in 2 hours and 10 minutes.

        Work out how long it took Nicky to get to the top of the hill.

This is a problem-solving question, so you need to make connections between different part of mathematics (in this case, speed, distance and time and simultaneous equations) and show your strategy clearly.

| | |
|---|---|
| $x + y = 22$     (1)<br><br>$\dfrac{x}{8} + \dfrac{y}{15} = 2\tfrac{1}{6}$    (2) | First, set up two simultaneous equations using the information given. |
| $15x + 8y = 260$   (3) | Multiply equation (2) by 120 (the lowest common multiple of 15, 8 and 6). |
| $8x + 8y = 176$   (4)<br><br>      $7x = 84$     (3) − (4) | Balance the coefficients (multiply (1) by 8) and subtract to eliminate $y$. |
| $x = 12$<br><br>The distance to the top of the hill is 12 km. | Solve the equation. |
| Time = 12 ÷ 8<br><br>       = 1 hour 30 minutes | Work out the time using distance ÷ speed. |

# Ready to progress?

I can solve inequalities such as $3x + 2 < 5$ and represent the solution on a number line.

I can solve linear simultaneous equations by balancing, substituting and elimination.

# Review questions

**1** Write each statement as an inequality. Use $x$ to represent the number of cucumbers on each stall.

  **a** Jennie's stall has more than 23 cucumbers.

  **b** Charlie's stall has 15 cucumbers at most.

  **c** Kayla's stall has fewer than 48 cucumbers.

  **d** Christine's stall has at least 6 cucumbers.

  **e** Kayden's stall has between 5 and 25 cucumbers inclusive.

**2** Solve each inequality and represent it on a number line.

  **a** $x + 2 < 9$        **b** $x - 6 > 9$        **c** $3x + 5 \leqslant 11$

  **d** $4x - 1 \geqslant 15$      **e** $\dfrac{x}{3} + 6 \geqslant 2$      **f** $\dfrac{x - 4}{2} < 2$

**3** Represent this inequality on a number line.      $-5 < x \leqslant 8$

**(PS) 4** A rectangle has sides of $(3x + 2)$ m and $(x - 5)$ m.

If $x$ is an integer and the perimeter of the rectangle is at least 66 m, work out the smallest possible value of $x$.

**(PS) 5** Sophie buys a clarinet from a music shop.

She pays a 30% deposit, leaving her no more than £238 to pay.

Calculate the maximum price of the clarinet.

**(PS) 6** My mother uses this formula to cook a turkey: $T = a + bW$

where $T$ is the cooking time (minutes), $W$ is the mass of the turkey (kilograms) and $a$ and $b$ are integers. She says it takes 4 hours 30 minutes to cook a 12 kg turkey, and 3 hours 10 minutes to cook an 8 kg turkey. How long will it take to cook a 5 kg turkey?

**(PS) 7** Two people went to the same shop to buy material to make Christingles. One bought 200 oranges and 220 candles at a cost of £65.60. The other bought 210 oranges and 200 candles at a cost of £63.30. They only needed 200 of each. How much should it have cost them?

**8** **a** Solve these inequalities.

**i** $-11 < 2x - 5 < -3$       **ii** $-3 < \dfrac{x-4}{2} < 0$       **iii** $21 < 3(x+8) < 30$

**b** Match two of your answers with the two representations shown below, and then draw a number line to represent the other solution.

**A**

**B**

(PS) **9** A pentagon has three angles of $x°$ and two of $y°$.

A hexagon has two angles of $x°$ and the rest are each $y°$.

Work out the values of $x$ and $y$.

(PS) **10** When you book Bingham Hall for a conference you pay a fixed booking fee plus a charge for each person attending. Jathika booked a conference for 65 people and was charged £192.50. Jasmine booked a conference for 40 people and was charged £180. James wants to book for 70 people. How much will he be charged?

(MR) **11** **a** Solve the simultaneous equations by balancing the coefficients of $x$ and subtracting.

$992x + 8y = 3992$

$8x + 992y = 3008$

**b** **i** Add together the equations $992x + 8y = 3992$ and $8x + 992y = 3008$. Divide the equation you get by its common factor.

**ii** Subtract the equation $8x + 992y = 3008$ from the equation $992x + 8y = 3992$. Divide the equation you get by its common factor.

**iii** Show that when you solve the simultaneous equations you got from parts **i** and **ii**, the answers are the same as those you got in **a**.

 **c** Solve these simultaneous equations without using a calculator:

$4576a + 10\,848b = 95\,424$

$5424a + 9152b = 94\,576$

(PS) **12** $x$ and $y$ are integers that satisfy the inequalities $3x + 2y > 11$ and $4x + y > 18$.

Work out the smallest possible values for $x$ and $y$.

(PS) **13** The racetrack shown is to be made with semicircles at each end, with an inner perimeter of 300 m and an outer perimeter of 320 m. How wide is the track?

| Hints and tips | Set up a pair of simultaneous equations. Remember that the formula for the circumference of a circle is $C = \pi d$. |

# 27 Algebra: Non-linear graphs

## This chapter is going to show you:

- how to interpret a distance–time graph
- how to draw and interpret graphs of the depths of a liquid as a container is filled
- how to interpret the gradients of straight lines on a velocity–time graph
- how to work out acceleration from a velocity–time graph
- how to draw quadratic graphs
- how to solve problems involving quadratic equations
- how to solve quadratic equations by factorisation
- how to solve quadratic equations graphically
- how to recognise and find the significant points of a quadratic graph
- how to recognise and draw cubic and reciprocal graphs.

## You should already know:

- how to substitute into simple algebraic functions
- how to collect together like terms
- how to multiply together two algebraic expressions
- how to solve simple linear equations
- how to draw linear graphs
- how to plot a graph from a given table of values using all four quadrants
- how to find the equation of a graph.

## About this chapter

The equations of straight-line graphs can be written in the form $y = mx + c$. Quadratic graphs are curved and their equations always include an $x^2$ term, usually a term in $x$ and a constant term. For example, $x^2 - 8x + 15 = 0$ is a quadratic equation. Like most mathematics, quadratic equations were first used in ancient Egypt.

Quadratic curves, also known as parabolas, can be used to model many situations in science and economics. If you hit a tennis ball or fire a cannonball, it will trace out a parabola. As gravity pushes the ball downwards, the ball rises at first but then falls to the ground. Strikers in football use maths to know how to kick the ball so that it will hit the target.

# 27.1 Distance–time graphs

This section will show you how to:

- interpret distance–time graphs
- draw a graph of the depth of liquid as a container is filled.

A **distance–time** graph gives information about how far someone or something has travelled.

When a plane travels at a constant speed, you can find the speed from the formula:

$$\text{speed} = \frac{\text{distance}}{\text{time}}$$

From the graph:

$$\text{speed} = \frac{500 \text{ km}}{2 \text{ h}}$$
$$= 500 \text{ km} \div 2 \text{ h}$$
$$= 250 \text{ km/h}$$

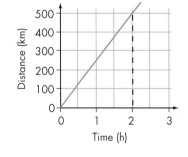

On a distance–time graph, the speed is the gradient of the graph.
The steeper the gradient, the greater the speed.

You can also find the **average speed** from a distance–time graph, using the formula:

$$\text{average speed} = \frac{\text{total distance travelled}}{\text{total time taken}}.$$

**Example 1**

Nottingham is 50 km from Barnsley. This distance–time graph represents a car journey from Barnsley to Nottingham, and back again.

a What can you say about points B, C and D?

b What can you say about the journey between points D and F?

c Work out the speed for each of the five stages of the journey.

d Find the average speed for the whole journey.

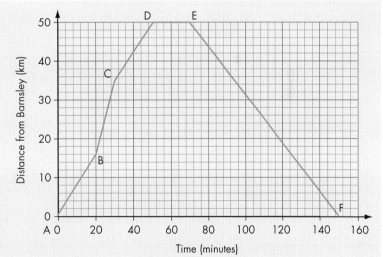

From the graph:

a At each point the speed changed.

At point B: After 20 minutes the car was 16 km away from Barnsley.

At point C: After 30 minutes the car was 35 km away from Barnsley.

At point D: After 50 minutes the car was 50 km away from Barnsley, so at Nottingham.

b The car stayed at Nottingham for 20 minutes, and then took 80 minutes for the return journey to Barnsley.

**c** Work out the speed for the five stages of the journey as follows.

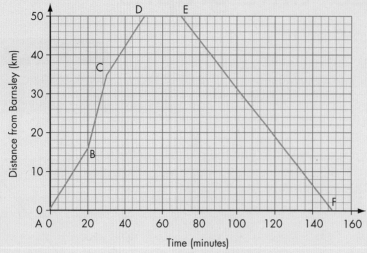

A to B represents 16 km in 20 minutes.

Since 20 minutes = $\frac{20}{60} = \frac{1}{3}$ of an hour, you could use the formula:

$$\text{speed} = \frac{\text{distance}}{\text{time}}$$

$$= 16 \div \frac{1}{3} \text{ km/h}$$

$$= 16 \times \frac{3}{1} \text{ km/h}$$

$$= 48 \text{ km/h}$$

B to C represents 19 km in 10 minutes. 10 minutes = $\frac{1}{6}$ of an hour, so:

$$\text{speed} = \frac{\text{distance}}{\text{time}}$$

$$= 19 \text{ km} \div \frac{1}{6} \text{ h}$$

$$= 114 \text{ km/h}$$

C to D represents 15 km in 20 minutes. 20 minutes = $\frac{1}{3}$ of an hour, so:

$$\text{speed} = \frac{\text{distance}}{\text{time}}$$

$$= 15 \text{ km} \div \frac{1}{3} \text{ h}$$

$$= 45 \text{ km/h}$$

D to E represents a stop, so the average speed is 0 km/h since no further distance was travelled.

E to F represents 50 km in 80 minutes. 80 minutes = $1\frac{1}{3}$ hours, or $\frac{4}{3}$ hours, so:

$$\text{speed} = \frac{\text{distance}}{\text{time}}$$

$$= 50 \text{ km} \div \frac{4}{3} \text{ h}$$

$$= 37\frac{1}{2} \text{ km/h}$$

**d** The total journey is 100 km and takes 150 minutes.

150 minutes = $2\frac{1}{2}$ hours or $\frac{5}{2}$, so:

$$\text{speed} = \frac{\text{distance}}{\text{time}}$$

$$= 100 \text{ km} \div \frac{5}{2} \text{ h}$$

$$= 40 \text{ km/h}$$

## Exercise 27A

**1** The distance–time graph shows Dan's journey from home to a church and back.

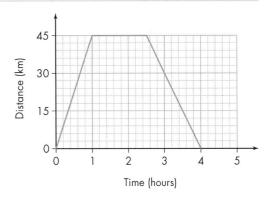

**a** How long did it take Dan to drive to the church?

**b** How long was Dan at the church?

**c** How long did it take Dan to drive home?

**d** What was Dan's speed on the way to the church (in kilometres per hour)?

**e** What was Dan's speed on the way home from the church (in kilometres per hour)?

**2** The distance–time graph shows Paul's journey to a meeting.

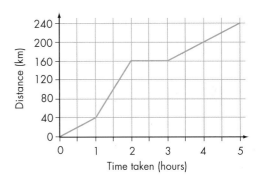

**a** How long after he set off did he:

  **i** stop for a break

  **ii** set off after his break

  **iii** reach his meeting place?

**b** At what speed was he travelling:

  **i** over the first hour

  **ii** over the second hour

  **iii** for the last part of his journey?

**3** For each graph, calculate the speed of the journey.

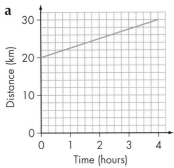 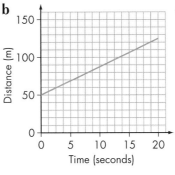 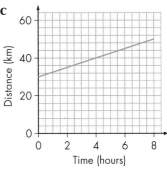

**4** A bus set off from Leeds to pick up Mike at 12:30 pm, his parents at 1:00 pm and his grandparents at 2:00 pm. The bus took them all to a hotel and then returned to Leeds.

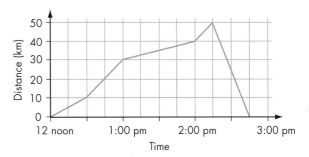

**a** How far from Leeds do Mike's parents live?

**b** How far from Leeds do Mike's grandparents live?

**c** How far from Leeds is the hotel at which Mike and his family stayed?

**d** What was the speed of the bus on its return journey to Leeds?

(MR) **5** Richard and Paul took part in a 5000-metre race.

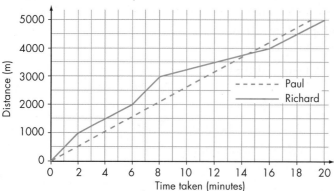

**a** Paul ran a steady race. What was his average speed in:

  **i** metres per minute

  **ii** kilometres per hour?

**b** What was Richard's highest speed?

**c** Who finished the race first? By how many minutes?

(CM) **6** Three friends, Patrick, Araf and Sean, ran a 1000-metre race.

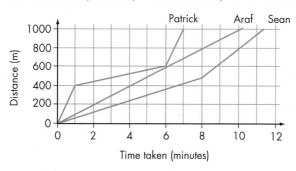

**a** Describe the race.

> Hints and tips | This means explain the important points for each of the three runners.

**b i** What was Araf's average speed in metres per second?

  **ii** What is this speed in kilometres per hour?

(PS) **7** A walker sets off at 9:00 am from point P.

He walks along a path at a steady speed of 6 km/h.

90 minutes later, a cyclist sets off from P on the same path.

She cycles at a steady speed of 15 km/h.

At what time does the cyclist overtake the walker?

> Hints and tips | Drawing a distance–time graph may help you.

(MR)  **8**  Two vehicles set off from Town $X$ at different times.

They both travelled to Town $Y$, then returned to Town $X$.

Vehicle 1 set off at 14:30.

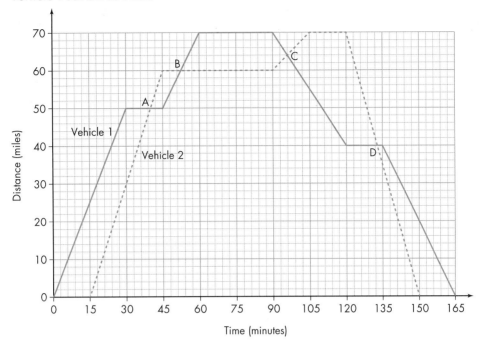

**a** What happened at point A?

**b** What happened at point B?

**c** What happened at point C?

**d** What happened at point D?

**e** At what time did Vehicle 1 return to Town $X$?

**f** Find the difference between the average speeds of the two vehicles. Give your answer in miles per hour, to 1 decimal place.

## Filling containers

This graph shows the change in the depth of water in a flat-bottomed flask, as it is filled at a steady rate.

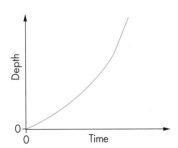

It shows that, at first, the depth of water increases slowly then speeds up as the flask gets narrower.

When the water reaches the neck, which has a constant cross-section, the depth increases at a constant rate up to the top of the neck.

Example 2

Draw a graph to show the change in depth of water in each container as they are filled at a steady rate.

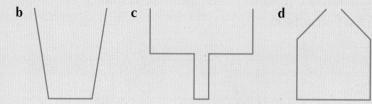

**a**   **b**   **c**   **d**

**a** The container has the same diameter from bottom to top. This means that the depth increases at a constant rate.

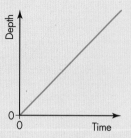

**b** The container gets wider from bottom to top. This means that at first the depth changes quickly, but it slows down as the container gets wider.

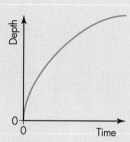

**c** The container is made from two parts. They both have a constant width, but the top half is much wider than the bottom half. The depth increases at a constant fast rate at first, then at a constant much slower rate.

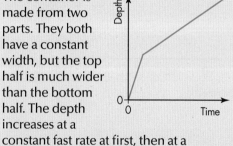

**d** The bottom of the container is a constant diameter so fills at a constant rate. Towards the top, the container gets narrower. This means the depth changes increasingly quickly.

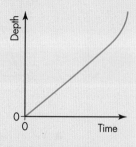

## Exercise 27B

  Maureen took a bath. The graph shows the depth of water in the bath from the time she started running the water to the time that the bath was empty again.

Explain what you think is happening for each part of the graph from **a** to **g**.

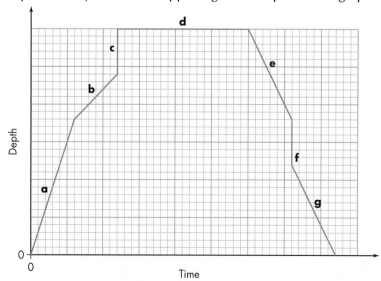

**2** **a** Liquid is poured at a steady rate into the bottle shown in the diagram.

The depth of the liquid in the bottle, $d$, changes as the bottle is filled.

Which of the four graphs shows the change in depth?

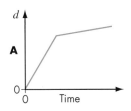 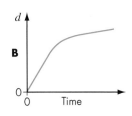 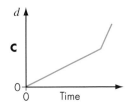 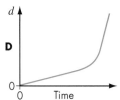

**b** Draw bottles for each of the other three graphs.

  **3** Draw a graph of the depth of water in each of these containers as it is filled steadily.

**a**      **b**      **c**

**d**      **e**      **f**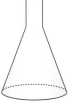

**g**

# 27.2 Velocity–time graphs

This section will show you how to:

- read information from a velocity–time graph
- work out acceleration from a velocity–time graph.

Velocity is the speed of someone or something in a certain direction. A **velocity–time graph** shows how the velocity changes during a journey. A positive gradient means the velocity is increasing. A **zero gradient**, when the line is horizontal, indicates a steady or *constant* velocity. A negative gradient means the velocity is decreasing.

**Key terms**

acceleration

deceleration

velocity–time graph

zero gradient

Look at the journey represented by this graph.

A to B takes 2 hours and the speed increases from 0 km/h to 10 km/h.

B to C takes 1 hour and the speed increases from 10 km/h to 40 km/h.

C to D takes 2 hours and the speed is constant at 40 km/h.

D to E takes 1 hour and the speed decreases from 40 km/h to 0 km/h.

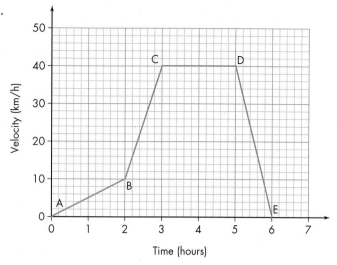

## Acceleration

The gradient of a velocity–time graph gives the rate at which the velocity is increasing or decreasing in a given time. If the gradient is positive, it is an **acceleration**. If the gradient is negative, it is a **deceleration**.

$$\text{Acceleration or deceleration} = \frac{\text{change in velocity}}{\text{time taken}}$$

The units for acceleration and deceleration are, for example, metres per second per second (m/s²) or kilometres per hour per hour (km/h²).

From this velocity–time graph:

**a** write down the initial velocity

**b** work out the acceleration.

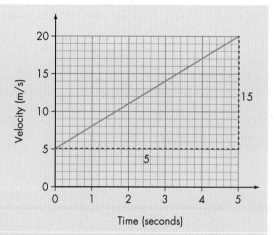

**a** Initial velocity is the velocity at the start of the journey.

Read this from the $y$-intercept.

Initial velocity = 5 m/s

**b** Acceleration = $\dfrac{\text{difference in velocity}}{\text{difference in time}}$

$= \dfrac{20 - 5}{5}$ m/s²

$= \dfrac{15}{5}$ m/s²

$= 3$ m/s²

An acceleration of 3 m/s² means that the velocity is increasing by 3 m/s every second.

# Exercise 27C 🖩

**1** The graph shows the journey of a car travelling between two sets of traffic lights.

   **a** Work out the acceleration of the car for the first part of the journey.

   **b** Work out the deceleration of the car for the last part of the journey.

   **c** For how long was the car travelling at a constant velocity?

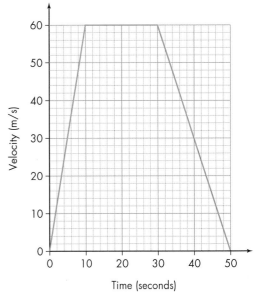

**2** The graph shows the journey of a train travelling between two towns.

Work out the acceleration or deceleration for each section of the graph.

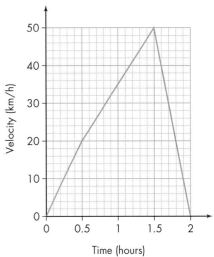

**PS** **3** The graph shows the velocity of a car between two junctions.

Work out the acceleration in the first 10 seconds, in terms of $v$.

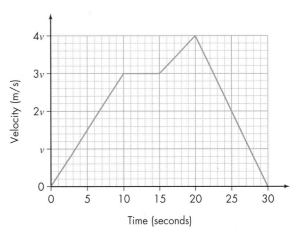

**4** The velocity–time graph shows a bus and a car travelling along the same road.

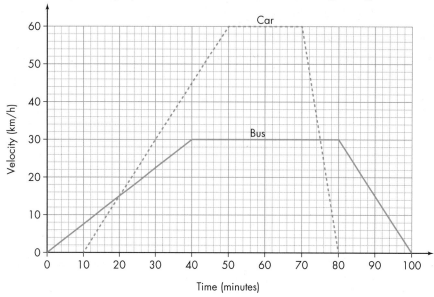

**a** Which vehicle set off first?

**b** Which vehicle had a greater initial acceleration?

**c** At what times were the bus and the car travelling at the same velocity?

# 27.3 Plotting quadratic graphs

## This section will show you how to:

- draw and read values from quadratic graphs.

A **quadratic** graph has an equation of the form $y = ax^2 + bx + c$. All of the following are quadratic equations.

$y = x^2$   $y = x^2 + 5$   $y = x^2 - 3x$   $y = x^2 + 5x + 6$   $y = 3x^2 - 5x + 4$

Each produces a quadratic graph, which is a smooth curve called a **parabola**.

To draw the graph of $y = x^2$ from $x = -3$ to $x = 3$, first make a table, as shown below.

| $x$ | −3 | −2 | −1 | 0 | 1 | 2 | 3 |
|-----|----|----|----|---|---|---|---|
| $y = x^2$ | 9 | 4 | 1 | 0 | 1 | 4 | 9 |

Next draw axes from $x = -3$ to $x = 3$ and from $y = 0$ to $y = 9$. Note that you could also write this as $-3 \leqslant x \leqslant 3$ and $0 \leqslant y \leqslant 9$.

Then plot the points from the table and join them to make a smooth curve.

This is the graph of $y = x^2$.

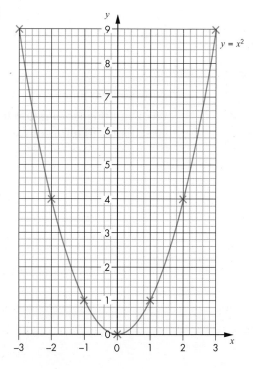

Example 4

**a** Draw the graph of $y = x^2 + 5x + 6$ for $-5 \leqslant x \leqslant 3$.

**b** Use your graph to find the value of $y$ when $x = 1.4$.

**c** Use your graph to solve the equation $x^2 + 5x + 6 = 4$.

**a** In a table, first work out the values of $x^2$, $5x$ and $6$ for each value of $x$. Then add them together to obtain the values of $y$.

| $x$ | –5 | –4 | –3 | –2 | –1 | 0 | 1 | 2 | 3 |
|---|---|---|---|---|---|---|---|---|---|
| $x^2$ | 25 | 16 | 9 | 4 | 1 | 0 | 1 | 4 | 9 |
| $+5x$ | –25 | –20 | –15 | –10 | –5 | 0 | 5 | 10 | 15 |
| $+6$ | 6 | 6 | 6 | 6 | 6 | 6 | 6 | 6 | 6 |
| $y$ | 6 | 2 | 0 | 0 | 2 | 6 | 12 | 20 | 30 |

Then plot the points from the table and join them with a smooth curve. Notice that the line joining the points $(-3, 0)$ and $(-2, 0)$ is curved, and so it goes below the $x$-axis.

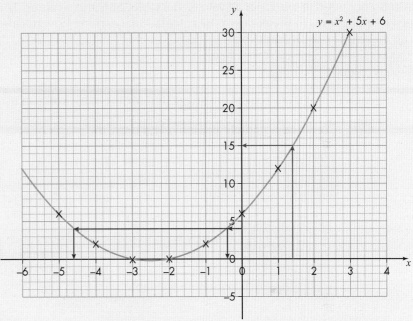

**b** To find the corresponding $y$-value for any value of $x$, start on the $x$-axis at that $x$-value, go up to the curve, across to the $y$-axis and read off the $y$-value. This procedure is marked on the graph with arrows.

Always show these arrows so you can double check your readings.

When $x = 1.4$, $y = 15$.

**c** To solve the equation $x^2 + 5x + 6 = 4$, you need to find the values of $x$ when $y = 4$. Start at 4 on the $y$-axis and read off the two $x$-values that correspond to a $y$-value of 4. Again, this procedure is marked on the graph with arrows.

When $y = 4$, $x = -4.6$ or $-0.4$.

## Drawing accurate graphs

Although it is difficult to draw accurate curves, you need to make sure that whatever you draw is the right shape and accurate enough to read off values. Try to avoid the following common errors:

- When the points are too far apart, a curve tends to 'wobble'.

- Drawing curves in small sections leads to 'feathering'.

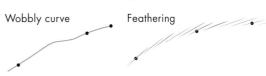

- The place where a curve should turn smoothly is drawn 'flat'.
- A line is drawn through a point that, clearly, has been incorrectly plotted.

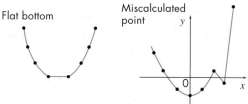

Here are some tips that will make it easier for you to draw smooth, curved lines.

- If you are *right-handed*, turn your paper or exercise book round so that you draw from left to right. Your hand is steadier this way than when you are trying to draw from right to left or away from your body. If you are *left-handed*, you should find drawing from right to left the more accurate way.
- Move your pencil over the points as a practice run without drawing the curve.
- Do one continuous curve and only stop at a plotted point.
- Use a *sharp* pencil and do not press too heavily, so that you can easily rub out mistakes.

You do not need to work out all values in a table, only the $y$-values. The other rows in the table are just working lines to break down the calculation. Learning how to calculate $y$-values with a calculator can make this process quicker.

## Exercise 27D 🖩

In this exercise, suitable ranges are suggested for the axes. You can use any type of graph paper.

**1** Copy and complete the table and draw the graph of $y = x^2$ for $-5 \leqslant x \leqslant 5$.

| $x$ | −5 | −4 | −3 | −2 | −1 | 0 | 1 | 2 | 3 | 4 | 5 |
|---|---|---|---|---|---|---|---|---|---|---|---|
| $y = x^2$ | 25 | 16 | | | 1 | 0 | | | | | |

> **Hints and tips** You may wish to plot other values such as $x = 4.5$ so that you can draw a more accurate graph. Note that you can use the symmetry of the curve to help complete the table.

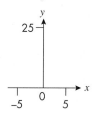

**2** Copy and complete the table for the graph of $y = 3x^2$ for values of $x$ from −3 to 3. Plot the graph using $-3 \leqslant x \leqslant 3$ and $0 \leqslant y \leqslant 30$ for your axes.

| $x$ | −3 | −2 | −1 | 0 | 1 | 2 | 3 |
|---|---|---|---|---|---|---|---|
| $y = 3x^2$ | 27 | | 3 | | | 12 | |

> **Hints and tips** Remember, you can work out other values if they will help you draw your graph more accurately.

**3 a** Copy and complete the table for the graph of $y = x^2 + 2$ for values of $x$ from −5 to 5. Plot the graph using $-5 \leqslant x \leqslant 5$ and $0 \leqslant y \leqslant 30$ for your axes.

| $x$ | −5 | −4 | −3 | −2 | −1 | 0 | 1 | 2 | 3 | 4 | 5 |
|---|---|---|---|---|---|---|---|---|---|---|---|
| $y = x^2 + 2$ | 27 | | 11 | | | | | 6 | | | |

**b** Where does the graph cross the $y$-axis?

**4** **a** Copy and complete the table for the graph of $y = x^2 - 3x$ for values of $x$ from $-5$ to $5$.

Plot the graph using $-5 \leqslant x \leqslant 5$ and $-4 \leqslant y \leqslant 40$ for your axes.

| $x$ | $-5$ | $-4$ | $-3$ | $-2$ | $-1$ | $0$ | $1$ | $2$ | $3$ | $4$ | $5$ |
|---|---|---|---|---|---|---|---|---|---|---|---|
| $x^2$ | 25 | | 9 | | | | | 4 | | | |
| $-3x$ | 15 | | | | | | | $-4$ | | | |
| $y$ | 40 | | | | | | | $-2$ | | | |

**b** Use your graph to find the value of $y$ when $x = 3.5$.

**c** Use your graph to solve the equation $x^2 - 3x = 5$.

**d** Where does the graph cross the $y$-axis?

**5** **a** Copy and complete the table or use a calculator to work out values for the graph of $y = x^2 - 4x - 5$ for values of $x$ from $-2$ to $6$. Plot the graph using $-2 \leqslant x \leqslant 6$ and $-10 \leqslant y \leqslant 7$ for your axes.

| $x$ | $-2$ | $-1$ | $0$ | $1$ | $2$ | $3$ | $4$ | $5$ | $6$ |
|---|---|---|---|---|---|---|---|---|---|
| $x^2$ | 4 | | | | 4 | | | | |
| $-4x$ | 8 | | | | $-8$ | | | | |
| $-5$ | $-5$ | | | | $-5$ | | | | |
| $y$ | 7 | | | | $-9$ | | | | |

**b** Use your graph to find the value of $y$ when $x = 0.5$.

**c** Use your graph to solve the equation $x^2 - 4x - 5 = -3$.

**Hints and tips** You need to find the values of $x$ when $y = -3$.

**6** **a** Copy and complete the table or use a calculator to work out the values for the graph of $y = x^2 + 6x + 8$ for values of $x$ from $-7$ to $1$. Plot the graph using $-7 \leqslant x \leqslant 1$ and $-1 \leqslant y \leqslant 15$ for your axes.

| $x$ | $-7$ | $-6$ | $-5$ | $-4$ | $-3$ | $-2$ | $-1$ | $0$ | $1$ |
|---|---|---|---|---|---|---|---|---|---|
| $x^2$ | | | | | 9 | | | | 1 |
| $+6x$ | | | | | $-18$ | | 6 | | |
| $8$ | | | | | 8 | 8 | | | |
| $y$ | | | | | $-1$ | | | | |

**b** Use your graph to find the $y$-value when $x = -2.5$.

**c** Use your graph to solve the equation $x^2 + 6x + 8 = 1$.

**d** On the same axes, draw the graph of $y = \frac{x}{2} + 2$.

**e** Where do the graphs $y = x^2 + 6x + 8$ and $y = \frac{x}{2} + 2$ cross?

**7** **a** Copy and complete the table or use a calculator to work out the values for the graph of $y = x^2 - 2x$ for values of $x$ from $-2$ to $4$. Plot the graph using $-2 \leqslant x \leqslant 4$ and $-1 \leqslant y \leqslant 8$ for your axes.

| $x$ | $-2$ | $-1$ | $0$ | $1$ | $2$ | $3$ | $4$ |
|---|---|---|---|---|---|---|---|
| $x^2$ | | | | $1$ | | | |
| $-2x$ | | | | $-2$ | | | |
| $y$ | | | | $-1$ | | | |

**b** Where do the graphs $y = x^2 - 2x$ and $y = x$ cross?

 **8** Shayla is writing out a table of values for the graph of $y = x^2 - 4x + 7$.

| $x$ | $-3$ | $-2$ | $-1$ | $0$ | $1$ | $2$ | $3$ | $4$ | $5$ | $6$ | $7$ |
|---|---|---|---|---|---|---|---|---|---|---|---|
| $y$ | | | | | $4$ | $3$ | $4$ | $7$ | $12$ | $19$ | $28$ |

**a** What do you notice about the value of $y$ when $x = 1$ and when $x = 3$?

**b** Complete the table without substituting any values. Explain how this is possible.

**9** **a** Copy and complete the table or use a calculator to work out the values for the graph of of $y = x^2 - 4$ for values of $x$ from $-4$ to $4$. Plot the graph using $-4 \leqslant x \leqslant 4$ and $-3 \leqslant y \leqslant 12$ for your axes.

| $x$ | $-4$ | $-3$ | $-2$ | $-1$ | $0$ | $1$ | $2$ | $3$ | $4$ |
|---|---|---|---|---|---|---|---|---|---|
| $y$ | $12$ | | | $-3$ | | | | $5$ | |

**b** Where does the graph cross the $x$-axis?

**c** Use your graph to find the $y$-value when $x = 1.5$.

**d** Use your graph to solve the equation $x^2 - 4 = 8$.

 **10** **a** Copy and complete the table or use a calculator to work out the values for the graph of $y = 4 - x^2$ for values of $x$ from $-4$ to $4$. Plot the graph using $-4 \leqslant x \leqslant 4$ and $-12 \leqslant y \leqslant 4$ for your axes.

| $x$ | $-4$ | $-3$ | $-2$ | $-1$ | $0$ | $1$ | $2$ | $3$ | $4$ |
|---|---|---|---|---|---|---|---|---|---|
| $y$ | $-12$ | | $3$ | | | | | $-5$ | |

**b** Where does the graph cross the $x$-axis?

**c** Use your graph to find the $y$-value when $x = 1.5$.

**d** Use your graph to solve the equation $4 - x^2 = -8$.

**e** Compare your answers for questions **9** and **10**.

# 27.4 Solving quadratic equations by factorisation

This section will show you how to:

- solve a quadratic equation by factorisation.

In Chapter 9, you learned how to factorise a quadratic expression, for example, $x^2 - 9x + 20 = (x - 5)(x - 4)$. This is the first step in solving a quadratic equation by factorisation.

To solve the equation $x^2 - 9x + 20 = 0$, first factorise the quadratic expression.

$$(x - 5)(x - 4) = 0$$

The expression $(x - 5)(x - 4)$ can only equal 0 if the value of one of the brackets is 0.

So either $(x - 5) = 0$ or $(x - 4) = 0$.

$$x - 5 = 0 \quad \text{or } x - 4 = 0$$

$$\rightarrow \qquad x = 5 \quad \rightarrow \quad x = 4$$

So, the solution is $x = 5$ or $x = 4$.

---

**Example 5**

Solve $x^2 - x - 6 = 0$ by factorisation.

First factorise the left-hand side of the equation. $\qquad (x + 2)(x - 3) = 0$

So either $(x + 2) = 0$ or $(x - 3) = 0$.

$$x + 2 = 0 \quad \text{or } x - 3 = 0$$

$$\rightarrow \qquad x = -2 \quad \rightarrow \quad x = 3$$

So, the solution is $x = -2$ or $x = 3$.

---

## EXERCISE 27E

**1** Solve these quadratic equations.

   **a** $(x + 2)(x + 5) = 0$      **b** $(y - 9)(y - 4) = 0$      **c** $(z + 6)(z - 3) = 0$

   **d** $(t + 3)(t + 1) = 0$      **e** $(a + 6)(a + 4) = 0$      **f** $(x + 3)(x - 2) = 0$

   **g** $(x + 1)(x - 3) = 0$      **h** $(t + 4)(t - 5) = 0$      **i** $(x - 1)(x + 2) = 0$

   **j** $(x - 2)(x + 5) = 0$      **k** $(a - 7)(a + 4) = 0$      **l** $(x - 3)(x - 2) = 0$

   **m** $(x - 1)(x - 5) = 0$      **n** $(a - 4)(a - 3) = 0$      **o** $(r + 7)(r - 5) = 0$

**2** First factorise and then solve these quadratic equations.

   **a** $x^2 + 5x + 6 = 0$      **b** $x^2 + 5x + 4 = 0$      **c** $x^2 - 6x + 8 = 0$

   **d** $x^2 - 3x - 10 = 0$      **e** $x^2 - 2x - 15 = 0$      **f** $t^2 + 3t - 18 = 0$

   **g** $x^2 - x - 2 = 0$      **h** $m^2 + 10m + 25 = 0$      **i** $a^2 - 14a + 49 = 0$

   **j** $x^2 + 11x + 18 = 0$      **k** $x^2 - 8x + 15 = 0$      **l** $t^2 + 4t - 12 = 0$

   **m** $x^2 + 4x + 4 = 0$      **n** $t^2 - 8t + 16 = 0$      **o** $t^2 + 8t + 12 = 0$

**3**  **a** Show by substitution that $x = 5$ is a solution of the equation $x^2 + 6x - 55 = 0$.

   **b** Find the other solution.

**4** First factorise and then solve these quadratic equations.

   **a** $v^2 + 5v - 50 = 0$      **b** $w^2 + w - 56 = 0$      **c** $r^2 - 19r + 18 = 0$

   **d** $z^2 + 13z + 22 = 0$      **e** $c^2 + 13c - 30 = 0$      **f** $c^2 - 13c + 30 = 0$

   **g** $s^2 + 1000s + 999 = 0$      **h** $k^2 - 2k + 1 = 0$      **i** $u^2 + 9u - 10 = 0$

(CM) **5** Fatimah says that the solution to the equation $x^2 + 8x + 12 = 0$ is $x = 2$ or $x = 6$.

Show why Fatimah is wrong.

(CM) **6** Tomisin says that the solution to the equation $x^2 + 5x - 6 = 0$ is $x = 2$ or $x = -3$.

Show why Tomisin is wrong.

**7** A rectangle has sides of $(x + 1)$ cm and $(x + 2)$ cm.

The area of the rectangle is 42 cm².

**a** Show that $x^2 + 3x - 40 = 0$.

**b** Solve the quadratic equation.

**c** Work out the perimeter of the rectangle.

**8** A rectangle has a base of $(x - 3)$ m and a height of $(x - 1)$ m.

The area of the rectangle is 48 m².

Work out the value of $x$.

# 27.5 The significant points of a quadratic curve

This section will show you how to:

- identify the significant points of a quadratic function graphically
- identify the roots of a quadratic function by solving a quadratic equation
- identify the turning point of a quadratic function.

**Key terms**

intercept

roots

turning point

A quadratic curve has four interesting points for a mathematician. These are the points labelled A to D on the diagrams below.

The curve crosses the $x$-axis at A and B. The $x$-values at A and B are called the **roots**.

C is the point where the curve crosses the $y$-axis (the $y$-**intercept**).

D is the **turning point**. This is the lowest or highest point of the curve.

## Exercise 27F

**1** Look at your graphs from Exercise 27D questions **5** to **7**.

- $y = x^2 - 4x - 5$
- $y = x^2 + 6x + 8$
- $y = x^2 - 2x$

**a** For each graph, state the coordinates of the point where the curve crosses the $y$-axis (the $y$-intercept).

**b** How does the $y$-intercept relate to the equation of the curve in each case?

**2** **a** Solve each equation in question **1** for $y = 0$.

**b** Each curve intersects the $x$-axis twice. Write down the coordinates of these two points for each curve.

**3** In each case in question **1**, the turning point is the lowest point on the graph. Write down the coordinates of the turning points.

## The roots

You can find the roots of a quadratic curve by putting the expression equal to zero and solving the quadratic equation. The roots are the same as the solutions to the quadratic equation. For example, the roots of the quadratic curve $y = x^2 - 4x - 5$ are $x = 5$ and $x = -1$, which are the same as the solutions to the quadratic equation $x^2 - 4x - 5 = 0$.

## The $y$-intercept

The curve crosses the $y$-axis when $x = 0$. The $y$-intercept is at $(0, c)$, where $c$ is the constant term of the equation $y = ax^2 + bx + c$.

## The turning point

Because a quadratic graph has a vertical line of symmetry passing through the turning point, the $x$-coordinate of the turning point is always halfway between the roots. You can find the $y$-value by reading from the graph or by substituting the $x$-value into the original equation.

**Note:** If the $x^2$ term is negative, the graph will be upside-down.

---

**Example 6**

**a** For the graph $y = x^2 + 2x - 15$, find the coordinates of:

   **i** the $y$-intercept   **ii** the roots   **iii** the turning point.

**b** Sketch the graph of $y = x^2 + 2x - 15$ including the points from part **a**.

**a** **i** The $y$-intercept is the point where $x = 0$. When $x = 0$, $y = -15$, so the $y$-intercept is at $(0, -15)$.

   **ii** To find the roots, solve the equation $x^2 + 2x - 15 = 0$.

     First factorise the left-hand side of the equation.   $(x + 5)(x - 3) = 0$

     So, either $x + 5 = 0$ or $x - 3 = 0$.     $\rightarrow$     $x = -5$ or $x = 3$

     The roots are at $(-5, 0)$ and $(3, 0)$.

   **iii** To find the turning point, find the value of $x$ halfway between the roots.

$$x = \frac{-5 + 3}{2}$$

$$= \frac{-2}{2}$$

$$= -1$$

     Then substitute to find the value of $y$.

     When $x = -1$, $y = (-1)^2 + 2 \times (-1) - 15$

$$= 1 - 2 - 15$$

$$= -16$$

     The turning point is at $(-1, -16)$.

**b** Sketch the graph, first marking the coordinates of the roots, $y$-intercept and turning point.

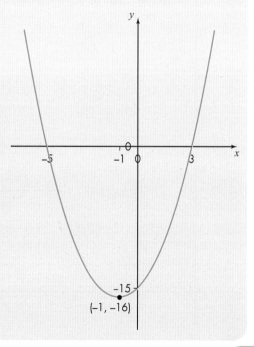

---

**1** From the graph, write down the coordinates of:

**i** the $y$-intercept      **ii** the points where the curve intersects the $x$-axis

**iii** the turning point.

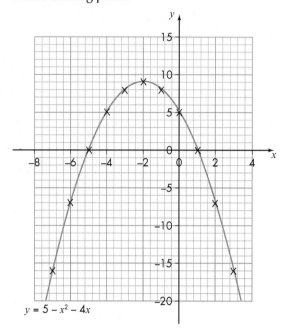

$y = 5 - x^2 - 4x$

**2** From the graph, write down the coordinates of:

**i** the $y$-intercept

**ii** the points where the curve intersects the $x$-axis

**iii** the turning point.

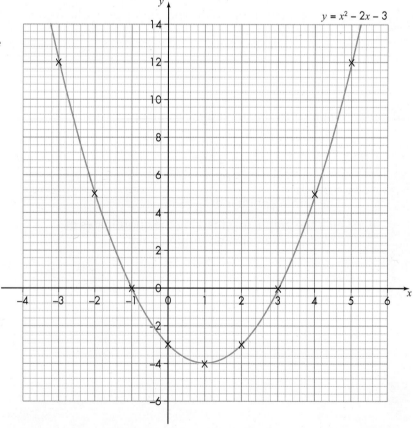

$y = x^2 - 2x - 3$

**3** Work out the roots and $y$-intercept of each graph.

    **a** $y = x^2 - 4$      **b** $y = x^2 - 6x$      **c** $y = x^2 - 2x - 3$    **d** $y = x^2 + 14x + 33$

**4** Write down the coordinates of the turning point of the graph of $y = x^2 - 4x + 3$.

 **5** For the graph of $y = 9 - x^2$, write down the coordinates of:

    **a** the points where the curve intersects the $x$-axis

    **b** the turning point.

**6** Without plotting the graph of $y = x^2 - 10x + 16$, write down the coordinates of:

    **i** the $y$-intercept        **ii** the points where the curve intersects the $x$-axis

    **iii** the turning point.

**(MR)** **7** Without plotting the graph of $y = x^2 + 2x - 15$, write down the coordinates of:

    **i** the $y$-intercept        **ii** the points where the curve intersects the $x$-axis

    **iii** the turning point.

**8** For the graph of $y = 2x^2 + 2x - 8$, write down the coordinates of:

    **a** the points where the curve intersects the $x$-axis

    **b** the turning point.

# 27.6 Cubic and reciprocal graphs

## This section will show you how to:

- recognise and plot cubic and reciprocal graphs.

**Key term**

cubic

### Cubic graphs

A **cubic** function or graph is one that contains a term in $x^3$. These are examples of cubic functions.

$y = x^3$      $y = x^3 + 3x$      $y = x^3 + x^2 + x + 1$

This is the graph of $y = x^3$.

It has a characteristic shape that you should learn to recognise.

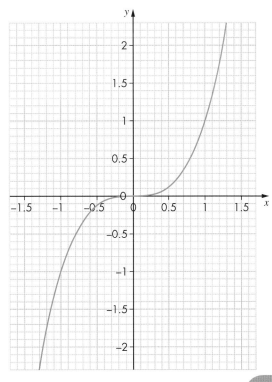

Example 7

Complete the table and draw the graph of $y = x^3 - 3$ for $-2 \leqslant x \leqslant 3$.

| x | -2 | -1 | 0 | 1 | 2 | 3 |
|---|----|----|---|---|---|---|
| y |    |    |   |   |   |   |

When $x = -2$,    $y = (-2)^3 - 3$

$= -8 - 3$

$= -11$

Complete the rest of the table in the same way.

| x | -2 | -1 | 0 | 1 | 2 | 3 |
|---|----|----|---|---|---|---|
| y | -11 | -4 | -3 | -2 | 5 | 24 |

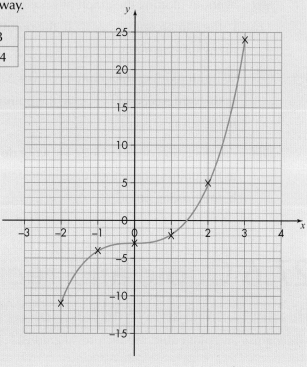

**Hints and tips**   Work out that when $x = 2.5$, $y = 12.6$ and plot (2.5, 12.6). Plotting an extra point at $x = 2.5$ will help you to draw a smooth curve.

You should use a calculator to work out the values of $y$ and round to 1 or 2 decimal places.

Note the difference between the shape of a positive cubic graph (one with $+x^3$) and a negative cubic graph (one with $-x^3$):

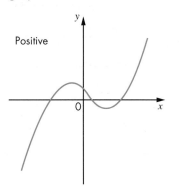

Positive

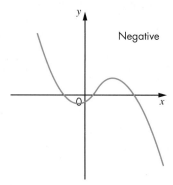

Negative

## Reciprocal graphs

A reciprocal function has the form $y = \dfrac{a}{x}$.

To draw the graph of $y = \dfrac{1}{x}$ for $-4 \leqslant x \leqslant 4$:

Start by finding the $y$-values for the integer $x$-values between $-4$ and $4$ (except 0, since $\frac{1}{0}$ is infinity).

Round values to 2 decimal places, as it is difficult to plot a value more accurately than this.

| $x$ | −4 | −3 | −2 | −1 | 1 | 2 | 3 | 4 |
|---|---|---|---|---|---|---|---|---|
| $y$ | −0.25 | −0.33 | −0.50 | −1.00 | 1.00 | 0.50 | 0.33 | 0.25 |

The graph plotted from these values is shown in graph A. This does not show the properties of the reciprocal function.

Find the $y$-values for $x$-values between −0.8 to 0.8 in steps of 0.2.

| $x$ | −0.8 | −0.6 | −0.4 | −0.2 | 0.2 | 0.4 | 0.6 | 0.8 |
|---|---|---|---|---|---|---|---|---|
| $y$ | −1.25 | −1.67 | −2.5 | −5 | 5 | 2.5 | 1.67 | 1.25 |

Plotting these points as well gives graph B.

**A** 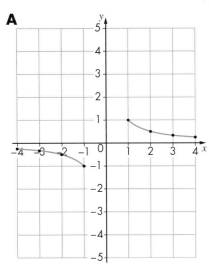  **B**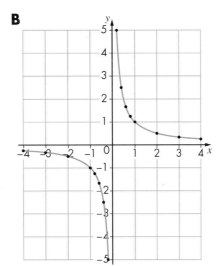

Graph B shows the characteristic properties of reciprocal graphs.

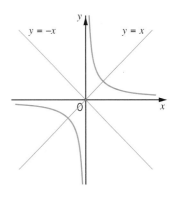

- The lines $y = x$ and $y = -x$ are lines of symmetry.
- The closer $x$ gets to zero, the nearer the graph gets to the $y$-axis.
- As $x$ increases, the graph gets closer to the $x$-axis.

Note the difference between the shape of a positive reciprocal graph and a negative reciprocal graph:

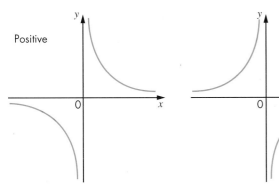

# Exercise 27H ▦

**1** Sketch the graph of $y = -x^3$.

**2** **a** Copy and complete the table and draw the graph of $y = x^3 + 1$ for $-3 \leqslant x \leqslant 3$ and $-30 \leqslant y \leqslant 30$.

| x | −3 | | −2 | | −1 | | 0 | | 1 | | 2 | | 3 |
|---|---|---|---|---|---|---|---|---|---|---|---|---|---|
| y | | | −7 | | | | 1 | | | | 9 | | |

**b** Use your graph to find the $y$-value when $x = 2.5$.

**3** **a** Draw the graph of $y = x^3 + 3$ for $-3 \leqslant x \leqslant 3$ and $-30 \leqslant y \leqslant 30$. Plot the $x$-values in steps of 0.5.

**b** Use your graph to find the $y$-value when $x = 1.2$.

**c** Find the coordinates of the point where the curve crosses the x-axis

**(CM)** **4** **a** Copy and complete the table and draw the graph of $y = \dfrac{1}{x}$ for $-20 \leqslant x \leqslant 20$.

| x | −20 | −10 | −5 | −4 | −2 | −1 | −0.5 | −0.4 | −0.2 | 0.2 | 0.4 | 0.5 | 1 | 2 | 5 | 10 | 15 | 20 |
|---|---|---|---|---|---|---|---|---|---|---|---|---|---|---|---|---|---|---|
| y | | −0.1 | | | | | | | | 5 | | 2 | | | | | | 0.05 |

**b** Explain why there is no value when $x = 0$.

**c** On the same axes, draw the graph of $y = 3 - x$.

**d** Use your graph to write down the $x$-values of any points where the graphs intersect.

**(MR)** **5** Write down whether each of these graphs is linear, quadratic, reciprocal, cubic or none of these.

**a**
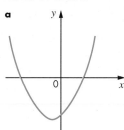

**b**
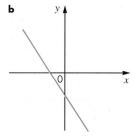

**c**
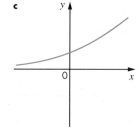

**d**
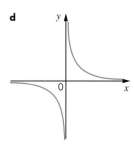

**e**
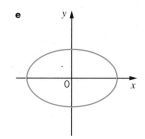

**f**
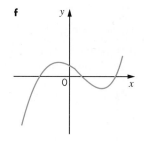

**g**
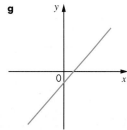

**h**
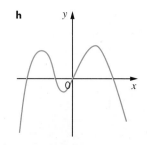

**i**
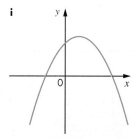

# Worked exemplars

(PS) **1** Tracy and Les both drove to the airport. The distance–time graphs of their journeys are shown below.

Given that 5 miles is approximately 8 km, calculate who drove faster.

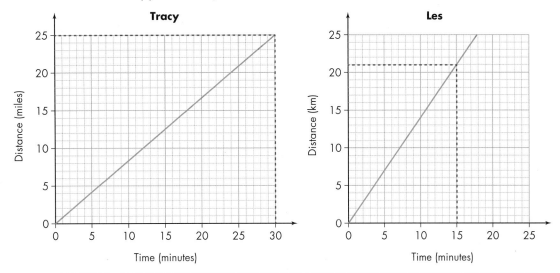

| This is a problem-solving question, so you need to make connections between different parts of mathematics, in this case distance–time graphs and converting between units. | |
|---|---|
| Tracy travels 25 miles in 30 minutes, so is travelling at 50 mph.<br><br>Les travels 21 km in 15 minutes, so is travelling at 84 km/h. | Start by converting the information from the graphs into speeds that can be used as a comparison.<br><br>These speeds are normally given per hour. |
| Tracy: $50 \times \frac{8}{5} = 80$ km/h<br><br>or<br><br>Les: $84 \times \frac{5}{8} = 52.5$ mph | 5 miles is equal to 8 km, so calculate Tracy's speed in km/h by multiplying her speed in mph by $\frac{8}{5}$. As a result, both Tracy's and Les' speeds are now in km/h.<br><br>Alternatively, multiply Les' speed by $\frac{5}{8}$ to convert it to mph. |
| So:<br><br>Tracy (80 km/h) is travelling more slowly than Les (84 km/h)<br><br>or<br><br>Tracy (50 mph) is travelling more slowly than Les (52.5 mph). | State your conclusion, comparing the two speeds. |

**2** Sketch the curve $y = x^2 + 8x - 20$, showing any significant points.

| This is a mathematical reasoning question so you need to show the logical steps to reach your conclusion. | |
| --- | --- |
| When $x = 0$, $y = 0 + 0 - 20$<br><br>The curve crosses the $y$-axis at $(0, -20)$. | Find the coordinates of the point where the curve crosses the $y$-axis by substituting $x = 0$ into the equation of the curve. |
| $0 = x^2 + 8x - 20$<br><br>$0 = (x + 10)(x - 2)$<br><br>$x = -10$ or $2$<br><br>The curve crosses the $x$-axis at $(-10, 0)$ and $(2, 0)$. | Find the coordinates of the points where the curve crosses the $x$-axis. Do this by substituting $y = 0$ into the equation of the curve and solving the equation by factorisation. |
| $\dfrac{-10 + 2}{2} = \dfrac{-8}{2} = -4$ | Because the quadratic graph is symmetrical, its turning point lies halfway between the points where it crosses the $x$-axis. |
| $y = (-4)^2 + 8(-4) - 20$<br><br>$\quad = 16 - 32 - 20$<br><br>$\quad = -36$<br><br>The turning point of the curve is at $(-4, -36)$. | Substitute $x = -4$ into the equation to find the value of $y$ for the turning point. |
| [graph showing y-axis from -40 to 10 and x-axis from -10 to 4, with points marked at (-10, 0), (2, 0), (0, -20) and (-4, -36)] | Draw an $x$-axis and a $y$-axis and plot the points $(0, -20)$, $(-10, 0)$, $(2, 0)$ and $(-4, -36)$. |

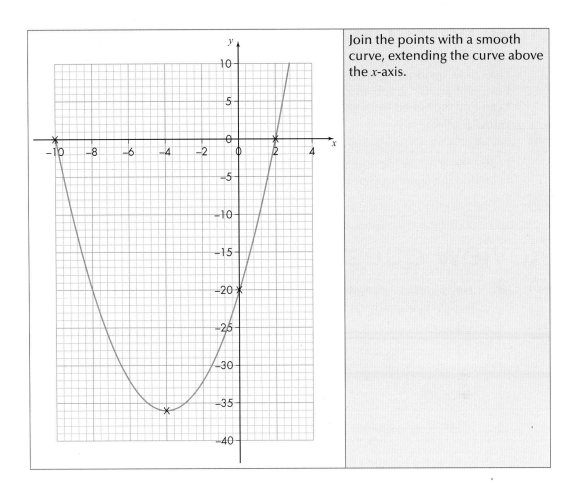

Join the points with a smooth curve, extending the curve above the $x$-axis.

# Ready to progress?

I can draw quadratic graphs from their tables of values.
I can plot a cubic graph and a reciprocal graph.
I can find the speed from a distance–time graph and the acceleration from a velocity–time graph.

I can solve a quadratic equation of the form $x^2 + ax + b = 0$.
I can find the significant points of a quadratic graph.

# Review questions

**MR** **1** The depth–time graph shows how the depth changes as a flask is filled at a steady rate.

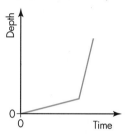

**a** Sketch a possible flask that could be represented by the graph.

**b** Draw a different flask which also has a graph made up of two straight-line sections.

**2** The velocity–time graph shows the journey of a car.

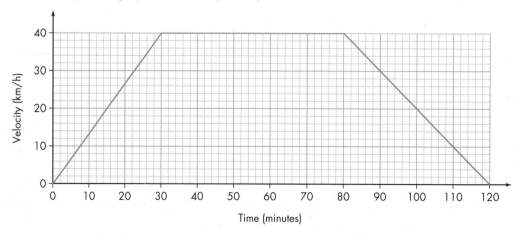

**a** Which part of the journey shows the car travelling at a constant velocity?

**b** What of these statements must be true from 80 minutes to 120 minutes?

- The car is travelling downhill
- The car is slowing down
- The car is travelling south-east

**3** Carlos left his apartment building by helicopter at 13:26. He flew to a hotel, collected some files and then returned to his apartment. The distance–time graph shows part of his journey. The missing section was completed at a speed of 40 km/h.

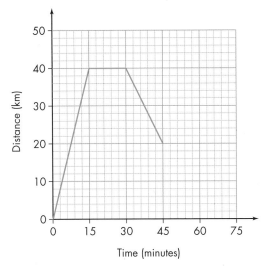

**a** Find Carlos' speed on the flight to the hotel.

**b** Find Carlos' initial speed on his return from the hotel.

**c** Copy and complete the distance–time graph.

**d** At what time did Carlos arrive back at his apartment?

**4** Solve $x^2 - 19x = -90$.

**5** Piravina is solving a quadratic equation by factorising but the coefficient of $x$ has been smudged.

$x^2 - \blacklozenge x - 24 = 0$

The two answers to the quadratic equation are both integers.

Write down all four possible values of the coefficient of $x$.

**6** Eric accelerated from a speed of 15 m/s to 23 m/s in 4 seconds.

Calculate Eric's acceleration.

**7** **a** Plot the graph of $y = x^2 - 3x + 1$ for $-2 \leqslant x \leqslant 5$ and $-1 \leqslant y \leqslant 11$.

**b** Use your graph to find two approximate solutions when:

   **i** $x^2 - 3x + 1 = 0$      **ii** $x^2 - 3x + 1 = 2$      **iii** $x^2 - 3x + 1 = 7$.

**8** A rectangle has edges of $(x + 10)$ cm and $(x - 3)$ cm. The area of the rectangle is 48 cm².

**a** Work out the value of $x$.

**b** Calculate the length of the perimeter.

**9** Plot the graph of $y = x^3 - 10$ for $-3 \leqslant x \leqslant 3$ and $-40 \leqslant y \leqslant 20$.

# Glossary

**3D shape**   A shape with three dimensions, length, width and height.

**acceleration**   The rate at which the velocity of a moving object increases.

**acute-angled triangle**   A triangle in which all the angles are acute.

**adjacent side**   The side that is between a given angle and the right angle, in a right-angled triangle.

**allied angles**   Interior angles that lie on the same side of a line that cuts a pair of parallel lines; they add up to 180°.

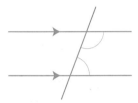

**alternate angles**   Angles that lie on either side of a line that cuts a pair of parallel lines; the line forms two pairs of alternate angles and the angles in each pair are equal.

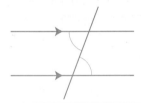

**angle bisector**   A line or line segment that divides an angle into two equal parts.

**angle of depression**   The angle between the horizontal line of sight of an observer and the direct line to an object that is viewed from above.

**angle of elevation**   The angle between the horizontal line of sight of an observer and the direct line to an object that is viewed from below.

**angle of rotation**   The angle through which an object is rotated, to form the image.

**angles around a point**   The angles formed at a point where two or more lines meet; their sum is 360°.

**angles on a straight line**   The angles formed at a point where one or more inclined (sloping) lines meet on one side of a straight line; their sum is 180°.

**annual rate**   A rate, such as interest, that is charged over a period of a year.

**apex**   The top point of a pyramid where all the edges of the sloping sides meet.

**approximate**   A value that is close but not exactly equal to another value, which can be used to give an idea of the size of the value; for example, a journey taking 58 minutes may be described as 'taking approximately an hour'; the ≈ sign means 'is approximately equal to'.

**approximation**   A calculated guess.

**arc**   Part of the circumference of a circle.

**arithmetic sequence**   A sequence of numbers in which the difference between one term and the next is constant.

**average**   A single value that represents a typical value for the whole set of data. The most common averages are the mode, the median and the mean.

**average speed**   The result of dividing the total distance travelled by the total time taken for a journey.

**balancing**   Doing the same thing to both sides of an equation.

**bar chart**   A diagram which is a series of bars or blocks of the same width to show frequencies.

**bearing**   The angle measured from North to define a direction.

**best buy**   The price that gives best value for money, the greatest quantity for the least price.

**better value**   The choice that gives more product per pound or penny.

**bias**   The property of a sample being unrepresentative of the population; for example, a dice may be weighted so that it gives a score of 5 more frequently than any other score.

**bisect**   Cut exactly in half.

**cancel**   When the numerator and denominator of a fraction have a common factor, this can be divided into both values to reduce the fraction to a fraction in its lowest terms.

**capacity**   The amount a container can hold.

**categorical**   Data that has non-numerical values.

**centilitre (cl)**   A measurement of capacity. A hundredth of a litre.

**centre of enlargement**   The point, inside, outside or on the perimeter of the object, on which an enlargement is centred; the point from which the enlargement of an object is measured.

**centre of rotation**   The point about which an object or shape is rotated.

**chord**   A straight line joining two points on the circumference of a circle.

**circumference**   The perimeter of a circle; every point on the circumference is the same distance from the centre, and this distance is the radius.

**class interval**   The range of a group of values in a set of grouped data.

**coefficient**   A number written in front of a variable in an algebraic term; for example, in $8x$, 8 is the coefficient of $x$.

**column method**   A method for multiplying large numbers, in which you multiply the units, tens and hundreds separately, then add the products together. This is also known as the *traditional method*.

**combination**   A way of selecting members from a group, when the order of selection does not matter. For example, given three letters ABC, there are three ways of choosing two letters: AB, AC and BC. The mathematical formula for a combination of $r$ items from a group of $n$ is $\frac{n!}{r!(n-r)!}$, which may be written as $_nC_r$.

**common factor**   A factor that divides exactly into two or more numbers; 2 is a common factor of 6, 8 and 10.

**common unit**   To enable you to compare quantities or simplify ratios, they must be expressed in the same or common units; for example, 2 m : 10 cm = 200 cm : 10 cm = 20 : 1.

**complement**   An event that does not happen. The probability that event A does not happen is written as P(A').

**composite bar chart**   A bar chart where each bar compares sets of related data.

**compound interest**   Interest that is paid on the amount in the account; after the first year interest is paid on interest earned in the previous years.

**compound shape**   A shape made up of a combination of two or more shapes.

**congruent**   Exactly the same shape and size.

**consecutive**   Numbers that follow each other continuously eg 3, 4, 5, 6 are consecutive whole numbers, 4, 9, 16, 25 are consecutive square numbers.

**consistency**   A way of comparing two or more sets of data. The data set with the smallest range is said to be more consistent.

**constant of proportionality**   If two variables are in direct proportion, you can write an equation, $y = kx$; if they are in inverse proportion, you can write $xy = k$. In either case, $k$ is the constant of proportionality.

**constant term**   A term that has a fixed value; in the equation $y = 3x + 6$, the values of $x$ and $y$ may change, but 6 is a constant term.

**construct**   Draw a shape by means of a ruler and a pair of compasses.

**continuous data**   Data, such as mass, length or height, that can take any value; continuous data has no precise fixed value.

**conversion graph**   A graph that can be used to convert from one unit to another.

**correlation**   A relationship or connection between two or more things showing how they vary in relation with each other, eg as the weather gets warmer, more ice creams are sold.

**corresponding angles**   Angles that lie on the same side of a pair of parallel lines cut by a line; the line forms four pairs of corresponding angles, and the angles in each pair are equal.

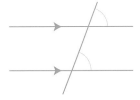

**cosine**   A trigonometric ratio related to an angle in a right-angled triangle, calculated as $\frac{\text{adjacent}}{\text{hypotenuse}}$.

**cover-up**   A method of solving equations by covering up one of the other terms.

**cross-section**   A cut across a 3D shape, or the shape of the face that is exposed when a 3D shape is cut. For a prism, a cut across the shape, perpendicular to its length.

**cubic** An expression where the highest power of the variable is 3.

**cylinder** A prism with a circular cross-section.

**data collection sheet** Any way of collecting data for a survey in written form. See *tally chart*.

**deceleration** The rate at which the velocity of a moving object decreases.

**decimal fraction** Another name for a decimal. It implies that any decimal can be written as a fraction with a numerator of 10, 100, etc. For example $0.7 = \frac{7}{10}$ or $0.94 = \frac{94}{100} = \frac{47}{50}$.

**decimal point** A symbol, usually a small dot, written between the whole-number part and the fractional part in a decimal number.

**denominator** The bottom number in a fraction.

**density** The mass of a substance divided by its volume.

**diameter** A chord in a circle that passes through the centre.

**difference** The result of a subtraction.

**difference of two squares** An expression of the form $x^2 - y^2$: the terms are squares and there is a minus sign between them.

**digit** Digits are the numbers from 0 to 9. 56 is a two-digit number.

**direct proportion** A relationship in which one variable increases or decreases at the same rate as another; in the formula $y = 12x$, $x$ and $y$ are in direct proportion.

**direct variation** Another name for direct proportion.

**direction** The line along which a vector such as force, weight or velocity acts.

**discrete data** Data that can only take certain values, such as a number of children; discrete data can only take fixed values.

**distance** The length between two points. For example, the distance from A to B is 10 centimetres.

**distance–time graph** A graph that represents a journey, based on the distance travelled and the time taken.

**dual bar chart** A bar chart to compare two sets of related data.

**edge** The line where two faces or surfaces of a 3D shape meet.

**element** Any member of a set.

**elevation** The view of a 3D shape when you view it from the side or the front.

**eliminate** Given a pair of simultaneous equations with two variables, you can manipulate one or both equations to remove or eliminate one of the variables by a process of substitution, addition or subtraction.

**enlargement** A transformation in which the object is enlarged to form an image.

**equally likely** The outcomes of an event are equally likely when each has the same probability of occurring. For example, when tossing a fair coin the outcomes 'Heads' and 'Tails' are equally likely.

**equation** A relation in which two expressions are separated by an equals sign with one or more variables. An equation can be solved to find one or more answers, but it may not be true for all values of $x$.

**equidistant** At equal distances.

**equilateral triangle** A triangle in which all the sides are equal and all the angles are 60°.

**equivalent** Exactly the same as, usually used in 'equivalent fraction'.

**equivalent fraction** Any fraction that can be made equal to another fraction by cancelling. For example $\frac{9}{12} = \frac{3}{4}$.

**error interval** The interval within which a rounded value can lie. For example, if $x = 25$ to the nearest whole number, the error interval for $x$ is $24.5 \leqslant x < 25.5$.

**estimate** A calculated guess.

**estimated mean** A mean that is estimated from grouped data, by multiplying the frequency by the mid-class value for each class, adding up the products and dividing by the total frequency.

**event** Something that happens in a probability problem, such as tossing a coin or predicting the weather.

**exhaustive** All possible outcomes of an event; the sum of the probabilities of exhaustive outcomes equals 1.

**expand** Multiply out (terms with brackets).

**expectation** Predicting the number of times you would expect an outcome to occur.

**experiment**   A method for collecting data, by carrying out a series of trials.

**experimental data**   Data that is collected by an experiment.

**experimental probability**   An estimate for the theoretical probability.

**expression**   A collection of numbers, letters, symbols and operators representing a number or amount; for example, $x^2 - 3x + 4$.

**exterior angle**   The angle formed outside a 2D shape, when a side is extended beyond the vertex.

**extrapolation**   To predict an outcome based on known facts or observations.

**face**   The area on a 3D shape enclosed by edges.

**factor**   A number that divides exactly (no remainder) into another number, for example, the factors of 12 are {1, 2, 3, 4, 6, 12}.

**factor pair**   A factor pair of a number is any pair of numbers whose product is the original number, for example, the factor pairs of 12 are 1 × 12, 2 × 6 and 3 × 4. Note that square numbers always have a number (the square root) that is its own 'pair', so 16 has a factor 'pair' of 4 × 4.

**factor tree**   A method of breaking down a number into its prime factors.

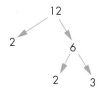

**factorial**   The product of the whole number $n$ and all the whole numbers less than $n$ down to 1. It is written as $n!$. For example $5! = 5 \times 4 \times 3 \times 2 \times 1 = 120$.

**factorisation**   The arrangement of a given number or expression into a product of its factors. (verb: factorise)

**flow diagram**   A diagram that splits a function into single steps.

**foot (ft)**   An imperial measurement of length. 3 feet = 1 yard

**formula**   A mathematical rule, using numbers and letters, which shows a relationship between variables; for example, the conversion formula from temperatures in Fahrenheit to temperatures in Celsius is: $C = \frac{5}{9}(F - 32)$.

**frequency**   The number of times each value occurs.

**frequency table**   A table that shows all the frequencies after all the data has been collected.

**frequency tree diagram**   A diagram that shows the frequencies when the probabilities of different events are known.

**function key**   See *shift key*.

**function**   An algebraic expression in which there is only one variable, often $x$.

**gallon (gal)**   An imperial measurement of volume. 1 gallon ≈ 4.55 litres

**geometric sequence**   A sequence in which each term is multiplied or divided by the same number, to produce the next term; for example, 2, 4, 8, 16, ... is a geometric sequence.

**gradient**   The slope of a line; the vertical difference between the coordinates divided by the horizontal difference.

**gradient-intercept**   A form for the equation of a line, written in terms of its gradient and the intercept on the vertical axis, $y = mx + c$ where $m$ is the gradient and $c$ is the $y$-intercept.

**grid method**   A method for multiplying numbers larger than 10, in which each number is split into its parts: for example, to calculate 158 × 67: 158 is 100, 50 and 8 and 67 is 60 and 7. These numbers are arranged in a grid and each part is multiplied by the others. This is also known as the *box method*.

| ×  | 100  | 50   | 8   |
|----|------|------|-----|
| 60 | 6000 | 3000 | 480 |
| 7  | 700  | 350  | 56  |

```
    6 0 0 0
    3 0 0 0
      4 8 0
      7 0 0
      3 5 0
  +     5 6
  ---------
  1 0 5 8 6
```

**grouped data**   Data arranged into smaller, non-overlapping sets, groups or classes, that can be treated as separate ranges or values, for example, 1–10, 11–20, 21–30, 31–40, 41–50; in this example there are equal class intervals.

**grouped frequency table**   A frequency table where the data has been collected using class intervals.

**highest common factor**   The largest number that is a factor common to two or more other numbers.

**hypothesis**   A statement that has to be proved true or false.

**identity**   Expressions either side of a ≡ sign with one or more variables, which is true for all values; for example, $3(x + 2) \equiv 3x + 6$ is an identity.

**image**   The result of a reflection or other transformation of an object.

**imperial**   Units commonly used in Britain that are gradually being phased out as we change to the metric system.

**improper fraction**   A fraction that has a numerator greater than the denominator.

**inch (in)**   An imperial measurement of length. 12 inches = 1 foot.

**included angle**   The angle made by two lines with a common vertex.

**inclusive inequality**   An inequality such as ⩽ or ⩾.

**index notation**   Expressing a number in terms of one or more of its factors, each expressed as a power.

**inequality**   A statement that one expression is greater or less than another, written with the symbol > (greater than) or < (less than) instead of = (equals).

**input**   The number that is put into a function.

**interpolation**   To insert, estimate or find an intermediate value.

**intercept**   The point where a line cuts or crosses the axis.

**interior angle**   The inside angle between two adjacent sides of a 2D shape, at a vertex.

**intersect**   To cross over.

**intersection**   The 'overlap', the set of elements that occur in two or more sets.

**inverse**   Going the other way.

**inverse flow diagrams**   A flow diagram that shows the reverse process.

**inverse operations**   An operation that reverses the effect of another operation; for example, addition is the inverse of subtraction, division is the inverse of multiplication.

**inverse proportion**   A relationship between two variables in which as one value increases, the other decreases; in the formula $xy = 12$, $x$ and $y$ are in inverse proportion.

**inverse variation**   Another name for inverse proportion.

**invert**   Turn upside down. Usually used when dividing by a fraction. The calculation is turned into a multiplication by inverting the dividing fraction.

**isometric grid**   A sheet with dots on to help draw a 3D representation.

**isosceles triangle**   A triangle in which two sides are equal and the angles opposite the equal sides are also equal.

**key**   A symbol that shows what each item represents.

**like terms**   Terms in which the variables are identical, but the coefficients may be different; for example, $2ax$ and $5ax$ are like terms but $5xy$ and $7y$ are not. Like terms can be combined by adding their numerical coefficients so $2ax + 5ax = 7ax$.

**line of best fit**   A straight line drawn on a scatter diagram where there is correlation, so that there are equal numbers of points above and below it; the line shows the trend of the data.

**line segment**   A line joining two points.

**linear graph**   A straight-line graph that represents a linear function.

**linear sequence**   A sequence or pattern of numbers in which the difference between consecutive terms is always the same.

**loci**   The plural of locus.

**locus**   The path of a point that moves, obeying given conditions.

**lowest common denominator**   When adding or subtracting fractions with different denominators, they must first be written with the same denominator. To avoid having to cancel this should be the lowest common multiple of the denominators.

**lowest common multiple (LCM)**   The lowest number that is a multiple of two or more numbers; 12 is the lowest common multiple of 2, 3, 4 and 6.

**magnitude**   The size of a quantity.

**mass**   The amount of matter in an object.

**metric**   The international standard for all measurement. Based on metres, kilograms and litres.

**mid-class value**   The mid-point value of each class interval.

**mirror line**   Another name for a line of symmetry.

**mixed number**   A number made up of a whole number and a fraction, for example $3\frac{3}{4}$.

**modal**   Any value that represents the mode.

**modal group**   In grouped data, the class with the highest frequency.

**multiple**   Any member of the times table, for example multiples of 7 are 7, 14, 21, 28, etc.

**multiplication table**   Any table with sets of numbers across the top and left hand side (usually 1 to 12) where the cells of the table are the products of the values in the top row and left hand column.

**multiplier**   A number that is used to find the result of increasing or decreasing an amount by a percentage.

**multiply out**   To multiply everything in a pair of brackets by the term in front of the brackets (or everything in one pair of brackets by everything in another pair of brackets).

**mutually exclusive**   Outcomes that cannot occur at the same time.

**negative**   Used in 'negative number' to mean a number less than zero.

**negative coordinates**   Coordinates such as (–2, 6), which contain one or more negative numbers.

**negative correlation**   A relationship between two sets of data, in which the values of one variable increase as the values of the other variable decrease.

**net**   A flat shape you can fold into a 3D shape.

**no correlation**   No relationship between two sets of data.

**$n$th term**   An expression in terms of $n$ where $n$ is the position of the term; it allows you to find any term in a sequence, without having to use a term-to-term rule.

**numerator**   The top number in a fraction.

**object**   The original or starting shape, line or point before it is transformed to give an image.

**observation**   A method for collecting data, by recording each item in a survey.

**obtuse-angled triangle**   A triangle containing an obtuse angle.

**opposite side**   The side that is opposite a given angle, in a right-angled triangle.

**order of rotational symmetry**   The number of times a 2D shape looks the same as it did originally when it is rotated through 360° about a central point. If a shape has no rotational symmetry, its order of rotational symmetry is 1, because every shape looks the same at the end of a 360° rotation as it did originally.

**ounce (oz)**   An imperial measurement of mass. 16 ounces = 1 pound

**outcome**   A possible result of an event in a probability experiment, such as the different scores when throwing a dice.

**outlier**   In a data set, a value that is widely separated from the main cluster of values.

**output**   The result when an input number is acted on by a function.

**parabola**   The shape of a quadratic curve.

**parallel**   Two lines which have the same gradient are called parallel lines.

**parallelogram**   A four sided shape where both pairs of opposite sides are parallel.

**partition method**   A method for long multiplication that requires the numbers to be written as separate digits across and down a grid. These are then multiplied and the values added diagonally to get the final answer. This is also known as 'Napier's Bones'.

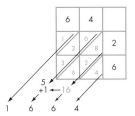

**pattern**   Numbers or objects that are arranged to follow a rule.

**per cent**   From Latin 'per centum' meaning out of a hundred.

**percentage**   Any fraction or decimal expressed as an equivalent fraction with a denominator of 100 but written with a percentage sign (%), for example $0.4 = \frac{40}{100} = 40\%$, $\frac{4}{25} = \frac{16}{100} = 16\%$.

**percentage change**   A change to a quantity, calculated as a percentage of the original quantity.

**percentage decrease**   A reduction or decrease to a quantity, calculated as a percentage of the original quantity.

**percentage increase**   An increase to a quantity, calculated as a percentage of the original quantity.

**percentage loss**   The loss on a financial transaction, calculated as the difference between the buying price and the selling price, calculated as a percentage of the original price.

**percentage profit**   The profit on a financial transaction, calculated as the difference between the selling price and the buying price, calculated as a percentage of the original price.

**perpendicular bisector**   A line that divides a given line exactly in half, passing through its midpoint at right angles to it.

**perpendicular height**   The shortest height from the base to a vertex.

**pi ($\pi$)**   The result of dividing the circumference of a circle by its diameter, represented by the Greek letter pi ($\pi$).

**pictogram**   A frequency table where the frequency for each type of data is shown by a symbol.

**pie chart**   A method of comparing discrete data. A circle is divided into sectors whose angles each represent a proportion of the whole sample.

**place value**   The position (place) of a digit in a number defines its value, so in 432.17, 4 represents 400 and 7 represents 7 hundredths.

**plan**   The view from directly above a solid shape.

**polygon**   A closed 2D shape with straight sides.

**population**   The complete data set in a survey.

**position-to-term**   A rule for generating a term in a sequence, depending on the position of the term within the sequence.

**positive**   Used in 'positive number' to mean a number greater than zero.

**positive correlation**   A relationship between two sets of data, in which the values of one variable increase as the values of the other variable increase.

**pound (lb)**   An imperial measurement of mass, approximately the same as a jar of jam.

**power**   The number of times you use a number or expression in a calculation; it is written as a small, raised number; for example, $2^2$ is 2 multiplied by itself, $2^2 = 2 \times 2$ and $4^3 = 4 \times 4 \times 4$.

**powers of 10**   A number that is produced by multiplying 10 by itself repeatedly.

**powers of 2**   A number that is produced by multiplying 2 by itself repeatedly.

**pressure**   The amount of force exerted divided by the area on which that force acts.

**primary data**   Data you have collected yourself.

**prime factor**   Any factor of a number that is a prime number. For example the factors of 12 are {1, 2, 3, 4, 6, 12}. The prime factors of 12 are {2, 3}.

**prime factorisation**   Breaking a number down into a product consisting of prime factors only. See *product of prime factors*.

**prime number**   A number with only two factors, 1 and itself.

**principal**   The amount invested or lent.

**prism**   A 3D shape that has the same cross-section wherever it is cut perpendicular to its length.

**probability**   The chance of an event happening.

**probability fraction**   The probability of an event happening, written as a fraction. Probabilities can also be written as decimals or percentages.

**probability scale**   A scale between 0 and 1 that shows the probability of an event happening.

**probability space diagram**   A diagram that shows all the outcomes for two events.

**product of prime factors**   A number written as a product of its prime factors, for example $12 = 2 \times 2 \times 3$.

**proper fraction**   A fraction that is less than one, with the numerator less than the denominator.

**pyramid**   A 3D shape with a base and sides rising to form a single point.

**Pythagoras' theorem**   The rule that, in any right-angled triangle, the square of the hypotenuse is equal to the sum of the squares of the other two sides.

**quadratic**   Having terms involving one or two variables, and constants, such as $x^2 - 3$ or $y^2 + 2y + 4$ where the highest power of the variable is two.

**quadratic expansion**   Multiplying out two pairs of brackets, leading to a quadratic expression.

**quadratic expression**   An expression in which the highest power of any variable is 2, such as $2x^2 + 4$.

**quadratic sequence**   A sequence in which the first differences are not constant, formed from a quadratic rule.

**quantity**   A measurable amount of something that can be written as a number, or a number with appropriate units; for example, the capacity of a milk carton.

**radius**   A straight line joining the centre of a circle to any point on the circumference.

**random** Chosen by chance, without looking; every item has an equal chance of being chosen.

**random sample** A sample in which every member of the population has an equal chance of being chosen.

**range** The difference between the highest and lowest values for a set of data.

**ratio** The ratio of A to B is a number found by dividing A by B. It is written as A : B. For example, the ratio of 1 m to 1 cm is written as 1 m : 1 cm = 100 : 1. Notice that the two quantities must both be in the same units if they are to be compared in this way.

**rational number** A number that can be written as a fraction, for example, $\frac{1}{4}$ or $\frac{10}{3}$.

**rearrange** Put into a different order, to simplify.

**reciprocal** The result of dividing a number into 1, so 1 divided by the number is its reciprocal.

**recurring decimal** A decimal number in which a digit or pattern of digits repeats for ever.

**reflection** The image formed when a 2D shape is reflected in a mirror line or line of symmetry; the process of reflecting an object.

**relative frequency** An estimate for the theoretical probability.

**representative** A value that is typical for the whole set of data.

**resultant vector** The result of combining two or more vectors.

**right-angled triangle** A triangle in which one angle is 90°.

**roots** The points on a graph where it crosses the x-axis.

**rotation** A turn about a central point, called the centre of rotation.

**rotational symmetry** A type of symmetry in which a 2D shape may be turned through 360° so that it looks the same as it did originally in two or more positions.

**round** The process of giving an estimate by changing the number of significant figures.

**sample** A selection taken from a larger data set, which can be researched to provide information about the whole population.

**sample size** The number of items of data collected when doing a survey or testing a hypothesis.

**sample space diagram** A diagram that shows all the outcomes of an experiment.

**scalar** A quantity such as mass that has quantity but does not act in a specific direction.

**scale** The number of squares that are used for each unit on an axis.

**scale drawing** A drawing that represents something much larger or much smaller, in which the lengths on the image are in direct proportion to the lengths on the object.

**scale factor** The ratio of the distance on the image to the distance it represents on the object; the number that tells you how much a shape is to be enlarged.

**scalene triangle** A triangle in which all sides are different lengths.

**scatter diagram** A graphical representation showing whether there is a relationship between two sets of data.

**secondary data** Data that has been collected by someone else.

**sector** A region of a circle, like a slice of a pie, enclosed by an arc and two radii.

**segment** A chord will divide the circle into two segments, one each side of the chord.

**sequence** A pattern of numbers that are related by a rule.

**set** A collection of objects or elements.

**shift key** The key on a calculator that enables you to use the alternative functions associated with the main keys.

**significant figure** In the number 12 068, 1 is the first and most significant figure and 8 is the fifth and least significant figure. In 0.246 the first and most significant figure is 2. Zeros at the beginning or end of a number are not significant figures.

**similar** Two shapes are similar if one is an enlargement of the other; angles in the same position in both shapes are equal to each other.

**simple interest** Money that a borrower pays a lender, for allowing them to borrow money.

**simplest form** A fraction written so that the numerator and denominator have no common factors.

**simplify** To make an equation or expression easier to work with or understand by combining like terms or cancelling; for example:
$4a - 2a + 5b + 2b = 2a + 7b$, $\frac{12}{18} = \frac{2}{3}$, $5 : 10 = 1 : 2$.

**simultaneous equations** Two equations that are both true for the same set of values for their variables.

**sine** A trigonometric ratio related to an angle in a right-angled triangle, calculated as $\frac{\text{opposite}}{\text{hypotenuse}}$.

**slant height** The length of the sloping side of a cone.

**solution** The answer for an equation; the method of finding the answer.

**speed** The rate at which an object moves. For example, the speed of the car was 40 miles per hour.

**sphere** A 3D shape that is the locus of a point that moves a fixed distance from a given point, the centre; a 3D shape that has a circular cross-section whenever it is cut through its centre.

**spread** A way of describing how a set of data is scattered for all the values. See *range*.

**square number** A number formed when any integer is multiplied by itself. For example, $3 \times 3 = 9$ so 9 is a square number.

**square root** A number that produces a specified quantity when multiplied by itself. For example, the square root of 16 is 4. Not all square roots are whole numbers. It uses the symbol $\sqrt{\phantom{x}}$, so $\sqrt{25} = 5$, and $\sqrt{7} = 2.645\ 751...$

**standard form** A way of writing a number as $a \times 10^n$, where $1 \leqslant a < 10$ and $n$ is a positive or negative integer.

**standard index form** See *standard form*.

**stem-and-leaf diagram** A diagram showing how discrete numerical data is distributed once the data is put in numerical order; the leaves are the unit's digits and the stems are the digits that occur before the units digits.

**stone (st)** An imperial measurement of mass, approximately equal to 6 kilograms.

**strict inequality** An inequality such as < or >.

**subject** The variable on the left-hand side of the equals (=) sign in a formula or equation.

**substitute** Replace a variable in an expression with a number and work out the value; for example,

if you substitute 4 for $t$ in $3t + 5$ the answer is 17 because $3 \times 4 + 5 = 17$.

**subtend** The joining of the lines from two points giving an angle.

**surface area** The total area of all of the surfaces of a 3D shape.

**survey** A method of collecting data by asking questions or observing.

**symbol** Symbols such as + and = are used to simplify expressions and equations.

**systematic counting** If you wanted to work out how many times the digit 6 was written when writing down all the numbers from 200 to 300 you would use a systematic counting strategy; for example, 206, 216, … 296 is 10 times plus 260, 261, … 269 which is 10 times so the digit 6 will be written 20 times. Note that if the question was how many numbers between 200 and 300 contain the digit 6, the answer would be 19 as 266 would be counted only once.

**tally chart** A data collection sheet where the data is collected using a tally.

**tangent** 1 A straight line that touches a circle just once.

2 A trigonometric ratio related to an angle in a right-angled triangle, calculated as $\frac{\text{opposite}}{\text{adjacent}}$.

**term** 1 A part of an expression, equation or formula. Terms are separated by + and − signs.

2 A number in a sequence or pattern.

**terminating decimal** A terminating decimal can be written down exactly. $\frac{33}{100}$ can be written as 0.33 so is a terminating decimal, but $\frac{1}{3}$ is 0.3333… with the 3s recurring forever.

**term-to-term** The rule that shows what to do to one term in a sequence, to work out the next term.

**theoretical probability** The exact or true probability of an event happening.

**three-figure bearing** The angle from north clockwise, generally given as a three-digit figure.

**time** A point or period of the day as measured in hours and minutes past midnight or noon. For example, the bus leaves at 10 am or the journey took 3 hours.

**time series graph** A line graph where the horizontal axis represents time.

**ton (T)** An imperial measurement of mass, approximately equal to a saloon car.

**tonne (t)** A metric measurement of mass, approximately equal to a saloon car.

**transformation** A change to a geometric 2D shape, such as a translation, rotation, reflection or enlargement.

**translation** A movement along, up, down or diagonally on a coordinate grid.

**trapezium** A four sided shape where only one pair of opposite sides are parallel.

**tree diagram** A diagram that is used to calculate the probability of combined events happening. All the probabilities of each single event are written on the branches of the diagram.

**trend** How data increases or decreases in a regular pattern.

**trial** A single experiment in a probability experiment.

**trigonometric functions** These are sine, cosine and tangent.

**trigonometric ratios** These are sine = $\frac{\text{opposite}}{\text{hypotenuse}}$, cosine = $\frac{\text{adjacent}}{\text{hypotenuse}}$ and tangent = $\frac{\text{opposite}}{\text{adjacent}}$.

**trigonometry** The study of the relationship between angles and sides in triangles.

**turning point** Any point on a graph where the gradient is zero; for a quadratic graph this is the lowest or highest point.

**two-way table** A table that records how two variables are linked.

**unbiased** The property of a sample being representative of the population, so that any member of the population may be chosen.

**union** The set of all the elements that occur in one or more sets.

**unique factorisation theorem** This states that every integer greater than 1 is a prime number or can be written as a product of prime numbers.

**unit cost** The cost of one unit, such as a kilogram, litre or metre, of something.

**unitary method** A method of finding best value by finding the price per unit, or the quantity per pound or penny.

**universal set** The set that contains all possible elements, usually represented by the symbol ξ.

**value for money** When comparing costs or offers of the same item, which offer gives the least unit cost.

**variable** A letter that stands for a quantity that can take various values.

**vector** A quantity such as velocity that has magnitude and acts in a specific direction.

**velocity–time graph** A graph in which distance travelled is plotted against time taken.

**Venn diagram** A diagram that shows the relationships between different sets.

**vertex** The point at which two lines meet, in a 2D or 3D shape.

**vertical height** The height of the top vertex of a 3D shape, measured perpendicular to the base.

**vertical line chart** Similar to a bar chart, but has vertical lines instead of bars.

**vertically opposite angles** The angles on the opposite side of the point of intersection when two straight lines cross, forming four angles. The opposite angles are equal.

**volume** The space taken up by a solid shape.

**$x$-values** The first number in a pair of coordinates, the input of a function.

**$y = mx + c$** The general equation of a straight line in which $m$ is the gradient of the line and $c$ is the intercept on the $y$-axis.

**yard (yd)** An imperial measurement; 1 yard, the approximate distance from your fingertip to your nose when you stretch out your arm.

**$y$-intercept** The point where a graph intersects the $y$-axis.

**$y$-values** The second number in a pair of coordinates, the output of a function.

**zero gradient** A line that is parallel to the horizontal axis has zero gradient.

# Index

# Notes

# Notes

# Notes

# Notes